Reliability of Power Electronic Converter Systems, 2nd Edition

Other related titles:

You may also like

- PBPO080 | Henry Shu-hung Chung, Huai Wang, Frede Blaabjerg, and Michael Pecht | Reliability of Power Electronic Converter Systems | 2016
- PBPO161| Marcelo Godoy Simões | Artificial Intelligence for Smarter Power Systems: Fuzzy logic and neural networks | 2021
- PBPO074 | Andrzej M. Trzynadlowski | Power Electronic Converters and Systems: Frontiers and applications | 2015
- PBPO082 | Keliang Zhou, Danwei Wang, Yongheng Yang, and Frede Blaabjerg | Periodic Control of Power Electronic Converters | 2016
- PBPO170 | Ahteshamul Haque, Frede Blaabjerg, Huai Wang, Yongheng Yang, and Zainul Abdin Jaffrey | Reliability of Power Electronics Converters for Solar Photovoltaic | 2021

We also publish a wide range of books on the following topics:
Computing and Networks
Control, Robotics and Sensors
Electrical Regulations
Electromagnetics and Radar
Energy Engineering Healthcare Technologies
History and Management of Technology
IET Codes and Guidance
Materials, Circuits and Devices Model Forms
Nanomaterials and Nanotechnologies
Optics, Photonics and Lasers
Production, Design and Manufacturing Security
Telecommunications Transportation

All books are available in print via https://shop.theiet.org or as eBooks via our Digital Library https://digital-library.theiet.org.

IET TRANSPORTATION SERIES 261

Reliability of Power Electronic Converter Systems, 2nd Edition

Edited by
Henry Shu-Hung Chung, Huai Wang, Frede Blaabjerg
and Michael Pecht

The Institution of Engineering and Technology

About the IET

This book is published by the Institution of Engineering and Technology (The IET).

We inspire, inform and influence the global engineering community to engineer a better world. As a diverse home across engineering and technology, we share knowledge that helps make better sense of the world, accelerate innovation and solve the global challenges that matter.

The IET is a not-for-profit organisation. The surplus we make from our books is used to support activities and products for the engineering community and promote the positive role of science, engineering and technology in the world. This includes education resources and outreach, scholarships and awards, events and courses, publications, professional development and mentoring, and advocacy to governments.

To discover more about the IET please visit https://www.theiet.org/.

About IET books

The IET publishes books across many engineering and technology disciplines. Our authors and editors offer fresh perspectives from universities and industry. Within our subject areas, we have several book series steered by editorial boards made up of leading subject experts.

We peer review each book at the proposal stage to ensure the quality and relevance of our publications.

Get involved

If you are interested in becoming an author, editor, series advisor, or peer reviewer please visit https://www.theiet.org/publishing/publishing-with-iet-books/ or contact author_support@theiet.org.

Discovering our electronic content

All of our books are available online via the IET's Digital Library. Our Digital Library is the home of technical documents, eBooks, conference publications, real-life case studies and journal articles. To find out more, please visit https://digital-library.theiet.org.

In collaboration with the United Nations and the International Publishers Association, the IET is a Signatory member of the SDG Publishers Compact. The Compact aims to accelerate progress to achieve the Sustainable Development Goals (SDGs) by 2030. Signatories aspire to develop sustainable practices and act as champions of the SDGs during the Decade of Action (2020-2030), publishing books and journals that will help inform, develop, and inspire action in that direction.

In line with our sustainable goals, our UK printing partner has FSC accreditation, which is reducing our environmental impact to the planet. We use a print-on-demand model to further reduce our carbon footprint.

Published by The Institution of Engineering and Technology, London, United Kingdom

The Institution of Engineering and Technology (the "**Publisher**") is registered as a Charity in England & Wales (no. 211014) and Scotland (no. SC038698).

First published 2024

The Institution of Engineering and Technology
Futures Place
Kings Way, Stevenage
Herts, SG1 2UA, United Kingdom

www.theiet.org

British Library Cataloguing in Publication Data
A catalogue record for this product is available from the British Library

ISBN 978-1-83953-941-1 (hardback)
ISBN 978-1-83953-942-8 (PDF)

Typeset in India by MPS Limited

Cover image credit: Christophe Lehenaff/Moment via Getty Images

Contents

About the editors

Henry Shu-Hung Chung is a chair professor at the Department of Electrical Engineering, City University of Hong Kong. He also serves as director of the Centre for Smart Energy Conversion and Utilization Research and is EiC or board member of key journals. His awards include the 2021 IEEE Power Electronics Society Award and the Outstanding Research Award of Hong Kong CityU 2020. He holds 80 patents. His research focus includes smart grids, renewable energy conversion, and application of AI in power electronics.

Huai Wang is a professor at AAU Energy at Aalborg University, Denmark. He participated in research stays at the Massachusetts Institute of Technology, United States of America, and Swiss Federal Institute of Technology in Zurich (ETH Zurich), Switzerland. He has received several academic awards, including Green Talents Award 2014. His research covers power electronics and transistors, inverters, power semiconductors and thermal resistance. He has already authored or co-authored 355 publications.

Frede Blaabjerg is a professor at AAU Energy at Aalborg University, Denmark. Prior he was with ABB-Scandia and joined Aalborg University in 1995. His current research interests include power electronics and its applications such as in wind turbines, PV systems, reliability, Power-2-X, and power quality. He has published more than 800 journal papers and edited or co-authored several key books. He was awarded 42 IEEE Prize Paper Awards, the Global Energy Prize in 2019, and the 2020 IEEE Edison Medal. In 2019–2020, he served as a president of the IEEE Power Electronics Society. He was nominated in 2014–2021 by Thomson Reuters to be among the 250 most-cited researchers in Engineering in the world.

Michael Pecht is a professor of Mechanical Engineering at Maryland University, USA. He is the founder and director of the Center for Advanced Life Cycle Engineering at UMD, funded by over 150 of the world's leading electronics companies. He is an IEEE fellow, an ASME fellow, an SAE fellow, and an IMAPS fellow and was editor-in-chief of Microelectronics Reliability for 16 years. He has written more than 40 books on his research areas of product reliability, development, use, and supply chain management and consults for 22 major international electronics companies. In 1999, he devised Pecht's Law providing an estimate of semiconductor device reliability trends and the requirements for accelerated testing.

Chapter 1

Reliability engineering in power electronic converter systems

Huai Wang[1], Frede Blaabjerg[1], Henry Shu-Hung Chung[2], Chang Lu[3,4] and Michael Pecht[5]

Power electronics has progressively gained important status in electrical power generation, transmission, distribution, and consumption. With more than 70% of electricity processed through power electronic systems, recent research has endeavored to improve its reliability performance to comply with more stringent constraints on cost of ownership, safety, and availability in various applications. This chapter serves to bridge the advancements between the disciplines of power electronics and reliability engineering and to give an overall introduction of the following chapters in this book. It briefly discusses the following topics: motivations for more reliable power electronic converter systems, the historical development of power electronics and reliability engineering, the physics-of-failure (PoF) of power electronics components, design for reliability (DFR), reliability and accelerated testing, strategies to improve reliability performance, and the challenges and opportunities in this multi-disciplinary area.

1.1 Performance factors of power electronic systems

Power electronic systems aim to best serve the needs of highly efficient generation and conversion of electrical energy. This section discusses the basic architecture of a power electronic system and its design objectives and performance factors.

[1]Department of Energy, Aalborg University, Denmark
[2]Centre for Smart Energy Conversion and Utilization, City University of Hong Kong, Hong Kong SAR, China
[3]Centre for Advances in Reliability and Safety, Hong Kong
[4]Department of Electrical and Electronic Engineering, The Hong Kong Polytechnic University, Hong Kong
[5]Department of Mechanical Engineering, Center for Advanced Life Cycle Engineering (CALCE), University of Maryland, USA

1.1.1 Power electronic converter systems

Electrical energy conversion by power electronic systems can be classified into the following four categories [1]:

1. Voltage conversion for both direct current (DC) and alternate current (AC)
2. Frequency conversion
3. Wave-shape conversion
4. Poly-phase conversion

The above-mentioned four types of conversions are used to meet the needs of industry sectors such as automotive, telecommunications, portable equipment, smart grids, High voltage DC (HVDC), flexible AC transmission systems (FACTs), traction, renewable energy, mining, electrical aircraft, and aerospace. The power level ranges from sub-W to multi-MW and GW, processed either by a single power converter or multiple power converters.

Figure 1.1 shows the general architecture of a typical power electronic converter system. The electrical energy in the input and output is represented in the

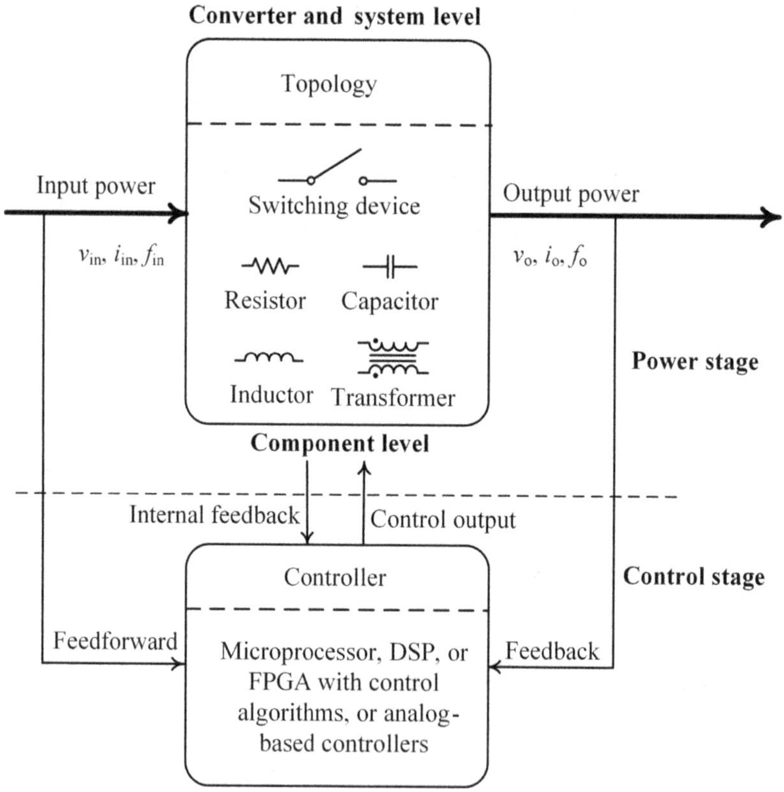

Figure 1.1 The basic architecture of a power electronic converter system

form of input voltage v_{in}, input current i_{in}, and input side frequency f_{in}, and output voltage v_o, output current i_o, and output side frequency f_o. The upper and lower blocks in Figure 1.1 show the power stage and control stage, respectively. The power stage is composed of switching devices and one or more kinds of passive components, connecting by a specific circuit topology. The switching devices are turned on and off at a frequency in the range of hundreds of Hz to hundreds of MHz, depending on the capability of the devices and the application requirements. The capacitors and inductors are used for energy storage and filtering purposes. The transformers are usually of the high frequency type and are used for galvanic isolation and step-up/down of voltage. Resistors are in fact not desirable in power electronic systems since they introduce power loss. However, in practical systems, there are parasitic resistances in components and resistors used for circuit snubbers, balancing circuits, and filter damping. The control stage receives conditioned low voltage signals from the power stage and sends back driven signals to control the on/off of the switching devices, including protection signals at the presence of abnormal operation. It can be implemented either in analog circuits, digital processors, or a hybrid way of both analog and digital parts.

1.1.2 Design objectives for power electronic converters

With the advancements in power switching devices and passive components, circuit topologies, control strategies, sensors, digital signal processors, and system integration technologies, there is a large variety of power electronic converter systems. The converter- or system-level performance is determined by the component-level performance, the applied circuit topology and control strategy, and the practical implementation and usage conditions. Besides the required functionality under specified conditions, power electronic converter design mainly considers the following five performance factors:

1. Cost
   Cost is usually the foremost consideration in most consumer and industrial applications such as lighting systems, photovoltaic plants, and wind turbines. For safety-critical applications, such as in aerospace, railway, and aircraft, other factors may weigh more than cost. A comprehensive cost analysis should include the design cost, manufacturing cost, operational cost, and recycle cost if applicable—that is, the life cycle cost.
2. Efficiency
   One of the distinctive features of power electronic converters is that they can convert and control electrical energy with high efficiency. Therefore, improving the efficiency is always an important design objective to push close to the limit of zero power loss. The widely used efficiency definitions are peak efficiency, rated power efficiency, and weighted efficiency under multiple loading conditions (e.g., European weighted efficiency for PV inverters). For power converters used for renewable energy applications, such as PV and wind power, the long-term total energy production is more useful since the power level could fluctuate frequently with the weather conditions. Therefore, the

energy efficiency defined by the annual output energy over the annual input power of a power converter provides much more insight. It takes into account the long-term environmental and operational conditions, as well as the impact of component degradation.

3. Power density (kW/L or kW/kg)
   A general trend in power electronics is toward increased power density in terms of reduced volume or weight for a given power rating. This can be achieved mainly by reducing passive components with the aid of increasing switching frequency of the power devices, and better thermal management and integration solutions.
4. Reliability
   The usual engineering definition of reliability is the probability that an item will perform a required function without failure under the stated conditions for a stated period of time [2]. Accordingly, a comprehensive reliability description includes five important aspects: definition of failure criteria, stress condition, reliability numbers (%), confidence level (%), and the time after which the reliability number and confidence level apply. A reliability number will vary by adjusting any one of the other four aspects, indicating the importance of understanding the background information behind a reliability number. As discussed in Section 1.1.3, more stringent reliability requirements and cost constraints are imposed on power electronic converters in both classical applications and emerging applications.
5. Manufacturability
   With the ever-increasing cost of labor involved in the manufacturing process, it is desirable to have power electronic design solutions that can be easily and economically implemented into final products. The manufacturability is largely dependent on the decisions made during the design phase [3]. When it comes to the power electronic converters, the modular design and integration of the component level, power module level, and system level can be accomplished to improve the manufacturability [4]. The emerging additive manufacturing technologies, including 3D printing, will provide new opportunities for power electronic converter design to have better manufacturability [5].

The performance requirements of power electronic products are increasingly demanding in terms of the above five performance factors. Of these, the reliability performance influences the safety, service quality, lifetime, availability, and life cycle cost of the specific applications.

1.1.3 Reliability requirements in typical power electronics applications

While targets concerning the efficiency of power electronic systems are within reach, the increasing reliability requirements create new challenges as discussed in [6,7]:

1. Mission profiles for critical applications (e.g., aerospace, military, avionics railway traction, automotives, data centers, medical electronics).

2. Emerging applications under harsh environments and long operation hours (e.g., onshore and offshore wind turbines, photovoltaic systems, air conditioners, and pump systems).
3. More stringent cost constraints, reliability requirements, and safety compliance requirements (e.g., demand for parts per million (ppm) level failure rates in future products).
4. Continuous need for higher power density in power converters and higher-level integration of power electronic systems that may invoke new failure mechanisms and thermal issues.
5. Uncertainty of reliability performance for new materials and packaging technologies (e.g., SiC and GaN devices).
6. Increasing complexity of electronic systems in terms of functions, number of components, and control algorithms.
7. Resource constraints (e.g., time, cost) for reliability testing and robustness validation due to time-to-market pressure and financial pressure.

Table 1.1 illustrates the industrial challenges from a reliability perspective of yesterday, today, and tomorrow. To meet the future application trends and customer expectations for ppm level failure rate per year, it is essential to have a better understanding of failure mechanisms of power electronic components and to explore innovative R&D approaches to build reliability in power electronic converter systems.

Table 1.2 summarizes the typical design target of lifetime in different applications. To meet those requirements, a paradigm shift is going on in the area of automotive electronics, avionics, and railway traction by introducing new reliability design tools and robustness validation methods [8,9].

Table 1.1 The reliability challenges in industry: past, present, and future [6,7]

	Yesterday	**Today**	**Tomorrow**
Customer expectations	• Replacement if failure • Years of warranty	• Low risk of failure • Request for maintenance	• Peace of mind • Predictive maintenance
Reliability target	• Affordable market returns (%)	• Low market return rates	• ppm market return rates
R&D approach	• Reliability test • Avoid catastrophes	• Robustness tests • Improve weakest components	• Design for reliability • Balance with field load / mission profile
R&D key tools	• Product operating and function tests	• Testing at the limits	• Understanding failure mechanisms, field load, root cause • Multi-domain simulation • ...

Table 1.2 Typical lifetime target in different power electronic applications [6,7]

Applications	Typical design target of lifetime
Aircraft	24 years (100,000 hours flight operation)
Automotive	15 years (10,000 operating hours, 300,000 km)
Industry motor drives	5–20 years (60,000 hours in at full load)
Railway	20–30 years (10 hours operation per day)
Wind turbines	20 years (24 hours operation per day)
Photovoltaic plants	5–30 years (12 hours per day)

In the applications listed in Table 1.2, the reality is that power electronic converters are usually one of the weakest links to limit the lifetime of the system. For example, with the increasing penetration of renewable energy sources and the increasing adoption of more efficient variable-speed motor drives [10,11], the failure of power electronic converters in wind turbines, photovoltaic systems, and motor drives is becoming an issue. Field experiences in renewables reveal that power electronic converters are usually one of the most critical assemblies in terms of failure level, lifetime, and maintenance cost [12]. For example, it shows that frequency converters caused 13% of the failures and 18.4% of the downtime of 350 onshore wind turbines in a recent study associated with 35,000 downtime events [13]. Another representative survey in [14] concludes that PV inverters are responsible for 37% of the unscheduled maintenance and 59% of the associated cost during five years of operation of a 3.5 MW PV plant. It should be noted that such statistics always look backward, as those designs are more than ten years old. The present technology could have different figures.

To fulfill future reliability requirements, multidisciplinary efforts devoted to both power electronics and reliability engineering are needed. Traditional academic research on power electronics focuses on improving the efficiency and power density, while reliability performance is usually not considered in the design phase. It is therefore necessary to better bridge the gap between the power electronics research in universities and the needs of industry.

The lifetime and robustness of the power electronics should be evaluated to ensure long-term performance, safety, and compliance with industry reliability requirements. One of the main approaches is reliability testing, including accelerated life testing, thermal cycling, vibration testing, and environmental stress screening, among others. In addition, specific requirements of sample sizes, pre- and post-stress tests, and definition of test failures should be considered for comprehensive reliability evaluation. Organizations like JEDEC (Joint Electron Device Engineering Council) and AEC (Automotive Electronics Council) develop detailed standards and guidelines for the reliability and qualification testing of electronic components and systems.

The common stress tests for evaluating power semiconductor devices' lifetime and intrinsic reliability are listed below.

1. High-temperature operating life (HTOL)
   This test stresses the device at high temperature and high voltage, under dynamic operational conditions over a period of time. The device should be biased at the maximum operating voltage, and the dynamic operational conditions should be set at a point where the device's junction temperature can remain within its maximum limit. The device's performance will usually be monitored during the testing. Figure 1.2 is a demonstration of the HTOL test setup. This reliability stress test is sometimes referred to as a lifetime test, device life test, or extended burn-in (BI) test.
2. High-temperature reverse bias (HTRB) and high-temperature gate bias (HTGB)
   HTRB and HTGB are the common static test methods for checking the robustness of the semiconductor's intrinsic structure. Figure 1.3 depicts the setups of HTRB and HTGB tests.

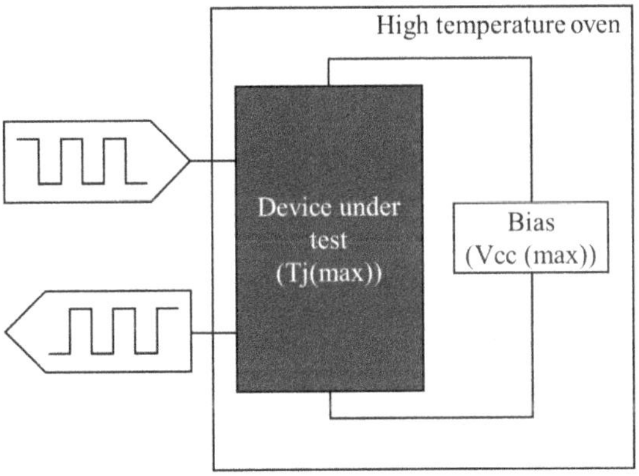

Figure 1.2 Demonstration of HTOL test setup

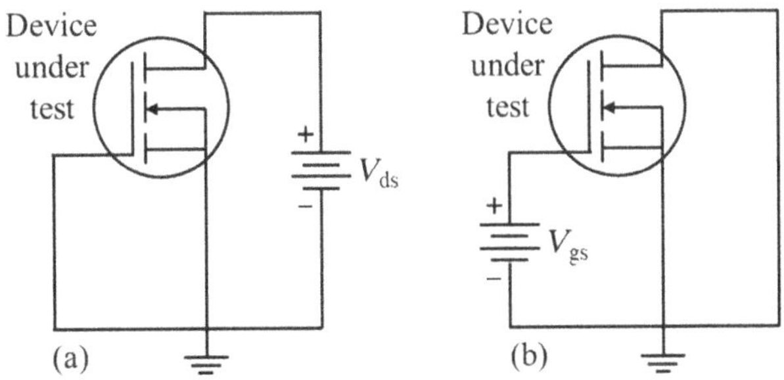

Figure 1.3 Demonstration of (a) HTRB and (b) HTGB test setups

HTRB stresses the device under high temperature and the maximum reverse bias condition between the drain and source terminals, and the drain to source leakage current should be monitored. This test aims to evaluate junction integrity and the robustness of field depletion structures of the die.

Similarly, HTGB stresses the device under the maximum bias condition across the gate oxide, and the gate leakage current should be monitored. This test is used to verify the quality of the oxide layer of the die.

The above-mentioned tests stress the device for a certain amount of time, for example, 500 or 1000 hours. During the testing, electrical characteristics are normally measured at intermediate intervals but should be within a limited time window.

In addition to testing the intrinsic characteristics and quality, the semiconductors' assembly and packaging reliability must be tested. Below are the methods used to qualify the assembly and packaging processes.

1. Bond wire pull test
   Bond wire quality is critical, given it is one of the major failure modes of power semiconductor devices. This test evaluates the bond strength of the wire-to-die, wire-to-substrate, or wire-to-package bonds. Any bond pulling that results in separation under an applied stress less than a required amount of force shall constitute a failure.
2. Shear test
   The shear test can be applied in multiple cases, such as die-attach, bond wires, solder balls, and the interconnections between surface-mounted components. This is a method of determining the maximum shear force of materials and procedures that can withstand before failure, and the failure criteria differ based on the attached area in between.
3. Solderability
   This test is to evaluate the ability of the component's terminations (e.g., leads, balls, pins) to be wetted and effectively soldered. The devices will first undergo preconditioning to simulate the long-term storage or exposure to environments and then be dipped into a molten solder bath for a specified duration. After the solder dip, the component's terminations are visually inspected for wetting, solder coverage, and the absence of defects.

Environmental stress tests are essential for evaluating the reliability of products, both at the device level and the system level. The following are some common test methods used for power electronic devices and systems.

1. Thermal cycling test
   This test is to evaluate the robustness of power electronic components under dynamic thermal stresses as shown in Figure 1.4. It helps assess the reliability of solder joints, wire bonds, and other interconnections under thermal expansion and contraction. There are two thermal cycling approaches, passive and active.

   The passive approach is called the temperature cycling test, in which the device under test is placed in a temperature cycling chamber or oven.

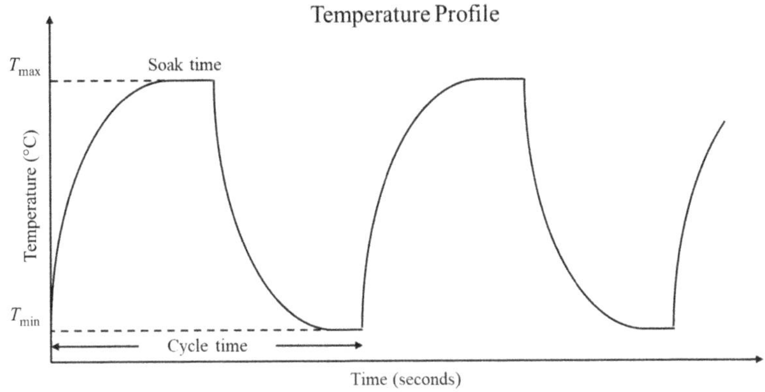

Figure 1.4 Temperature profile of thermal cycling test

This approach has simpler test setup and equipment requirements, the temperature cycling profile can be controlled, and it is suitable for testing a large number of samples simultaneously.

In the active approach, the device under test is electrically powered on and off and actively generates heat at the junction during the testing so it is called the power cycling test. This method is useful for evaluating the reliability of power devices because it can better simulate the actual electrical operating conditions as well as the thermal stresses experienced by the device.

2. Thermal shock

   Thermal shock testing also stresses the device with dynamic temperature changes, but the changes are more abrupt and faster than those of the thermal cycling test. This test determines whether the device under test is capable of withstanding sudden temperature changes. In real applications, power devices may experience rapid temperature changes due to self-heating, and the system-level circuits may experience rapid temperature changes during hand or wave soldering.

   The temperature profile of thermal shock is similar to thermal cycling but with a faster ramp rate and shorter soak time.

3. Highly accelerated temperature and humidity stress test (HAST)

   HAST test subjects the device or system to high-temperature, high-humidity conditions under certain electrical bias conditions. It assesses the product's resistance to moisture-induced failures such as corrosion, electrical leakage, and delamination. This test is commonly used for evaluating the reliability of packaging and material interfaces of the semiconductor devices.

   Sometimes, the devices may be stressed under high temperature and high humidity conditions without electrical bias, and these scenarios are called uHAST.

4. High humidity, high temperature, and high voltage reverse bias (H3TRB)

   This test evaluates the insulation ability and reliability of the semiconductor transistor's package. In power electronic applications like offshore wind turbines, the power devices are exposed to high temperature and humidity environments and need to withstand high voltages as well. The multiple stressors will accelerate the degradation process and unexpected failures might occur. H3TRB test can simulate extremely humid and high-temperature conditions that may create chemical reactions from the edge and surface of the passivation layer inside the device package.

5. Vibration test
   This test exposes the device or system to controlled vibrations, mimicking the mechanical stresses encountered during transportation, installation, or operation. It evaluates the robustness of the product's mechanical structure and interconnections. The appropriate vibration test methods, such as sinusoidal vibration, random vibration, or a combination of both, will be applied to the device under test. The electrical characteristics and the physical changes will be monitored during the vibration test. Abnormal behavior, such as resonance frequencies, excessive vibration amplification, or performance degradation, will be documented.
6. Electrical overstress test
   This test exposes the device or system to electrical conditions beyond the normal operating limits such as overvoltage, overcurrent, or electrostatic discharge (ESD). It evaluates the product's robustness and protection mechanisms against electrical failures.

   ESD testing is performed to test the susceptibility of the devices or systems to ESD. For the device-level ESD testing, based on JEDEC standards, there are two main testing models, the human body model (HBM) and the charged device model (CDM). These methods are to simulate the different ESD scenarios that may induce different failure mechanisms. The HBM and CDM scenarios are illustrated in Figure 1.5.

   The test pulses for different ESD test models will be applied to each pin of the device's package. After the ESD stress, the devices under test will be measured for functional and parametric failures. ESD testing can also be applied to system-level reliability evaluation. In system-level tests, ESD events are applied to the I/O pins that are externally exposed, outside cases or shielding, an arc through a vent hole or a seam of the PCB, or a discharge event inside the system. Compared to device-level testing, the ESD pulse waveform for system-level testing requires higher voltage and current values. Figure 1.6

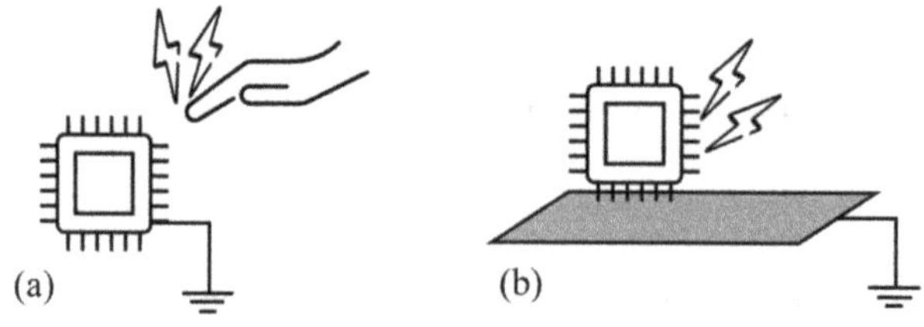

Figure 1.5 (a) HBM and (b) CDM scenarios

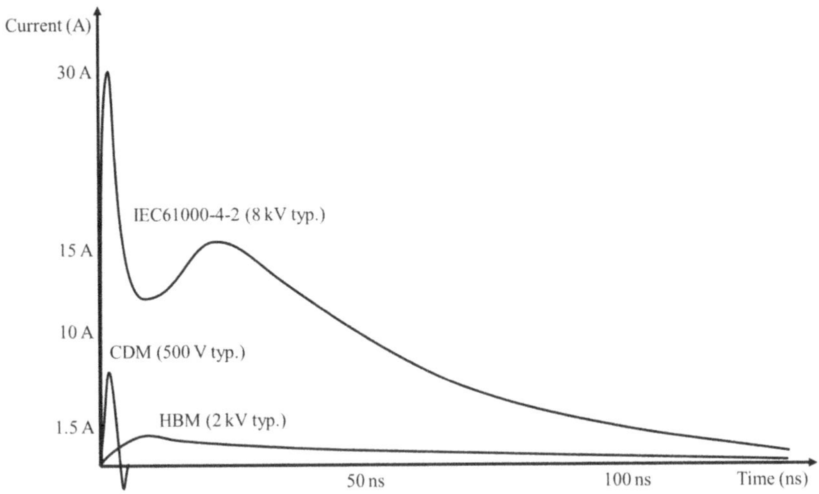

Figure 1.6 Comparison of CDM, HBM, and IEC62000-4-2 test pulse models

compares the three ESD test pulse models, CDM, HBM, and IEC6100-4-2 which is the system-level testing model.

1.2 Reliability engineering in power electronics

This section will start with the key terms and metrics that are widely used in reliability engineering. Then the historical development of both power electronics and reliability engineering will be discussed. After that, a brief presentation on the topics that are correlated to the following chapters in this book will be given. It covers the reliability of power electronic components, DFR in power electronics, accelerated testing, and strategies to improve the reliability of power electronic converter systems.

1.2.1 Key terms and metrics in reliability engineering

1.2.1.1 Failure distribution

A failure distribution shows the frequency histogram of the failure occurrence, modeled as a kind of probability density function (pdf) $f(x)$. The variable x could be time, distance, cycles, or something else depending on the parameter of interest. Figure 1.7 shows an example of the failure distribution of a group of capacitors for power electronics applications. By defining $F(x)$ as the cumulative distribution function, reliability is shown as

$$R(x) = 1 - F(x) = 1 - \int_0^x f(x)dx \tag{1.1}$$

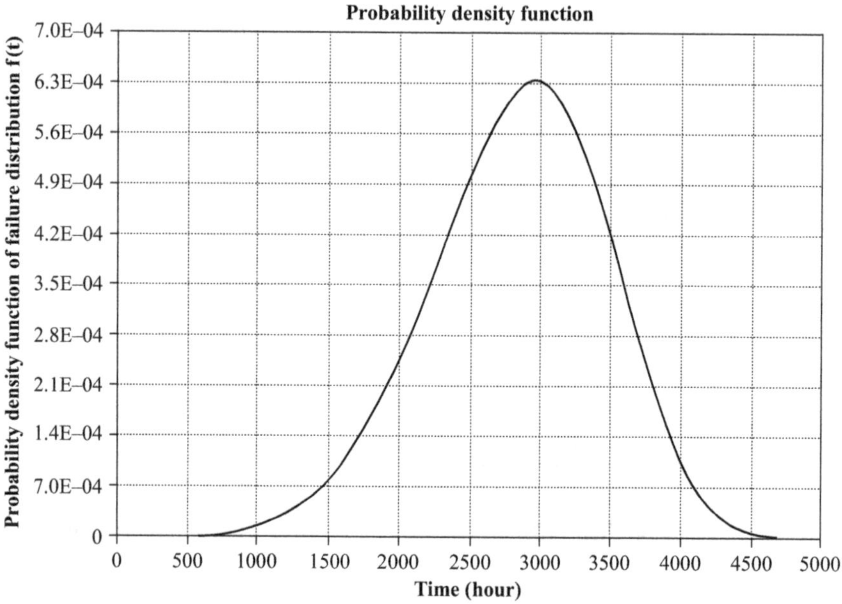

Figure 1.7 An example of a failure distribution of power electronics capacitors

The hazard rate $h(x)$ is defined as the conditional probability of failure in the interval x to $(x+\Delta x)$ [2]; that is

$$h(x) = \frac{f(x)}{R(x)} \tag{1.2}$$

There are numerous failure distributions as discussed in [2]. In this chapter, the Weibull distribution is discussed. The exponential distribution is not discussed as it describes a scenario of constant hazard rate, also called the constant failure rate, λ, which rarely occurs in practice.

The Weibull distribution was introduced by Walloddi Weibull in 1951 [15]. Its pdf function, reliability function, and hazard rate are defined as

$$f(x) = \frac{\beta}{\eta^{\beta}} x^{\beta-1} \exp\left[-\left(\frac{x-\gamma}{\eta}\right)^{\beta}\right] \tag{1.5}$$

$$R(x) = \exp\left[-\left(\frac{x-\gamma}{\eta}\right)^{\beta}\right] \tag{1.6}$$

$$h(x) = \frac{\beta}{\eta^{\beta}} x^{\beta-1} \tag{1.7}$$

where β is the shape parameter and η is the scale parameter, or characteristic life, which is the life at which 63.2% of the population will have failed. γ is the location parameter, called the failure free period. The distribution presented in (5) is a three-parameter Weibull distribution. In many practical applications with failure occurring from time zero, γ is zero and (5) becomes a two-parameter Weibull distribution accordingly. The Weibull distribution can be applied to model a wide range of life distributions of engineered products, since with different values of β, a Weibull distribution is equivalent or approximated to other kinds of distributions. For example, when $\beta = 1$, it results in an exponential distribution with a constant hazard rate; and when $\beta = 3.5$, it approximates to a normal distribution. When $\beta < 1$, the hazard rate is decreasing; and when $\beta > 1$, the hazard rate is increasing.

1.2.1.2 Lifetime and percentile life

Lifetime is the time to which an item reaches its failure criteria. The criteria could be a complete loss of function, a certain level of degradation, or the stage of being uneconomic.

In practice, another term—percentile life—is more widely used to present the lifetime of a population of items. It is the time by which a certain percentage of the items might have failed. For example, B10 lifetime corresponds to the time by which 10% of the items have failed, that is, when the reliability is equal to 90%. Figure 1.8 describes the relationship between the reliability and percentile life based on the example shown in Figure 1.7. The B1 lifetime and B10 lifetime in the example are 1277 hours and 2003 hours, respectively.

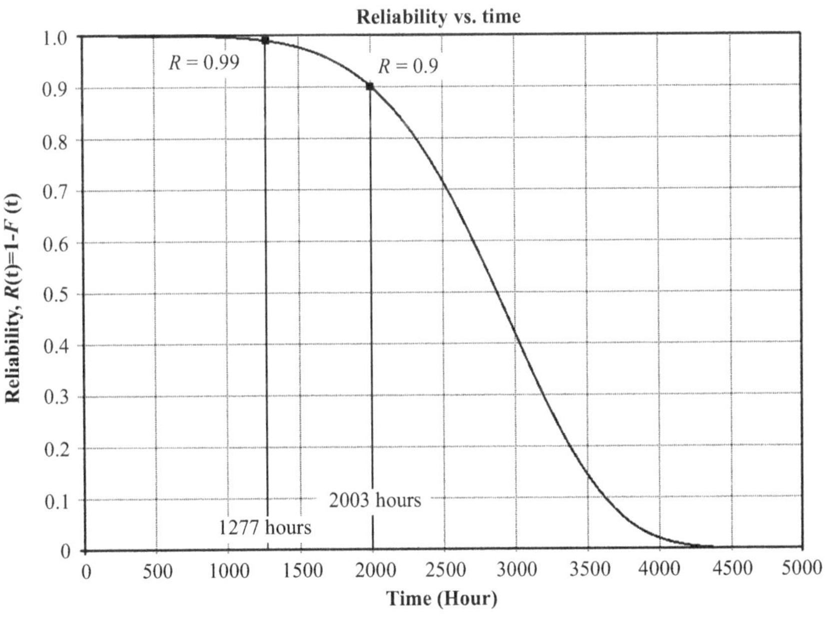

Figure 1.8 An example of reliability and percentile lifetime based on Figure 1.7

1.2.1.3 Bathtub curve

The bathtub curve [16] shown in Figure 1.9 is commonly used to illustrate the different forms of the hazard rate change during the entire life of an electronic component or system, although electronics rarely have this profile There are three distinct intervals as follows:

Interval I – The early failure is dominant due to quality control issues or early wear out failures, with a decreased hazard rate (i.e., $\beta < 1$).

Interval II – The random failure is dominant, for example, catastrophic failure due to a single event of overvoltage, overcurrent, or overheating, human error. It is widely assumed that the hazard rate is constant in this time interval (i.e., $\beta = 1$)

Interval III – The end-of-life of components due to degradation is dominant, with an increased hazard rate (i.e., $\beta > 1$).

It should be noted that the hazard rate in Interval II is rarely constant in practical applications. Moreover, the degradation (e.g. wearout) of power electronic components usually starts immediately when in use or even in storage, which is much earlier than what is shown in Figure 1.9.

1.2.1.4 MTTF and MTBF

Mean-time-to-failure (MTTF) and mean-time-between-failure (MTBF) are two classical metrics that are widely appeared in the literature and in product manuals. They are used for non-repairable items and repairable items, respectively. In statistics, it is the expected value of the failure distribution function $f(x)$ and is applicable for any type of distribution. In reliability engineering, they are more often applied for the case of exponential distribution. MTBF (and MTTF) is

$$\mathrm{MTBF} = \frac{1}{\lambda} \tag{1.8}$$

The fundamental assumption of (8) is that the hazard rate is constant throughout the entire life, which is not valid for most of the durable components and systems in industrial applications [12] and [17,18]. Moreover, it should be

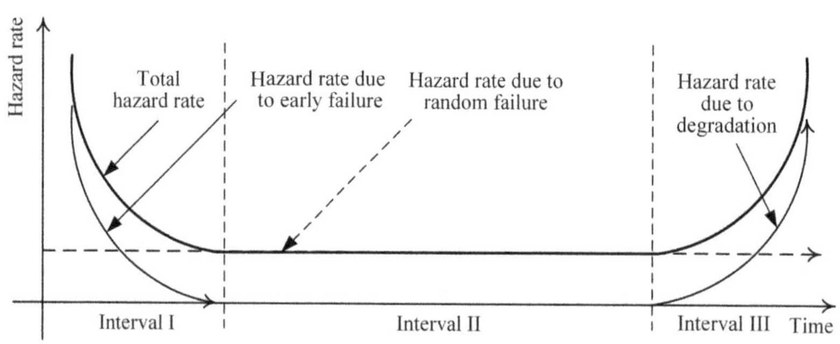

Figure 1.9 Bathtub curve: a widely assumed hazard rate curve for electronic components and systems

noted that MTTF or MTBF corresponds to the time when 63% of the items have failed and the reliability is 37%. Therefore, it is irrelevant to the lifetime or percentile life (except for B63) discussed before. The value of MTTF or MTBF provides very limited insights for reliability design and reliability performance comparison. Many power electronics users care most about the time during which the reliability is 90% or above.

1.2.1.5 Mean cumulative function (MCF) curve

As discussed above, the hazard rate over operational time is usually not constant, and MTTF and MTBF in these cases are not recommended in order to avoid misleading results. An alternate technique to present the failure level and time is the MCF curve [19]. When analyzing repairable systems, the MCF curve graphs the number of failures versus time since installation. It is also possible to represent the behavior of the group of systems by the average number of failures versus time. The MCF curve is the integration of hazard rate with time. The customer will be the person who sees the accumulated failure level of all random failures and failures due to degradation. More details on the MCF curve can be found in [6].

1.2.1.6 Accelerated factor (AF)

The equation for AF is as follows [2]:

$$AF = \frac{L_{\text{Field}}}{L_{\text{Test}}} \tag{1.9}$$

where L_{Field} is the product lifetime at the field stress level and L_{Test} is the lifetime at the test stress level. The lifetime L can be calculated by failure models such as the Arrhenius equation and the Eyring equation.

To calculate the AF for the thermally activated degradation process, the Arrhenius equation is commonly used as the failure model, which is expressed as

$$k = Ae^{\frac{-E_a}{k_B T}} \tag{1.10}$$

where k is the rate constant (frequency of collisions resulting in a reaction), T is the device field/test temperature in degrees Kelvin, A is the Arrhenius factor (a temperature-independent constant), and E_a is the thermal activation energy in electron volts (0.7eV), k_B is the Boltzmann's constant 8.62×10^{-5} eV/K.

HTOL is a useful stress test method for calculating the AF of failures exhibiting a constant failure rate. Applying the Arrhenius equation (10) to (9), the AF of T_{Test} to T_{Field} can be calculated by

$$\text{AF}(T_{\text{Test}}\ to\ T_{\text{Field}}) = e^{\frac{E_a}{k_B}\left(\frac{1}{T_{\text{Field}}} - \frac{1}{T_{\text{Test}}}\right)}. \tag{1.11}$$

For example, if the T_{Field} is 55 °C, and the T_{Test} is 125 °C for the HTOL stress test, the AF will be 78.6, which means that every hour of the HTOL stress test at 125 °C is equivalent to 78.6 hours of use at 55 °C in the field application.

1.2.1.7 Failure-in-time (FIT)

FIT is the unit to describe the number of failures per 10^9 device hours. 1 FIT means one failure in 10^9 device hours.

The FIT calculation can be based on AF, and the equation is expressed as follows:

$$FIT = \frac{failure \cdot 10^9}{hours \cdot sample\ size \cdot AF}$$

Using the example for AF calculation, if 3 out of 80 samples fail after 2000 hours of the 125°C HTOL stress test, the FIT will be 239, indicating there are estimated 239 failures after 10^9 hours of operation.

The above seven terms and metrics will be frequently used in the following chapters of this book. Moreover, it is worth mentioning the definitions of the following three terms as discussed in detail in [8]:

1. Mission profile: a representation of all relevant conditions that a specific item will be exposed to in all of its intended applications throughout its entire life cycle.
2. Robustness: insensitivity to noise (i.e., variation in operating environment, manufacture, distribution, and all factors and stresses in the life cycle).
3. Robustness validation: a process to demonstrate that a product performs its intended function(s) with sufficient margin under a defined mission profile for its specified lifetime.

1.2.2 Historical development of power electronics and reliability engineering

The invention of the practical transformer and the poly-phase AC system in the 1880s brought about the demand for better rectifying devices, which were the initial enablers of the emergence of power electronics. The introduction of the thyristor in 1957 is accepted as the beginning of modern power electronics. Since then, the historical development of power electronics is device-driven, as shown in Figure 1.10. The advancement of power semiconductor devices enables higher

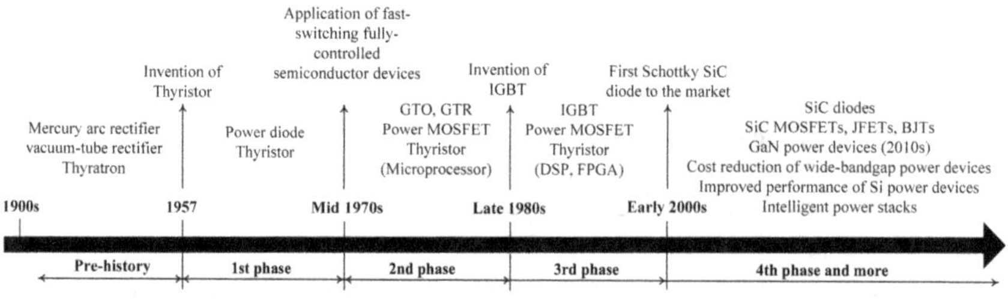

Figure 1.10 Key milestones in the advancement of power electronic semiconductors

switching speed, wider power and temperature range, and better efficiency and reliability of power electronic systems.

The birth of statistics in 1654 and the adoption of mass production in 1913 are the essential ingredients of reliability engineering [20]. After World War I, the US Department of Defense initiated the study of the failures of vacuum tubes and these efforts along the years, eventually gave birth to a new discipline. In the same year, 1957, as the era of modern power electronics began, the Advisory Group on Reliability of Electronic Equipment (AGREE) report was published. It is when the reliability engineering became a distinct discipline. Since then, much pioneer work has been devoted to various reliability topics, as shown in Figure 1.11. One of the main streams is quantitative reliability prediction based on empirical data and various handbooks released by military and industry [17]. Another stream of the discipline focuses on identifying and modeling the physical causes of component failures, which was the initial concept of PoF presented in 1962 [21]. The PoF approach is a methodology based on root cause failure mechanism analysis and the impact of materials, defects, and stresses on product reliability [22]. However, until the 1980s, the handbook-based constant failure rate models (e.g., Military-Handbook-217 series [23]) have been predominantly applied for describing the useful life of electronic components. Since the 1990s, with the increased complexity of electronic systems and especially the application of integrated circuits, more evidence was suggesting that constant failure rate models are wrong and the 217 approach flawed [24]. The Military-Handbook-217F was therefore officially cancelled in 1995. Similar handbook documents, including FIDES and Telcordia, were also determined flawed and of no value. In its place, the PoF approach has been accepted as key to reliability engineering.

Intensive PoF research has been continuously conducted since the 1990s in microelectronics, as discussed in [24] and [25]. It changes the analysis of a system from a box of components to a box of failure mechanisms. The traditional handbook-based reliability prediction provides failure rate models for various components. The PoF approach analyzes and models each failure mechanism induced by environmental and usage stresses. For a given component, there could be multiple failure mechanisms which should be identified individually. Moreover, failure mechanisms are not limited to the component level. As discussed in the standard ANSI/VITA 51.2 [25], there are various failure mechanisms in component level (i.e., single transistor level), package level, and Printed Circuit Board (PCB) level. From this perspective, it is challenging to apply PoF to a complex system of

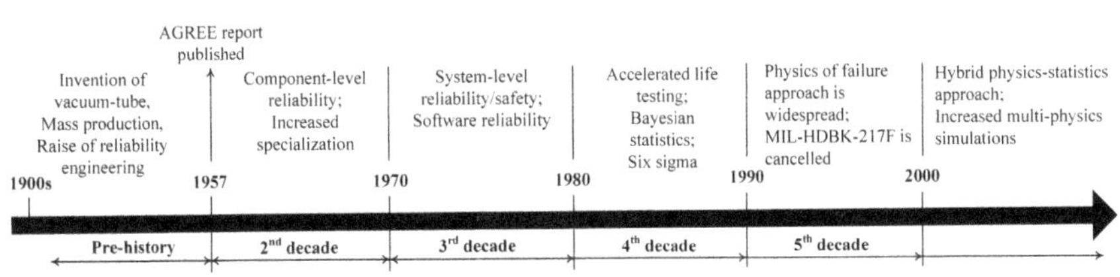

Figure 1.11 Key milestones in the advancement of reliability engineering

which limited number of models and their associated parameters are available [25]. Therefore, it is important to identify and focus on the critical failure mechanisms in specific applications.

With the transition from pure empirical based methods to more scientific based approaches, the paradigm shift in reliability research is going on from the following aspects [6]:

1. From components to failure mechanisms
2. From constant failure rate λ to MCF curve
3. From reliability prediction to comprehensive robustness validation [8]
4. From microelectronics to also power electronics

In power electronic applications, reliability has been and will continue to be one of the important performance factors in many applications, as discussed in Section 1.1.3. To address the challenges, power electronic engineers and scientists have started to apply various reliability tools for reliability prediction and reliability-oriented design of power electronic converter systems. Several literature reviews on field experiences [26], strategies to improve reliability of power electronic systems [27] and DFR for power electronic systems [28] have been presented in the last few years. Respective research in different applications is also discussed in various studies, such as three-phase converters for aircraft [29], power inverters for railway traction [30], inverters for hybrid electric vehicles [31], high-power variable-speed motor drives [32], and pulsed power converters for industrial process control [33]. Besides these applications, the last decade also saw much pioneering work on the reliability of power converters for wind turbines [34,35] and inverters for photovoltaic systems [36]. This reveals that, unlike the case in microelectronics, conventional handbook methods are still dominantly applied nowadays for reliability prediction in those studies.

While the pace of power electronics toward the PoF approach is relatively slower than that of microelectronics, the need for this paradigm shift has been well recognized in the automotive industry [8] and in other sectors. Especially, much interesting work from the semiconductor side investigates the failure mechanisms of IGBT modules [37] and physics-based lifetime models [38]. More realistic thermal stress analyses of Si- and SiC-based devices under long-term mission profiles are studied in [39,40], respectively. The level of technology and scientific understanding is still rapidly evolving. The research in microelectronics could provide a very important foundation for ongoing and future work in power electronics, especially from the methodology point of view. Nevertheless, it should be noted that most of the physics-based models are not scalable for power electronic components. System-level reliability issues (e.g., active thermal stresses, interconnections among components, interaction of different components) are still of interest to be investigated.

From this perspective, opportunities exist for power electronics to expand its role in dealing with efficient and reliable power processing in different kinds of applications. Nearly four decades ago, the scope of power electronics was defined by Newell as three of the major disciplines of electrical engineering, as shown in

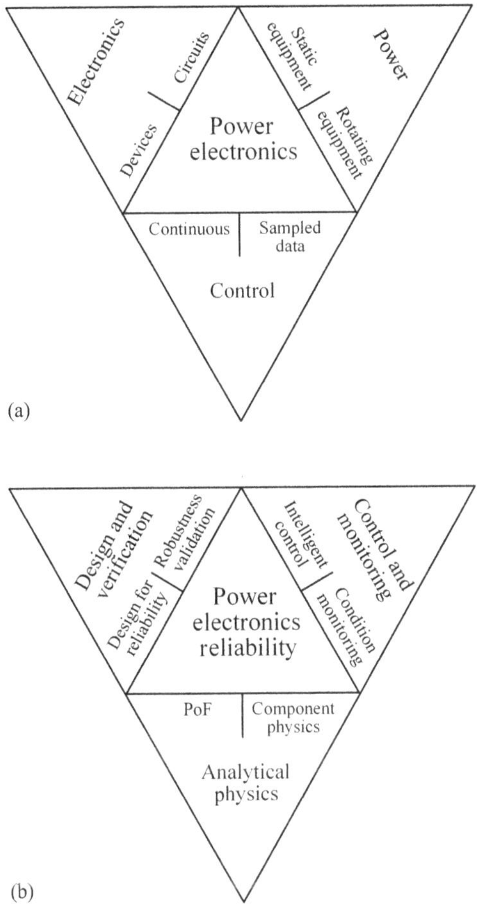

Figure 1.12 Defined scope in (a) power electronics by Newell in the 1970s [41] and (b) power electronic reliability research needs seen from today

Figure 1.12(a) [41]. Likewise, the future reliability research in power electronics that involves multidisciplinary knowledge is defined here in Figure 1.12(b). It covers three major aspects: analytical analysis to understand the nature of why and how power electronic products fail; DFR and a robustness validation process to build in reliability and sufficient robustness in power electronic products during each development process; and intelligent control and condition monitoring to ensure reliable field operation under specific mission profiles.

1.2.3 Physics of failure of power electronic components

The failure of a power electronic components is explained in Figure 1.13. A component fails when the applied load L exceeds the design strength S. The load L here refers to a kind of stress (e.g., voltage, cyclic load, temperature); and strength

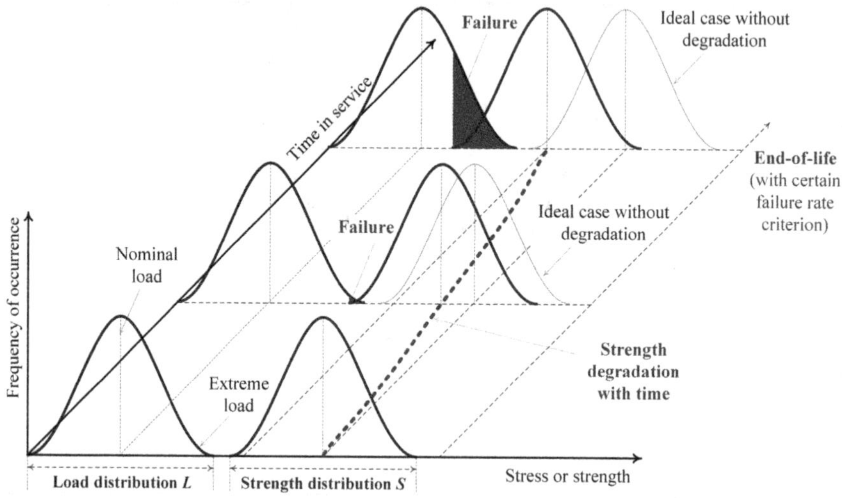

Figure 1.13 Load-strength analysis to explain overstress failure and wear-out failure

S refers to any resisting physical property (e.g., hardness, melting point, adhesion) [2]. Figure 1.13 presents a typical load-strength interference evolving with time. The load and strength of power electronic components are allocated within certain ranges, which can be described by specific probability density functions. Moreover, the strength of a material or device could be degraded with time. This implies also that failure could be reduced or eliminated within service life by either design with an increased strength (i.e., an increased design margin) or with a reduced load by control (i.e., stress control or load management).

As shown in Figure 1.12(b), understanding the reliability physics of power electronic components is the most fundamental aspect. In power electronic converter systems, power semiconductor devices (e.g., Si and SiC IGBTs and MOSFTs, GaN devices), capacitors, connectors, and fans are considered as the most vulnerable components. They are considered as the reliability critical components in power electronic converters, especially the IGBT modules in medium- to high-power applications and capacitors for AC filtering and DC-link applications. Therefore, Chapter 3 and Chapter 9 in this book discuss the reliability of capacitors and IGBT modules, respectively.

A focus point matrix (FPM), as described in [8], is a useful way to analyze the critical stressors that will cause the components to fail. Based on the accumulated industrial experiences and future research needs, Table 1.3 shows the critical stressors for different components in power electronic systems. Steady-state temperature, temperature swings, humidity, voltage, and vibration have different levels of impact on semiconductor devices, capacitors, inductors, and low-power control boards. The FPM provides information for determining the critical failure mechanisms. The interactions among different stressors are also of interest. More details on the failure mechanisms of respective components will be covered in later chapters of this book.

Table 1.3 Focus points matrix in reliability of power electronic components

Load			**Focus points**									
Climate + Design ⇒ Stressor			**Active power components**			**Passive power components**		**Control circuitry, IC, PCB, connectors…**				
Ambient	**Product design**	**Stressors**	**Die**	**LASJ**	**Wire-bond**	**Cap.**	**Ind.**	**Solder Joint**	**MLCC**	**IC**	**PCB**	**Connectors**
Relative humidity – *RH(t)* Temperature – *T(t)*	– thermal system – operation point – ON/Off – power *P*(*t*)	Temperature swing ΔT	**X**	**X**	**X**			**X**				
		Average Temperature *T*	X	X	X	*X*		X	X	*x*	*x*	*x*
		dT/dt	*x*	*x*	*x*	*x*						
		water								**X**	**X**	*x*
		Relative Humidity	*x*	*x*	*x*	X	*x*	*x*	*x*	*X*	*X*	*x*
Pollution	Tightness	Pollution						*x*			*x*	
Mains	Circuit	Voltage	*x*	*x*	*x*	*X*	X		*x*	*x*	*x*	*x*
Cosmic	Circuit	Voltage	*x*									
Mounting	Mechanical	Chock/vibration	*x*			*x*	*x*	*x*	*x*			*x*

LASJ—large area solder joint, MLCC—multi-layer ceramic capacitor, IC—integrated circuit, PCB—printed circuit board, Cap.—capacitor, Ind.—inductor, Level of importance (from high tolow): **X**-*X*-X-*x*.

Table 1.4 Failure components of power electronic systems [42,43]

Components	Percentage of total failure	Major failure modes
Capacitors	30	Short circuit (dendrite formation) Open circuit (high voltage overstress)
PCBs	26	Self-heating (bad layout design) Trace cracking (vibration or torsion)
Power semiconductors	21	Open circuit (bond wire lift-off/crack) Short circuit or high leakage current (oxide breakdown) Die attach delamination
Solder joints or connections	13	Open circuit/intermittent (fatigue or cracking) Surface contamination or oxidation
Connectors	3	Open circuit (mechanical damage or corrosion)
Others	7	Microelectronics digital logic error Electro-optical component degradation Resistors open circuit

Identifying the failure modes and understanding the root causes are critical to the physics of failure study. The failure of the overall power electronic system might be caused by one or more sub-components failing. Studies have shown that power semiconductors, driver circuits, capacitors, and PCBs are the major failures inside the system [42]. Table 1.4 summarizes the common failure components inside power electronic systems and their corresponding failure modes [42,43].

Failure analysis is essential for identifying the root cause and understanding the physics of failure. Below are some common failure analysis methods including non-destructive and destructive techniques.

1. Non-destructive failure analysis techniques
   (a) X-ray imaging
   X-ray imaging, including techniques like X-ray radiography and 3D X-ray computed tomography (CT), which are for the non-destructive examination of the internal structure and features of electronic devices. This method can reveal hidden defects in solder joints, metal traces, or wires without physically disassembling the device. Defects like cracks, voids, or misalignments are all possible to be spotted by the X-ray inspection technique.
   (b) Scanning acoustic microscopy (SAM)
   SAM has been a common method of checking the adhesion between two layers inside semiconductor devices [44] such as the interface between die and die-attach and the connection between the bond wire and bond pad. It is specifically useful for detecting the delamination of layers. It is worth noting that the device needs to be soaked in water for the SAM inspection. A proper dry-up process might be considered after the inspection.

2. Destructive failure analysis techniques
   (a) Decapsulation
   Decapsulation involves physically removing the encapsulation material from an electronic device to reveal the internal structures for further inspection and analysis. There are various methods of decapsulation such as chemical dissolution and laser etching. This is a necessary step to take for further destructive failure analysis on semiconductor devices.
   (b) Cross-section
   Instead of inspecting from the bird's eye view after decapsulation, cross-section is to cut the electronic device or component vertically to expose its internal structure and check for any cracks or breakdowns inside the layers.
   (c) Emission microscopy (EMMI)/optical beam-induced resistance change (OBIRCH)
   EMMI and OBIRCH are different methods for precisely locating the "hot spot" where current is leaking, and excessive heat is generated. Although these methods are non-destructive, the decapsulation of the device is often required for spotting the short-circuit location.
   Both methods are able to reveal the defective spots at the micrometer level, but they rely on different working principles and are sensitive to varying usage scenarios, as discussed in detail in [44].
   (d) Scanning electron microscopy (SEM) and energy-dispersive X-ray spectroscopy (EDX)
   SEM uses a focused beam of electrons to generate high-resolution images of the surface and internal structures of electronic devices at the nanometer scale. EDX is a complementary technique to SEM, which allows for the elemental analysis of the material composition at the area of interest.
   The combination of SEM and EDX can provide detailed information about the morphology, composition, and defects of integrated circuits, electronic components, and printed circuit boards.
   (e) Focused ion beam (FIB) analysis
   FIB analysis uses a finely focused beam of ions to selectively remove or mill material from the sample. Similar to SEM, FIB is also useful for analyzing the microstructure of electronic components at the nanometer scale, for example, defects like electromigration, dielectric breakdown, and interconnects and vias at the die level.
   Usually, to perform a satisfying FIB analysis, a dual-beam FIB configuration is needed, which allows material cutting/removal from one beam while observing the cutting area from the other.

1.2.4 Design for reliability of power electronic converter systems

The second aspect of power electronics reliability is to build reliability and sufficient robustness into the system design through the DFR process. Industries have advanced the development of reliability engineering from traditional testing for reliability to

DFR [2]. DFR is the process conducted during the design phase of a component or system that ensures them to be able to achieve the required level of reliability. It aims to understand and fix the reliability problems up front in the design process.

Due to the difference in chosen reliability tools and specified requirements of products, the DFR process varies with industry sectors; however, the generic form usually covers the processes of identify, design, analyze, verify, validate, and control [2]. A systematic DFR procedure specifically applicable to the design of power electronic converter systems is shown in Figure 1.14. By implementing the procedure, reliability is well considered and treated in each development phase, especially in the design phase. The design of power electronic converters is mission profile-based by taking into account parametric variations (e.g., temperature swings, solar irradiance level changes, wind speed fluctuations, load changes, manufacturing process). A mission profile-based case study will be discussed in

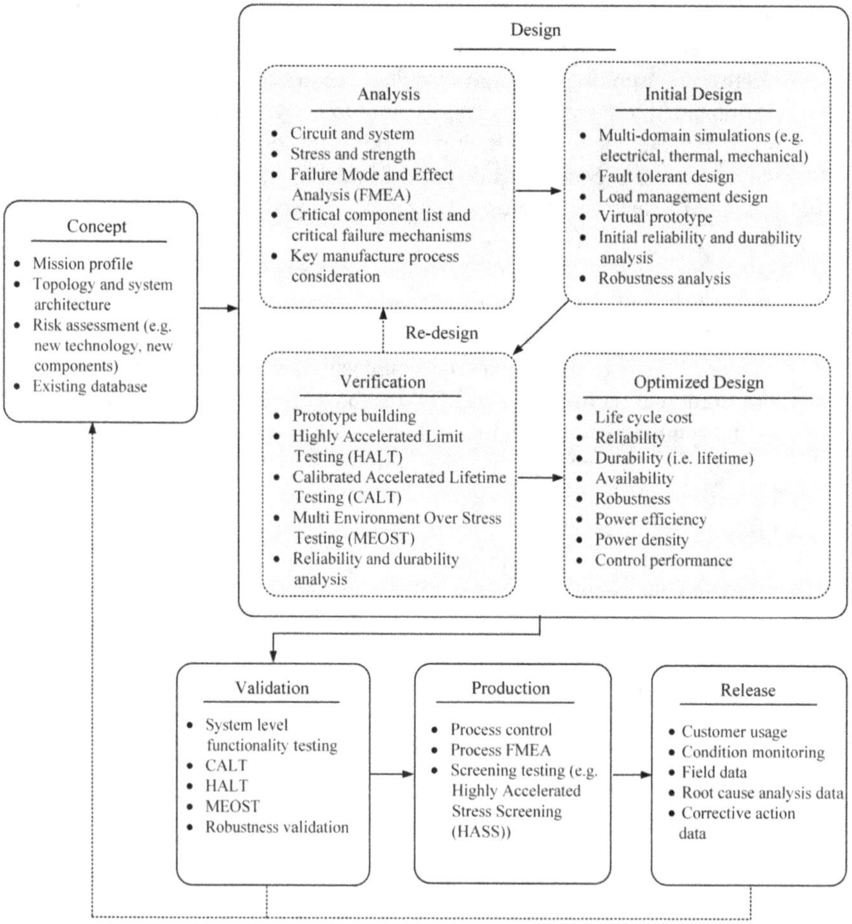

Figure 1.14 State-of-the-art reliability design procedure for power electronics

Chapter 13. Detailed discussions of the procedures can be found in [28]. Here the reliability prediction toolbox is discussed, and various kinds of accelerated testing concepts will be discussed in Section 1.2.5.

Reliability prediction is an important tool to quantify the lifetime, failure level, and design robustness based on various sources of data and prediction models. Figure 1.15 presents a generic prediction toolbox. It includes statistical models and lifetime models and various sources of available data (e.g., manufacturer testing data, simulation data, and field data) for the reliability prediction of individual components and the overall system. The statistical models are well presented in [2], while the number of physics-based lifetime models available for power electronic components is still limited. Research efforts on both accelerated testing and advanced multi-disciplinary simulations will be beneficial for obtaining those lifetime models.

To map the reliability from the component level to the system level [45], the reliability block diagram (RBD), fault-tree analysis (FTA), and state-space analysis (e.g., Markov analysis) are widely applied, as summarized in Table 1.5.

It should be noted that the three tabulated methods are conventionally applicable to constant failure rate cases. The PoF-based system-level reliability prediction is still an open research topic, even in the area of microelectronics [23,25]. Interactions among different failure mechanisms will bring additional complexity for the analysis. Moreover, it should be noted that the system reliability depends not only on components but also on packaging, interconnects, manufacturing process, and human errors. The latter also needs to be treated properly for a more accurate reliability assessment.

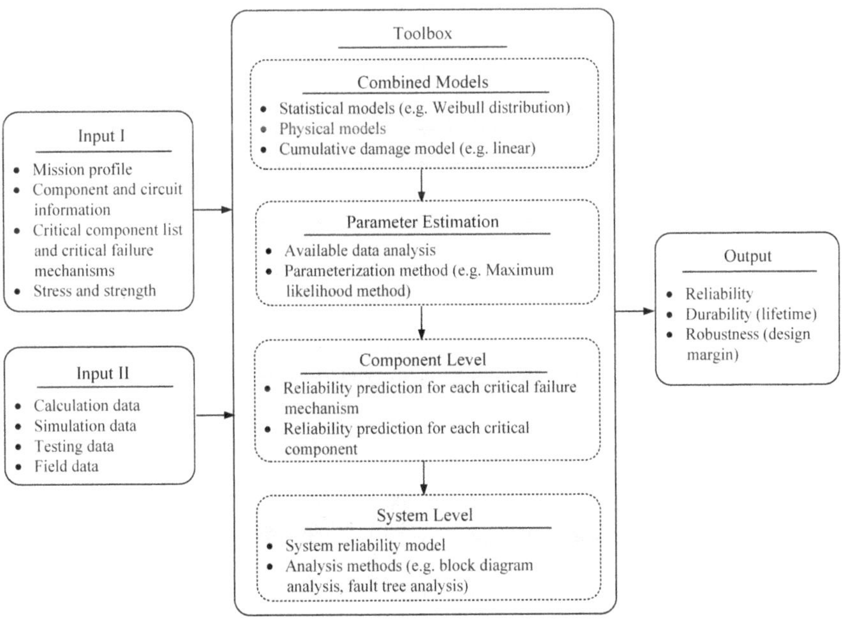

Figure 1.15 Reliability prediction toolbox for power electronic systems

Table 1.5 Summary of system-level reliability prediction methods

	Reliability block diagram (RBD)	**Fault tree analysis (FTA)**	**Markov analysis (MA)**
Concepts	RBD is an analytical technique graphically representing the system components and their reliability-wise connections (from simple series-parallel to complex) by a logic diagram based on the system characteristics.	FTA is an analytical technique using a top-down approach to analyze various system combinations of hardware, software, and human failures (i.e., sub-events) that could cause the system failure (i.e., top event).	MA is a dynamic state-space analytical technique presenting all possible system states (i.e., functioning or failed) and the existing transitions between these states.
Elements	• Rectangle blocks • Direction lines • Failure level and time of the component/subsystem represented by each blocks	• Events (i.e., initiating fault events, intermediate events, and top event) • Logic gates (e.g., AND, OR and more complex ones) • Probability of each event	• States (i.e., functioning or failed) • Transitions between states • Transition rates based on failure rates and repair rates of components/ subsystems
Outcome	• System-level reliability	• System-level reliability • Identified all possible faults (similar to the results from FMEA)	• System-level reliability • System availability
Applications	For non-repairable systems • Without redundancy • With redundancy	For non-repairable systems • Without redundancy • With redundancy	Mainly for repairable systems • Without redundancy • With redundancy
Advantages	• Simplicity and ease of application	• All factors including human factors could be taken into account • Useful also for identifying failure causes and design problems	• Dynamic (i.e., represent state of every component at any time and the dependences among them) • Applicable for repairable systems
Disadvantages/ Limitations	• Limitations in considering external events (e.g., human factor) and priority of events • Dependencies among components or subsystems are not well treated	• Dependencies among components/subsystems are not well treated	• State-based models easily get large (e.g. maximum 2^n states with n components) • Primarily applicable for constant failure rate and constant repair rate (which works in theory only)

1.2.5 Accelerated testing concepts in reliability engineering

The aim of an accelerated testing lies in twofold:

1. Quantitatively yield desired information on product life or performance under normal use
2. Qualitatively identify the weakest points of products to improve the design and manufacturing

The basic idea of accelerated testing is to shorten the life of products or to hasten the degradation of their performance by testing them at an accelerated stress level compared to that in normal use. The basic accelerated lifetime testing (ALT) and the corresponding statistical models, test plan, and data analysis methods are well discussed in [46]. Depending on the testing conditions and testing sample size, there are various other accelerated testing concepts, such as calibrated accelerated lifetime testing (CALT) [47], multiple environment over-stress tests [48], and highly accelerated limit testing (HALT) [49]. The basic concept of CALT and HALT will be briefly discussed below.

CALT is a sequential method of quantitative ALT that can be used under a constrained test time and with a minimal sample size of six, to arrive at a useful estimate of life [47]. There are three groups of accelerated testing, which are Test I, Test II, and Test III, as indicated in Figure 1.16. The testing procedure is as follows:

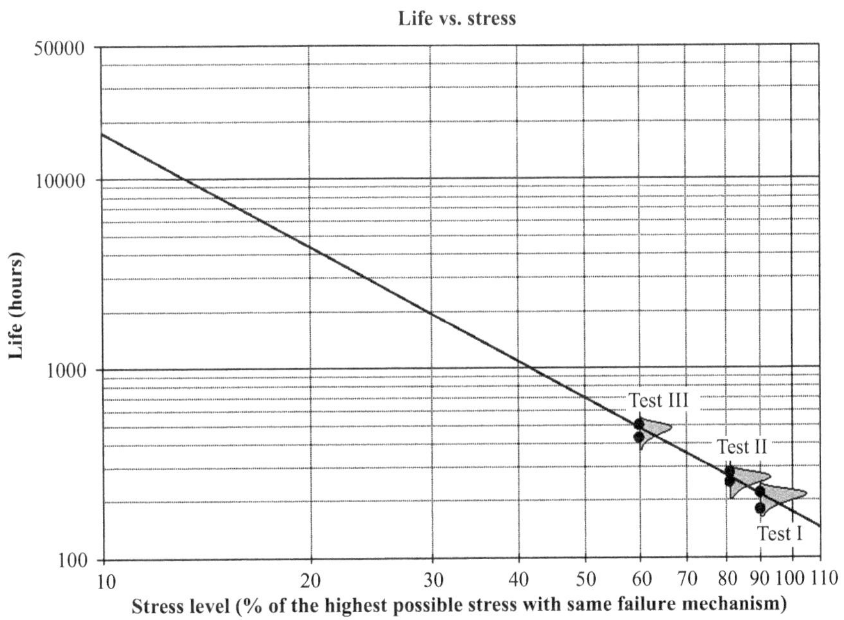

Figure 1.16 An example for illustration of CALT concept

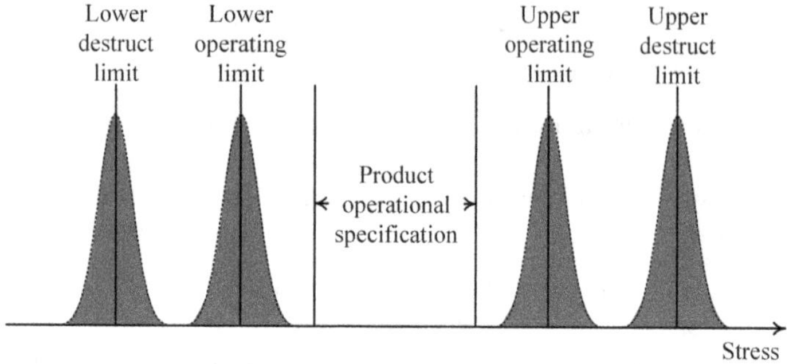

Figure 1.17 Illustration of the testing conditions of HALT

1. Identify the destruct limit of testing samples without altering the failure mechanism
2. Test I: reduce the stress level to 90% of destruct limit and test two parts to failure at this stress level
3. Test II: reduce further the stress level to 90% of that of Test I, that is 81% of the destruct limit, and test two parts to failure at this stress level
4. Test III: according to the time constraints, identify the lowest possible stress level that will result in failures and test two or more parts at this lowest stress level.

The aim of HALT is to identify the operating limit, destruct limit, new failure mechanisms, or design weakest points. It tests components or systems under accelerated conditions close to the destruct limit, as shown in Figure 1.17. HALT is a qualitative testing method that does not intend to predict the reliability. Detailed discussions on HALT can be found in [49].

1.2.6 Strategies to improve the reliability of power electronic converter systems

After power electronic converter systems have been designed, their reliability could be further improved through control and condition monitoring. This is the third important aspect shown in Figure 1.12. Among many options, three main actions can be taken to increase the reliability of power electronic systems: prognostics health management (PHM); active thermal control for reducing temperature and temperature swing, which are the main killing factors of power device modules; and fault-tolerant operation to continue operating the system even in case of failures. The last can be considered as an alternative measure with respect to the first two or like the last attempt to make the system operate if it was not possible to predict failures or to avoid them. Of course, all these actions entail important investments in terms of devices, sensors, and control actions, and even request redundancies. All of them should be evaluated in terms of cost with respect to the specific application.

1. Prognostics health management

   The Electronic Prognostics and Health Management Research Center at the University of Maryland has categorized the main approaches as: use of fuses and canary devices, built-in-test (BIT), monitoring and reasoning of failure precursors, and modeling accumulated damage based on measured life-cycle loads [50]. Chapter 2 will discuss them in detail. Moreover, the online monitoring of IGBT module wear out status by means of thermo-sensitive electrical parameters (TSEPs) will be discussed in Chapter 10.
2. Active thermal control

   The thermal analysis of power converters, especially in case of more complex structures, such as multilevel or multi-cell ones, reveals that some of the power semiconductor devices can be more stressed with respect to others, and this difference can be even more evident in some particular conditions, such as those caused by system faults. Hence, the possibility to modify the modulation and control of the power converter using as feedback the junction temperature of the most stressed device is an appealing possibility. Figure 1.18 gives the general block diagram for active thermal control of the power semiconductors once the junction temperature is measured or estimated. More details on active thermal control will be discussed in Chapter 8.
3. Fault tolerant control

   Working outside the safe operating area leads to damage of power electronic components. Taking an example of power semiconductor switches, the main failure causes are: fault currents—either over-current, short-circuit current or earth fault current; over-voltages; over-temperature; and cosmic radiation. Other problems may arise because of the driver of the power semiconductor: malfunctioning of the driver board, auxiliary power supply failure or *dv/dt* disturbance. As a consequence, five main types of faults can be identified:

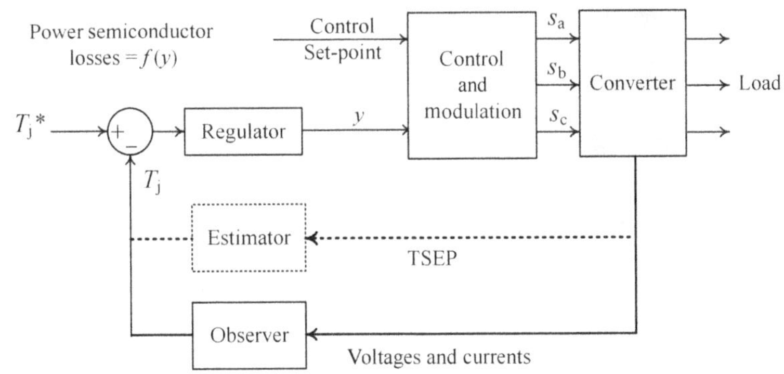

Figure 1.18 Active thermal control of the power semiconductor junction temperature T_j by means of y (switching frequency, reactive power, or any other quantity that can modify the power semiconductor losses). T_j is obtained by using an estimator based on TSEP or an observer using measured voltages and currents

single switch short-circuit (power semiconductor is de-saturated, working as current source, or has a physical short-circuit), phase-leg short-circuit, single-switch open-circuit, single-phase open-circuit, intermittent gate-misfiring. Chapter 12 will present fault tolerant strategies for power semiconductor switches in adjustable speed drive applications.

1.3 Challenges and opportunities in research on power electronics reliability

Reliability is an important performance index of power electronic systems. The status and future trends of DFR in power electronics are presented in this chapter. The traditional reliability research on power electronics has left simple handbook based on constant failure rate for the PoF approach and DFR process. Joint efforts from engineers and scientists in multiple disciplines are required to fulfill the research needs and promote a paradigm shift in reliability research. The major challenges and opportunities in research on reliability for power electronic systems are addressed as below.

1.3.1 Challenges in power electronics reliability research

Power electronics can be used across a wide range of applications, the uncertainties and variations in mission profiles make it significantly challenging to implement pervasive and fast reliability evaluations with all kinds of environmental exposures. Apart from that, the research is often plagued by outdated paradigms, unclear failure mechanisms and failure modes of reliability critical components, and the lack of understanding of DFR process specific to power electronics.

With the expansion of renewable energy applications, the electrical and electronic contents are increasing, and the power electronic systems are evolving rapidly with growing complexities. Due to these changes, the traditional system-level reliability prediction methods, which are based on constant failure rates, are not accurate enough to characterize the degradation process. However, the PoF-based component-level reliability prediction results in varying failure levels with time. In addition, the traditional testing methods for reliability prediction and robustness validation from components to entire systems are resource-consuming, but the results still cannot achieve zero defects in the case of high-volume production. The mass-manufactured power electronic products still end up with ppm-level return rates from the field.

The new generation products can operate in higher temperature (e.g., with wide band-gap devices), which challenges the reliability and lifetime of the packaged products and the overall system. The traditional packaging and assembly solution limits the performance of the advanced products and needs to be updated. The highly integrated devices introduced more digital controllers in power electronic systems, which require more complex software control strategies; thus, software reliability becomes a potential issue and should be treated adequately. Moreover, the design changes for cost reduction and physical size requirements for some applications (particularly consumer products) have not been taken into account.

1.3.2 Opportunities in power electronics reliability research

Research in microelectronics provides an important foundation for the ongoing and future work in power electronics, especially from the methodologies' point of view. The outcomes contributing to the PoF approach for power semiconductor reliability study have provided insights into avoiding failures in power electronic components, circuits, and systems. In the meantime, emerging semiconductor and capacitor technologies enable more reliable power electronic components and systems, which can be further researched.

The new generation of power semiconductors is the game changers, but new reliability issues have also been created, such as silicon carbide body diode bipolar degradation, which is yet to be solved. At the product or system level, new designs for high-level power integration (packaging) or hybridization, like 3D packaging, trends of the modular design of power converters, system smart de-rating operation, and the standardization of power electronic components and packaging technologies are gaining traction. For the advancement in design software, computer-aided automated design software tools can be improved to save time and cost for the development process.

Innovative controlling and monitoring techniques are also possible directions of development, for example, active thermal control by controlling the power flow in power electronic circuits and smart condition monitoring and fault tolerant design allowing extended lifetime and reduced failure rate. Consequently, more and more mission profiles and online monitoring data from the field are available and accessible for analysis and research. An adequate amount of data can also support the decision-making for tradeoffs among the cost, expected service time, and reliability of the system.

Additionally, with better understandings of failure mechanisms in power electronics, more failure mechanism–specific accelerated life testing could be designed, leading to improved reliability predictions for targeted applications. Various application-oriented reliability testing concepts and facilities are introduced in [51].

For more comprehensive discussions on the opportunities in power electronics reliability research, please refer to [52].

References

[1] P. T. Krein, *Elements of Power Electronics*, 1st edn, New York: Oxford University Press; 1998, p. 10.

[2] P. O'Connor and A. Kleyner, *Practical Reliability Engineering*, 5th edn, West Sussex: Wiley; 2012.

[3] J. Bralla, *Design for Manufacturability Handbook*, 2nd edn, New York: McGraw-Hill Professional; 1998.

[4] J. D. van Wyk and F. C. Lee, "On a future for power electronics," *IEEE Journal of Emerging and Selected Topics in Power Electronics*, vol. 1, no. 2, pp. 59–72, 2013.

[5] H. Ke and D. C. Hopkins, "3D packaging for high density and high performance GaN-based circuits," Tutorial Presentation at the *IEEE Applied Power Electronics Conference and Exposition*, Charlotte, 2015.
[6] H. Wang, M. Liserre, F. Blaabjerg, *et al.*, "Transitioning to physics-of-failure as a reliability driver in power electronics," *IEEE Journal of Emerging and Selected Topics in Power Electronics*, vol. 2, no. 1, pp. 97–114, 2014.
[7] K. Ma, Y. Yang, H. Wang, and F. Blaabjerg, "Design for reliability of power electronics in renewable energy systems," in M. A. Sanz-Bobi, (ed.), *Use, Operation and Maintenance of Renewable Energy Systems: Experiences and Future Approaches*, ch. 3, Berlin: Springer; 2014.
[8] ZVEI Robustness Validation Working Group, *Handbook for Robustness Validation of Automotive Electrical/Electronic Modules*, revised version, Frankfurt: Zentralverband Elektrotechnik- und Elektronikindustrie e. V., 2013.
[9] T. Jomier, *Final public MOET technical report*, EU FP6 project on More Open Electrical Technologies, Dec. 2009. [Online]. Available: https://www.yumpu.com/en/document/read/27200319/moet-technical-report (accessed: 25 January 2025).
[10] F. Blaabjerg, Z. Chen, and S. B. Kjaer, "Power electronics as efficient interface in dispersed power generation systems," *IEEE Transactions on Power Electronics*, vol. 19, no. 4, pp. 1184–1194, 2004.
[11] S. B. Kjaer, J. K. Pedersen, and F. Blaabjerg, "A review of single-phase grid connected inverters for photovoltaic modules," *IEEE Transactions on Industry Applications*, vol. 41, no. 5, pp. 1292–1306, 2005.
[12] H. Wang, M. Liserre, and F. Blaabjerg, "Toward reliable power electronics – Challenges, design tools and opportunities," *IEEE Industrial Electronics Magazine*, vol. 7, no. 2, pp. 17–26, 2013.
[13] RELIAWIND, "Reliability-focused research on optimizing wind energy system design, operation and maintenance: tools, proof of concepts, guidelines & methodologies for a new generation," Final Report Summary, 2011. [Online]. Available: https://cordis.europa.eu/project/id/212966/reporting (accessed: 25 January 2025).
[14] L. M. Moore and H. N. Post, "Five years of operating experience at a large, utility-scale photovoltaic generating plant," *Journal of Progress in Photovoltaics: Research and Applications*, vol. 16, no. 3, pp. 249–259, 2008.
[15] W. Weibull, "Statistical distribution function of wide applicability," *ASME Journal of Applied Mechanics*, vol. 18, no. 3, pp. 293–297, 1951.
[16] G. A. Klutke, Peter C. Kiessler, and M. A. Wortman, "A critical look at the Bathtub curve," *IEEE Transactions on Reliability*, vol. 52, no. 1, pp. 125–129, 2003.
[17] M. G. Pecht and F. R. Nash, "Predicting the reliability of electronic equipment," *Proceedings of IEEE*, vol. 82, no. 7, pp. 992–1004, 1994.

[18] M. Krasich, "How to estimate and use MTTF/MTBF would the real MTBF please stand up?" in *Proceedings of IEEE Annual Reliability and Maintainability Symposium*, 2009, pp. 353–359.

[19] W. B. Nelson, *Recurrent-Events Data Analysis for Repairs, Disease Episodes, and Other Applications*, Philadelphia, PA: Society for Industrial and Applied Mathematics; 2003.

[20] J. H. Saleh and K. Marais, "Highlights from the early (and pre-) history of reliability engineering," *Reliability Engineering and System Safety*, vol. 91, no. 2, pp. 249–256, 2006.

[21] K. Chatterjee, M. Modarres, and J. B. Bernstein, "Fifty years of physics of failure," *Journal of the Reliability Information Analysis Center*, vol. 20, no. 1, pp. 2–7, 2012.

[22] M. Pecht and A. Dasgupta, "Physics-of-failure: An approach to reliable product development," in *Proceedings of the International Integrated Reliability Workshop*, 1995, pp. 1–4.

[23] *Military Handbook: Reliability Prediction of Electronic Equipment*, MIL-HDBK-217F, Dec. 2, 1991.

[24] M. White and J. B. Bernstein, *Microelectronics Reliability: Physics-of-Failure Based Modeling and Life Evaluation*, La Cañada Flintridge, CA: Jet Propulsion Laboratory; 2008.

[25] ANSI/VITA 51.2 Standard, *Physics of Failure Reliability Predictions*, Oklahoma City, OK: VEMbus International Trade Association; 2011.

[26] S. Yang, A. T. Bryant, P. A. Mawby, D. Xiang, L. Ran, and P. Tavner, "An industry-based survey of reliability in power electronic converters," *IEEE Transactions on Industry Applications*, vol. 47, no. 3, pp. 1441–1451, 2011.

[27] Y. Song and B. Wang, "Survey on reliability of power electronic systems," *IEEE Transactions on Power Electronics*, vol. 28, no. 1, pp. 591–604, 2013.

[28] H. Wang, K. Ma, and F. Blaabjerg, "Design for reliability of power electronic systems," in *Proceedings of the IEEE Industrial Electronics Society Annual Conference*, 2012, pp. 33–44.

[29] R. Burgos, C. Gang, F. Wang, D. Boroyevich, W. G. Odendaal, and J. D. V. Wyk, "Reliability-oriented design of three-phase power converters for aircraft applications," *IEEE Transactions on Aerospace and Electronic Systems*, vol. 48, no. 2, pp. 1249–1263, 2012.

[30] X. Perpiñà, X. Jordà, M. Vellvehi, J. Rebollo, and M. Mermet-Guyennet, "Long-term reliability of railway power inverters cooled by heat-pipe-based systems," *IEEE Transactions on Industrial Electronics*, vol. 58, no. 7, pp. 2662–2672, 2011.

[31] D. Hirschmann, D. Tissen, S. Schroder, and R. W. De Doncker, "Reliability prediction for inverters in hybrid electrical vehicles," *IEEE Transactions on Power Electronics*, vol. 22, no. 6, pp. 2511–2517, 2007.

[32] P. Wikstrom, L. A. Terens, and H. Kobi, "Reliability, availability, and maintainability of high-power variable-speed drive systems," *IEEE Transactions on Industry Applications*, vol. 36, no. 1, pp. 231–241, 2000.

[33] F. Carastro, A. Castellazzi, J. Clare, and P. Wheeler, "High-efficiency high-reliability pulsed power converters for industrial processes," *IEEE Transactions on Power Electronics*, vol. 27, no. 1, pp. 37–45, 2012.

[34] S. S. Smater and A. D. Dominguez-Garcia, "A framework for reliability and performance assessment of wind energy conversion systems," *IEEE Transactions on Power Systems*, vol. 26, no. 4, pp. 2235–2245, 2011.

[35] K. Fischer, F. Besnard, and L. Bertling, "Reliability-centered maintenance for wind turbines based on statistical analysis and practical experience," *IEEE Transactions on Energy Conversion*, vol. 27, no. 1, pp. 184–195, 2012.

[36] S. Harb and R. S. Balog, "Reliability of candidate photovoltaic module-integrated-inverter (PV-MII) topologies – A usage model approach," *IEEE Transactions on Power Electronics*, vol. 28, no. 6, pp. 3019–3027, 2013.

[37] M. Ciappa, "Selected failure mechanisms of modern power modules," *Journal of Microelectronics Reliability*, vol. 42, no. 4–5, pp. 653–667, 2002.

[38] L. Yang, P. A. Agyakwa, and C. M. Johnson, "Physics-of-failure lifetime prediction models for wire bond interconnects in power electronic modules," *IEEE Transactions on Device and Materials Reliability*, vol. 13, no. 1, pp. 9–17, 2013.

[39] Y. Yang, H. Wang, F. Blaabjerg, and K. Ma, "Mission profile based multi-disciplinary analysis of power modules in single-phase transformerless photovoltaic inverters," in *Proceedings of the European Conference on Power Electronics and Applications*, 2013, pp. 1–10.

[40] N. C. Sintamarean, F. Blaabjerg, H. Wang, F. Iannuzzo, and P. de Rimmen, "Reliability oriented design tool for the new generation of grid connected PV inverters," *IEEE Transactions on Power Electronics*, vol. 30, no. 5, pp. 2635–2644, 2015.

[41] W. E. Newell, "Power electronics – Emerging from limbo," *IEEE Transactions on Industry Applications*, vol. IA–10, no. 1, pp. 7–11, 1974.

[42] J. Falck, C. Felgemacher, A. Rojko, M. Liserre, and P. Zacharias, "Reliability of power electronic systems: An industry perspective," *IEEE Industrial Electronics Magazine*, vol. 12, no. 2, pp. 24–35, 2018.

[43] S. Yang, D. Xiang, A. Bryant, P. Mawby, L. Ran and P. Tavner, "Condition monitoring for device reliability in power electronic converters: A review," *IEEE Transactions on Power Electronics*, vol. 25, no. 11, pp.2734–2752, 2010.

[44] P. Jacob, "Defect- and structure-weakness-localization on power semiconductors using OBIRCH (optical beam induced resistivity change)," in *Proceedings of the 9th International Symposium on the Physical and Failure Analysis of Integrated Circuits* (Cat. No.02TH8614), Singapore, 2002, pp. 152–156.

[45] *IEEE Standard Framework for the Reliability Prediction of Hardware*, IEEE Std. 1413, New York: IEEE; 2009.

[46] W. B. Nelson, *Accelerated Testing – Statistical Models, Test Plans, and Data Analysis*, Hoboken, NJ: Wiley; 2004.

[47] General Motors Corporation Handbook GMW8758, *Calibrated Accelerated Life Testing*, 2004.
[48] K. Bhote and A. Bhote, *World Class Reliability: Using Multiple Environment Overstress Tests to Make It Happen*, New York: AMACOM; 2004.
[49] G. K. Hobbs, *Accelerated Reliability Engineering: HALT and HASS*, Hoboken, NJ: Wiley; 2000.
[50] M. G. Pecht, *Prognostics and Health Management of Electronics*, Hoboken, NJ: Wiley; 2008.
[51] H. Wang, F. Iannuzzo, A. S. Bahman, *et al.*, "Application-oriented reliability testing of power electronic components and converters," *IEEE Power Electronics Magazine*, vol. 9, no. 4, pp. 22–31, 2022.
[52] H. Wang and F. Blaabjerg, "Power electronics reliability: State of the art and outlook," *IEEE Journal of Emerging and Selected Topics in Power Electronics*, vol. 9, no. 6, pp. 6476–6493, 2021.

Chapter 2

Anomaly detection and prognostics for power electronics

Chang Lu[1,2] and Michael Pecht[3]

2.1 Introduction

Power electronics is a core competence within the industry, and there is an increasing number of critical systems—uninterruptible power supplies, electric power transmission, automotive electronics, motor drives, and energy storage. According to a 2014 market research report, the power electronics market is expected to boom in the near future. The power electronics market in 2014 was valued at $36.86 billion and was expected to grow at a compound annual growth rate of 7.74% from 2014 to 2020 [1].

As the reliance on power electronics continues to rise, electronic systems have been evolving and have become more sophisticated. However, with increasing operating time, the aging of electronic systems occurs, and so the reliability decreases gradually with time.

Failures of power electronic devices often lead to disastrous accidents, which can cause the loss of manpower, material resources, and financial resources. To avoid the weak points of traditional timing maintenance or breakdown maintenance, monitoring technology and real-time capturing of the status of power electronics should be improved to identify problems and take appropriate countermeasures. Failures can be effectively prevented before they occur, and some serious failures can be controlled and excluded. As a result, the maintenance range and workload and failure rates can be reduced, maintenance costs can be saved, and equipment uptime can be improved. Thus, maintenance can change from passive to active.

Prognostics and health management (PHM) is a method that permits the reliability of a system to be evaluated in its actual life-cycle conditions to determine the advent of failure and mitigate system risks. Performing diagnostics and

[1]Centre for Advances in Reliability and Safety, Hong Kong
[2]Department of Electrical and Electronic Engineering, The Hong Kong Polytechnic University, Hong Kong
[3]Center for Advanced Life Cycle Engineering (CALCE), University of Maryland, USA

prognostics are the critical steps in implementing PHM to power electronic products and systems.

The term "diagnostics" refers to the detection and isolation of anomalies or failures. Anomalies are items, events, or observations that do not conform to an expected pattern or other items in a data set. Anomalies are also referred to as outliers, novelties, noise, deviations, or exceptions. Anomaly detection is to detect unusual or unexpected (anomalous) behavior of the product by identifying deviations from nominally healthy behavior.

The term "prognostics" refers to the process of predicting a future state (of reliability) based on current and historic conditions. Remaining useful life (RUL) is the useful life left for an asset at a particular time of operation. The prediction of RUL estimates the time at which a system or component will no longer perform its intended function to aid in online decision-making activities such as fault mitigation and mission planning. Prognostics refers to predicting a product's RUL within appropriate confidence intervals, which often requires additional information that is not traditionally provided by sensors such as maintenance history, past and future operating profiles, and environmental factors.

To perform diagnostic and prognostic of electronic devices, the physics-of-failure (PoF)-, data-driven-, canary-, and fusion-based approaches utilizing either physical laws or machine learning (ML) techniques or both have been studied. A PoF approach is based on the physical laws of degradation linked with mathematical failure models to perform reliability modeling, design, and assessment. The approach to implementing PoF into PHM can be based on the failure models and a failure mode, mechanism, and effect analysis (FMMEA), which are further discussed in Sections 2.2 and 2.3.

Data-driven approaches use data analytics and ML techniques to determine anomalies and make predictions about the reliability of electronic devices, systems, and products based on internal and/or external covariates [2]. In Section 2.4, different data-driven methods for anomaly detection, classification, and prediction are introduced.

A canary is a component or device designed to be more sensitive to the operating environment factors and degrade earlier than the actual devices. These canary devices (also called prognostic cells) are embedded in the same system as the actual ones experiencing the same stresses and provide early incipient failure warnings before actual system or component failure. Examples of canary methods are given in Section 2.5.

The fusion approach is a method that combines the advantages of the PoF-based and data-driven approaches introduced in this chapter. The PoF models can be used to determine the variables to monitor, assess the in-situ degradation of the system, and support failure definition. The ML techniques are beneficial for feature extraction from the monitored variables and the early warning and prediction of failures.

Safety-critical mechanical systems and structures—such as propulsion engines, aircraft structures, bridges, buildings, roads, pressure vessels, rotary equipment, and gears—have benefited from advanced sensor systems developed specifically for in-situ fault diagnosis (condition monitoring) and health and usage monitoring [3–8]. As a result, a considerable body of knowledge exists on PHM of mechanical systems, with research conducted in establishing failure precursors (such as changes in vibration signatures of roller bearings and variations in acoustic levels due to wear) and developing reasoning algorithms. Degradation in electronics is more difficult to detect and inspect compared to most mechanical systems and structures, due to the micro- to nano-scale size and the complex architecture of most electronic products. Since faults in electronic products may not necessarily lead to failure or loss of designated electrical performance or functionality, it is difficult to quantify product degradation and the progression from faults to final failure.

In addition, there is a significant shortage of knowledge about the failure precursors in electronics [9]. Consequently, it can be more difficult to implement diagnostic and prognostic systems that can directly monitor the faults or conditions in which faults occur in electronics.

2.2 Failure models

During the life cycle of a power electronic device, the device will be subject to various complex conditions—temperature, humidity, vibration, and so forth. Field failures related to the operating environments of electronic equipment show that about 55% of failures are due to high temperatures and temperature cycling, 20% are related to vibration and shock, and another 20% are due to humidity [10]. The modeling of stress and damage in electronic parts and structures utilizing exposure conditions to compute accumulated damage is one of the main approaches for prognostic implementation that Vichare and Pecht summarized [11]. Currently, in the field of power electronics, the most common failure models include time-dependent dielectric breakdown (TDDB) for electrical stress induced failures, and energy-based and thermal cycling models for thermo-mechanical stress related failures.

2.2.1 TDDB models

TDDB refers to the physical process where a dielectric stored under a constant electric field, which is less than the material's breakdown strength, will break down with time [12]. TDDB is a failure mechanism in metal-oxide-silicon field effect transistors (MOSFETs), wherein the gate oxide breaks down as a result of the long-term application of a relatively low electric field (as opposed to immediate breakdown, which is caused by a strong electric field). The breakdown is caused by the formation of a conducting path through the gate oxide to the substrate due to electron tunneling current when MOSFETs are operated close to or beyond their specified operating voltages.

There are three important oxide breakdown models employed by reliability engineers: the thermochemical (E) model, the classic anode hole injection ($1/E$) model, and the $\sqrt{E}$ model based on the Poole–Frenkel conduction mechanism.

McPherson [13] proposed the E model to describe the process in which the oxide creates traps as an electric field is applied to the gate. The temperature will accelerate this process. The model is shown as follows:

$$\tau = A\exp\left(\frac{E_a}{kT}\right)\exp(-\gamma E_f) \tag{2.1}$$

where τ is the mean time to failure, A is a constant, E_a is the activation energy, k is Boltzmann's constant, T is temperature, γ is the field acceleration factor, and E_f is the electric field strength in the oxide. McPherson [14] used the E model to describe the low-field TDDB in SiO_2 thin films.

Chen *et al.* [15] proposed the 1/E model to explicate TDDB in thin gate oxides. Schuegraf and Hu [16] gave a modified model in 1994. The 1/E model is shown in (2.2):

$$\tau = \frac{A\exp\left(\frac{E_a}{kT}\right)\exp\left(-\frac{\gamma}{E_f}\right)}{E_f^2} \tag{2.2}$$

where A, E_a, T and γ are the same as in (2.1). This can happen at either high or low energies, but it can be observed most clearly with electrons which have very high energy [17].

The $\sqrt{E}$ model is based on assumptions of the charge to breakdown (QBD) hypothesis and the Poole–Frenkel leakage mechanism [18]. The corresponding equation is:

$$t_{BD} \infty \frac{Q_{BD}}{E}\exp\left(\frac{q\left[\Phi_B - \sqrt{\frac{qE}{\pi\varepsilon_0\varepsilon_\infty}}\right]}{kT}\right) \tag{2.3}$$

where t_{BD} is the lifetime of the dielectric under stress until the dielectric breaks down. E is an external field; the electric field is the same as in the previous two models; Q_{BD} is the critical charge or the breakdown charge; q is the elementary charge; Phi_B is the trap depth; ε_0 is permeability in a vacuum; and ε_∞ is permeability in the dielectric [18].

Although there are many methods employed to combat the effects of TDDB, TDDB is still one of the main failure mechanisms for MOSFETs [19–21].

2.2.2 *Energy-based fatigue models*

Fatigue failures will occur when the component experiences cyclic stresses and strains that produce permanent damage. Power electronic devices are subjected to thermal cycling as a result of turning them on and off or varying the load conditions in their applications. These repeated material expansion and contraction cycles can induce stresses which may lead to material fatigue at the solder joints over extended periods. As the device and interconnect sizes continues to shrink, the reliability of solder joints inside the device package has become critical to the overall quality of power electronics. The necessity of modeling the fatigue behavior of solder joints has been recognized, and several models have been proposed.

Energy-based models are the newest models today used to predict fatigue failure based on a hysteresis energy term or a type of volume-weighted average stress-strain history [22]. Based on the total strain energy, Akay [23] has presented the fatigue model on LLCC leaded package fatigue behavior as follows:

$$N_f = \left(\frac{\Delta \overline{W}_{\text{total}}}{W_o}\right)^{\frac{1}{k}} \tag{2.4}$$

where N_f is the mean cycles to failure, W_o and k are the fatigue coefficients, and $\Delta \overline{W}_{\text{total}}$ is the total strain energy.

Liang *et al.* [24] proposed an integrated fatigue life prediction methodology for optimum design and reliability assessment of solder interconnections and calculated on an energy-based fatigue failure criterion, as shown in (2.5):

$$\overline{N}_f = C(W_{ss})^{-m} \tag{2.5}$$

where W_{ss} is the stress–strain hysteresis energy density, and C and m are temperature-dependent material constants derived from low-cycle fatigue tests. The model was verified by the testing of ball grid array (BGA) packages.

Jung *et al.* [25] performed fatigue analysis for crack initiation in plastic ball grid array (PBGA) packages. They gave (2.6) for 62Sn/36Pb/2Ag solder joints:

$$N_0 = 7,860(\Delta W_0)^{-1.00} \tag{2.6}$$

where N_0 is the number of cycles to crack initiation and ΔW_0 (psi) is the viscoplastic strain energy density.

To estimate solder fatigue life, Wu *et al.* [26] used Heinrich's energy model to describe the number of cycles to crack initiation of the fine pitch BGA packages. N_0 can be expressed as follows:

$$N_0 = 18,083\Delta W^{-1.46} \tag{2.7}$$

where N_0 is the number of cycles to crack initiation and ΔW is the viscoplastic strain energy density per cycle (psi).

Based on Darveau's research, the number of cycles to crack initiation (N_0) and energy density (ΔW) can be expressed as follows:

$$N_0 = 54.2\Delta W^{-1.00} \tag{2.8}$$

The crack growth $da\backslash dN$ [m/cycle] is described by:

$$\frac{da}{dN} = 3.49 \cdot 10^{-7} \cdot \Delta W^{1.13} \tag{2.9}$$

Equations (2.8) and (2.9) represent primary (p) and secondary (s) crack initiations growing toward each other with two different crack growth rates.

Based on the above research, Gustafsson [27] has reported another energy-based fatigue model, as shown as follows:

$$N_{aw} = N_{0s} + \frac{a - \left(N_{0s} - N_{0p}\right)\frac{da_p}{dN}}{\frac{da_s}{dN} + \frac{da_p}{dN}} \tag{2.10}$$

where N_{aw} is the number of cycles to failure and a is the total possible crack length.

The benefit of an energy-based fatigue model, as compared to strain-based or creep-based fatigue models, is the ability to capture test conditions with more accuracy. The limitation of energy-based fatigue models is their inability to predict the actual number of cycles to failure [22].

2.2.3 Thermal cycling models

Among many environmental accelerated testing methodologies for assessing the reliability of electronic systems, thermal cycling is the most commonly used one for the characterization of devices as well as interconnections. Thermal cycling is one important reason that leads to power electronics fatigue, which is a key failure mechanism for many power electronic products.

The reliability of electronic assemblies is highly dependent on the quality of the solder joints. Ghaffarian [28] used two-parameter Weibull distributions to fit the Ceramic Ball Grid Array 625 (CBGA 625) thermal cycles to failure data. The equation is:

$$F(N) = 1 - \exp(-(N/N_0)^m) \tag{2.11}$$

where $F(N)$ is the cumulative failure distribution function; N is the number of thermal cycles; N_0 is a scale parameter that commonly is referred to as characteristic life and is the number of thermal cycles with 63.2% failure occurrence, and m is the shape parameter. A large value for m is approximately inversely proportional to the coefficient of variation (CV) by 1.2/CV. The Coffin–Manson model predicts the number of cycles needed to drive to fail a given percentage of parts when exposed to temperature variations. The Coffin–Manson model is given by the following equation:

$$N_f = \frac{\delta}{(\Delta t)^{\alpha_1}} \tag{2.12}$$

where N_f is the number of cycles to failure, Δt is the cycle temperature range, and δ and α_1 are characteristics of the material and test setup, respectively. This power-rule relationship explains the effect that temperature range has on thermal-fatigue life. A model of this type, known as the (modified) Coffin–Manson model, has been used successfully to model crack growth in solder and other metals due to repeated temperature cycling as equipment is turned on and off [29]. This model takes the form of:

$$N_{\mathrm{f}} = C \cdot f^{\alpha} \cdot \Delta T^{-\beta} \cdot G_{\mathrm{Tmax}} \tag{2.13}$$

where N_f is the number of cycles to failure, f is the cycling frequency, ΔT is the cycle temperature range, $G_{T\max}$ is an Arrhenius term evaluated at the maximum temperature reached in each cycle, α is the cycling frequency exponent (typical value 0.33), and β is the temperature range exponent (typical value of 1.9–2.0). Another thermal cycling model for solder fatigue behavior is the Norris–Landzberg model for thermal fatigue of solder interconnections, which Norris and Landzberg first proposed [30]:

$$\mathrm{AF} = \frac{N_a}{N_t} = \left(\frac{f_a}{f_t}\right)^{\frac{1}{3}} \left(\frac{\Delta T_t}{\Delta T_a}\right)^2 \Phi(T_{\max}) \tag{2.14}$$

where

$$\Phi(T_{\max}) = \frac{\mathrm{life}(T_{\max a})}{\mathrm{life}(T_{\max t})} \tag{2.15}$$

AF is the calculated acceleration factor, $T_{\max}$ is the maximum solder joint temperature, f is the cyclic frequency, and a, t are two stress conditions.

2.3 FMMEA to identify failure mechanisms

To predict the RUL of a product, it is essential to understand the root cause behind the damage to the product and how the damage is manifested in the product. To address the root cause of failure, it is necessary to know not only the failure mode but also the failure mechanism that causes the failure. If the failure mechanisms and modes are not known, then the sensors for monitoring, the location of monitoring, and the models to analyze the collected data may be selected erroneously.

Precursor failures are events that indicate impending failure. A precursor indication is usually a change in a measurable variable that can be associated with subsequent failures. For example, a shift in the output voltage of a power supply would suggest impending failure due to a damaged feedback regulator and optoisolator circuitry. Failures can then be predicted by using a causal relationship between a measured variable that can be correlated with subsequent failure for PoF assessment. If the identification of a precursor parameter is not based on the fundamental understanding of the failure mechanisms of a product, then an erroneous parameter may be monitored. Monitoring such a parameter may not provide an appropriate precursor and may lead to faulty prediction and, therefore, improper or delayed corrective actions.

Born and Boenning [31] and Pecht *et al.* [32] proposed several measurable parameters which can be used as failure precursors for electronic switching power supplies, cables, connectors, CMOS-integrated circuits (CMOS-IC), and voltage-controlled high-frequency oscillators (see Table 2.1). Testing was conducted to demonstrate the potential of select parameters to be viable for the detection of incipient failures in electronic systems.

Table 2.1 Potential failure precursors for electronics

Electronic subsystem	Failure precursor parameter
Switching power supply	• DC output (voltage and current levels) • Ripple • Pulse width duty cycle • Efficiency • Feedback (voltage and current levels) • Leakage current • RF noise
Cables and connectors	• Impedance changes • Physical damage • High-energy dielectric breakdown
CMOS IC	• Supply leakage current • Supply current variation • Operating signature • Current noise • Logic level variations
Voltage-controlled oscillators	• Output frequency • Power loss • Efficiency • Phase distortion • Noise
FET	• Gate leakage current/resistance • Drain-source leakage current/resistance
Ceramic chip capacitors	• Leakage current/resistance • Dissipation factor • RF noise
General purpose diodes	• Reverse leakage current • Forward voltage drop • Thermal resistance • Power dissipation • RF noise
Electrolytic capacitors	• Leakage current/resistance • Dissipation factor • RF noise
RF power amplifier	• Voltage standing wave ratio • Power dissipation • Leakage current

The FMMEA is a method based on assessing the root cause of failure mechanisms of a given product [33]. A potential failure mode is the manner in which a failure manifests itself in a product. Failure mechanisms are the processes

Table 2.2 Examples of failure mechanisms [34]

Categories	Overstress failure mechanisms	Wear-out failure mechanisms
Mechanical	Delamination, fracture	Fatigue, creep
Thermal	Excessive heating of component past glass transition temperature	Diffusion voiding, creep
Electrical	Dielectric breakdown, electrostatic discharge, electrical overstress	Electromigration
Radiation	Single-expose failure	Embrittlement, charge trapping
Chemical	Etching	Dendrites, whiskers, corrosion

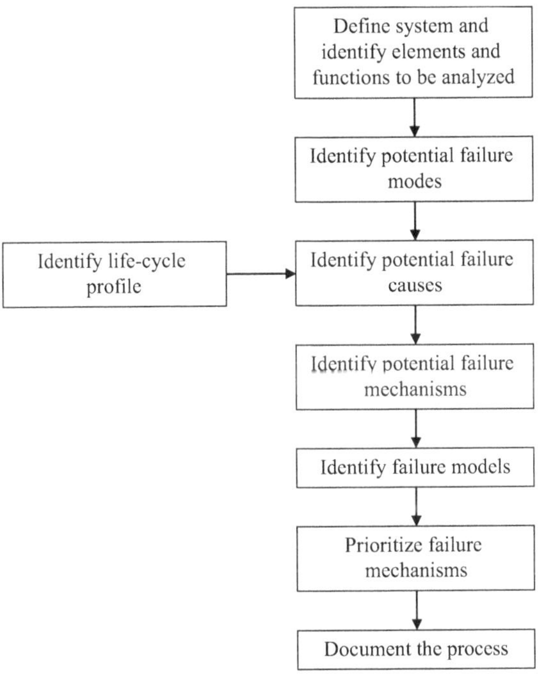

Figure 2.1 FMMEA methodology [35]

by which physical, electrical, chemical, and mechanical stresses induce failures individually or in combination. Failure mechanisms are categorized as either overstress or wear-out mechanisms—examples of which are shown in Table 2.2.

FMMEA is based on an understanding of the relationships between product requirements and the physical characteristics of the product (and their variations in the production process), the interactions of product materials with loads (stresses at application conditions), and their influence on the product's susceptibility to failure with respect to the use conditions [35]. A schematic diagram reflecting the steps in FMMEA is shown in Figure 2.1.

The potential failure identification step involves using the geometry and material properties of the product, together with the measured lifecycle loads acting on the product, to identify the potential failure modes, mechanisms, and failure sites in the product. This task is best performed through virtual qualification, which is a simulation-based methodology used to identify and rank the potential failure mechanisms. The reliability assessment step involves the identification of appropriate PoF models for the specified failure mechanisms. A load-stress analysis is conducted using material properties, product geometry, and the lifecycle loads. With the computed stresses and the failure models, an analysis is conducted to determine the cycles to failure and then the accumulated damage is estimated using a damage model. The PoF methodology can provide a systematic approach to reliability assessment early in the design process [36].

Ganesan *et al.* [35] have provided a detailed description of the FMMEA methodology. FMMEA uses lifecycle environmental and operating conditions and the duration of the intended application with knowledge of the active stresses and potential failure mechanisms. The purpose of FMMEA is to identify potential failure mechanisms and models for all the potential failure modes and to prioritize the failure mechanisms. Table 2.3 provides a list of potential failure modes, mechanisms, and the corresponding models of an electronic system [2]. The T, H, V, M, J, and S in the table indicate temperature, humidity, voltage, moisture, current density, and stress, respectively, and Δ and ∇ mean cyclic range and gradient, respectively.

To ascertain the criticality of the failure mechanisms, a risk priority number (RPN) is calculated for each mechanism. The higher the RPN, the higher it is ranked among the failure mechanisms. Figure 2.2 shows the prioritization of failure mechanisms based on the calculated RPN for each mechanism. The RPN is the product of the severity (S),

Table 2.3 Examples of failure mechanisms, loads, and failure models in electronics via FMMEA

Failure mechanisms	**Failure sites**	**Relevant loads**	**Failure models**
Fatigue	Die attach, wirebond/TAB, solder leads, bond pads, traces, vias/PTHs, interfaces	ΔT, T_{mean}, dT/dt, dwell time, ΔH, ΔV	Nonlinear power law (Coffin–Manson, Basquin)
Corrosion	Metallization	M, ΔV, T	Eyring (Howard)
Electromigration	Metallization	T, J	Eyring (Black)
Conductive filament formation	Between metallization	M, ΔV	Power law (Rudra)
Stress-driven diffusion voiding	Metal traces	S, T	Eyring (Okabayashi)
Time-dependent dielectric breakdown	Dielectric layers	V, T	Arrhenius (Fowler–Nordheim)

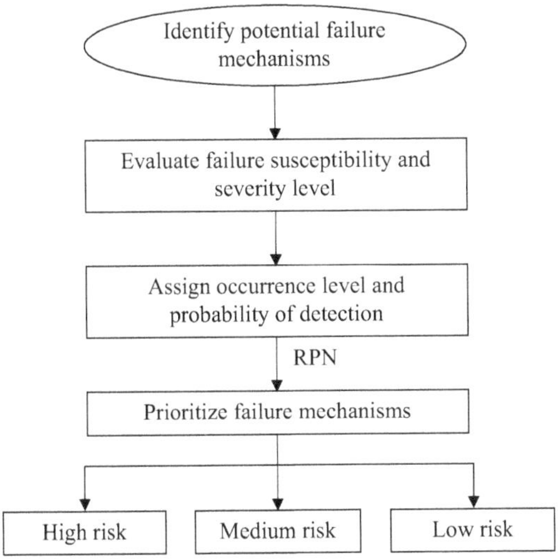

Figure 2.2 Prioritization of failure mechanisms

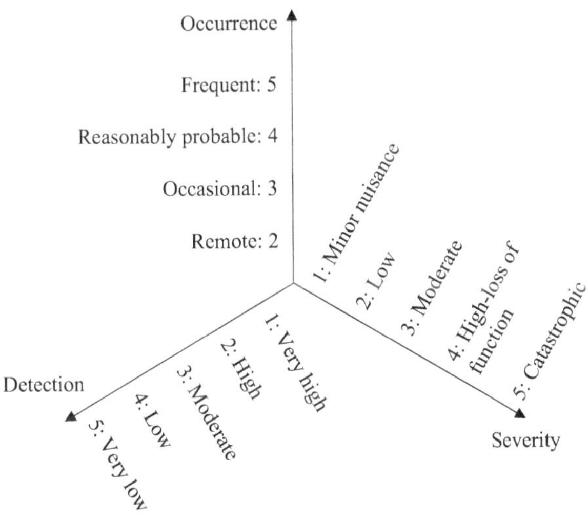

Figure 2.3 Risk matrix

occurrence (O), and detection rankings (D), the equation is RPN $= S \times O \times D$. Severity describes the seriousness of the effect of the failure caused by a mechanism. Occurrence describes how frequently a failure mechanism is expected to result in failure. Detection describes the probability of detecting the failure modes associated with the failure mechanism.

Figure 2.3 shows the axis of a three-dimensional risk matrix. From the estimation of the critical/dominant failure mechanisms that affect a product, the appropriate environmental and operational loads and performance parameters can be selected for health monitoring of the product.

FMMEA is a major improvement over traditional design for reliability methods since it internalizes the concept of failure mechanisms at every step of the decision-making process. The utilization of failure mechanisms as the basis of reliability assessment has been accepted in standards by major technical organizations such as IEEE [37], EIA/JEDEC [38–43], and SEMATECH [44–47].

2.4 Data-driven methods for anomaly detection and prognostics

Data-driven approaches are used for both the diagnosis and prognosis stages in PHM, often based on statistical and ML techniques. Data-driven approaches involve learning statistical relationships and patterns from sensor data to provide valuable decision-making information. They are based on the assumption that the statistical characteristics of the system data remain relatively unchanged until a fault occurs in the system. In this approach, in-situ monitoring of the environmental and operational loads and system parameters is performed. The data is then analyzed using a variety of techniques for anomaly detection followed by prediction of RUL. Anomaly detection techniques are used for diagnostics to detect deviations from the nominally healthy behavior of the system that may lead to system malfunction or failure. The results from anomaly detection can provide advanced warnings of failure, referred to as failure precursors. For prognostic purposes, trends in parameter values, features, or changes in probabilities of the system state are then used to estimate the time to failure of the system using prediction algorithms. This chapter discusses several approaches that have been developed for the prognostics of electronics.

2.4.1 The variable reduction methods

When power electronics fail, many parameters are sometimes measured that result in a huge data set consisting of a large number of variables [48]. Some variables in the data set are strongly correlated with the others causing data redundancy. While some variables are irrelevant and cannot provide useful information to the diagnostic or prognostic results. Using this collected data set may cause several problems in any kind of statistical analysis. Data reduction methods in diagnostics and prognostics can free up storage space, data-logger CPU load, and align with life prediction models. The efficiency measures of data reduction methods should consider gains in computing speed and testing time, the ability to condense load histories without sacrificing important damage characteristics and the estimation of the error introduced by omitting data points or variables [36]. The dimensionality reduction technique, also known as feature extraction, is used in PHM to simplify the data variables without losing too much information. There are many methods to reduce variables. Below we introduce some of the main methods for feature extraction and selection.

2.4.1.1 Principal component analysis

Principal component analysis (PCA) is one of the main methods for reducing the dimensionality of the data set without disturbing the main features of the whole data set [49]. PCA is an unsupervised feature extraction method that transforms the original data set. The number of principal components is less than or equal to the number of original variables. This transformation is defined in such a way that the first principal component has the largest possible variance, and each succeeding component in turn has the highest variance possible under the constraint that it is orthogonal to (i.e., uncorrelated with) the preceding components. Principal components are guaranteed to be independent if the data set is jointly normally distributed. PCA is sensitive to the relative scaling of the original variables. Thus, the features need to be on the same scale; otherwise, variables on larger measurement scales will be emphasized more. The PCA algorithm is as follows:

The standardization of original data:

1. Collect n samples $x_j = (x_{j1}, x_{j2}, \ldots, x_{jp})^T$, $j = 1, 2, \ldots, n$, $n > p$, of the p-dimensional vector $\mathbf{x} = (x_1 + x_2, \ldots, x_p)^T$ to construct a sample matrix. Do the standardized transformation of every element of the sample matrix,

$$Z_{ij} = \frac{x_{ij} - \bar{x}_j}{s_j}, i = 1, 2, \ldots, n; j = 1, 2, \ldots, p \tag{2.16}$$

where $\bar{x}_j = \frac{\sum_{i=1}^{N} x_{ij}}{n}, s_j^2 = \frac{\sum_{i=1}^{n} (x_{ij} - \bar{x}_j)^2}{n-1}$, to get the standardized matrix Z.

2. Calculate the correlative coefficient matrix of the standardized matrix Z:

$$R = [r_{ij}]_p xp = \frac{Z^T Z}{n - 1} \tag{2.17}$$

where $r_{ij} = \frac{\sum z_{kj} \cdot z_{kj}}{n-1} i, j = 1, 2, \ldots, p$.

3. Solve the characteristic equation $|R - \lambda I_p| = 0$ of the sample correlation matrix R for p-characteristic roots to determine the principal component.

   According to $\frac{\sum_{j=1}^{m} \lambda_j}{\sum_{j=1}^{p} \lambda_j} \geq 0.85$, m is determined to make the information utilization rate over 85%. For each $\lambda_j, j = 1, 2, \ldots, m$, solve the equations $Rb = \lambda_j b$, to get the unit characteristic vector b_j^o.

4. Transform the standardized indicator variables into principal components.

$$U_{ij} = z_i^T b_j^o, j = 1, 2, \ldots, m \tag{2.18}$$

   U_1 is called the first principal component, U_2 is the second principal component, and U_p is the p principal component.

5. Conduct a comprehensive evaluation of the m principal components.

Calculate the weighted sum of the m principal components to get a final evaluation value. The e weight is the contribution ratio of each principal component variance.

2.4.1.2 Linear discriminant analysis

As mentioned in the last section, PCA is an unsupervised ML technique, whereas linear discriminant analysis (LDA) is a supervised technique. This means that PCA finds directions (the so-called principal components) of maximum variance regardless of class labels, LDA finds directions that maximize the class separability. Figure 2.4 illustrates the difference between PCA and LDA in feature extraction tasks.

Listed below are the five general steps for performing LDA [50]:

1. Compute the d-dimensional mean vectors for the different classes from the data set matrix X.

$$m_i = \frac{1}{n_i}\sum_{i=1}^{n_i} x_i \tag{2.19}$$

2. After calculating the mean vectors, the scatter matrices (in-between-class and within-class scatter matrices) can be calculated. Within-class scatter matrix S_W is computed by the following equation:

$$S_W = \sum_{i=1}^{c} S_i \tag{2.20}$$

where S_i is the scatter matrix for a specific class c and m_i is the mean vector for that class.

Alternatively, the class-covariance matrices can be computed by adding the scaling factor $\frac{1}{n-1}$ to the within-class scatter matrix.

$$\sum\nolimits_i = \frac{1}{N_i - 1}\sum_{x \in D_i}^{n} (x - m_i)(x\_m_i)^T \tag{2.21}$$

$$S_W = \sum_{i=1}^{c} (N_i - 1)\Sigma_i \tag{2.22}$$

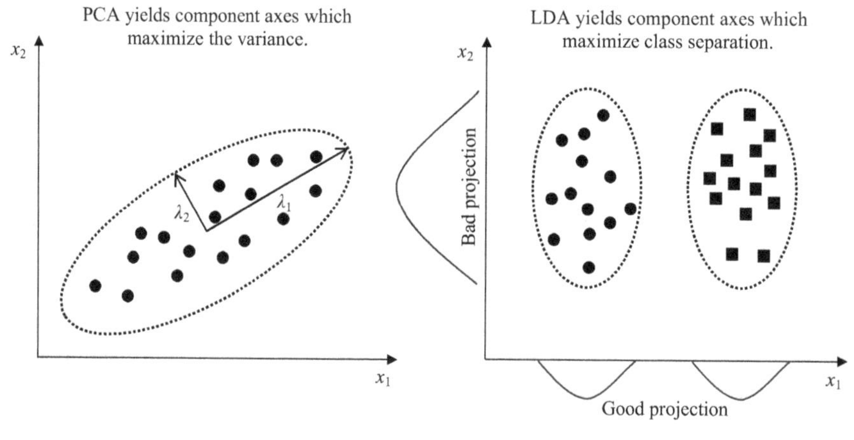

Figure 2.4 Pictorial illustration of PCA versus LDA

Between-class scatter matrix S_B is computed by the following equation:

$$S_B = \sum_{i=1}^{c} N_i (m_i - m)(m_i - m)^T \quad (2.23)$$

where m is the overall mean, m_i is the mean of the respective class, and N_i is the sample size of that class.

3. Compute the eigenvectors and corresponding eigenvalues for the scatter matrices. Solve the generalized eigenvalue problem for the matrix $S_W^{-1} S_B$ to obtain the linear discriminants.

$$S_W^{-1} S_B w_i = \lambda_i w_i \quad (2.24)$$

where $\lambda_i = J(w_i) = \frac{w^T S_B w_i}{w^T S_W w_i} = \text{scalar and } i = 1, 2, \ldots, c-1$

4. After calculating the eigenvectors and eigenvalues, the eigenvectors w_i are sorted from the highest to lowest depending on their corresponding eigenvalue λ_i and then choose the top k eigenvectors, where k is the number of dimensions to keep.
5. After selecting the k eigenvectors, the resulting $d \times k$ dimensional eigenvector matrix W can be used to transform data onto the new subspace such that

$$Y = X \times W \quad (2.25)$$

where X is a $n \times d$-dimensional matrix representing the n samples, and Y are the transformed $n \times k$-dimensional samples in the new subspace.

As shown, PCA and LDA are both powerful techniques for data dimensionality reduction. Although it might seem intuitive that LDA is superior to PCA when the class labels are known, this might not always be the case. For example, comparisons between accuracies in fault diagnosis applications after using PCA or LDA show that PCA tends to outperform LDA if the number of samples per class is relatively small [51]. It is also not uncommon in practice to use both LDA and PCA in combination: for instance, PCA for dimensionality reduction followed by an LDA.

2.4.1.3 Minimum redundancy maximum relevance

The minimum redundancy maximum relevance (mRMR) is a feature selection method that uses mutual information, correlations, and distance or similarity scores as feature selection criteria [52]. mRMR is a robust feature selection method that can rank the features based on their relevance to the target and also exclude redundant features. To summarize, this combined method selects the features with maximum relevance and minimum redundancy. Given two discrete variables, X and Y, the mutual information of these two variables can be defined as:

$$I(X;Y) = \sum_{y \in Y} \sum_{x \in X} p(x,y) \log\left(\frac{p(x,y)}{p(x)p(y)}\right) \quad (2.26)$$

where $p(x)$,$p(y)$ are the marginal probability distribution function. The relevance between the features and class variables can be achieved by:

$$D(S,c) = \frac{1}{|S|}\sum\nolimits_{f_i \in S} I(f_i;c) \tag{2.27}$$

The redundancy of all of the features takes the form:

$$R(S) = \frac{1}{|S|^2}\sum\nolimits_{f_i,f_j \in S} I(f_i;f_j) \tag{2.28}$$

Finally, the mRMR criterion is the combination of $D(S,c)$ and $R(S)$:

$$\max\{D(S,c) - R(S)\} \tag{2.29}$$

However, in certain situations, mRMR can lead to poor performance [53] when the features are individually useless but are useful when combined (a pathological case is found when the class is a parity function of the features). It is an instance of a large class of filter methods that tradeoff between relevancy and redundancy in different ways.

2.4.2 Approaches to anomaly detection

Various approaches and algorithms are available for anomaly detection, depending on the type of data available from the system under consideration. Some commonly used techniques include distance-based methods (e.g., Mahalanobis Distance (MD), *k*-Nearest Neighbors (KNNs)), clustering-based methods (e.g., *k*-means), and classification-based methods (e.g., support vector machines (SVMs) and neural networks (NNs)) [54]. Some popular methods are introduced here for handling different types of data sets.

When healthy data from a system is available and the method is selected, anomaly detection can be implemented by determining a detection threshold based on the healthy data to identify outliers. Threshold detection is an important step in diagnostics to have advanced warning of failure. Threshold values are typically defined based on expert knowledge of known fault conditions and economic factors such as the need to reduce the number of false alarms. These diagnostic approaches may not be able to detect anomalies when a priori knowledge of faults is not available. Therefore, it is useful to implement a generalized probabilistic approach to determine thresholds for anomaly detection [55].

2.4.2.1 MD approach to anomaly detection

MD is a distance measure that is used in applications such as anomaly detection, pattern recognition, and process control [56]. In electronics, MD has been used for detecting anomalies in notebook computers [55] and multilayer ceramic capacitors [57]. Kumar *et al.* [55] developed a probabilistic technique for determining detection thresholds using MD, followed by a power transformation of the MD values, which are always positive, so they follow the normal distribution. Statistical thresholds were then developed based on the mean and standard deviation of the transformed data.

The MD approach to anomaly detection involves distinguishing between healthy and anomalous data using the distance measure, thus reducing multivariate data to univariate data. MD is sensitive to changes between various parameters monitored as it takes into account the correlation between the different parameters. Additionally, MD is not sensitive to the differing scales of the parameters monitored, since MD values are calculated using normalized parameters.

Monitored data that is known to be healthy is used to calculate the mean and standard deviation for normalization. Further, this healthy data is used to compute the correlation matrix. With the mean, the standard deviation, and the correlation matrix obtained from the healthy data, the MD is calculated for every test data point. The MD values calculated from the healthy data are transformed into the normal distribution via the Box–Cox power transformation. The collection of MD values for a healthy system is known as the Mahalanobis space (MS). A detection threshold is then calculated based on the mean and standard deviation of the transformed healthy MD data. The calculations are then repeated for all test data using the mean and standard deviation, correlation matrix, and Box–Cox transformation parameter learned from the healthy data. Using this approach, anomalies are detected when the transformed MD for a test data point crosses the detection threshold.

The anomaly detection approach as shown in Figure 2.5 starts with performance parameter monitoring. For a test product, the MD value for each observation is calculated using the performance parameters' mean and standard deviation and a correlation coefficient matrix that is obtained from the training data (Figure 2.6). The calculated MD value is then compared with a threshold MD value τ, which is established from a baseline to classify the product as being healthy or unhealthy. Then, if the product were to be classified as unhealthy, further processing would be performed to isolate the faulty parameter(s) to establish reasons for the fault.

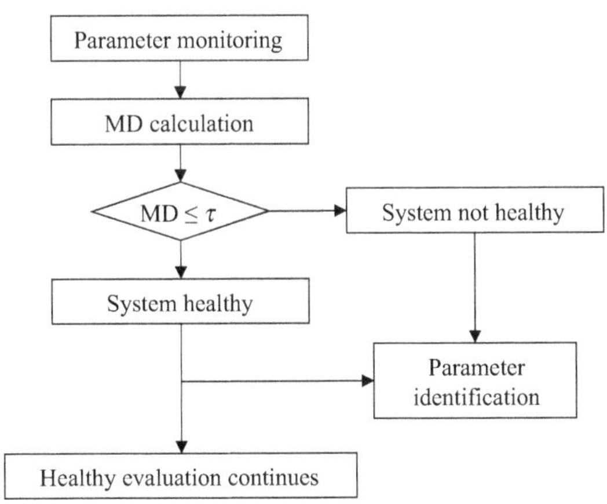

Figure 2.5 Fault detection approach

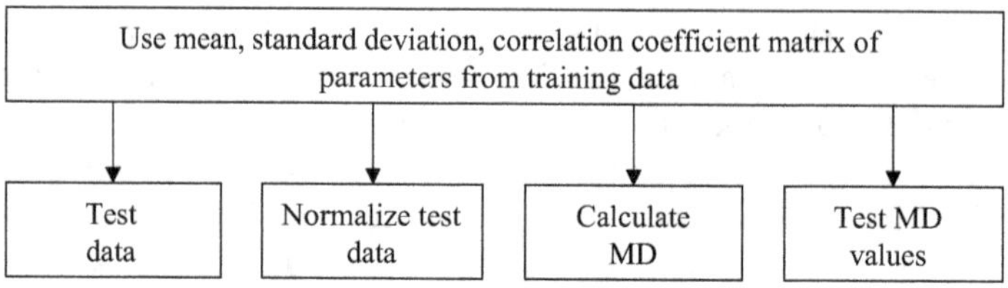

Figure 2.6 MD calculation using test data

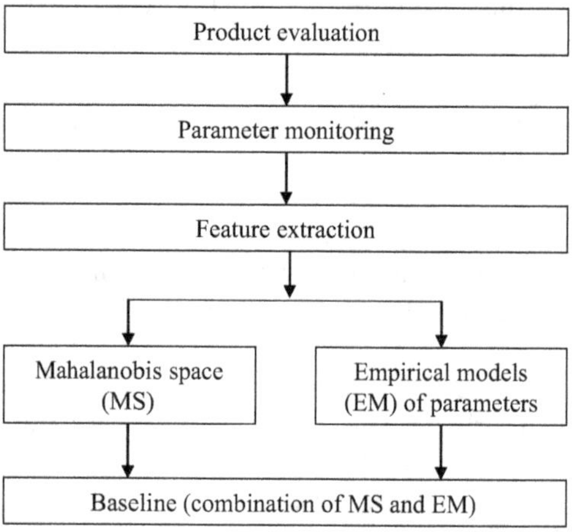

Figure 2.7 Baseline establishment methodology

A product's performance range is defined by measurements made of its performance parameters under different operating conditions. The combination of performance parameters can be summarized by a distance measure. A data set formed by measuring the performance parameters of a healthy product is used as training (or baseline) data. A baseline consists of an MD profile, a threshold MD value, and empirical models of performance parameters. The process of constructing a baseline is shown in Figure 2.7.

2.4.2.2 MD for threshold determination

A probabilistic approach is presented to determine two types of threshold MD values. First, a generic threshold for detecting any type of fault or anomaly present in a product based on the MDs obtained from the training data is determined. Second, a fault-specific threshold for detecting the presence of a particular fault based on historical data related to a particular fault is determined. The second threshold can be considered as a second-tier fault isolation process.

Generic threshold determination

An approach for determining a generic threshold—an MD value—for fault diagnosis is shown in Figure 2.8.

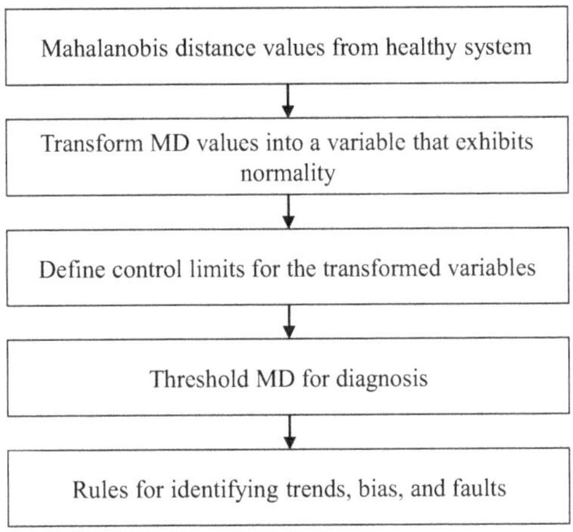

Figure 2.8 Approach for defining the threshold MD value

The MDs are always positive, but they do not generally follow a normal distribution. The Box–Cox power transformation can be used to transform a variable which has positive values and does not follow a normal distribution into a normally distributed transformed variable [58]. The Box–Cox transformation is defined as follows:

$$x(\lambda) = \frac{(x_\lambda - 1)}{\lambda}, \lambda \neq 0 \quad x(\lambda) = \ln(x), \lambda = 0 \tag{2.30}$$

where the vector of data observations is $x = x_1, x_2, \ldots, x_n$, and $x(\lambda)$ is the transformed data. The power λ is obtained by maximizing the logarithm of the likelihood function:

$$f(x,\lambda) = -\frac{n}{2}\ln\left[\sum_{i=1}^{n}\frac{(x_i(\lambda) - \bar{x}(\lambda))^2}{n}\right] + (\lambda - 1)\sum_{i=1}^{n}\ln(x_i) \tag{2.31}$$

where

$$\bar{x}(\lambda) = \frac{1}{n}\sum_{i=1}^{n} x_i(\lambda) \tag{2.32}$$

The normality of $x(\lambda)$, which is a transformed variable, is confirmed by plotting it into a normal plot. The mean (μ_x) and standard (σ_x) deviations of the transformed variable are used to determine the control limits of an x-bar chart. A threshold value corresponding to the warning limit $(\mu_x + 2\sigma_x)$ and a threshold value corresponding to a fault alarm $(\mu_x + 3\sigma_x)$ are defined. Since higher MD values are of concern from an "unhealthiness" perspective, the upper portion of the control chart is of importance for identifying changes in system health. Rules from quality control, including bias and variance identification, can be used [59].

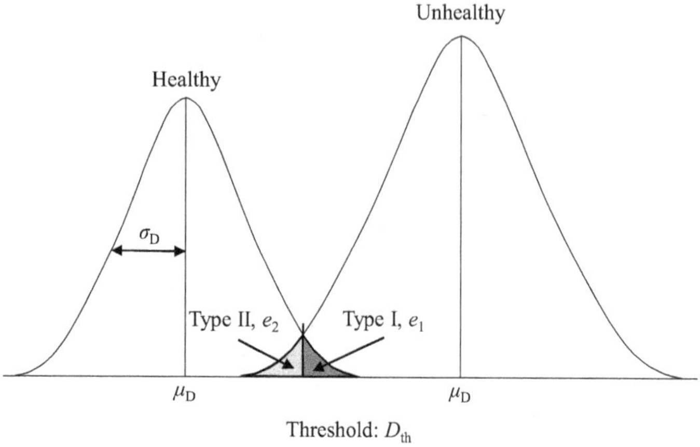

Figure 2.9 Threshold value calculation

Fault-specific threshold determination

A normally distributed transformed variable, which corresponds to MD values, can be used to determine Type-I and Type-II errors [60]. A Type-I error, which is often referred to as a false positive, is a statistical error made in testing the health of a product, in which the product is healthy but is incorrectly determined to be unhealthy. A Type-II error, which is often referred to as a false negative, is a statistical error made in testing the unhealthiness of a product, in which a product is determined to be healthy when it is not. Figure 2.9 shows Type-I and Type-II errors using a variable's distribution for a healthy and an unhealthy system, where the healthy distribution is defined from the training data and the unhealthy distribution is defined from the data representing a specific fault in a system. For a known fault, an optimal transformed variable can be defined such that the combined error (i.e., the sum of Type-I and Type-II errors) remains minimal (i.e., the shaded region in Figure 2.8), and an MD value corresponding to the optimal transformed variable x is calculated.

For a healthy product, the probability of having MD values higher than the threshold value is the number of observations that produce an MD value higher than the threshold MD value divided by the total number of observations for a healthy product. Similarly, for an unhealthy product, the probability of having an MD value less than the threshold value is the number of observations that produce MD values less than the threshold MD value divided by the total number of observations for an unhealthy product. The threshold value τ_x of a transformed variable for detecting a known anomaly is established using the following error function ε:

$$\varepsilon(\tau_x) = \frac{e_1}{n_h} + \frac{e_2}{n_u} \tag{2.33}$$

where τ is the threshold, e_1 is the number of observations classified as unhealthy in the healthy population n_h and e_2 is the number of observations classified as healthy in the unhealthy population n_u. The threshold value is obtained by minimizing the error function (i.e., by choosing a different value for τ_x).

2.4.2.3 *k*-nearest neighbors approach to anomaly detection

k-nearest neighbor (KNN) is a technique in ML wherein a new data point is classified based on its proximity to other data points belonging to known classes; in this case, healthy and faulty classes [61,62]. Traditionally, classification by KNN is done by a majority vote of the nearest neighboring points, where the class with the most number of neighbors wins the vote and classifies the new point as its own class. In this study, however, the KNN classification method is slightly modified by selecting three nearest neighbors from each class and then calculating the distance of the new point to the centroid of the neighbors of each class. An illustration of this distance-based KNN algorithm is shown in Figure 2.10. The advantage of using a distance-based KNN instead of a majority-vote-based KNN is that the information obtained on the trajectory of degradation in the feature space as the change tracked in the distance. The criterion for anomaly used in this study is when the distance of a point to the faulty class is closer than the distance to the healthy class, which suggests that the IGBT is now behaving more like a faulty device. This criterion can be summarized by the following statement:

$$\textit{Anomaly exists if}: \ (\textit{Distance to Healthy Centroid} - \textit{Distance to Faulty Centroid}) > 0$$

Figures 2.11 and 2.12 show the result of the KNN anomaly detection algorithm applied to two IGBT samples. The two figures show that the algorithm successfully detects anomalies before failure when the distance curve crosses above the zero line.

2.4.2.4 *k*-means approach to anomaly detection

Clustering-based methods aim to group similar instances together based on their intrinsic patterns or similarities. Anomalies in clustering-based methods are detected as data points which do not belong to any of the identified clusters or that are assigned to small, sparse, or outlier clusters.

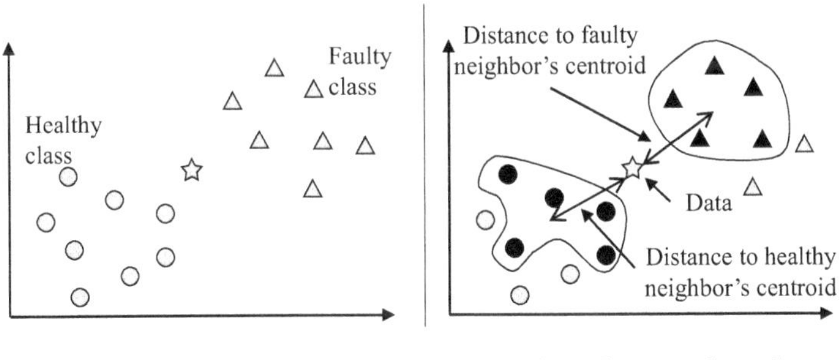

Figure 2.10 Illustration of a distance-based KNN algorithm

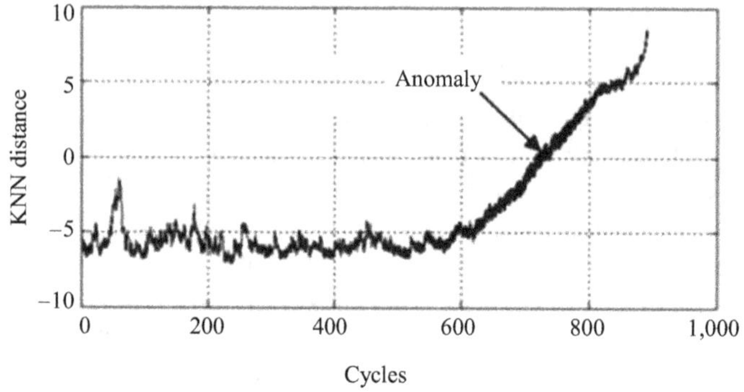

Figure 2.11 Anomaly detection for IGBT test sample #1. Anomaly detected at 717 cycles. The sample failed at 891 cycles. Experiment loading conditions: 1 kHz switching frequency, 50% duty cycle, 100 °C temperature swing.

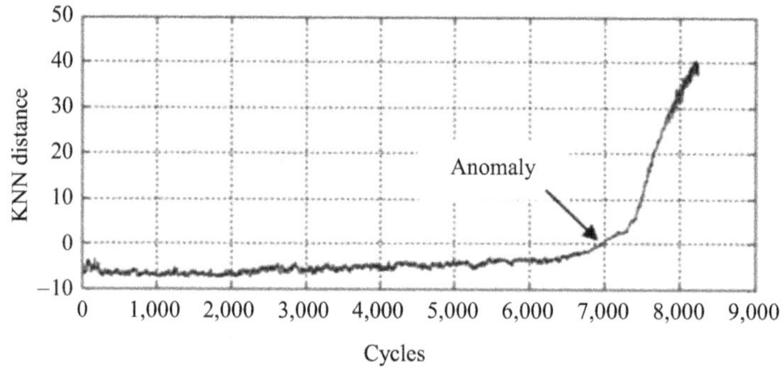

Figure 2.12 Anomaly detection for IGBT test sample #2. Anomaly detected at 6,973 cycles. The sample failed at 8,249 cycles. Experiment loading conditions: 1 kHz switching frequency, 50% duty cycle, 50 °C temperature swing.

The key difference from distance-based methods is that clustering-based methods evaluate each observation with respect to the cluster it belongs to, whereas distance-based methods analyze each observation with respect to its local neighborhood. Clustering-based methods can be effective for dealing with reference observations collected under varying operating conditions of an object system by partitioning them into clusters and using each as a reference [2].

k-means is one of the clustering methods for pattern recognition and ML. Figure 2.13 illustrates the procedure of k-means clustering—this begins with a decision of the value of k, which is a parameter to specify the number of clusters to

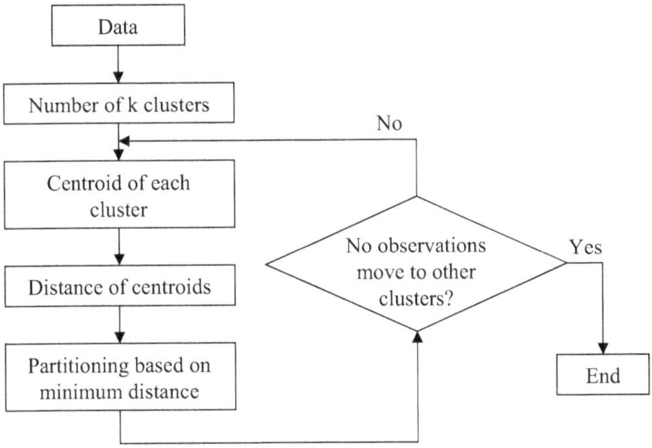

Figure 2.13 Procedure of k-means clustering

be partitioned. The next step is to define k centroids, one for each cluster. One of the simplest ways is to randomly choose and use k data points in the given data set as cluster centroids. In fact, the centroids need to be placed strategically due to the fact that they cause clustering results; the better choice is to place cluster centroids as far away from each other as possible. The following step is to take each instance in the data set and compute the distance between the data and each cluster centroid. Although Euclidean distance is a common distance metric in k-means clustering, other distance metrics (e.g., MD) can be used. Then, each data instance is assigned to the closest cluster based on the distances calculated in the previous step and each cluster centroid is updated by calculating a mean of all data instances in the same cluster. The aforementioned steps repeat iteratively until a predetermined termination condition is achieved. For example, if no data instances (or centroids) move to other clusters, the clustering process is terminated. That is, the k-means clustering algorithm aims to minimize an objective function J, in this case a squared error function:

$$J = \sum_{j=1}^{k} \sum_{i=1}^{N} ||x_i^{(j)} - c_j||^2 \tag{2.34}$$

where x_i is the ith data instance, $i = 1, 2, \ldots, N = 1$ N is the number of data instances in the given data set, and c_j is the centroid of cluster j, respectively.

2.4.2.5 Support vector machine approach to anomaly detection

SVM [63] is a supervised learning algorithm commonly used for one-class classification tasks and can also be adapted for anomaly detection. In the context of anomaly detection, SVM aims to learn the characteristics of normal instances and identify anomalies as instances that fall on the other side of the hyperplane, away from the region of normal instances. Its primary focus is on learning decision boundaries and separating data points based on their class labels or anomaly status, rather than directly relying on distance-based calculations.

SVM operates by mapping the input data points into a higher-dimensional feature space using a kernel function and then finding the hyperplane that maximally separates the classes. The decision boundary is defined by the distances between the support vectors (which are the data points closest to the hyperplane) and the hyperplane.

Given a set of input-output pairs $D = \{(x_1, y_1), (x_2, y_2), \ldots, (x_n, y_n)\}$, where $x_i \in \mathbb{R}^d$ is the ith input data instance, $y_i \in \{-1, 1$ is the corresponding output to x_i, n is the number of data instances, and $\mathbb{R}^d$ indicates the d-dimensional feature space.

The hyperplane equation can be represented as follows:

$$y_i(w^T x_i + b) \geq 1, \quad i \in [1,n] \tag{2.35}$$

where $w \in \mathcal{F}$ and $b \in \mathcal{R}$. The hyperplane determines the margin between the classes; that is, data instances for a class (i.e., $y_i = -1$) are on one side, whereas data instances for another class (i.e., $y_i = 1$) are on the other:

$$\begin{cases} w^T x_i + b \leq -1, & \text{if} \quad y_i = -1 \\ w^T x_i + b \geq 1, & \text{if} \quad y_i = 1 \end{cases} \tag{2.36}$$

The distance from the closest instances (i.e., support vectors) from each class to the hyperplane is equal to $\frac{margin}{2}$; that is, the constructed hyperplane searches for the maximal margin between the classes, as shown in Figure 2.14, where $r = \frac{w^T x_i + b}{\|w\|}$ indicates the distance of any data instance x_i to the hyperplane.

If all data instances are off the margin, this is called a "hard-margin SVM." As depicted in Figure 2.15, there are two well-known issues with hard-margin SVMs. If data instances are not linearly separable, hard-margin SVMs will fail, creating hyperplanes (see the left-hand side of Figure 2.15). Likewise, hard-margin SVMs are very sensitive to noisy data (see the right-hand side of Figure 2.15).

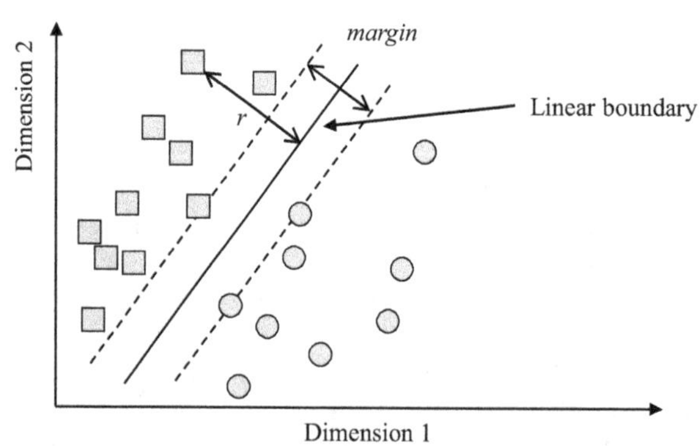

Figure 2.14 Classification margin

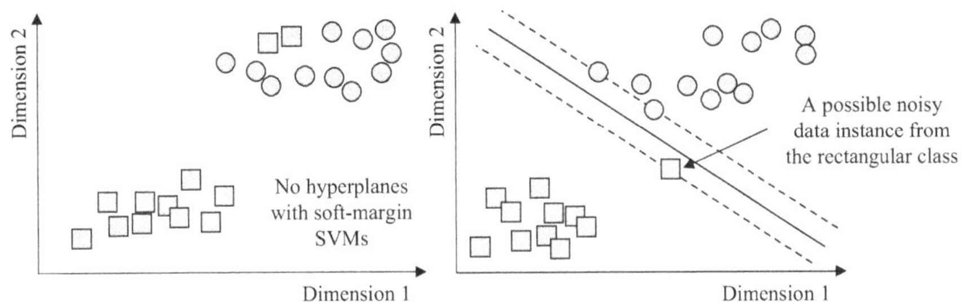

Figure 2.15 Issues in hard-margin SVMs

Accordingly, to avoid such issues, it is preferable to use a more flexible model yielding a good balance between keeping the margin as large as possible and limiting the margin violations. This is called a "soft-margin SVM."

Slack variables ξ_i are introduced to the previous equation, which can be rewritten as:

$$y_i(w^T x_i + b) \geq 1 - \xi_i \tag{2.37}$$

If $\xi_i = 0$, the point can be considered as correctly classified. If $\xi_i > 0$, the point i violates the margin and is incorrectly classified. ξ_i is considered as an error term and the constant $C > 0$ controls the trade-off between maximizing the margin and the number of training data instances within that margin. The SVM's objective function can be defined as the following minimization formulation:

$$\min_{w,b,\xi_i} \frac{||w||^2}{2} + C\sum_{i=1}^{n} \xi_i \tag{2.38}$$

subject to $y_i(w^T x_i + b) \geq 1 - \xi_i$ and $\xi_i \geq 0$ for all $i = 1, 2, \ldots, n$

This formulation is called the soft margin technique. For large values of C, a smaller-margin hyperparameter will be chosen if that hyperplane performs well in classifying all training data instances correctly. In contrast, a very small value of C will choose a larger-margin hyperplane. SVMs were originally designed for binary classification, but they can be extended to solve multiclass classification problems by combining many binary classification SVMs such as one-against-one SVM, one-against-all SVM, and directed acyclic graphic SVM [64].

2.4.3 Techniques for prognostics

Prognostics of the RUL can be carried out based on the knowledge of the physics of the degradation process. For those systems that are too complicated to have an explicit physics model to describe the degradation process, a data-driven method is preferable to automate the diagnosis and prognosis. The changes and anomalies in the precursors data can be detected using several ML techniques as introduced, and the data trends can then be estimated using the techniques covered in the following sections.

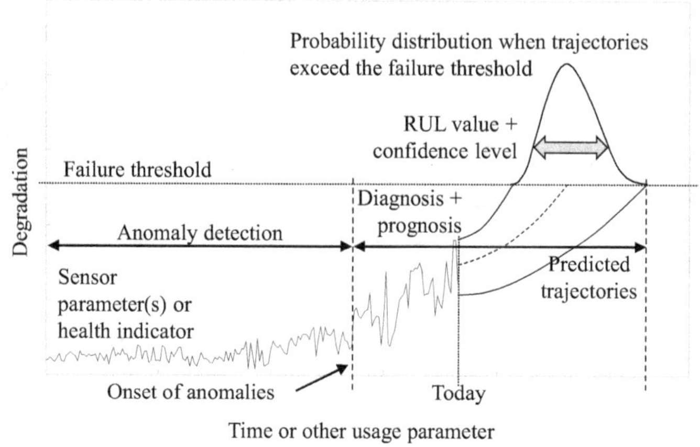

Figure 2.16 Prognostics concept

The key role of prognostics is to forecast the RUL of a system under its usage conditions. Figure 2.16 illustrates the concept of prognostics [2]. The in-situ monitoring of health indicators or sensor measurements is performed during the degradation. Using techniques discussed in Section 2.4.2, the onset of anomalies is identified. Once anomalies are detected, RUL estimation can be made, where "today" is the time at which RUL is estimated. As a prognostic result, a RUL value with a certain confidence interval will be provided. Hence, the end goal of prognostics is to develop a method(s) to provide not only the prior history of the system's health status but also diagnostics of failure modes/mechanisms, enabling the application of the correct model (e.g., PoF and data-driven) for a prediction of RUL. To select the predictive modeling technique for prognostics tasks, one may want to consider the data type, the model complexity, and the interpretability of the model's output. Some models work better for certain types of data than others. For example, if there is numerical data and explainable AI is preferred, techniques such as regression analysis and filtering may fit better. If the prediction task is based on image data, more complex models such as deep learning NNs may be considered.

2.4.3.1 Statistical regression

Regression models are used to predict a continuous numerical value based on one or more input variables. The goal of regression predictive modeling is to identify the best mapping function from the input variables to a continuous output variable. A linear regression model describes the relationship between a dependent variable and one or more independent variables. The case of one explanatory variable is called simple linear regression. Suppose n data pairs are observed and called $\{(x_i, y_i), i = 1, \ldots, n\}$, they can be expressed using the following equation involving the noise ε:

$$y_i = \beta_0 + \beta_1 x_i + \varepsilon \tag{2.39}$$

The more general case is multiple linear regression where a model is created for more than one independent variable. The model is expressed using the following equation:

$$y_i = \beta_0 + \beta_1 x_{i1} + \beta_2 x_{i2} + \ldots + \beta_p x_{ip} + \varepsilon_i, \quad i = 1, \ldots, n \tag{2.40}$$

where n is the number of observations. y_i is the ith response. β_k is the kth coefficient, where β_0 is the constant term. x_{ij} is the ith observation on the jth predictor variable $j = 1, \ldots, p$. ε_i is the ith error term (noise). Nonlinear regression models represent the relationship between a continuous response variable and one or more continuous predictor variables in the form

$$y = h(x)\beta + \varepsilon \tag{2.41}$$

where y is an n-by-1 vector of observations of the response variable. h is any function of x and β that evaluates each row of x along with the vector β to compute the prediction for the corresponding row of y. x is an n-by-p matrix of predictors. β is a p-by-1 vector of unknown parameters to be estimated. Assume the noise $\varepsilon \sim N(0,\sigma^2)$ follows an independent, identically distributed Gaussian distribution with zero mean and variance σ^2. A Gaussian process regression (GPR) model [65] explains the response by introducing latent variables, $f(x_i), i = 1, 2, \ldots, n$, from a Gaussian process (GP). The covariance function of the latent variables captures the smoothness of the response. A GP is a set of random variables, such that any finite number of them have a joint Gaussian distribution. If $\{f(x), x \in \mathbb{R}^d\}$ is a GP, then given n observations $x_1, x_2, \ldots, x_n$, the joint distribution of the random variables $f(x_1), f(x_2), \ldots, f(x_n)$ is Gaussian. A GP is defined by its mean function $m(x)$ and covariance function, $k(x, x')$. That is, if $\{f(x), x \in \mathbb{R}^d\}$ is a GP, then

$$E(f(x)) = m(x) \tag{2.42}$$

and

$$\mathrm{Cov}[f(x_0), f(x')] = E[\{f(x) - m(x)\}\{f(x') - m(x')\}] = k(x, x') \tag{2.43}$$

Now consider the following model.

$$h(x)^T\beta + f(x) \tag{2.44}$$

where $f(x) \sim GP(0, k(x, x'))$, that is $f(x)$ are from a zero mean GP with covariance function, $k(x, x')$. $h(x)$ are a set of basis functions that transform the original feature vector x in $\mathbb{R}^d$ into a new feature vector $h(x)$ in $\mathbb{R}^p$. β is a p-by-1 vector of basis function coefficients. This model represents a GPR model. An instance of response y can be modeled as:

$$P(y_i|f, N) \sim N(y_i|h(x_i)^T\beta + f(x_i), \sigma^2) \tag{2.45}$$

Hence, a GPR model is a probabilistic model. There is a latent variable $f(x_i)$ introduced for each observation x_i, which makes the GPR model nonparametric. A GPR model has the ability to provide uncertainty representation and facilitate accurate prediction without physical models. The use of GPR for state of health estimation of lithium-ion batteries is a good example of model-based learning [66].

2.4.3.2 Bayesian filtering

A power electronics device is usually viewed as a black box because of its unmeasured state, while its electromechanical properties are nonlinear. The application of the Bayesian techniques to state the estimations as in particle filters (PF) provides a powerful tool to integrate the diagnosis and prognosis of complex systems (such as batteries [67]). PF accurately models the underlying dynamics of a physical system with nonlinearity and non-Gaussianity. Moreover, it is important to process data online and, hence, rapidly adapt to changing signal characteristics [68].

Bayesian filtering aims to eliminate or reduce noise and then to restore the actual signal or the optimal estimation of the system state [68]. The basic idea is to establish a state-space model that involves the hidden state variables, random noises, and observed measurements:

$$x_k = f_k(x_{k-1}, v_k) \leftrightarrow p(x_k|x_{k-1}) \tag{2.46}$$

$$y_k = h_k(x_k, w_k) \leftrightarrow p(y_k|x_k) \tag{2.47}$$

where (2.46) and (2.47) are possibly nonlinear functions of the state and measurement, respectively; v_k is an independent and identically distributed (i.i.d.) process noise sequence; w_k is the i.i.d. measurement noise; $p(x_k|x_{k-1})$ is the transition density; and $p(y_k|x_k)$ is the observation density. Recursive Bayesian estimation provides a generic approach for estimating the posterior expectation and posterior probability density function (pdf) based on observed data. Prediction and update are the two steps in processing the filtering. First, the state is predicted with the probability density one step ahead of the system model. Then, the current measurement is used to modify the predictive result and determine the estimation of the current final state. The specific procedure is described below.

Let $x_{0:k}$ and $y_{1:k}$ denote the set of all states and the set of observed data, respectively, of the signal and the observations up to time k. It is useful to estimate the posterior distribution and expectations recursively in time.

1. Prediction:

$$p(x_{0:k}|y_{1:k-1}) = \int p(x_k|x_{k-1})p(x_{0:k-1}|y_{1:k-1})dx_{0:k-1} \tag{2.48}$$

   where $p(x_{0:k}|y_{1:k-1})$ is the a priori state pdf.
2. Update:

$$p(x_{0:k}|y_{1:k}) \propto p(y_k|x_k)p(x_k|x_{k-1})p(x_{0:k-1}|y_{1:k-1})dx_{0:k-1} \tag{2.49}$$

$$I(f_k) \int f_k(x_{0:k})p(x_{0:k}|y_{1:k})dx_{0:k} \tag{2.50}$$

   where $p(x_{0:k}|y_{1:k})$ is the posterior state pdf.

This recursive thinking led to the emergence of the Kalman filter (KF). However, the KF is only available for linear filtering whose posterior density is Gaussian at every step [62]. Thus, based on the fundamental principle of KF, the extended KF (EKF) adopts the local linearization to the nonlinear function around the estimates of the state through first-order terms of the Taylor series expansion. Similar to the EKF, the unscented KF (UKF) was created based on the assumption that the state distribution is still Gaussian. The advantage of UKF is that the sample points are deterministically selected from the Gaussian approximation and propagated through a truly nonlinear system [69]. Thus, it is more accurate to capture the true mean and covariance to the second-order term of the Taylor series expansion. Although EKF and UKF promote recursive filtering to the nonlinear system, the state variables estimated from both are subject to Gaussian distribution. In fact, the non-Gaussianity of the true density is more common. In this case, particle filtering was derived to solve the nonlinear and non-Gaussian problem based on Monte Carlo (MC) simulation.

2.4.3.3 Particle filtering

PFs are sequential MC methods that are coupled with recursive Bayesian estimation [70]. The principle is to utilize the set of weighted samples (particles) to represent the probability densities, which are available for any state-space model. As the number of samples (N_s) becomes very large, the MC characterization becomes an equivalent representation of the usual functional description of the posterior pdf [71]. $\{x^i_{0:k}, w^i_k\}$ is introduced to denote a random measure that characterizes the posterior pdf $p(x_{0:k}|y_{1:k})$, where $\{x^i_{0:k}, i = 0, \ldots, N_s\}$ is a set of support points with associated weights $\{w^i_{0:k}, i - 0, \ldots, N_s\}$. The weights are normalized as $\sum_i w^i_k = 1$. Thus, the posterior filtered pdf at the k step can be approximated as:

$$p(x_k|y_{1:k}) \approx \sum_{i=1}^{N_s} w^i_k \delta(x_k - x^i_k) \tag{2.51}$$

The normalized weights w^i_k are based on the principle of importance sampling [70,72]. Here, $\delta(\cdot)$ is a Dirac function, and $x^i_{0:k}$ is drawn from an importance density $q(x_{0:k}|y_{1:k})$. Through recursive relation, the weights are given by:

$$w^i_k \propto w^i_{k-1} \frac{p(y_k|x^i_k)p(x^i_k|x^i_{k-1})}{q(x^i_k|x^i_{0:k-1}, y_{1:k}} \tag{2.52}$$

Currently, two problems still exist. One is that large numbers of sample points (particles) are needed to satisfy high accuracy, adding to the complexity of the algorithm. The other problem is that it is impossible to avoid the degeneracy of the particles for sequential importance sampling [68,72]. After several iterations, the possible extreme situation is that only one particle is retained, while others are neglected because of low weight. Thus, the effective sample size, N_{eff}, is introduced for resampling to solve the degeneracy of the particles.

$$N_{eff} = \frac{N_s}{1 + \mathrm{Var}(w^{*i}_k)}, \quad \text{where } w^{*i}_k = \frac{p(x^i_k|y_{1:k})}{q(x^i_k|x^i_{k-1}, y_k)} \tag{2.53}$$

Since the true weight w_k^{*i} cannot be exactly evaluated [17], an estimate N_{eff} is approximated by

$$N_{eff} = \frac{1}{\sum_{i=1}^{N_s} (w_k^i)^2} \tag{2.54}$$

where w_k^i is given by (2.52). The basic idea of resampling is to eliminate particles that have small weights and to concentrate on particles with large weights [61] (i.e., when N_{eff} falls below the threshold N_T, the resampling process will be invoked).

2.4.3.4 Prognostics using particle filtering

The objective is to estimate the j-step ahead of the prediction of the state pdf $p(x_{k+j}|y_{0:k}), (j \geq 2 \in \mathbf{N}^*), j = 1, \ldots, T-k$ where T is the time horizon for the end-of-life of the device under study. If no information is available about the degradation model, then it is unavailable for estimating the likelihood of the degrading state following the future path, $x_{k+1:k+j}$. Thus, the initial condition $p(x_k|y_{0:k})$ should be projected among all possible future paths. The probability from the state x_{t-1} to x_t is $p(x_t|x_{t-1})dx$. The prediction distribution is given by:

$$p(x_{k+j}|y_{0:k}) = \int p(x_{k+j}|x_{k+j-1})p(x_{k+j-1}|y_{0:k})dx_{k+j-1} \tag{2.55}$$

Based on (2.51), (2.55) can be approximated as:

$$p(x_{k+j}|y_{0:k}) \approx \int \sum_{i=1}^{N} w_{k+j-1}^i \delta_{x_{k+j-1}^i}(dx_{k+j-1})p(x_{k+j}|x_{k+j-1})dx_{k+j-1} \tag{2.56}$$

2.4.3.5 Prognostics using deep learning methods

Deep learning, a subset of ML, has gained significant attention in prognostics. Deep neural networks, such as recurrent neural networks (RNNs), long short-term memory (LSTM) networks, and convolutional neural networks (CNNs), can capture complex temporal dependencies and patterns in the data. These models can be trained to predict the RUL or health condition of a system based on historical sensor data (usually time series in nature) and do not rely on explicit knowledge of the system's physics.

The LSTM RNN is designed to learn the long-term dependencies and is widely used for time series data predictions. The network architecture of the LSTM RNN predictor is schematically shown in Figure 2.17 [73]. All elements of the LSTM RNN are enumerated and described below. Note that the subscript c of each vector represents the neuron, and the superscript (t) represents the moment t. $x^{(t)}$ is the input layer at the current moment t, and $h^{(t-1)}$ is the hidden layer at the previous moment $t-1$. "Moment" is used here for the generality of LSTM RNN applications, and for the RUL prediction of the device under study, "moment" amounts to the "cycle" of the cyclic degradation test.

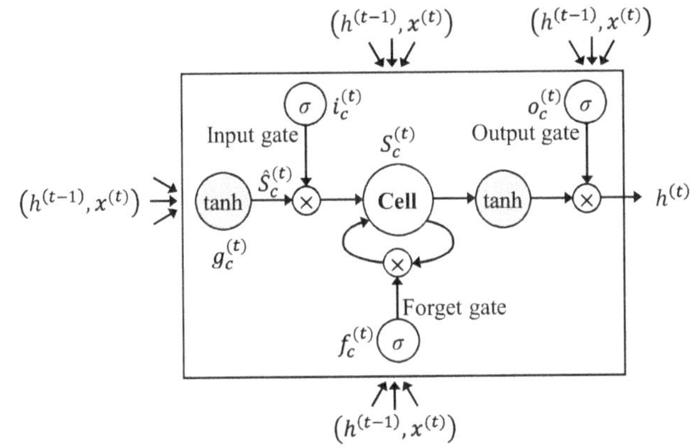

Figure 2.17 Network architecture of the LSTM RNN predictor

LSTM is able to remember information for long periods of time via the introduced gates. Gates are a distinctive feature of the LSTM approach and are able to optionally decide which information should pass through. A gate is a sigmoidal unit that is activated from the current input layer $x^{(t)}$ and from the hidden layer at the previous time step. The forget gate is able to discard redundant information; the input gate is able to select key information to be stored in the internal state; and the output gate is used to determine output information. In this case, LSTM RNN is able to store and update the key information efficiently over a long time without gradients vanishing. The below equations give the full algorithm for a modern LSTM.

$$f^{(t)} = \sigma(W^{fX}x^{(t)} + W^{fh}h^{(t-1)}b_f) \tag{2.57}$$

$$i^{(t)} = \sigma(W^{iX}x^{(t)} + W^{ih}h^{(t-1)} + b_i) \tag{2.58}$$

$$g^{(t)} = \tanh(W^{gX}x^{(t)} + W^{gh}h^{(t-1)} + b_g) \tag{2.59}$$

$$s^{(t)} = g^{(t)*}i^{(t)} + s^{(t-1)*}f^{(t)} \tag{2.60}$$

$$o^{(t)} = \sigma(W^{oX}x^{(t)} + W^{oh}h^{(t-1)} + b_o) \tag{2.61}$$

$$h^{(t)} = \tanh(s^{(t)})^{*}o^{(t)} \tag{2.62}$$

where W and b values are the layer weights and biases, respectively.

Zhang *et al.* [73] investigated deep-learning-enabled battery RUL prediction using LSTM RNN. The LSTM RNN was adaptively optimized using the resilient mean square backpropagation (RMSprop) method, and a dropout technique was used to address the overfitting problem. The developed MC simulation is combined to generate a probabilistic RUL prediction with confidence level. Figure 2.18 shows the features and capabilities of the constructed LSTM RNN.

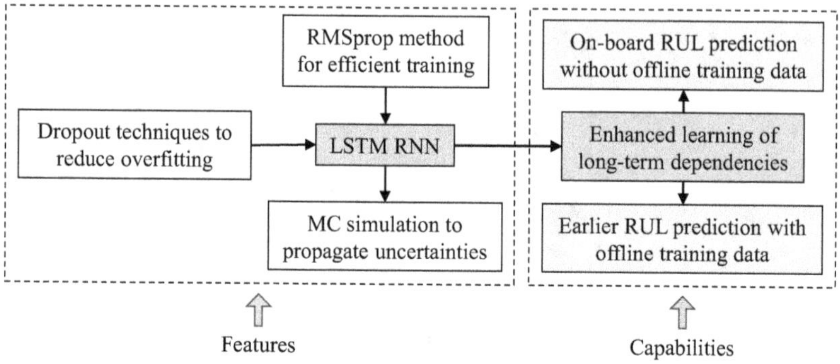

Figure 2.18 Features and capabilities of the constructed LSTM RNN

2.5 Examples of anomaly detection and prognostics for electronic circuits

Interest in non-invasive circuit-level anomaly detection and prognostics methods has recently increased as such methods are often simpler to implement from a hardware perspective. These methods are focused on data-driven analysis of the output signal of a device in response to a known input. This section provides examples of data-driven anomaly detection and prognostics. In addition, the circuit-level canary methods in which a prognostics chip is designed to fail, thus providing early warning of the impending failure of a monitored system, are also discussed.

2.5.1 Data-driven anomaly detection and prognostics for electronic circuits

Data-driven methods for anomaly detection and prognostics do not require information pertaining to the failure mechanisms or materials used in a monitored device. This is advantageous for circuit-level monitoring where exact failure mechanisms are not known. Multiple studies have shown that it is possible to monitor the characteristics of the output of a device to detect anomalies and make RUL predictions. For analog circuits, it is possible to complete a frequency sweep of the circuit using a larger bandwidth than that of the circuit and perform an analysis of the output to determine the behavioral characteristics of the device.

When analyzing the frequency sweep of the circuit, there are two main categories of features that are extracted: time domain features and frequency domain features. In typical frequency analysis of a time domain response, a Fourier transform is conducted; however, the Fourier transform of the signal assumes a stationary signal as time domain transients simply become integrated into the frequency signal. Therefore, wavelet transforms are used to reveal break points and discontinuities now observable in a Fourier transform. In addition to wavelet features, time domain statistical features, such as mean, variance, skewness, and kurtosis, can be extracted from the signal to be used in diagnosis.

Extracting multiple features over long periods of time will generate large data sets, which can lead to significant computational expenses and longer time for estimation. Variable reduction methods (e.g., PCA), as discussed previously, can be used to make such data more manageable. A health or fault indicator (FI) must be determined from the features to track health for anomaly detection and perform prognostics.

One study used both the frequency domain features and the time domain features to create a health operating region for the circuit [74]. The FI was calculated based on the well-known Mahalanobis–Taguchi (MT) methodology and started with the collection of the two feature sets (wavelet and statistical features) under a no-fault condition. For both feature sets, two individual MS were constructed using their normalized feature elements and correlation coefficients. The MD values obtained from the two feature sets are plotted on a 2-D plane as shown in Figure 2.19, the fault conditions are separable from the no-fault condition. The figure also illustrates a typical trajectory in the 2-D space representing circuit degradation.

Following the creation of the FI, the circuits under test were seeded with various faults to create training data for a kernel-based ML fault detection and prognostic system. The FI was able to summarize the degradation level in each of the circuit's critical components using the features extracted from the circuit response. The study showed that the value of FI increased with respect to the amount of deviation in the faulty component from its nominal value. Thus, the anomaly detection threshold could be defined based on a certain FI value when a component value drifted outside its tolerance range. The developed fault diagnostic method was applied to the band-pass and the low-pass filter circuits. The result showed that the method was able to detect faulty circuit conditions and isolate faulty components with an accuracy of 100% and 96%, respectively.

FI was also developed to perform prognostics for analog circuits. Once an anomaly was detected by the diagnostic routine, it triggered the prognostic module. A PF approach was employed for the FI values prediction and the remaining useful

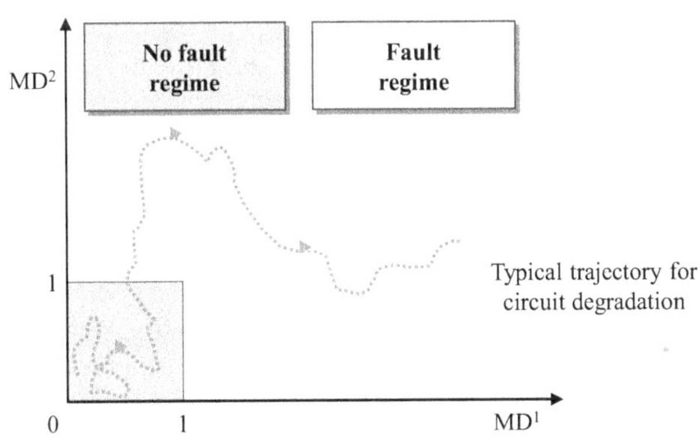

Figure 2.19 Illustration of fault propagation on MS

performance (RUP) estimation. The RUP was calculated based on the time index at which the predicted FI value crosses the failure threshold. The circuit FI is also compatible with future changes in the extracted features. Thus, even if new features (for circuit fault diagnosis) are introduced in the future, RUP estimates can still be obtained using the developed FI for the circuit.

An advantage of complex systems is that they often incorporate a data bus that can be easily and non-invasively probed and then used for in-situ monitoring. From this data bus, specific parameters can be monitored and trended to update the FI. One such parameter is the voltage ripple of the device, which increases as the power supply filter fails. Health data can be monitored for this, and then a threshold can be determined to provide classification for the FI.

2.5.2 Canary methods for anomaly detection and prognostics for electronic circuits

Another strategy for detecting circuit-level faults is to integrate expendable devices into circuits to monitor the health of the components within. Such devices are often called canaries or prognostic cells. These devices are designed in such a way that they receive enhanced loading compared to the devices they are mirroring. There are two ways to accomplish the early warning feature. First, the load on a device can be amplified and sent to the prognostic cell where it stresses the components in the prognostic cell. The second and more common method is to make the devices in the prognostic cell less robust by changing the geometry or the material of which the device is composed. There can also be a combination of the two. Thus, under the same environmental and operational loading conditions, the canary health monitors wear out faster to indicate the impending failure of the actual product.

For example, to create a canary for a solder joint, the stress concentration factor can be increased by thinning or sharpening the corners of the pad. The prognostic cell is then isolated and placed in the same package as the circuit where it sees the same loading but enhanced stress and will fail a certain time before the devices in the circuit are given an RUL estimation. It is essential that prognostic cells fail by the same mechanism as the board components for adequate anomaly detection. The time between the failure of the canary and the failure of the circuit is called the prognostic distance (see Figure 2.20), and these devices can be designed to yield the desired magnitude of this distance.

One study demonstrated an implementation of a prognostic cell to monitor TDDB of the gate oxide of MOSFET [75]. Here, the electric field observed by the gate oxide of the MOSFET die was amplified by the designed circuit (see Figure 2.21) and applied to the gate oxide of the MOSFET in the prognostic cell. Therefore, the acceleration of the breakdown of the gate oxide can be achieved. A feedback module is designed into the circuit that shuts off power to the prognostic cell once a failure has been observed. When the prognostic cell failed, a certain fraction of the circuit lifetime was consumed. The fraction of consumed circuit life was dependent on the amount of the overvoltage applied and could be estimated from the known failure models of TDDB.

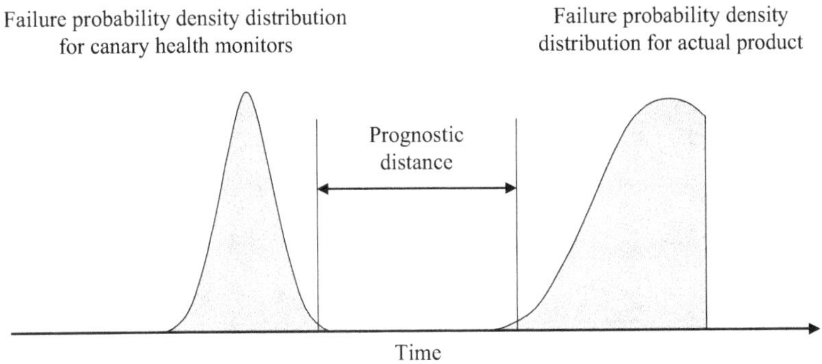

Figure 2.20 Advanced warning of failure using canary structures

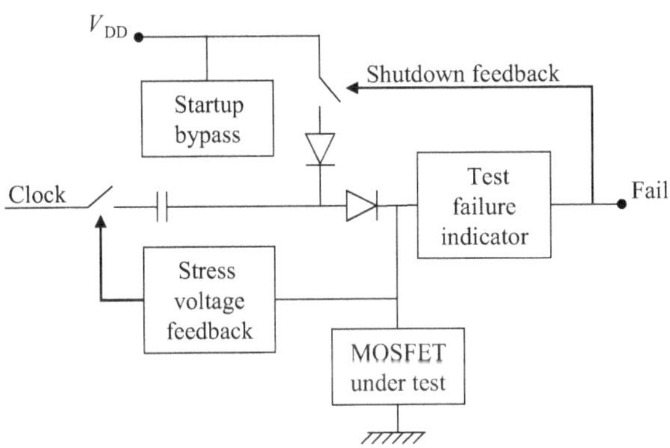

Figure 2.21 Schematic circuit for the TDDB prognostic circuit

A prognostic cell must be triggered before the accompanying circuit fails. This means that the cumulative failure probability of the prognostic cell must be close to unity before the cumulative failure probability of the circuit becomes appreciable, as shown in Figure 2.22. This can be realized by operating the prognostic cell at a higher oxide field than the circuit as described above.

The probability of correctly predicting circuit failure using a prognostic cell is equal to the probability that the prognostic cell triggers (fails) before the circuit. If the following variables are defined: $p(t)$ is the trigger (failure) pdf of the prognostic cell, $c(t)$ is the failure pdf of the circuit, then the probability of correct prediction is given by:

$$p(\text{correct}) = \int_{\infty}^{0} c(t)\left(\int_{0}^{t} p(u)du\right)dt \tag{2.63}$$

The probability that the prognostic cell does not correctly predict failure of the circuit is equal to the probability that the circuit fails before the prognostic cell

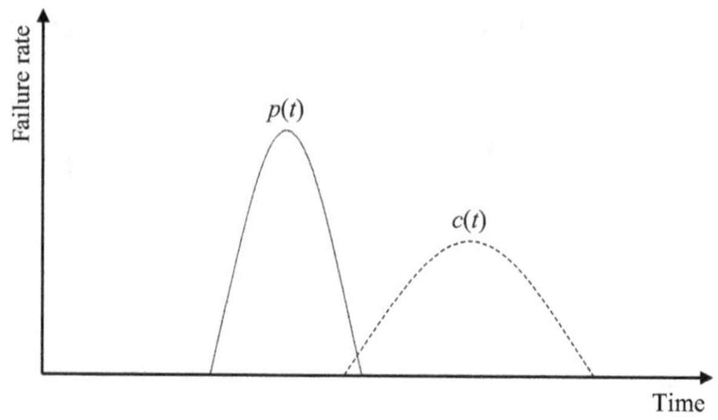

Figure 2.22 The failure probability density of the circuit c(t) and the trigger rate of the prognostic cell p(t) as a function of time

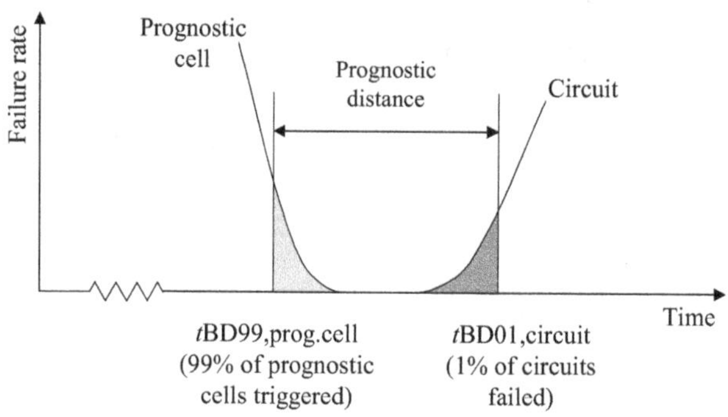

Figure 2.23 Prognostic distance in terms of cumulative failure probabilities $t_{BD01,circuit}$ and $t_{BD99,prog.cell}$

fails. This is given by:

$$p(\text{incorrect}) = \int_{\infty}^{0} p(t)\left(\int_{0}^{t} c(u)du\right)dt \tag{2.64}$$

The prognostic distance is the time between the 1% cumulative failure probability point of the circuit and the 99% cumulative failure probability point of the prognostic cell, as shown in Figure 2.23. Where: t_{BDF} = time-to-breakdown with a failure fraction of F%.

The design in this example attempts to achieve a prognostic distance of 10%. This means that before 90% of the time has elapsed at which 1% of the circuits have failed, 99% of the prognostic cells must have triggered. The reliability requirement can be stated in terms of time-to-breakdown of the circuit and prognostic cell:

$$\frac{9}{10} t_{\text{BD01,circuit}} = t_{\text{BD99,prog.cell}} \tag{2.65}$$

where t_{BDF} can be expressed by the accumulative distribution equation derived from the E model:

$$t_{\mathrm{BDF}} = \frac{C'\exp(-\gamma E_{\mathrm{ox}})\exp\left(\frac{E_a}{kT}\right)\left[\ln\left(\frac{1}{1-F}\right)\right]^{1\beta}}{A^{1\beta}} \tag{2.66}$$

Applying the accumulative distribution equation for $t_{\mathrm{BD01,circuit}}$ and $t_{\mathrm{BD01,circuit}}$ results in the basis for a design equation for a prognostic cell.

$$\frac{9}{10}\frac{C'\exp\left(-\gamma E_{\mathrm{ox,circuit}}\right)\exp\left(\frac{E_a}{kT}\right)\left[\ln\left(\frac{1}{1-0.01}\right)\right]^{1\beta}}{A_{\mathrm{circuit}}^{1\beta}} = \frac{C'\exp\left(-\gamma E_{\mathrm{ox,prog.cell}}\right)\exp\left(\frac{E_a}{kT}\right)\left[\ln\left(\frac{1}{1-0.99}\right)\right]^{1\beta}}{A_{\mathrm{prog.cell}}^{1\beta}} \tag{2.67}$$

From this, the amount of overvoltage required is obtained:

$$\begin{aligned}\Delta V_G &= \Delta E t_{\mathrm{ox}} = \left(E_{\mathrm{ox,prog.cell}} - E_{\mathrm{ox,circuit}}\right)t_{ox} \\ &= \frac{t_{\mathrm{ox}}}{\gamma}\ln\left[\frac{9}{10}\left(\frac{A_{\mathrm{circuit}}}{A_{\mathrm{prog.cell}}}\right)^{1\beta}\left(\frac{\ln(1-0.99)}{\ln(1-0.01)}\right)^{1\beta}\right]\end{aligned} \tag{2.68}$$

Equation (2.68) is a design equation for the TDDB prognostic cell. The equation gives the excess stress voltage (ΔV_G) that is required for a given prognostic distance—in this case, 10%.

2.6 Summary

There is growing interest among industry, government, and academia to monitor the ongoing reliability, or health, and predict the RUL of power electronic systems because most complex systems today contain significant power electronics content.

This chapter provides a basic understanding of PHM for power electronics. It introduces the failure model of power electronics, including TDDB, energy-based models, and thermal cycling models. The chapter also discusses the PoF prognostic approach. The PoF-based prognostic method can provide an estimate of damage for given loading conditions and failure mechanisms and identify the components that are critical to system reliability.

Data-driven prognostics use statistics and probability for analyzing current and historical data to estimate the RUL of a product. Data-driven methods do not require product-specific knowledge, such as material properties, constructions, and failure mechanisms. Data-driven methods can capture complex relationships and learn trends available in the data without the need for specific failure models.

This chapter also provides three methods (PCA, LDA, and mRMR) to reduce variable dimension. Additionally, various anomaly detection methods including distanced-based (MD and KNN), clustering based (k-means) and classification-based (SVMs) are introduced and the method of defining failure threshold based on MD is described in detail. For prognostics, examples using statistical regression analysis, PF and deep learning methods (LSTM) are given to predict RUL for power electronics.

Examples of data-driven and canary methods for anomaly detection and prognostics for electronic circuits are given. The calculation of FI for failure prognosis based on the MT methodology is introduced. The canary method uses the prognostic cell to provide advanced warning of failure based on the defined prognostic distance.

The advanced techniques of anomaly detection and prognostics can enhance system reliability, safety, and performance while reducing downtime and costs. Nevertheless, the continued research and improvements are necessary for further advancements in power electronic converter systems.

Acknowledgments

The authors would like to thank the Center for Advanced Life Cycle Engineering (CALCE) at the University of Maryland, College Park, and 100-plus companies and organizations that sponsor CALCE annually. Furthermore, the authors would like to express great gratitude to the Centre for Advances in Reliability and Safety (CAiRS) admitted under AIR@InnoHK Research Cluster for providing tremendous supports and contributions to this chapter.

References

[1] Marketsandmarkets. *Power Electronics Market by Substrate Wafer Technology (GaN, SiC, and Others), Devices (Power IC, Power Module & Power Discrete), Applications, and Geography—Analysis & Forecast to 2014–2020*. Pune, India: Marketsandmarkets Research Private Ltd.; May 2014. Report No. SE 2434.

[2] Pecht MG, and Kang M (eds.). *Prognostics and Health Management of Electronics: Fundamentals, Machine Learning, and the Internet of Things*. New York: Wiley; 2018.

[3] Tumer I, and Bajwa A. A survey of aircraft engine health monitoring systems. In: *Proc. AIAA*; 1999. pp. 1–6.

[4] Carden EP, and Fanning P. Vibration based condition monitoring: A review. *Structural Health Monitoring*. 2004;3(4):355–377.

[5] Chang PC, Flatau A, and Liu SC. Review paper: Health monitoring of civil infrastructure. *Structural Health Monitoring*. 2003;2(3):257–267.

[6] Krok MJ, and Goebel KF. Prognostics for advanced compressor health monitoring. In: Willett PK, and Kirubarajan T (eds.). *System Diagnosis and Prognosis: Security and Condition Monitoring Issues III. Vol. 5107*. Bellingham, WA: International Society for Optics and Photonics, SPIE; 2003. pp. 1–12.

[7] Kacprzynski GJ, Roemer MJ, Modgil GA, *et al.* Enhancement of physics-of-failure prognostic models with system level features. In: *Proc. IEEE Aerospace Conference*. vol. 6; 2002. pp. 2919–2925.

[8] Xie J, and Pecht MG. Applications of in-situ health-monitoring and prognostic sensors. In: *Proc. 9th Pan Pacific Microelectronics Symposium Exhibits & Conference*. Oahu, HI, USA; February 2004.

[9] Vichare N, Rodgers P, Eveloy V, *et al.* Environment and usage monitoring of electronic products for health assessment and product design. *Quality Technology & Quantitative Management*. 2007;4(2):235–250.

[10] Steinberg DS. *Vibration Analysis for Electronic Equipment*, 3rd edn. New York: Wiley; 2000.

[11] Vichare N, and Pecht M. Enabling electronic prognostics using thermal data. In: *Proc. 12th International Workshop on Thermal Investigation of ICs and Systems*. Nice, France; September 2006.

[12] McPherson JW. Time dependent dielectric breakdown physics – Models revisited. *Microelectronics Reliability*. 2012;52(9–10):1753–1760.

[13] McPherson JW, and Baglee DA. Acceleration factors for thin gate oxide stressing. In: *23rd International Reliability Physics Symposium*; 1985. pp. 1–5.

[14] McPherson JW, and Mogul HC. Underlying physics of the thermochemical E model in describing low-field time-dependent dielectric breakdown in SiO_2 thin films. *Journal of Applied Physics*. 1998;84(3):1513–1523.

[15] Chen IC, Holland SE, and Hut C. A quantitative physical model for time-dependent breakdown in SiO_2. In: *Proc. 23rd International Reliability Physics Symposium*. Orlando, FL, USA; 1985. pp. 24–31.

[16] Schuegraf KF, and Hu C. Hole injection SiO_2 breakdown model for very low voltage lifetime extrapolation. *IEEE Transactions on Electron Devices*. 1994;41(5):761–767.

[17] Henderson CL. Time dependent dielectric breakdown. *Semiconductor Reliability*. 2002.

[18] Allers KH. Prediction of dielectric reliability from *I–V* characteristics: Poole–Frenkel conduction mechanism leading to $\sqrt{E}$ model for silicon nitride MIM capacitor. *Microelectronics Reliability*. 2004;44(3):411–423.

[19] Swartz GA. Gate oxide integrity of NMOS transistor arrays. *IEEE Transactions on Electron Devices*. 1986;33(11):1826–1829.

[20] Strong AW, Wu EY, Vollertsen RP, *et al. Reliability Wearout Mechanisms in Advanced CMOS Technologies*. Piscataway, NJ: Wiley-IEEE Press; 2009.

[21] McPherson JW. *Reliability Physics and Engineering*. New York: Springer; 2010.

[22] Lee WW, Nguyen LT, and Selvaduray GS. Solder joint fatigue models: Review and applicability to chip scale packages. *Microelectronics Reliability*. 2000;40(2):231–244.

[23] Akay HU, Zhang H, and Paydar NH. Experimental correlations of an energy-based fatigue life prediction method for solder joints. In: *Proc. Pacific Rim/ASME Int. Intersociety Electronic and Photonic Packaging Conf. INTERpack'97*. vol. 2; 1997. pp. 1567–1574.

[24] Liang J, Gollhardt N, Lee PS, *et al.* An integrated fatigue life prediction methodology for optimum design and reliability assessment of solder interconnections. In: *Proc. Pacific Rim/ASME Int. Intersociety Electronic and Photonic Packaging Conf. INTERpack'97*. vol. 2; 1997. pp.1583–1592.

[25] Jung W, Pao YH, and Lau JH. Nonlinear analysis of full-matrix and perimeter plastic ball grid array solder joints. In: *Nepcon West'97*. California, USA; February 1997. pp. 1076–1095.

[26] Wu SX, Chin J, Grigorich T, *et al.* Reliability analysis for fine pitch BGA package. In: *Proc. 48th Electronic Components and Technology Conference (Cat. No.98CH36206)*; June 1998. pp. 737–741.

[27] Gustafsson G. Solder joint reliability of a lead-less RF-transistor. In: *Proc. 48th Electronic Components and Technology Conference (Cat. No. 98CH36206)*; June 1998. pp. 87–91.

[28] Ghaffarian R. Accelerated thermal cycling and failure mechanisms for BGA and CSP assemblies. *Journal of Electronic Packaging*. 2000;122 (4):335–340.

[29] Cui H. Accelerated temperature cycle test and Coffin-Manson model for electronic packaging. In: *Proc. Annual Reliability and Maintainability Symposium*; 2005. pp. 556–560.

[30] Norris KC, and Landzberg AH. Reliability of controlled collapse interconnections. *IBM Journal of Research and Development*. 1969;13 (3):266–271.

[31] Born FH, and Boenning RA. Marginal checking-a technique to detect incipient failures. In: *Proc. IEEE National Aerospace and Electronics Conference*; May 1989. pp. 1880–1886.

[32] Pecht MG, Radojcic R, and Rao G. *Guidebook for Managing Silicon Chip Reliability*. Boca Raton, FL: CRC Press; 1999.

[33] Pecht MG, and Dasgupta A. Physics-of-failure: An approach to reliable product development. *Journal of the Institute of Environmental Sciences*. 1995;38:30–34.

[34] Snook I, Marshall JM, and Newman RM. Physics of failure as an integrated part of design for reliability. In: *Annual Reliability and Maintainability Symposium*; 2003. pp. 46–54.

[35] Ganesan S, Eveloy V, Das D, *et al.* Identification and utilization of failure mechanisms to enhance FMEA and FMECA. In: *Proc. IEEE Workshop on Accelerated Stress Testing & Reliability (ASTR)*; 2005. pp. 2–5.

[36] Gu J, and Pecht M. Prognostics and health management using physics-of-failure. In: *2008 Annual Reliability and Maintainability Symposium*. Piscataway, NJ: IEEE; 2008. pp. 481–487.

[37] IEEE Std 1413.1-2002, *IEEE Guide for Selecting and Using Reliability Predictions Based on IEEE 1413*. New York: IEEE Standard; 2003.

[38] JESD659-A: Failure-mechanism-driven reliability monitoring. EIA/JEDEC Standard; September 1999.

[39] JEP143A: Solid-state reliability assessment and qualification methodologies. JEDEC Publication; May 2004.

[40] JEP150: Stress-test-driven qualification of and failure mechanisms associated with assembled solid state surface-mount components. JEDEC Publication; May 2005.
[41] JESD74: Early life failure rate calculation procedure for electronic components. JEDEC Standard; April 2000.
[42] JESD94: Application specific qualification using knowledge based test methodology. JEDEC Standard; January 2004.
[43] JESD91A: Method for developing acceleration models for electronic component failure mechanisms. JEDEC Standard; August 2003.
[44] SEMATECH, #00053955A-XFR: Semiconductor device reliability failure models. SEMATECH Publication; May 2000.
[45] SEMATECH, #00053958A-XFR: Knowledge-based reliability qualification testing of silicon devices. SEMATECH Publication; April 2000.
[46] SEMATECH, #04034510A-TR: Comparing the effectiveness of stress-based reliability qualification stress conditions. SEMATECH Publication; April 2004.
[47] SEMATECH, #99083810A-XFR: Use condition based reliability evaluation of new semiconductor technologies. SEMATECH Publication; August 1999.
[48] Sharma S. *Applied Multivariate Techniques*. Toronto: Wiley; 1996.
[49] Rencher AC. *Methods of Multivariate Analysis*. New York: Wiley; 1995.
[50] Tanner G. Linear discriminant analysis (LDA)[online]. Machine learning explained; 2021. Available from https://ml-explained.com/blog/linear-discriminant-analysis-explained [accessed: 1 February 2024].
[51] Martinez AM, and Kak AC. PCA versus LDA. *IEEE Transactions on Pattern Analysis and Machine Intelligence*. 2001;23(2):228–233.
[52] Peng H, Long F, and Ding C. Feature selection based on mutual information criteria of max-dependency, max-relevance, and min-redundancy. *IEEE Transactions on Pattern Analysis and Machine Intelligence*. 2005;27 (8):1226–1238.
[53] Brown G, Pocock A, Zhao MJ, *et al.* Conditional likelihood maximisation: A unifying framework for information theoretic feature selection. *The Journal of Machine Learning Research*. 2012;13:27–66.
[54] Chandola V, Banerjee A, and Kumar V. Anomaly detection: A survey. *ACM Computing Surveys*. 2009;41(3):1–58.
[55] Kumar S, Chow TW, and Pecht M. Approach to fault identification for electronic products using Mahalanobis distance. *IEEE Transactions on Instrumentation and Measurement*. 2009;59(8):2055–2064.
[56] De Maesschalck R, Jouan-Rimbaud D, and Massart DL. The Mahalanobis distance. *Chemometrics and Intelligent Laboratory Systems*. 2000;50(1):1–18.
[57] Nie L, Azarian MH, Keimasi M, *et al.* Prognostics of ceramic capacitor temperature-humidity-bias reliability using Mahalanobis distance analysis. *Circuit World*. 2007;33(3):21–28.
[58] Box GE, and Cox DR. An analysis of transformations. *Journal of the Royal Statistical Society Series B: Statistical Methodology*. 1964;26(2):211–243.
[59] Nelson LS. Technical aids: Notes on the Shewhart control chart. *Journal of Quality Technology*. 1984;16(4):238–239.

[60] Schabenberger O, and Pierce FJ. *Contemporary Statistical Models for the Plant and Soil Sciences*. Boca Raton, FL: CRC Press; 2001.

[61] Fix E, and Hodges J JL *Discriminatory Analysis-Nonparametric Discrimination: Consistency Properties*. Randolph Field, TX: USAF School of Aviation Medicine; February 1951. Report Number 4, Project Number 21-49-004.

[62] Hand DJ, Mannila H, and Smyth P. *Principles of Data Mining*. Cambridge, MA: MIT Press; 2001.

[63] Cortes C, and Vapnik V. Support-vector networks. *Machine Learning*. 1995; 20:273–297.

[64] Hsu CW, and Lin CJ. A comparison of methods for multiclass support vector machines. *IEEE Transactions on Neural Networks*. 2002;13(2):415–425.

[65] Gaussian Process Regression Models [online]. The MathWorks, Inc.; 2023. Available from https://ww2.mathworks.cn/help/stats/gaussian-process-regression-models.html [accessed: 1 February 2024].

[66] Liu D, Pang J, Zhou J, *et al.* Prognostics for state of health estimation of lithium-ion batteries based on combination Gaussian process functional regression. *Microelectronics Reliability*. 2013;53(6):832–839.

[67] Saha B, and Goebel K. Uncertainty management for diagnostics and prognostics of batteries using Bayesian techniques. In: *2008 IEEE Aerospace Conference*. Piscataway, NJ: IEEE; 2008. pp. 1–8.

[68] Xing Y, Ma EW, Tsui KL, *et al.* A case study on battery life prediction using particle filtering. In: *Proc. IEEE 2012 Prognostics and System Health Management Conference (PHM-2012 Beijing)*. Piscataway, NJ: IEEE; 2012. pp. 1–6.

[69] Wan EA, and Van Der Merwe R. The unscented Kalman filter. In: Haykin S (ed.), *Kalman Filtering and Neural Networks*. New York: Wiley; 2001. pp. 221–280.

[70] Doucet A, Godsill SJ, and Andrieu C. On sequential Monte Carlo sampling methods for Bayesian filtering. *Statistics and Computing*. 2000;10:197–208.

[71] Arulampalam MS, Maskell S, Gordon N, *et al.* A tutorial on particle filters for online nonlinear/non-Gaussian Bayesian tracking. *IEEE Transactions on Signal Processing*. 2002;50(2):174–188.

[72] Lee DJ. *Nonlinear Bayesian Filtering with Applications to Estimation and Navigation*. College Station, TX: A&M University; 2005.

[73] Zhang Y, Xiong R, He H, *et al.* Long short-term memory recurrent neural network for remaining useful life prediction of lithium-ion batteries. *IEEE Transactions on Vehicular Technology*. 2018;67(7):5695–5705.

[74] Sai Sarathi Vasan A, Long B, and Pecht M. Diagnostics and prognostics method for analog electronic circuits. *IEEE Transactions on Industrial Electronics*. 2013;60(11):5277–5291.

[75] Goodman D, Vermeire B, Ralston-Good J, *et al. A Board-Level Prognostic Monitor for MOSFET TDDB*. Tucson, AZ: Ridgetop Group, Inc.; 2005.

Chapter 3

Reliability of DC-link capacitors in power electronic converters

Huai Wang[1]

DC-link capacitors are an important component in a majority of power electronic converters. They contribute to the cost, size, and failure rate of power electronic converters on a considerable scale. From the viewpoint of capacitor users, this chapter presents a review of the improvement of the reliability of DC-link capacitors in power electronic converters from two aspects: (1) reliability-oriented DC-link design solutions; and (2) conditioning monitoring of DC-link capacitors during operation. Failure mechanisms, failure modes, lifetime models of capacitors, and capacitor reliability testing are also discussed as a basis. This chapter serves to provide an overall picture of the state-of-the-art research in this area and to identify the corresponding challenges and future research directions for capacitors and their DC-link applications. Parts of the contents of this chapter are excerpted from [1,2].

3.1 Capacitors for DC links in power electronic converters

DC-link capacitors are widely used in power electronic converters to balance the instantaneous power difference between the input source and output load and to minimize voltage variation in the DC link. In some applications, they are also used to provide energy storage. Figure 3.1 shows some typical configurations of power electronic conversion systems with capacitive DC links. Such configurations cover a wide range of power electronics applications, for example, power factor corrections, wind turbines, photovoltaic systems, motor drives, electric vehicles, and lighting systems. It should be noted that a capacitive DC link discussed here does not necessarily consist of capacitors only. There could also be some inductive components (e.g., DC choke) in the DC bus in some of the above applications.

[1]Department of Energy, Aalborg University, Denmark

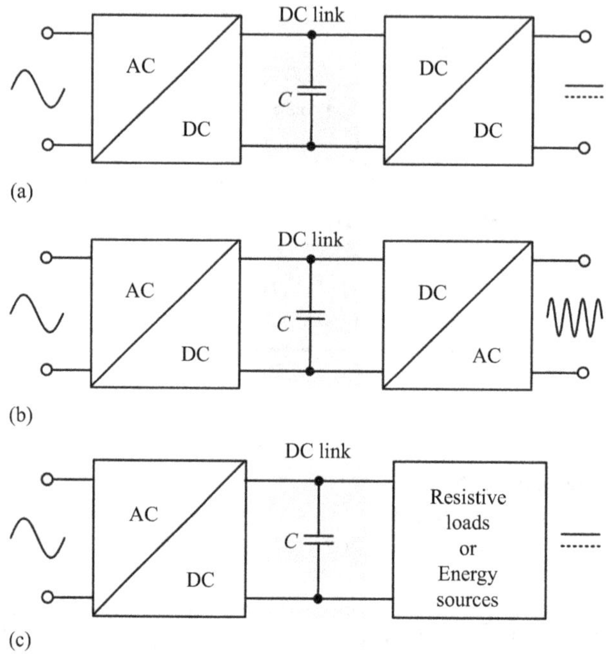

Figure 3.1 *Configurations of power electronic converters with capacitive DC links. (a) AC–DC–DC or DC–DC–AC power converters with a capacitive DC link. (b) AC–DC–AC power converters with a capacitive DC link. (c) AC–DC or DC–AC power converters with a capacitive DC link.*

3.1.1 The type of capacitors used for DC links

The type of capacitors used for DC links depends on the required capacitance value, voltage rating, maximum operating temperature, frequency characteristics, cost, size, and reliability. Figure 3.2 illustrates the ranges of capacitance and voltage ratings of various types of capacitors for power electronics applications. Of these, the electric double-layer capacitors are for energy storage purposes if used in DC links. They have the highest achievable capacitance value and low voltage rating. There are two types of electrolytic capacitors: aluminum electrolytic capacitors (Al-Caps) and solid tantalum capacitors. Solid tantalum capacitors are mainly used for low DC-link voltage applications below 100 V. Film capacitors and ceramic capacitors are also used for DC links in some applications such as the metallized polypropylene film capacitors (MPPF-Caps) and multi-layer ceramic capacitors (MLC-Caps).

3.1.2 Comparison of different type of capacitors for DC links

This chapter focuses on the discussion of Al-Caps, MPPF-Caps, and MLC-Caps for DC links, since they represent capacitors with a relatively wide range of voltage

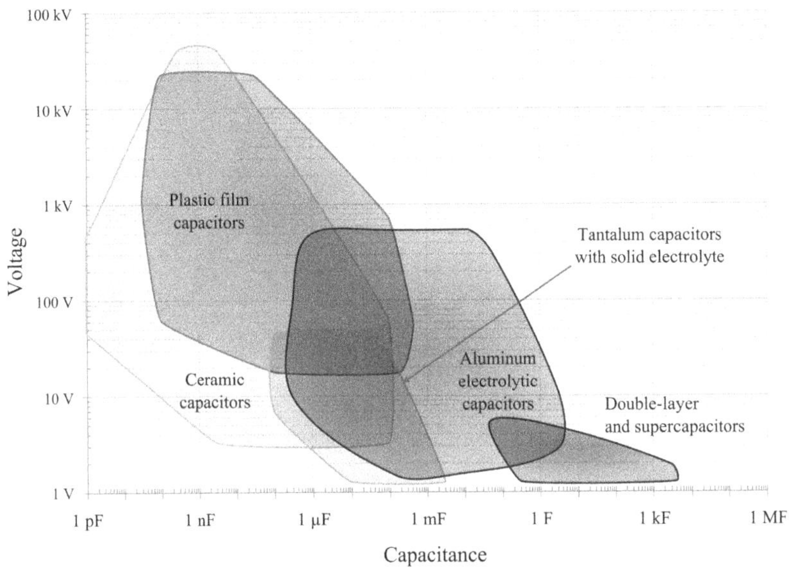

Figure 3.2 Typical capacitance and voltage ranges of different types of capacitors applied in power electronic systems

and capacitance. The selection and sizing of capacitors require the matching of available capacitor characteristics and parameters to the specific application needs under various environmental, electrical, and mechanical stresses.

Figure 3.3(a) shows a lumped electro-thermal model of capacitors. In the electrical domain model, C, R_s, and L_s are the capacitance, equivalent series resistance (ESR), and equivalent series inductance (ESL), respectively. The dissipation factor (DF) is $\tan\delta = 2\pi f R_s C$. R_p is the insulation resistance, where f is the frequency of interest. R_d is the dielectric loss due to dielectric absorption and molecular polarization, and C_d is the inherent dielectric absorption [3]. The widely used simplified capacitor model is composed of C, R_s, and L_s. It should be noted that their values vary with temperature, voltage stress, frequency, and time (i.e., operating conditions). The absence of the consideration into these variations may lead to improper analysis of the electrical stresses and thermal stresses. In the thermal domain model, $P_{c,loss}$ is the power loss, T_a, T_c, and T_h are the ambient temperature, case temperature, and hot spot temperature, respectively. R_{thhc} and R_{thca} are the thermal resistances from hot spot to case and from case to ambient, respectively. The value of R_{thca} depends on the thermal management of capacitors in power electronic systems. R_{th} is the equivalent thermal resistance from the hot spot to ambient.

Figure 3.3(b) plots the impedance characteristics of capacitors, which can be divided into three distinct frequency regions. The impedance is dominated by capacitance, ESR, and ESL in the Region I, Region II, and Region III, respectively. The resonant frequency, f_r, of a capacitor is corresponding to the minimum

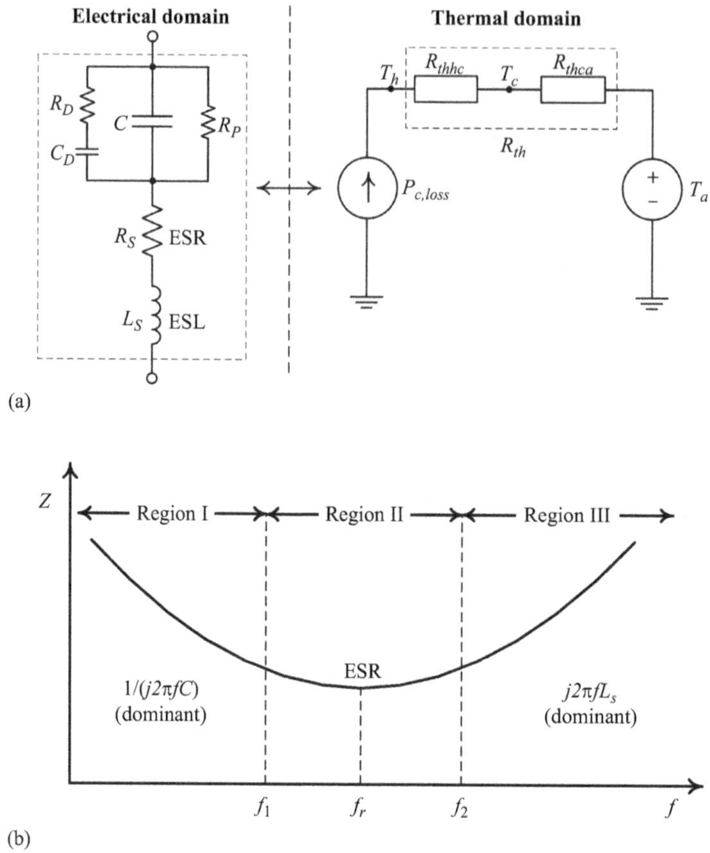

Figure 3.3 An equivalent model and impedance characteristics of capacitors: (a) a lumped equivalent electro-thermal model of capacitors (ESR—equivalent series resistance; ESL—equivalent series inductance), (b) impedance characteristics of capacitors

impedance value of ESR, which defines the maximum operational frequency of the capacitor. Otherwise, the capacitor will become an inductive component at frequencies above f_r.

The property of dielectric materials is a major factor that limits the performance of capacitors. Al_2O_3, polypropylene, and ceramics are the materials used in Al-Caps, MPPF-Caps, and MLC-Caps, respectively. According to the discussion in [4], Al_2O_3 has the highest energy density due to high field strength and high relative permittivity. The theoretical limit is in the range of 10 J/cm^3 and the achieved energy density in commercial products is about 2 J/cm^3. Ceramics could have a much higher dielectric constant than Al_2O_3 and film; however, it suffers from low field strength, resulting in similar energy density as that of film.

Table 3.1 Performance comparison of three types of capacitors for DC links

	Al-Caps	MPPF-Caps	MLC-Caps
Capacitance	+ + +	+ +	+
Voltage	+ +	+ + +	+
Ripple current	+	+ + +	+ + +
Equivalent series resistance	+	+ + +	+ + +
Dissipation factor	+	+ + +	+ + +
Frequency range	+	+ +	+ + +
Capacitance stability	+ +	+ + +	+
Overvoltage capability	+ +	+ + +	+
Temperature range	+ +	+	+ + +
Energy density	+ + +	+	+ +
Reliability under electro-thermal stresses	+	+ + +	+ + +
Cost per Joule	+ + +	+ +	+

+ + + superior, + + intermediate, + inferior.

The three types of capacitors therefore exhibit specific advantages and shortcomings, as shown in the comparison in Table 3.1. Al-Caps could achieve the highest energy density and lowest cost per Joule, but with relatively high ESRs, low ripple current ratings, and wear-out issues due to the evaporation of electrolyte. MLC-Caps have a smaller size, wider frequency range, and higher operating temperatures up to 200 °C. However, they suffer from higher cost and mechanical sensitivity. The recent release of CeraLink series ceramic capacitors [5] is of interest to extend the scope of MLC-Caps for DC-link applications. It is based on new ceramic materials with antiferroelectric behavior and strong positive bias effect (i.e., capacitance versus voltage stress). MPPF-Caps provide a well-balanced performance for high voltage applications (e.g., above 500 V) in terms of cost and ESR, capacitance, ripple current, and reliability. Nevertheless, they have the shortcomings of large volume and moderate upper operating temperature.

3.1.3 Reliability challenges for capacitors in power electronic converters

With more stringent reliability constraints brought by automotive, aerospace, and energy industries, the design of DC links encounters the following challenges: (a) capacitors are one kind of stand-out component in terms of failure rate in the field operation of power electronic systems [6,7]; (b) cost reduction pressure from global competition dictates the minimum design margin of capacitors without undue risk; (c) capacitors are to be exposed to more harsh environments (e.g., high ambient temperature, high humidity) in emerging applications; and (d) constraints on volume and thermal dissipation of capacitors with the trends for high power density power electronic systems.

The efforts to overcome the above-mentioned challenges can be divided into three categories: (a) advance the capacitor technology with improved and pre-determined reliability built in; (b) optimal DC-link design solutions based on the present capacitors to achieve proper robustness margin and cost-effectiveness; and (c) implementations of condition monitoring to ensure reliable field operation and preventive maintenance. By taking advantage of the progress in new dielectric materials and innovative manufacturing processes, leading capacitor manufacturers have been continuously releasing new generations of products with improved reliability and cost performance. The proper application of these capacitors for specific DC-link design is equally important as the operating conditions (e.g., temperature, humidity, ripple current, voltage) could significantly influence the reliability of the capacitors. Compared to the first category, the latter two are more relevant from the power electronic designers' perspective, which therefore will be discussed in this chapter. Moreover, the failure modes, failure mechanisms, corresponding critical stressors, and lifetime models of them are mapped in the next section.

3.2 Failure mechanisms and lifetime models of capacitors

As discussed in Chapter 1 of this book, the understanding of failure mechanisms is essential to the reliability prediction and reliability improvement of power electronic components and systems. In this section, the failure mechanisms and lifetime models of Al-Caps, MPPF-Caps, and MLC-Caps will be discussed.

3.2.1 Failure modes, failure mechanisms, and critical stressors of DC-link capacitors

DC-link capacitors can fail due to intrinsic and extrinsic factors such as design defect, material wear-out, operating temperature, voltage, current, moisture, mechanical stress, and so on. Generally, the failures can be divided into catastrophic failures due to single-event overstress and wear-out failures due to the long-term degradation of capacitors. The major failure mechanisms have been presented in [8–11] for Al-Caps, [12–15] for MPPF-Caps, and [16–18] for MLC-Caps. Based on these prior research results, Table 3.2 shows an overview of the failure modes, failure mechanisms, and corresponding critical stressors of the three types of capacitors.

Table 3.3 shows the comparison of failure and self-healing capability of Al-Caps, MPPF-Caps, and MLC-Caps. Electrolyte vaporization is the major wear-out mechanism of small-size Al-Caps (e.g., snap-in type) due to their relatively high ESR and limited heat dissipation surface. For large-size Al-Caps, the wear-out lifetime is dominantly determined by the increase of leakage current, which is relevant to the electrochemical reaction of oxide layer [19]. The most important reliability feature of MPPF-Caps is their self-healing capability [14]. Initial dielectric breakdowns (e.g., due to overvoltage) at local weak points of a MPPF-Cap will be cleared, and the capacitor regains its full ability except for a negligible capacitance reduction. With the increase of these isolated weak points, the

Table 3.2 Overview of failure modes, critical failure mechanisms, and critical stressors of the three main types of DC-link capacitors (with emphasis on the ones relevant to the design and operation of power converters)

Cap. type	Failure modes	Critical failure mechanisms	Critical stressors
Al-Caps	Open circuit	Self-healing dielectric breakdown	V_C, T_a, i_C
		Disconnection of terminals	Vibration
	Short circuit	Dielectric breakdown of oxide layer	V_C, T_a, i_C
	Wear-out: electrical parameter drift (C, ESR, tanδ, I_{LC}, R_p)	Electrolyte vaporization	T_a, i_C
		Electrochemical reaction (e.g., degradation of oxide layer, anode foil capacitance drops)	V_C
MPPF-Caps	Open circuit (typical)	Self-healing dielectric breakdown	V_C, T_a, dV_C/dt
		Connection instability by heat contraction of a dielectric film	T_a, i_C
		Reduction in electrode area caused by oxidation of evaporated metal due to moisture absorption	Humidity
	Short circuit (with resistance)	Dielectric film breakdown	V_C, dV_C/dt
		Self-healing due to overcurrent	T_a, i_C
		Moisture absorption by film	Humidity
	Wear out: electrical parameter drift (C, ESR, tanδ, I_{LC}, R_p)	Dielectric loss	V_C, T_a, i_C, humidity
MLC-Caps	Open circuit	Severe cracking (e.g., due to temperature excursions)	T_a, i_C, vibration
	Short circuit (typical)	Dielectric breakdown	V_C, T_a, i_C
		Cracking; damage to capacitor body	Vibration
	Wear-out: electrical parameter drift (C, ESR, tanδ, I_{LC}, R_p)	Oxide vacancy migration; dielectric puncture; insulation degradation; micro-crack within ceramic	V_C, T_a, i_C, vibration

V_C: capacitor voltage stress, i_C: capacitor ripple current stress, i_{LC}: leakage current, T_a: ambient temperature.

Table 3.3 A comparison of failure and self-healing capability of capacitors

	Al-Caps	MPPF-Caps	MLCC-Caps
Dominant failure modes	Open circuit	Wear-out Open circuit	Short circuit
Dominant failure mechanisms	Electrolyte vaporization; electrochemical reaction	Moisture corrosion; dielectric loss	Insulation degradation; flex cracking
Most critical stressors	T_a, V_C, i_C	T_a, V_C, humidity	T_a, V_C, vibration
Self-healing capability	Moderate	Good	No

capacitance of the capacitor is gradually reduced to reach the end-of-life. The metallized layer in MPPF-Caps is typically less than 100 nm [20], which is susceptible to corrosion due to the ingress of atmospheric moisture. In [21], the corrosion mechanism is well studied. Unlike that of Al-Caps and MPPF-Caps, the dielectric material of MLC-Caps is expected to last for thousands of years at use-level conditions without showing significant degradation [19]. Therefore, wear-out of ceramic capacitors is typically not an issue. However, an MLC-Cap could degrade much more quickly due to the "amplifying" effect from the large number of dielectric layers [17]. In [22], it has been shown that a modern MLC-Cap could wear out within ten years due to increasing miniaturization through the increase of the number of layers. Moreover, the failure of MLC-Caps may induce severe consequences to power converters due to the short circuit failure mode. The dominant failure causes of MLC-Caps are insulation degradation and flex cracking. Insulation degradation due to the decrease of the dielectric layer thickness results in increased leakage currents. Under high voltage and high temperature conditions, avalanche breakdown (ABD) and thermal runaway (TRA) could occur, respectively. ABD features with an abrupt burst of current leading to an immediate breakdown, while TRA exhibits a more gradual increase of leakage current [16].

3.2.2 Lifetime models of DC-link capacitors

Lifetime models are important for lifetime prediction and benchmark of different capacitor solutions. The most widely used lifetime model for capacitors is shown in (1), which describes the influence of temperature and voltage stress:

$$L = L_0 \times \left(\frac{V}{V_0}\right)^{-n} \times \exp\left[\left(\frac{E_a}{K_B}\right)\left(\frac{1}{T} - \frac{1}{T_0}\right)\right] \tag{3.1}$$

where L and L_0 are the lifetime under the use condition and testing condition, respectively. V and V_0 are the voltages at use condition and test condition, respectively. T and T_0 are the temperatures in Kelvin at use condition and test condition, respectively. E_a is the activation energy, K_B is Boltzmann's constant 8.62×10^{-5} eV/K, and n is the voltage stress exponent. Therefore, the values of E_a and n are the key parameters to be determined in the above model.

In [23], the E_a and n are found to be 1.19 and 2.46, respectively, for high dielectric constant ceramic capacitors. In [22], the ranges of E_a and n for MLC-Caps are 1.3–1.5 and 1.5–1.7, respectively. The large discrepancies could be attributed to the ceramic materials, dielectric layer thickness, testing conditions, etc. With the trend for smaller size and thinner dielectric layer, the MLC-Caps will be more sensitive to the voltage stress, implying a higher value of n. Moreover, under different testing voltages, the value of n might be different as discussed in [24].

For Al-Caps and film capacitors, a simplified model from (1) is popularly applied as follows:

$$L = L_0 \times \left(\frac{V}{V_0}\right)^{-n} \times 2^{\frac{T_0 - T}{10}} \tag{3.2}$$

The derivation of (2) from (1) is discussed in [25]. The model presented by (2) is corresponding to a specific case of (1) when $E_a = 0.94$ eV and T_0 and T are substituted by 398 K. For MPPF-Caps, the exponent n is from around 7 to 9.4 used by leading capacitor manufacturers [26]. For Al-Caps, the value of n typically varies from 3 to 5 [27]. However, the voltage dependency of lifetime for Al-Caps depends on the voltage stress level. In [9], instead of a power law relationship, a linear equation is found to be more suitable to describe the impact of voltage stress. Moreover, the lifetime dependence on temperature presented in (2) is an approximation only [28]. In [28,29], a lifetime model of electrolytic capacitors is proposed based on the ESR drift due to electrolyte evaporation and loss. The estimation of the ESR is based on the electrolyte pressure and the reduction of the electrolyte volume. The prediction results fit well with the lifetime–temperature relationship shown in (1) (i.e., Arrhenius equation). To obtain the physical explanations of the lifetime model variants from different capacitor manufacturers, a generic model is derived in [30] as follows:

$$\frac{L}{L_0} = \begin{cases} \left(\frac{V_0}{V}\right) \times \exp\left[\left(\frac{E_a}{K_B}\right)\left(\frac{1}{T} - \frac{1}{T_0}\right)\right] & \text{(low voltage stress)} \\ \left(\frac{V}{V_0}\right)^{-n} \times \exp\left[\left(\frac{E_a}{K_B}\right)\left(\frac{1}{T} - \frac{1}{T_0}\right)\right] & \text{(medium voltage stress)} \\ \exp[a_1(V_0 - V)] \times \exp\left[\frac{E_{a_0} - a_0\xi}{K_B T} - \frac{E_{a_0} - a_0\xi_0}{K_B T_0}\right] & \text{(high voltage stress)} \end{cases} \tag{3.3}$$

where a_0 and a_1 are constants describing the voltage and temperature dependency of E_a. ξ and ξ_0 are stress variables (i.e., voltage and/or temperature) under operation and test, respectively. E_{a0} is the activation energy under test. It can be noted that the influence of voltage stress is modeled as linear, inverse power law, and exponential equations, respectively, for low voltage stress, medium voltage stress, and high voltage stress. Another important observation is that the activation energy, E_a, varies with voltage and temperature, especially under high voltage stress conditions. It is in agreement with the observations in [28] that the equivalent values of E_a/K_B are varying under different temperature ranges.

It should be noted that the above lifetime models for DC-link capacitors consider the impact of voltage and temperature (i.e., including both ambient temperature and internal temperature rise due to ripple currents) only. Nevertheless, humidity and vibration are also critical stressors for MPPF-Caps and MLC-Caps, respectively, as shown in Table 3.3. A humidity-related lifetime model has been discussed in [31], which will also be discussed in Section 3.2.3.

3.2.3 Capacitor reliability testing to investigate failure mechanisms and obtain lifetime models

3.2.3.1 Example of accelerated lifetime testing of film capacitors under humidity conditions

An example of accelerated lifetime testing (ALT) of DC-link capacitors is given in this part [31]. The ALT concept has been presented in Section 1.2.5 of Chapter 1 in this book. The purpose of this testing example is to study the degradation and lifetime of a type of 1100 V/40 μF MPPF-Caps. The testing is performed by a capacitor testing system shown in Figure 3.4, which is composed of a climatic chamber with a temperature range from −70 °C to 180 °C and a relative humidity level from 10% to 95% (within a certain temperature range), ripple current testers, and parameter characterization equipment. The system allows the testing of a wide range of film capacitors for the DC-link applications in power electronics.

Figures 3.5 and 3.6 present the testing results of ten samples of a specific type of 1,100 V/40 μF MPPF-Caps. The capacitance drops along with the 2,000 hours of testing are plotted in Figure 3.5. The end-of-life criteria of the capacitors are 5% capacitance drop from its initial values. It can be noted that all the ten capacitor

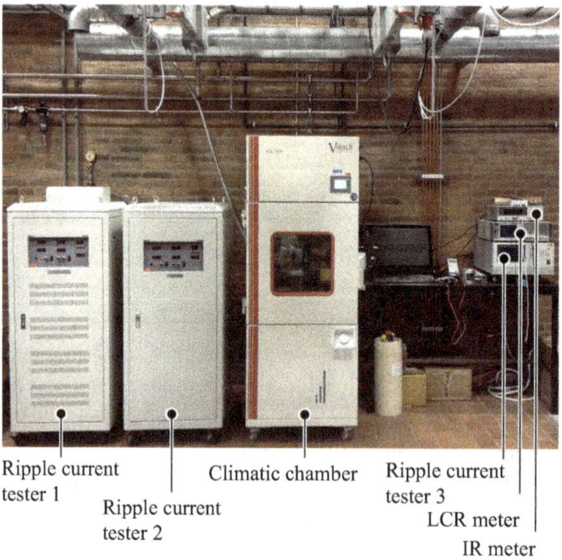

Figure 3.4 Photo of a capacitor degradation testing system

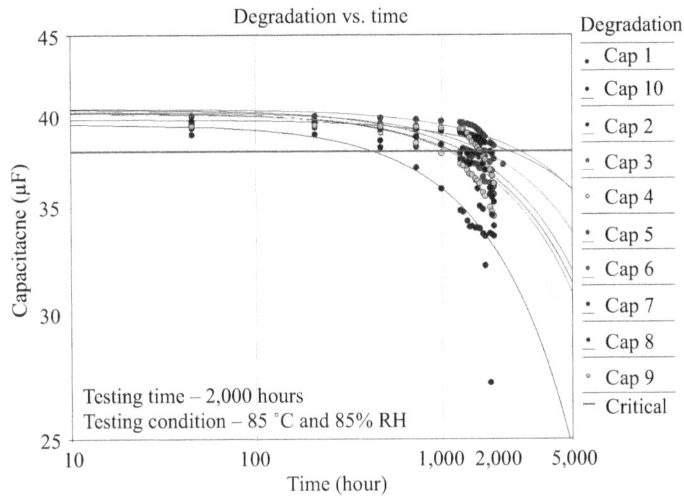

Figure 3.5 Degradation of the ten capacitor testing samples in terms of capacitance drop under 85 °C and 85% relative humidity (RH)

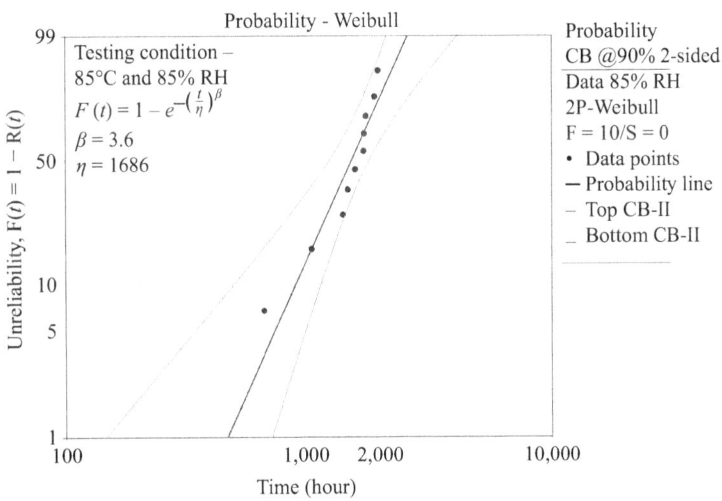

Figure 3.6 Unreliability curve in a Weibull plot based on the time-to-failure data of the ten capacitor samples under 85 °C and 85% relative humidity (RH)

samples reach the end-of-life before 2000 hours, while the lifetime indicated in its datasheet is more than 100,000 hours under a condition of 85 °C operation temperature and no humidity. Therefore, the testing results reveal a significant impact of humidity on the degradation and lifetime of the MPPF-Caps. Figure 3.6 shows the unreliability of the capacitors in a Weibull plot based on the time-to-failure

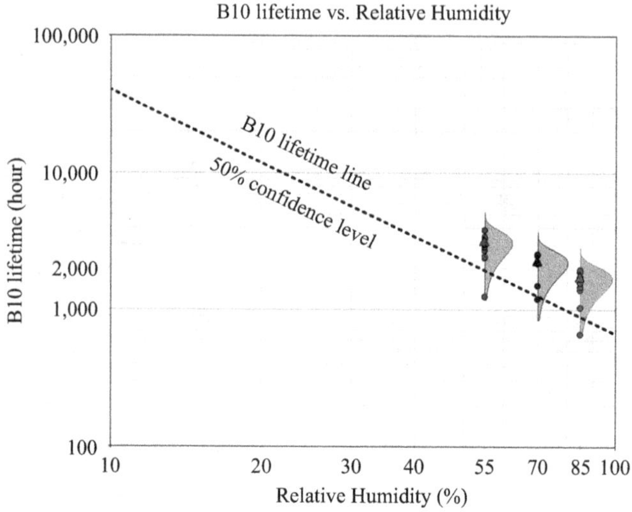

Figure 3.7 B10 lifetime under 85 °C and different humidity levels based on three groups of capacitor testing under relative humidity levels of 85%, 70%, and 55% (ten capacitor samples for each group of testing)

values of the ten capacitors. The B10 lifetime is 899 hours at a 50% confidence level. The information on Weibull analysis can be found in Section 1.2.1 of Chapter 1 in this book, which therefore will not be discussed here.

Figure 3.7 plots the B10 lifetime of the capacitor under different relative humidity levels. The results are based on the above-accelerated testing under 85% humidity level, and another two group of testing under 70% humidity level and 55% humidity level.

3.2.3.2 Testing to obtain the safe operating area (SOA) of DC-link film capacitors

Besides the degradation behavior and lifetime model, the operation limits of capacitors are also important information used for decision-making in the design phase. The following example introduces a safe operating area (SOA) concept for capacitors, inspired by the power semiconductor field. A theoretical model to derive the capacitor SOA in terms of voltage, ripple current, and ambient temperature is presented in [32]. The derived model considers temperature-dependent power loss coming from both ESR and R_p shown in Figure 3.3(a). Figure 3.8 shows the testing results of two out of the ten samples, with rated voltage of 1.25 kV, capacitance of 645 μF, and rated surge voltage of 1.875 kV. The two samples, Sample 5 and Sample 6, are tested with an initial ambient temperature of 96 °C, voltage of 1540 V and ripple current of 20 A. Sample 5 survives at the end of testing, while Sample 6 has an over-temperature failure. Especially, the hotspot temperature increases faster after it reaches 125 °C, due to that it triggers strong dependence between temperature and leakage current. This type of short-term overstress testing helps to obtain the capacitor SOA, as discussed in detail in [32].

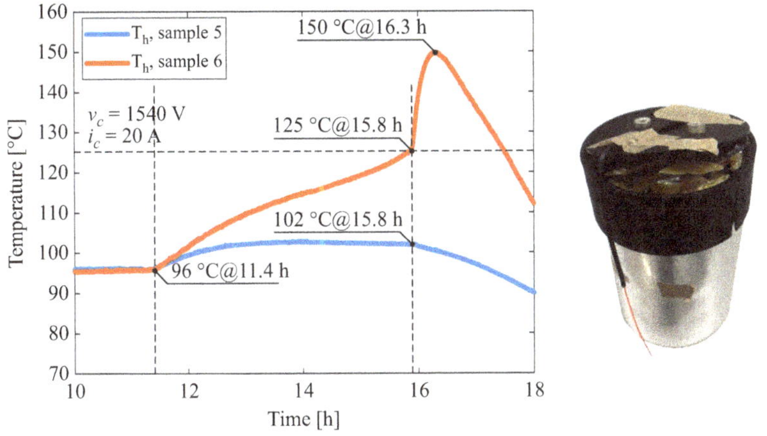

Figure 3.8 Testing results of two capacitor samples (Sample 5 and Sample 6 out of ten samples) [32]. The picture of Sample 6 at the end of testing is shown on the right-hand side.

3.2.3.3 Application-oriented testing method of capacitors for high-power applications

The existing commercially available ripple current testers used for capacitor testing either have limited voltage or current capability. In applications such as wind turbines, railways, and high-voltage-direct-current (HVDC) applications, capacitors withstand both high voltage and high ripple current. To address the limitation, a robust capacitor stress emulation stress method is proposed in [33] and generalized in [34] which can generate application-oriented voltage and current stresses of capacitors. The circuit architecture is shown in Figure 3.9(a), which includes a low-voltage power supply (LV), a high-voltage power supply (HV), a three-phase DC/AC power stack, and AC side inductors (L). HV is used to provide the DC bias voltage. LV is used to generate the AC voltage by the power stack and the AC side inductors and provide the power consumed by the testing system and the capacitors under test. With the circuit, the maximum required current of LV is dependent on the total power loss only, and the current of the HV is negligible (e.g., in the range of tens of mA). An implemented testing system is shown in Figure 3.9(b), which can provide up to 10 kV DC voltage, 1.8 kV AC voltage, and 1.2 kV ripple current concurrently, while the rated current of HV and LV are 1 A and 30 A, respectively. One application example of the testing system is presented in [35] for capacitors used in the AC filters of wind turbine converters.

3.3 Reliability-oriented design for DC links

As the DC-link capacitors contribute to cost, size, and failure of power electronic converters on a considerable scale [1], research efforts have been devoted to either

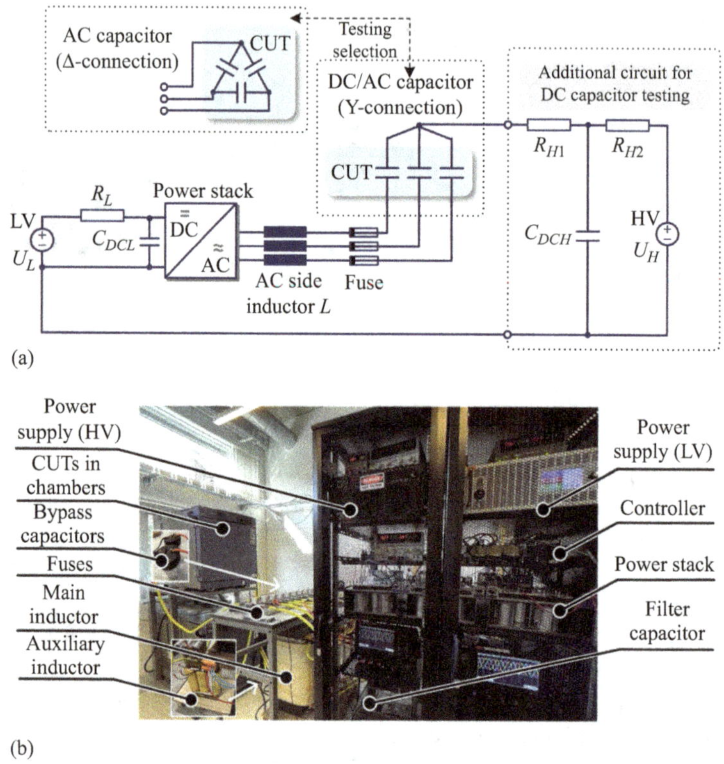

Figure 3.9 Capacitor stress emulation system for high power applications with minimum required power supplies [33,34]. (a) Circuit architecture for capacitor stress emulation. (b) Developed capacitor testing system for high-power applications (testing capability: concurrently up to 10 kV DC voltage, 1.8 kV AC voltage, and 1.2 kA ripple current).

optimal design of a DC-link capacitor bank [36,37] or to the reduction of the DC-link capacitance requirement [38]. This section will discuss six types of DC-link design solutions and a reliability-oriented procedure to select and size DC-link capacitors.

3.3.1 Six types of capacitive DC-link design solutions

Figure 3.10 shows the main types of capacitive DC-link design solutions. The most widely applied one is shown in Figure 3.10(a), where either Al-Caps, MPPF-Caps, or MLC-Caps are applied. A hybrid design solution composed of both Al-Caps and MPPF-Caps is presented in [39], as illustrated in Figure 3.10(b). A DC link composed of a 40 mF Al-Caps bank and a 2 mF MPPF-Cap is designed for a 250 kW inverter by taking advantage of their different frequency characteristics. By adopting this solution, the reliability of the Al-Caps bank is to be improved due to

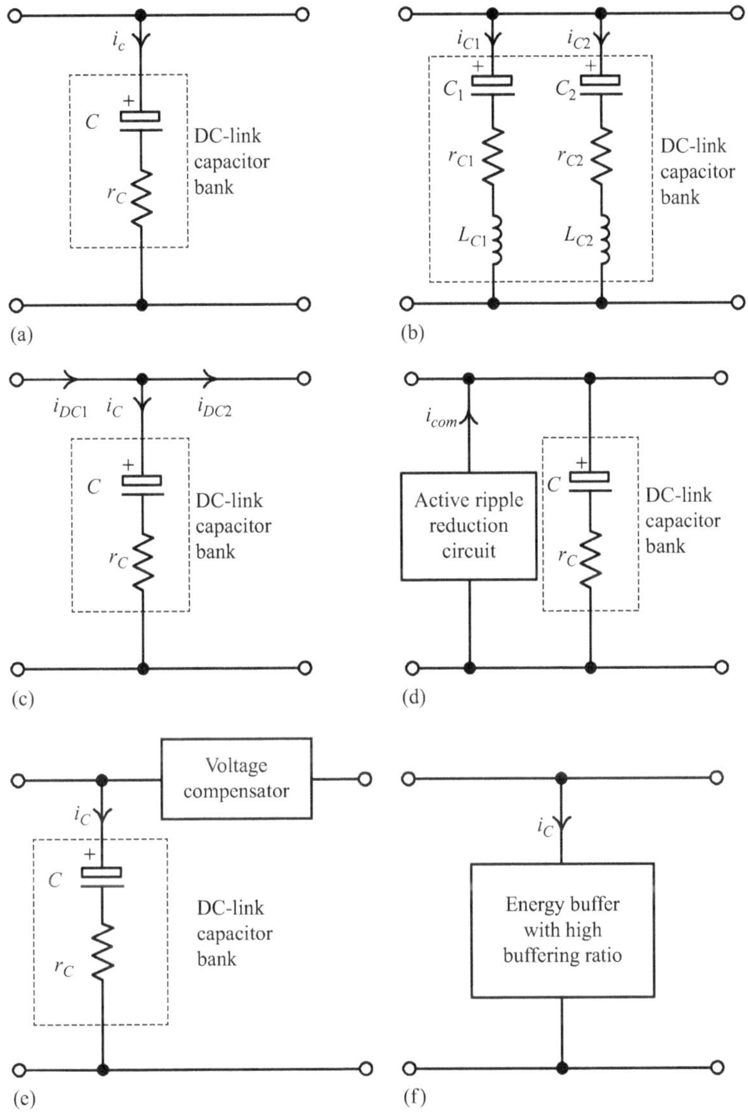

Figure 3.10 Six types of capacitive DC-link design solutions. (a) Convensional design, (b) hybrid DC-link bank design, (c) by an additional control scheme, (d) a parallel-connected active circuit, (e) a series-connected active circuit, and (f) replacement by an active circuit.

reduced current stresses. Another research direction is to reduce the energy storage requirement in the DC link so that Al-Caps could be replaced by MPPF-Caps to achieve a higher level of reliability without considerably increasing the cost and volume. For example, the concept of Figure 3.10(c) is to synchronize the current

i_{DC1} and i_{DC2} by an additional control scheme to reduce the ripple current flowing through the DC-link capacitor [40]. This solution is especially applicable for the application when there is a specific relationship in the operating frequency between the two converters connected to the DC link. The concept of Figure 3.10(d) and (e) is to introduce an additional ripple power port apart from the DC link [41,42]. These two solutions could reduce the overall energy storage requirement of the DC link. The advantage of the series voltage compensator solution in [42] is that the power capacity of the compensator is much lower than that of the parallel circuit shown in Figure 3.10(d). It is due to very low voltage stresses on the active devices inside the compensator. Figure 3.10(f) shows the sixth type of DC-link solution, of which the conventional DC-link capacitors are directly replaced by an energy buffer with a high energy buffering ratio. The energy buffering ratio is defined as the ratio of the energy that can be injected and extracted from the DC link in one cycle to the total energy stored in the DC link [43]. A variant of Figure 3.10(f) is presented in [44,45], which the DC link is implemented by a capacitor connected in series with an auxiliary circuit. The auxiliary circuit processes only the voltage ripple and ripple current of the passive capacitor. A two-terminal active capacitor concept is proposed in [45] which does not require any external feedback signal and power supply, making it with only two power terminals same as a passive capacitor. More detailed discussions on capacitance reduction in power electronic converters can be found in Chapter 6.

The active DC-link solutions shown in Figure 3.10(d)–(f) break the classical trade-offs among cost, reliability, and power density of capacitors and bring the opportunities in improving at least one aspect of them. A comprehensive comparison of different capacitive DC link solutions is presented in [38].

3.3.2 A reliability-oriented design procedure of capacitive DC links

In addition to the possibilities brought by innovative DC-link solutions, a reliability-oriented design procedure could provide further potential to build reliability into DC links. Figure 3.11 presents a reliability-oriented design procedure for DC links.

The key steps shown in Figure 3.11 are discussed as follows.

3.3.2.1 Higher-level definition

The DC-link design depends on the converter-level specifications (e.g., power rating, voltage level, and lifetime target), circuit topologies, control methods, and design constraints on other components. For voltage source converters or inverters, the DC link is capacitive and composed of capacitors, while for current source converters or inverters, the DC link is inductive and mainly composed of inductors [46]. The capacitive type is more widely used than the inductive one due to the popularity of voltage source converters and inverters in various applications. One of the reasons is that capacitors generally have higher energy density than that of inductors. The selections of other components also affect the sizing of DC-link

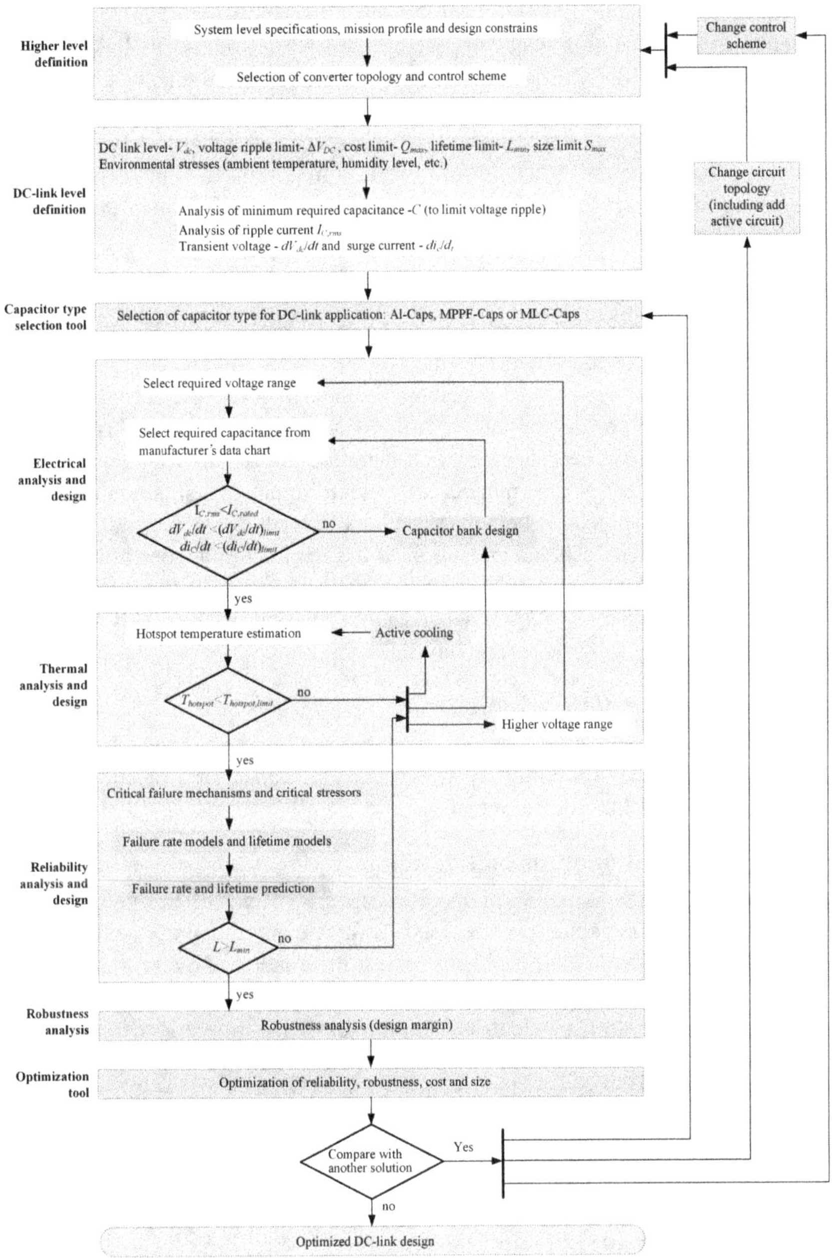

Figure 3.11 A reliability-oriented design procedure for capacitive DC links

capacitors. For example, the choice of the input side inductor in an AC variable-frequency drive has a significant impact on the lifetime of its DC-link capacitor bank, as studied in [29]. It reveals that a higher inductor value (i.e., a higher line impedance) is beneficial to the improvement of the capacitor bank lifetime or to the reduction of the required capacitance. Therefore, it is essential to have a system-level scope. From the reliability perspective, it is important to allocate the system-level reliability target to each important components, including the DC-link capacitors.

3.3.2.2 DC-link level definition

Based on the converter-level specifications, the major design constraints of DC link are DC-link voltage level, limit of DC-link voltage ripple, volume, cost, and lifetime. Another important aspect of the definition is the environmental conditions (e.g., ambient temperature profile, humidity profile). Based on the above information, the ripple current stress can be calculated, and therefore the required minimum capacitance can be preliminarily determined. An accurate ripple current stress analysis of DC-link capacitors is crucial to both the selection of proper capacitors and the lifetime prediction of them. The detailed derivation of the ripple current spectra for a three-phase inverter and for general voltage source inverters is presented, respectively, in [46,47]. The ripple current stress is dependent on mission profiles such as dynamic solar irradiances [46] and grid voltage unbalances [47].

3.3.2.3 Capacitor type selection

Depending on the application and the calculated ripple current stress and required minimum capacitance, a preliminary selection of the capacitor type and the corresponding DC-link solution can be made.

3.3.2.4 Electrical analysis and design

This step involves the selection of specific capacitors and the design of the DC-link bank if multiple capacitors are used [36]. A capacitor bank design with a low parasitic inductance is desirable to reduce the chance of overvoltage of both the capacitors and the relevant switching devices [48]. Moreover, it is important to consider the variation of electrical parameters with time and with operation conditions. For example, a lumped capacitor impedance model is presented in [49] which differentiates the three major sources of ESR. Therefore, the ESR variation with ripple current frequency and temperature can be considered the circuit-level simulations, allowing for a more accurate thermal stress estimation in the next step.

3.3.2.5 Thermal analysis and design

As shown in Table 3.3, temperature is one of the most important stressors that influence the reliability of capacitors. Therefore, in addition to the electrical stress analysis [36], thermal stress analysis is equally important to the choice of capacitors and the design of DC-link banks. The connection between the electrical stress analysis and thermal stress analysis is the thermal impedance network of the capacitor of interest. The thermal impedance of a single electrolytic capacitor and a numerical heat transfer model of the capacitor banks have been investigated in [50,51], respectively. In [51], the heat transfer dependence on capacitor spacing and capacitor location (i.e., center capacitors and side capacitors) is also studied. In [52], the impact of the dynamics of capacitor ambient temperature is studied and modeled. It shows the necessity to consider the ambient temperature dynamics in operations with dynamic mission profiles due to the relatively large thermal capacitance of capacitors.

3.3.2.6 Reliability analysis and design

This step covers the lifetime prediction of the pre-selected capacitors based on the failure mechanisms and corresponding lifetime models, as discussed in Section 3.2. An additional example of motor drive application is discussed in [53].

3.3.2.7 Robustness analysis and optimization

The final steps of the design procedure are the design margin (i.e., robustness) analysis [54] and multi-objective optimization on reliability, robustness, cost, and size of the DC-link solution. Different DC-link design solutions, alternative topologies, and control schemes may also need to be evaluated and compared to reach the final design solution.

While the above design procedure provides a systematic way to select the DC-link capacitors with optimized cost, size, and lifetime, it may still not be easy to be applied due to its high level of complexity as well as the need for further research in some of the steps shown in Figure 3.11. Among others, the research effort needed is to develop user-friendly software tools that can implement the procedure so that the power electronic designers can practically apply the procedure in a much easier way.

3.4 Condition monitoring of DC-link capacitors

Condition monitoring is another reliability engineering tool to achieve reliable operation of DC-link capacitors, besides the reliability-oriented design. Figure 3.12 shows a summary of the condition monitoring methods for various types of capacitors, which can be classified into online and offline implementations. The degradation of capacitors can be indirectly estimated by a few physical parameters such as capacitance, ESR, insulation resistance, voltage ripple, weight, and volume. There are three methodologies to obtain one or more of these degradation indicators: (a) capacitor current sensor–based method, (b) circuit model–based method, and (c) data and advanced algorithm–based method.

Among the indicators shown in Figure 3.12, capacitance, ESR, and insulation resistance are the three widely used ones. Table 3.4 shows the associated typical end-of-life criteria of Al-Caps, MPPF-Caps, and MLC-Caps.

Impressive research work has been done on the condition monitoring of Al-Caps [55–59]. As shown in Figure 3.3(b), in the low frequency range ($f < f_1$), the

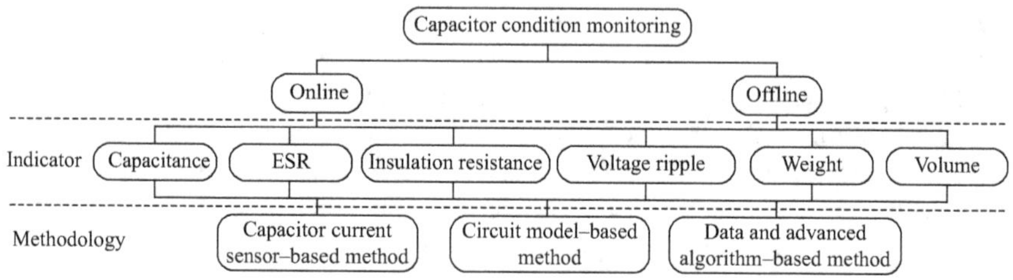

Figure 3.12 A classification of condition monitoring methods for capacitors

Table 3.4 Typical end-of-life criteria and condition monitoring parameters

	Al-Caps	MPPF-Caps	MLC-Caps
Failure criteria	C: 20% reduction ESR: 2 times	C: 5% reduction DF: 3 times	C: 10% reduction $R_p < 10^7\ \Omega$ DF: 2 times
Degradation indicators	C or ESR, or both	C	C, R_p

DF: dissipation factor; R_p: insulation resistance.

impedance is approximated to $2\pi fC$. In the medium frequency range ($f_1 < f < f_2$), the impedance is dominated by the ESR. Therefore, by extracting the voltage and/ or current information in the respective frequency ranges, the capacitance and ESR can be estimated.

There are two main principles for ESR estimation:

1. $ESR = V_C/I_C$ where V_C and I_C are the root-mean-square (RMS) values of the capacitor voltage and capacitor current in the ohmic region (i.e., $f_1 < f < f_2$) [55–57]. The case temperature of the capacitor is usually measured to compensate the temperature dependence of ESR. This method requires two band-pass filters, which should have sufficient bandwidths to extract the frequency components of interest. At the same time, the frequency components below f_1 shall be rejected sufficiently.
2. $ESR = P_C/I_C^2$ where P_C is the average power dissipated in the capacitor and I_C is the RMS current of the capacitor [58,59]. This method does not require specific band-pass filters. The introduction of the sensor in the capacitor current path may not be desirable in practical applications due to its stray inductance.

The main principle applied to estimate the capacitance of both Al-Caps and MPPF-Caps is $C = \left(\int i_C\, dt\right)/\Delta v_C$, where i_C is the capacitor current and the Δv_C is the capacitor voltage ripple. In [60], the online condition monitoring of MPPF-Caps for an aerospace drive application is presented. To avoid the use of a current sensor in series with the DC-link capacitor, the DC-link current i_C is calculated based on

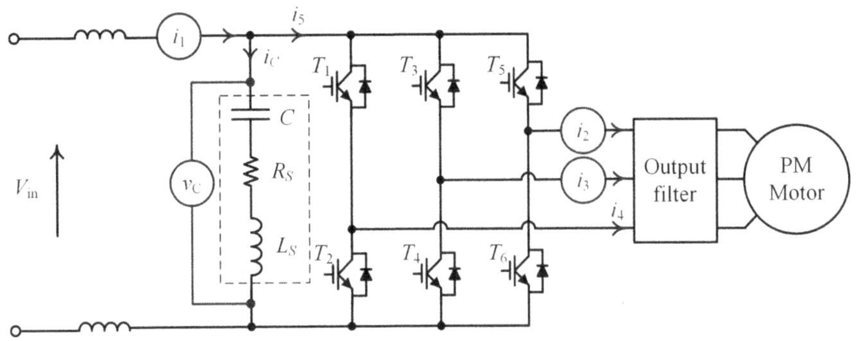

Figure 3.13 An example of condition monitoring of DC-link capacitor based on the circuit model [60]

the circuit operation model and the available signals in the control circuits. As shown in Figure 3.13, the electrical parameters i_1, i_2, i_3, and v_C are measured by three existing current sensors and one voltage sensor, respectively, for the control purpose of the converter. The value of i_4 is estimated by assuming that the three phase output currents i_2, i_3, and i_4 are balanced, and $i_2 + i_3 + i_4 = 0$. Then the value of i_5 can be obtained from the information of i_2, i_3, i_4, and the switching modulation signals of the six switches T_1 to T_6. Therefore, the capacitor ripple current i_C is calculated as the difference between the input current, i_1, and the current flowing to the inverter, i_5. The measurement system should have a wide bandwidth to capture all the harmonics of the DC-link voltage ripple and have a fast sampling rate.

In [61], an offline prognostic method for MLC-Caps is presented in which the insulation resistance and capacitance are measured. The method is based on the parameter residual generated by the difference between the measured capacitance and its estimation.

A few remarks are given below on the condition monitoring of DC-link capacitors:

1. Extensive academic research has been performed in the last three decades on various capacitor condition monitoring methods. Industry application demands are emerging in recent years. Nevertheless, cost, complexity, and new risk associated with condition monitoring solutions remain the barriers. Especially, there is a lack of analysis and verification of the new risk due to the failure of hardware, software, or both needed for condition monitoring and the false alarm due to incorrect monitoring results.
2. Methods having no or minimum additional hardware requirements are favorable to minimize the cost and reduce the complexity.
3. While majority of the research focuses on online monitoring, it is worthwhile to considering offline or special operation modes for capacitor condition monitoring if it is possible for the specific application, especially when the goal is to monitor the degradation level, which is a slow process. For example, a hand-carry unit is proposed in [62] to estimate the DC-link filter parameters of

traction inverter during the routine checks. Another example discussed in [63] leverages the discharge operation mode, which is an existing operation mode for safety reason, after parking electric vehicles to estimate the DC-link capacitance based on existing information.

4. Besides the possible health indicators shown in Figure 3.12, it is worthwhile to considering new health indicators. For example, a health indicator based on strain sensing on the surface of capacitor is proposed in [64], making it possible to have a solution without electrical connection with the converter to be monitored.

References

[1] H. Wang and F. Blaabjerg, "Reliability of capacitors for DC-link applications in power electronic converters – an overview," *IEEE Transactions on Industry Applications*, vol. 50, no. 5, pp. 3569–3578, 2014.

[2] H. Wang and F. Blaabjerg, "Reliability of capacitors for DC-link applications—an overview," in *Proceedings of the IEEE Energy Conversion Congress and Exposition*, 2013, pp. 1866–1873.

[3] B. W. Williams, *Power Electronics: Devices, Drivers, Applications, and Passive Components*, 2006. Available: https://personal.eee.strath.ac.uk/barry.williams/book.html.

[4] M. Marz, A. Schletz, B. Eckardt, S. Egelkraut, and H. Rauh, "Power electronics system integration for electric and hybrid vehicles," in *Proceedings of International Conference on Integrated Power Electronics Systems (CIPS)*, 2010, pp. 1–10.

[5] TDK datasheet, *CeraLink capacitors for fast-switching semiconductors*. Online: https://www.tdk-electronics.tdk.com/inf/20/10/ds/B58033_SP.pdf (accessed: 25 January 2025).

[6] H. Wang, M. Liserre, and F. Blaabjerg, "Toward reliable power electronics - challenges, design tools and opportunities," *IEEE Industrial Electronics Magazine*, vol. 7, no. 2, pp. 17–26, 2013.

[7] S. Yang, A. T. Bryant, P. A. Mawby, D. Xiang, L. Ran, and P. Tavner, "An industry-based survey of reliability in power electronic converters," *IEEE Transactions on Industry Applications*, vol. 47, no. 3, pp. 1441–1451, 2011.

[8] R. S. Alwitt and R. G. Hills, "The chemistry of failure of aluminum electrolytic capacitors," *IEEE Transactions on Parts Materials and Packaging*, vol. PMP-I, no. 2, pp. 28–34, 1965.

[9] C. Dubilier, *Application guide, aluminum electrolytic capacitors*. Online: http://www.cde.com/catalogs/AEappGUIDE.pdf (accessed: 25 January 2025).

[10] TDK, *Aluminum electrolytic capacitors, general technical information*, August 2022. Online: https://www.tdk-electronics.tdk.com/download/530704/e724fb43668a157bc547c65b0cff75f8/pdf-generaltechnicalinformation.pdf (accessed: 25 January 2025).

[11] Nippon Chemi-con, *Technical note on judicious use of aluminum electrolytic capacitors*. 2024. Online: https://www.chemi-con.co.jp/products/relatedfiles/capacitor/catalog/al-technote-e.pdf (accessed: 25 January 2025).
[12] A. Ritter, *Capacitor reliability issues and needs*, Presentation at the Sandia National Laboratories Utility-Scale Grid-Tied PV Inverter Reliability Technical Workshop, January 2011.
[13] Q. Sun, Y. Tang, J. Feng, and T. Jin, "Reliability assessment of metallized film capacitors using reduced degradation test sample," *Journal of Quality and Engineering International*, vol. 29, no. 2, pp. 259–265, 2013.
[14] TDK, *Film capacitors – general technical information*. June 2018. Online: https://www.tdk-electronics.tdk.com/download/530754/480aeb04c789e45ef5bb9681513474ba/pdf-generaltechnicalinformation.pdf (accessed: 25 January 2025).
[15] Electronicon, *Capacitors for power electronics*, May 2024. Online: https://www.electronicon.com/fileadmin/inhalte/PDF-Dateien/Kataloge_und_Broschueren/200.003-020021_App-notes.pdf (accessed: 25 January 2025).
[16] B. S. Rawal and N. H. Chan, *Conduction and Failure Mechanisms in Barium Titanate Based Ceramics Under Highly Accelerated Conditions*. Greenville, SC: AVX Corporation Technical Information; 1984.
[17] D. Liu and M. J. Sampson, "Some aspects of the failure mechanisms in BaTiO3 – Based multilayer ceramic capacitors," in *Proceedings of CARTS International*, 2012, pp. 59–71.
[18] M. J. Cozzolino, "Electrical shorting in multilayer ceramic capacitors," in *Proceedings of CARTS International*, 2004, pp. 57–68.
[19] J. L. Stevens, J. S. Shaffer, and J. T. Vandenham, "The service life of large aluminum electrolytic capacitors: effects of construction and application," *IEEE Transactions on Industry Applications*, vol. 38, no. 5, pp. 1441–1446, 2002.
[20] R. M. Kerrigan, "Metallized polypropylene film energy storage capacitors for low pulse duty," in *Proceedings of CARTS USA*, 2007, pp. 97–104.
[21] R. W. Brown, "Linking corrosion and catastrophic failure in low-power metallized polypropylene capacitors," *IEEE Transactions on Device Material Reliability*, vol. 6, no. 2, pp. 326–333, 2006.
[22] C. Hillman, *Uprating of ceramic capacitors*, DfR Solution white paper.
[23] W. J. Minford, "Accelerated life testing and reliability of high *K* multilayer ceramic capacitors," *IEEE Transactions on Components, Hybrids, and Manufacturing Technology*, vol. CHMT-5, no. 3, pp. 297–300, 1982.
[24] N. Kubodera, T. Oguni, M. Matsuda, H. Wada, N. Inoue, and T. Nakamura, "Study of the long term reliability for MLCCs," in *Proceedings of CARTS International*, 2012, pp. 1–9.
[25] S. G. Parler, "Deriving life multipliers for electrolytic capacitors," *IEEE Power Electronics Society Newsletter*, vol. 16, no. 1, pp. 11–12, 2004.
[26] Emerson Network Power, *Capacitors age and capacitors have an end of life*, White paper.

[27] A. Albertsen, *Electrolytic capacitor lifetime estimation*, Jianghai capacitor technical note. Online: https://jianghai-america.com/uploads/technology/JIANGHAI_Elcap_Lifetime_-_Estimation_AAL.pdf (accessed: 25 January 2025).

[28] M. L. Gasperi, "Life prediction model for aluminum electrolytic capacitors," in *Proceedings of IEEE Industry Applications Society Annual Meeting*, 1996, pp. 1347–1351.

[29] M. L. Gasperi, "Life prediction modeling of bus capacitors in AC variable-frequency drives," *IEEE Transactions on Industry Applications*, vol. 41, no. 6, pp. 1430–1435, 2005.

[30] H. Wang, K. Ma, and F. Blaabjerg, "Design for reliability of power electronic systems," in *Proceedings of the IEEE Industrial Electronics Society Annual Conference*, 2012, pp. 33–44.

[31] H. Wang, P. D. Reigosa, and F. Blaabjerg, "A humidity-dependent lifetime derating factor for DC film capacitors," in *Proceedings of the IEEE Energy Conversion Congress and Exposition*, 2015, pp. 3064–3068.

[32] S. Liu, Z. Shen, and H. Wang, "Safe operating area of DC-link film capacitors," *IEEE Transactions on Power Electronics*, vol. 36, no. 10, pp. 11014–11018, 2021.

[33] B. Yao, Q. Wang, H. Wang, K. Hasegawa, and H. Wang, "A robust testing method for DC and AC capacitors with minimum required power supply," *IEEE Transactions on Power Electronics*, vol. 37, no. 5, pp. 4942–4946, 2022.

[34] B. Yao, H. Wang, Q. Wang, and H. Wang, "A unified capacitor stress emulation method for high power converter applications," *IEEE Transactions on Power Electronics*, vol. 38, no. 8, pp. 10213–10226, 2023.

[35] B. Yao, X. Wei, Y. Zhang, *et al.*, "Accelerated degradation testing and failure mechanism analysis of metallized film capacitors for AC filtering," *IEEE Transactions on Power Electronics*, vol. 39, no. 5, pp. 6256–6270, 2024.

[36] P. Pelletier, J. M. Guichon, J. L. Schanen, and D. Frey, "Optimization of a DC capacitor tank," *IEEE Transactions on Industry Applications*, vol. 45, no. 2, pp. 880–886, 2009.

[37] H. Wang, C. Li, G. Zhu, Y. Liu, and H. Wang, "Model-based design and optimization of hybrid DC-link capacitor banks," *IEEE Transactions on Power Electronics*, vol. 35, no. 9, pp. 8910–8925, 2020.

[38] H. Wang, H. Wang, G. Zhu, and F. Blaabjerg, "An overview of capacitive DC links – topology derivation and scalability analysis," *IEEE Transactions on Power Electronics*, vol. 35, no. 2, pp. 1805–1829, 2020.

[39] M. A. Brubaker, D. El Hage, T. A. Hosking, H. C. Kirbie, and E. D. Sawyer, "Increasing the life of electrolytic capacitor banks using integrated high performance film capacitors," in *Proceedings of PCIM Europe*, 2013, pp. 1–8.

[40] I. S. Freitas, C. B. Jacobina, and E. C. Santos, "Single-phase to single-phase full-bridge converter operating with reduced AC power in the DC-link

capacitor," *IEEE Transactions on Power Electronics*, vol. 25, no. 2, pp. 272–279, 2010.

[41] R. X. Wang, F. Wang, D. Boroyevich, *et al.*, "A high power density single-phase PWM rectifier with active ripple energy storage," *IEEE Transactions on Power Electronics*, vol. 26, no. 5, pp. 1430–1443, 2011.

[42] H. Wang, H. S. H. Chung, and W. Liu, "Use of a series voltage compensator for reduction of the dc-link capacitance in a capacitor-supported system," *IEEE Transactions on Power Electronics*, vol. 29, no. 3, pp. 1163–1175, 2014.

[43] M. Chen, K. K. Afridi, and D. J. Perreault, "Stacked switched capacitor energy buffer architecture," *IEEE Transactions on Power Electronics*, vol. 28, no. 11, pp. 5183–5195, 2013.

[44] S. Qin, Y. Lei, C. Barth, W. C. Liu, and R. Pilawa-Podgurski, "A high power density series-stacked energy buffer for power pulsation decoupling in single-phase converters," *IEEE Transactions on Power Electronics*, vol. 32, no. 6, pp. 4905–4924, 2017.

[45] H. Wang and H. Wang, "A two-terminal active capacitor," *IEEE Transactions on Power Electronics*, vol. 32, no. 8, pp. 5893–5896, 2017.

[46] N. C. Sintamarean, F. Blaabjerg, H. Wang, F. Iannuzzo, and P. de Rimmen, "Reliability oriented design tool for the new generation of grid connected PV inverters," *IEEE Transactions on Power Electronics*, vol. 30, no. 5, pp. 2635–2644, 2015.

[47] H. Wang, P. Davari, H. Wang, D. Kumar, F. Zare, and F. Blaabjerg, "Lifetime estimation of DC-link capacitors in adjustable speed drives under grid voltage unbalances," *IEEE Transactions on Power Electronics*, vol. 34, no. 5, pp. 4064–4078, 2019.

[48] M. C. Caponet, F. Profumo, R. W. Doncker, and A. Tenconi, "Low stray inductance bus bar design and construction for good EMC performance in power electronic circuits," *IEEE Transactions on Power Electronics*, vol. 17, no. 2, pp. 225–231, 2002.

[49] M. L. Gasperi, "A method for predicting the expected life of bus capacitors," in *Proceedings of IEEE Industry Applications Society Annual Meeting*, 1997, pp. 1042–1047.

[50] T. Huesgen, "Thermal resistance of snap-in type aluminum electrolytic capacitor attached to heat sink," *in Proceedings of IEEE Energy Conversion Congress and Exposition*, 2012, pp. 1338–1345.

[51] H. Wang and H. Wang, "Analytical modeling and design of capacitor bank considering thermal coupling effect," *IEEE Transactions on Power Electronics*, vol. 36, no. 3, pp. 2629–2640, 2021.

[52] H. Wang, R. Zhu, H. Wang, M. Liserre, and F. Blaabjerg, "A thermal modeling method considering ambient temperature dynamics," *IEEE Transactions on Power Electronics*, vol. 35, no. 1, pp. 6–9, 2020.

[53] H. Wang, S. Huang, D. Kumar, X. Deng, G. Zhu, and H. Wang, "Lifetime prediction of DC-link capacitors in multiple drives system based on

simplified analytical modeling," *IEEE Transactions on Power Electronics*, vol. 36, no. 1, pp. 844–860, 2021.

[54] ZVEI Robustness Validation Working Group, *Handbook for Robustness Validation of Automotive Electrical/Electronic Modules*, Frankfurt: Zentralverband Elektrotechnik- und Elektronikindustrie e. V., 2008.

[55] K. Harada, A. Katsuki, and M. Fujiwara, "Use of ESR for deterioration diagnosis of electrolytic capacitor," *IEEE Transactions on Power Electronics*, vol. 8, no. 4, pp. 355–361, 1993.

[56] K. P. Venet, F. Perisse, M. H. El-Husseini, and G. Rojat, "Realization of a smart electrolytic capacitor circuit," *IEEE Industry Application Magazine*, vol. 8, no. 1, pp. 16–20, 2002.

[57] A. M. Imam, T. G. Habetler, R. G. Harley, and D. Divan, "Failure prediction of electrolytic capacitors using dsp methods," in *Proceedings of IEEE Applied Power Electronics Conference*, 2005, pp. 965–970.

[58] M. A. Vogelsberger, T. Wiesinger, and H. Ertl, "Life-cycle monitoring and voltage-managing unit for DC-Link electrolytic capacitors in PWM converters," *IEEE Transactions on Power Electronics*, vol. 26, no. 2, pp. 493–503, 2011.

[59] E. Aeloiza, J. H. Kim, P. Enjeti, and P. Ruminot, "A real time method to estimate electrolytic capacitor condition in PWM adjustable speed drives and uninterruptible power supplies," in *Proceedings of IEEE Power Electronics Specialists Conference*, 2005, pp. 2867–2872.

[60] A. Wechsler, B. C. Mecrow, D. J. Atkinson, J. W. Bennett, and M. Benarous, "Condition monitoring of DC-link capacitors in aerospace drives," *IEEE Transactions on Industry Applications*, vol. 48, no. 6, pp. 1866–1874, 2012.

[61] J. Sun, S. Cheng, and M. Pecht, "Prognostics of multilayer ceramic capacitors via the parameter residuals," *IEEE Transactions on Device and Materials Reliability*, vol. 12, no. 1, pp. 49–57, 2012.

[62] B. Yao, Y. Peng, Y. Zhang, H. Wang, and H. Wang, "An estimation method of high-order LC circuits in power electronic converters," *IEEE Transactions on Industrial Electronics*, vol. 71, no. 5, pp. 5274–5284, 2024.

[63] X. Wei, B. Yao, Y. Peng, K. Wang, Y. Sun, and H. Wang, "An improved discharge profile-based DC-link capacitance estimation for traction inverters in electric vehicle applications," *IEEE Transactions on Power Electronics*, vol. 39, no. 7, pp. 8696–8708, 2024.

[64] B. Yao, S. Zhao, Y. Zhang, and H. Wang, "A health indicator of aluminum electrolytic capacitors based on strain sensing," *IEEE Transactions on Power Electronics*, vol. 38, no. 7, pp. 7982–7987, 2023.

Chapter 4

Minimization of DC link capacitance in power electronic converter systems

Henry Shu-Hung Chung[1]

Power electronics is a pervasive part of our daily lives. The main aims of power electronic systems are to control, convert, and condition electrical power flow from one form to another through the use of solid-state electronics. Regardless of the application, electricity is processed through power conditioning systems with multiple power conversion units for meeting source and load requirements. Those units are typically interconnected through a DC link, and the DC-link voltage is typically stabilized by a bank of capacitors. Among different types of capacitors, electrolytic capacitors are a popular choice for capacitor banks due to their high capacitance in a small form factor and also their cost effectiveness. However, their life expectancy is comparatively short and is reduced dramatically at elevated ambient temperatures. Thus, capacitor banks have become a bottleneck for the reliability of power electronic converter systems. For high-voltage, high-power converter systems, DC-link capacitors are periodically replaced and monitored for reliable and safe operation, leading to substantial maintenance costs and electronic waste. To enhance reliability and lifetime, many high-performance capacitors have been used as replacements for electrolytic capacitors. However, the benefits of using those capacitors are to some degree compromised by their reduced volumetric efficiency. To lessen the dependency on DC-link capacitors, many techniques have been proposed to minimize DC-link capacitance or make systems less dependent on DC-link capacitance. This chapter gives an overview of various strategies and their implementation.

4.1 Introduction

The major function of a DC-link capacitor is to stabilize the DC-link voltage by buffering the instantaneous power difference among various units connected to the DC link. For the sake of simplicity and without loss of generality, a simplified

[1]Centre for Smart Energy Conversion and Utilization Research, City University of Hong Kong, Hong Kong SAR, China

model with two units, Systems A and B, connected together through a DC link, is illustrated (see Figure 4.1). The DC link is supported by a capacitor C.

Systems A and B can be different kinds of electrical systems such as power electronic converters, voltage or current sources, and electrical loads. The DC-link voltage is v_{dc}. In the following discussion, assume that System A delivers a current i_A to the DC link and that System B absorbs a current i_B from the DC link. The two currents are composed of DC and AC components. Mathematically, they can be expressed as

$$i_A(t) = I_A + \Delta i_A(t) \tag{4.1}$$

$$i_B(t) = I_B + \Delta i_B(t) \tag{4.2}$$

where I_A and I_B represent the DC component of i_A and i_B, respectively, and Δi_A and Δi_B represent the AC component of i_A and i_B, respectively.

As the DC-link capacitor C allows AC current only to flow, the DC components of i_A and i_B should be the same, and the capacitor current i_C equals the difference between i_A and i_B. Thus,

$$I_A = I_B \tag{4.3}$$

$$i_C(t) = \Delta i_A(t) - \Delta i_B(t) \tag{4.4}$$

The capacitor voltage v_C consists of DC and AC components. The magnitude of the AC component depends on the magnitude of i_C and the value of the capacitor C,

$$\frac{d}{dt} v_C(t) = \frac{1}{C} i_C(t) \tag{4.5}$$

As the DC-link voltage, v_{dc}, is the capacitor voltage, v_C, its AC ripple is also governed by (4.5). Many applications require a minimum DC-link voltage such as a grid-connected inverter to maintain the output power quality or sustain operation.

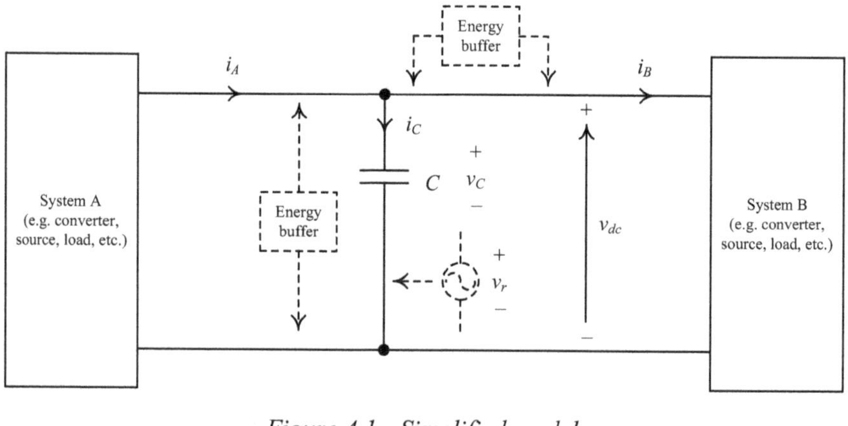

Figure 4.1 Simplified model

The value of the DC-link capacitor is chosen such that the ripple voltage is smaller than the designed value.

To reduce the DC-link capacitance or make the whole system less dependent on the value of the DC-link capacitor, many techniques have been proposed. They are mainly based on the following strategies:

1. *Performance tradeoff*—The requirement of the ripple on the DC-link voltage is relaxed. Thus, a larger voltage ripple is allowed, and the required DC-link capacitance can then be reduced.
2. *Reduction of the capacitor current*—Since the capacitor current is a function of Δi_A and Δi_B (4.4), it can be reduced by reducing the magnitudes of Δi_A and Δi_B, or synchronizing Δi_A and Δi_B with the same frequency, phase, and magnitude.
3. *Energy storage in systems A and B*—Systems A and B can have local energy storage to stabilize the input and/or output performance. Thus, the entire system will become less dependent on the DC-link capacitance.
4. *External energy buffer*—An external parallel- or series-type energy buffer can be introduced, so that the energy storage unit in the energy buffer can absorb energy from or release energy to the DC link in order to stabilize the DC-link voltage, or the voltage and/or current supplied to System B.
5. *Ripple cancellation*—An external voltage source can be added in series with the DC-link capacitor. It generates a voltage that cancels the ripple voltage on the capacitor. Thus, the DC-link voltage has a small ripple.

The strategies can be realized by different methods. As outlined in Figure 4.2, they are classified into three main categories: (1) performance tradeoff, (2) passive approach, and (3) active approach. The performance tradeoff is based on sacrificing the overall system performance to lessen the requirement of the DC-link capacitance. The passive approach is based on using passive components to buffer part of the ripple power or to cancel the DC-link ripple. The active approach is based on using active devices to decouple the energy storage on the DC link, form a new circuit structure, or cancel DC-link ripple. It also includes modification of the

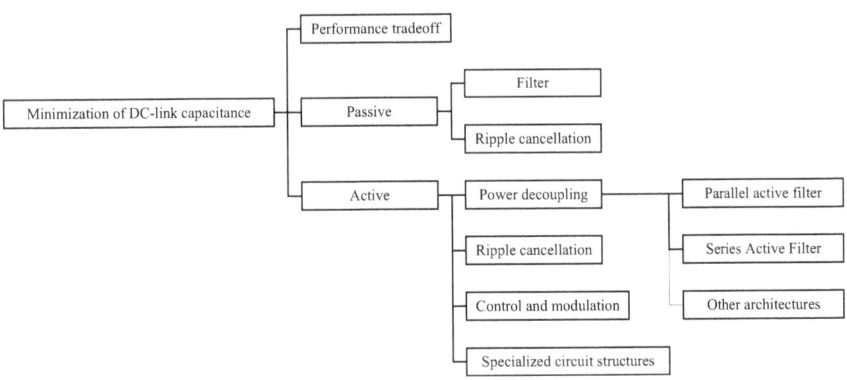

Figure 4.2 Classification of DC-link capacitance minimization techniques

operations of the units connected to the DC link. The implementation methods of each category are briefly described as follows.

4.2 Performance tradeoff

The value of the DC-link capacitance increases as the difference between the energy delivered by System A and the energy absorbed by System B increases. Such difference includes instantaneous power difference and period of variation. However, the variation of the power delivery to the DC link or the absorption from the DC link is sometimes determined by the input and output requirements of the connected systems. An example is an LED lamp with power factor correction (PFC). As illustrated in Figure 4.3(a), the input voltage and input current are both sinusoidal and in phase. Thus, the power delivered to the DC link, which is equal to the product of the input voltage and current, is time-varying and the variation is double that of the line frequency. However, the power utilized by the LEDs is fairly constant to avoid possible lamp flickering. Thus, the DC-link capacitance buffers the power difference between the time-varying power delivered by the PFC circuit and the power utilized by the LEDs.

When the DC-link capacitance is reduced, the ripple voltage on the DC-link will increase. However, in order to maintain the normal operation of the PFC circuit, the input and DC-link voltages have to meet some requirements. In [1], the

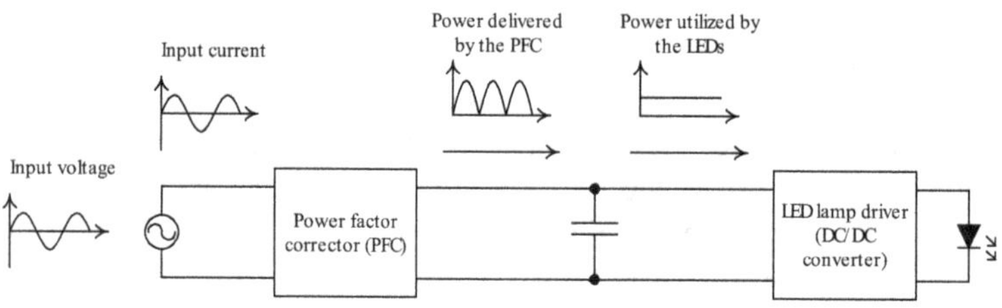

(a)

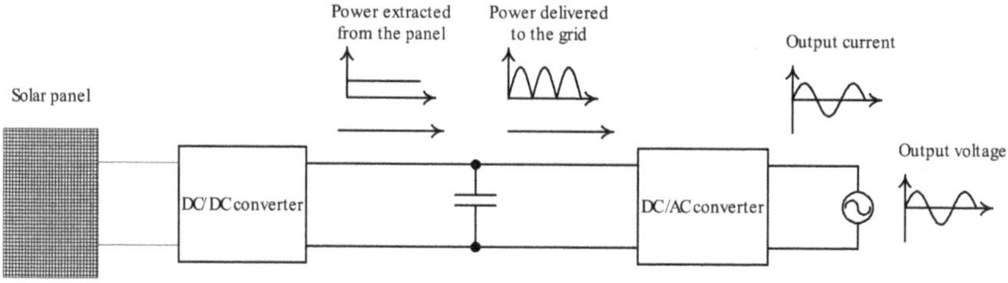

(b)

Figure 4.3 Illustrations of various systems with DC-link. (a) LED lamp with input power factor correction and (b) inverter.

DC-link capacitance is reduced to a value where the DC-link voltage is higher than the input voltage over a line cycle in the boost-type PFC circuit. In order to further reduce the DC-link capacitance, third and fifth harmonic currents [1,2] are injected, so that the profile of the input power is modified to be more even over the line cycle. In [3], the limit of the input current distortion is estimated so that the DC-link capacitance can be reduced and the statutory standard can be met.

Another example is a grid-connected inverter. The structure is similar to the one in Figure 4.3(a), except that the input is DC and the output is AC. The front-stage DC/DC converter is used to stabilize the operating point of the panel and extract the maximum power from the panel. The output DC/AC converter converts the DC power into AC power and injects it into the grid. Thus, the power delivered by the panel to the DC link is fairly constant, and the power delivered to the grid is time-varying. The function of the DC-link capacitance is to absorb the power difference between the input and output power. Similarly, the instantaneous DC-link voltage has to meet a requirement in order to ensure the normal operation of the DC/AC converter. As discussed in [4], the increased ripple on the DC link with small DC-link capacitance will affect the design of the input and output filters, thus ensuring a constant input current and sinusoidal output current. In [5] and [6], control methods for regulating the DC-link voltage and modulating the switches in the DC/AC converter are used to ensure that the output current is of low distortion.

4.3 Passive approach

Power electronic systems typically deal with an AC grid. As illustrated in Figure 4.3, the ripple voltage on the DC link is typically double the line frequency. There are two basic techniques to mitigate the ripple on the DC link: passive filtering and ripple cancellation techniques.

4.3.1 Passive filtering techniques

In order to stabilize the DC-link voltage, a straightforward way is to use a passive filter to eliminate the dominant frequency component on the DC link. As shown in Figure 4.4(a), a passive series resonant LC filter can be placed in parallel with the DC link [7]. Its resonant frequency is tuned at the dominant frequency (i.e., double the line frequency) so that a short circuit path for the chosen frequency component is established. However, this will increase the order of the system and the complexity of the system dynamics [8], and thus will increase difficulties for the voltage control design [7].

Depending on the characteristics of the application, the DC-link filter can have different forms. For example, a DC-link inductor and a parallel resonant LC filter are placed in series with the output load. The structure is shown in Figure 4.4(b). The inductor L acts as a current source, which is effectively an energy buffer, to drive the load. In [9], the load is an LED, which is current-driven device. Similar to the series resonant LC filter in Figure 4.4(a), the parallel resonant filter formed by L_r and C_r is used to reduce the low-frequency ripple. It resonates at double the line

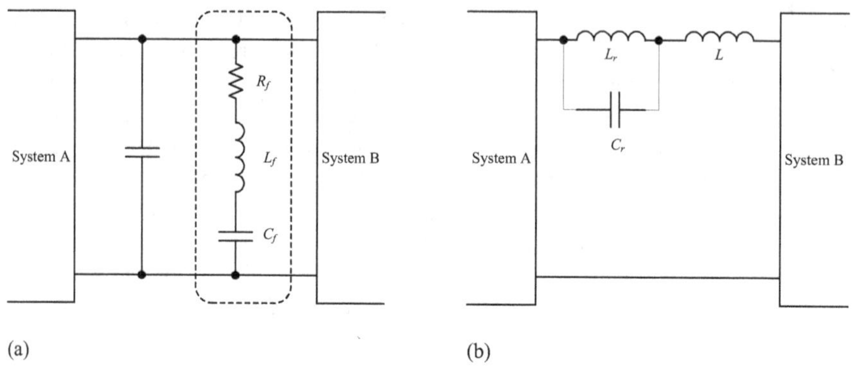

Figure 4.4 Reduction of DC-link capacitance with DC-link filter. (a) With series resonant LC filter. (b) With parallel resonant LC filter.

frequency. Thus, the required value and the physical size of the inductor can be reduced. Such a configuration allows a large ripple at the output of System A without affecting the current driving function for the load.

4.3.2 Ripple cancellation techniques

Another technique is based on using an auxiliary source to compensate the current ripple on the DC link for the load. A low-frequency ripple cancellation method, which is based on using a coupled inductor [10] to eliminate ripple current, is studied in [11] for driving LEDs. The coupled inductor is connected in series with the load. When the input current is higher than the mean value, the capacitor will be charged by the coupled inductor. Conversely, when the input current is lower than the average value, the capacitor will be discharged to the load through the coupled inductor.

In order to improve the input power factor of the system and give a relatively stable voltage to the coupled inductor, a valley-fill circuit [12] is used to provide output to the load.

An extension of the ripple cancellation technique for high-frequency applications is investigated in [13]. The frequency response of the filter can be tuned by using adjusting the operating point of the saturable inductor, L_{sat}. As shown in Figure 4.5(b), it is applied to a buck converter with ripple correlation control so that the overall function of the structure is considered as an active filter using passive components.

The use of passive components has the merit of circuit simplicity. However, as the resonance frequency of the filters is low, e.g., 100 Hz for a 50 Hz grid, the main challenge lies in the physical size and weight of the passive components required. The voltage across the capacitor in the resonant LC filter in Figure 4.4(a) can also be significantly higher than the maximum DC-link voltage. In addition, in order to allow the tolerance of the component values and deviation of the grid frequency, the designed value of the capacitance in the DC link would be conservative, imposing the constraint of increasing the power density of the filter.

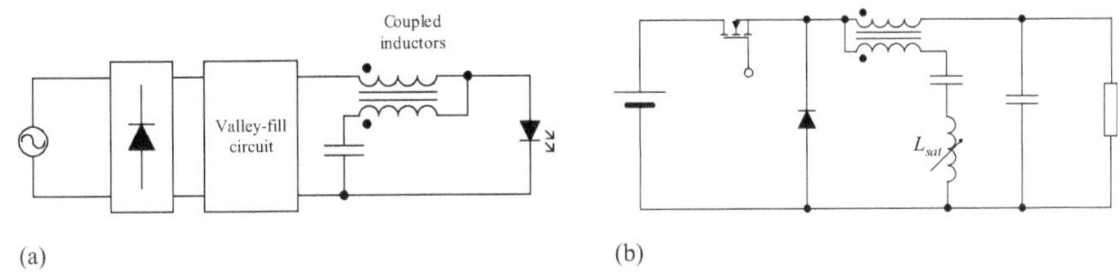

Figure 4.5 Ripple cancellation methods. (a) Application for driving an LED lamp. (b) Extended method with tunable resonant frequency.

4.4 Active approach

The active approach is mainly based on introducing active circuits or control and modulation methods to minimize the ripple power handled by the DC-link capacitor. As shown in Figure 4.1, the objective can be realized by various methods, including a power decoupling circuit that uses a separate energy storage unit to share the energy handled by the DC-link capacitor, an active ripple cancellation device to counteract the capacitor voltage variation, and control and modulation of Systems A and B to control the transfer of energy between them. The methods are briefly described below.

4.4.1 Power decoupling techniques

The concept of power decoupling is based on using a separate energy storage medium to minimize the ripple power handled by the DC-link capacitor. The energy storage medium can be a capacitor or an inductor that allows large variation in its voltage or current level. Different power decoupling circuits have been proposed. They can be classified into several forms, including parallel module, series module, and circuit-specific structures [14].

4.4.1.1 Parallel active filter

The parallel module is connected in parallel with the DC-link capacitor. It is operated as a DC-side parallel active filter [14,15]. The basic structure is shown in Figure 4.6(a). It is composed of a bidirectional power converter and an energy storage medium [16,17]. The energy storage medium can be an inductor or a capacitor, or a part of the whole system. There are many circuit topologies for implementing a bidirectional power converter. In [14], the bi-directional power converter is an inverter for handling ripple energy.

Figure 4.6(b) shows an inductor [18] that is used in traction drives. There are four modes of operation. In the first mode, the two switches, S_1 and S_2, are conducting, and current is fed from the DC link to the inductor. The inductor absorbs extra energy to stabilize the DC link. In the second mode, the switches are off and the two diodes are on. Current is fed into the DC link by the inductor. The inductor releases energy to stabilize the DC link. In the third and fourth modes, the current is

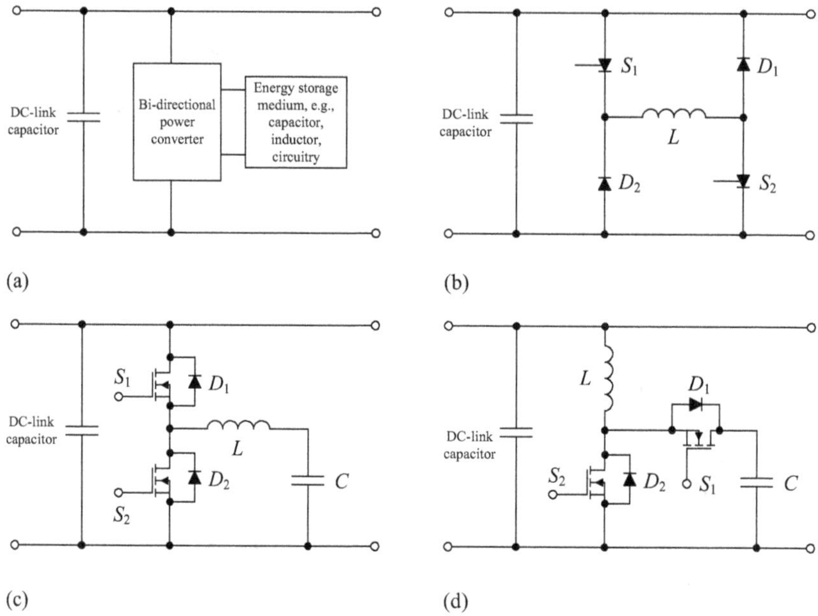

Figure 4.6 Parallel active filter. (a) Basic concept. (b) Use of inductor. (c) Two-quadrant DC/DC converter I. (d) Two-quadrant DC/DC converter II.

in a freewheeling stage wherein either one of the switches is on. For example, when S_1 is on, the inductor current will be freewheeling through D_1. When S_2 is on, the inductor current will be freewheeling through D_2.

In Figure 4.6(c), a two-quadrant DC/DC converter is used for the bidirectional power converter [19,20]. It is implemented by the switches S_1 and S_2 and the inductor L. The capacitor C is the energy storage element. When it is necessary to transfer the ripple energy from the DC link to the capacitor, S_1, D_2, and L form a buck-type converter. When it is necessary to transfer the ripple energy from the capacitor to the DC link, S_2, D_1, and L form a boost-type converter. The concept is used in a battery charger [17] and an AC/DC converter [20].

In Figure 4.6(d), another two-quadrant DC/DC converter is used for a bidirectional power converter [21]. Its operation is similar to the circuit in Figure 4.6(c), but the buck and boost functions are different. When it is necessary to transfer the ripple energy from the DC link to the capacitor, S_2, D_1, and L form a boost-type converter. When it is necessary to transfer the ripple energy from the capacitor to the DC link, S_1, D_2, and L form a buck-type converter. The voltage on the energy storage capacitor in this structure is higher than the voltage of the circuit in Figure 4.6(c). The concept is used in an LED driver [16], battery charger [17], AC/DC converter [21], and DC/AC converter [22].

Some variants have features included such as a soft-switching function into the bidirectional power converter [23]. Instead of using a separate module, the architecture shown in Figure 4.6(a) can be integrated or extended into the system design

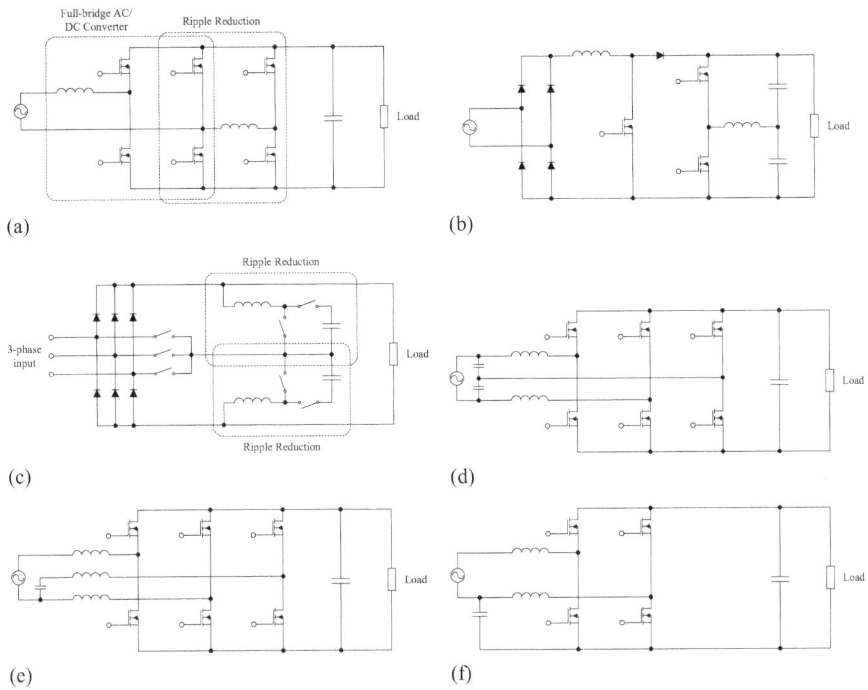

Figure 4.7 *Extension of the power decoupling methods. (a) Full-bridge AC/DC converter and parallel active filter in Figure 4.6(b). (b) Use of DC-link capacitors as the energy storage medium. (c) Two series-connected parallel active filter modules for diode rectifier. (d) Power decoupling on the AC-side capacitors. (e) Power decoupling using an added capacitor on the AC side. (f) With an AC-side decoupling capacitor only.*

so as to simplify the circuit design. For example, the circuit shown in Figure 4.6(b) is combined with the full-bridge AC/DC converter in [24] and [25]. It is also applied to stabilize the DC link in [26]. The structure is shown in Figure 4.7(a). By synchronizing the operation of the full bridge and the ripple reduction circuit with a specialized control method, one of the legs in the full bridge can be shared for both power factor correction and ripple reduction.

In [27], the DC-link capacitor is split into two series-connected capacitors, as shown in Figure 4.7(b). The function of the energy-storage capacitor in Figure 4.6(c) is performed by the DC-link capacitors, where the two capacitors are undergoing charging and discharging in the antiphase. That is, while the upper capacitor is charging, the lower capacitor will be discharging, and vice versa. Thus, the DC-link voltage can be stabilized.

By using two modules in Figure 4.6(c) and connecting them in series on the DC side, a structure for a three-phase diode rectifier is shown in Figure 4.7(c).

The switches are operated so that charging and discharging of the capacitors in the modules are synchronized with the operation of the diode bridge [28].

Instead of using an explicit device on the DC side, the ripple energy can also be handled by another part of the circuit or another component in the system. An example is shown in Figure 7(d) and (e). The ripple energy is used to charge and discharge the input capacitors on the AC side of the AC/DC converter [29,30]. A similar concept is studied for an inverter application in [31]. The structure shown in Figure 4.7(f) has an auxiliary AC-side capacitor added into it [32].

4.4.1.2 Series active filter

Another possibility is to place the energy buffer in series with the DC link. This is illustrated in Figure 4.8. The series module generates an AC voltage counteracting the ripple voltage on the DC-link capacitor [33]. Its DC component is zero. Thus, the series module only handles reactive power. The module is a DC/AC converter with a full-bridge circuit and an output LC filter. A half-bridge circuit can also be used if the output current is unidirectional. The DC side of the full bridge is connected to a DC voltage source, such as a capacitor or an external voltage source. The switches in the full-bridge circuit are pulse-width-modulated in order to make the output voltage constant. Compared with the parallel module, the series module requires low-voltage switching devices, as it only produces low voltage and handles reactive power.

A series energy buffer for a series resonant converter is shown in Figure 4.9, in which the modulation of the switches in it acts as an equivalent blocking voltage source in series with the line voltage for a photovoltaic inverter [34].

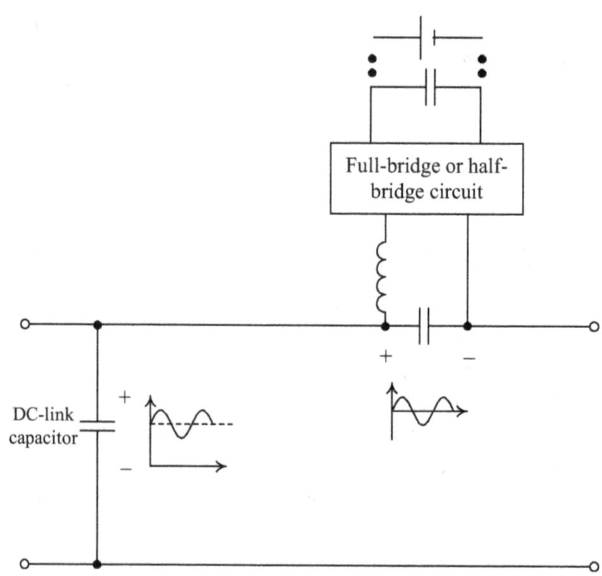

Figure 4.8 Use of series module to compensate voltage ripple

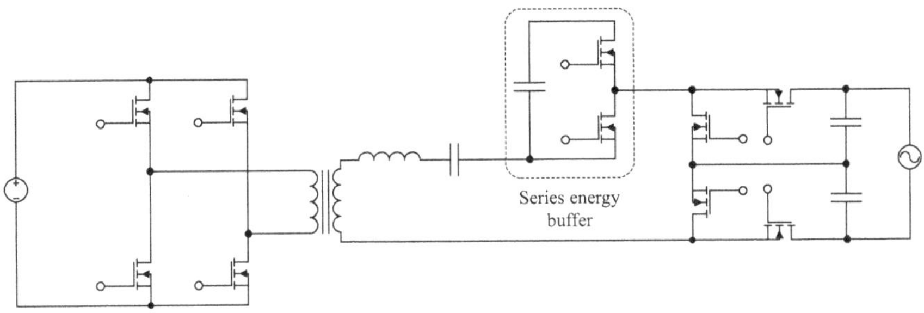

Figure 4.9 Series energy buffer for a resonant converter

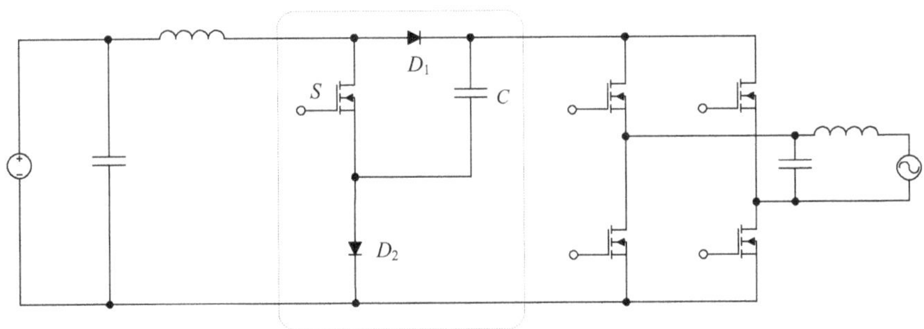

Figure 4.10 Switched-capacitor-based active energy buffer

4.4.1.3 Other architectures

Apart from applying the concept of the parallel- and series-type active filters, there are also DC-link capacitance minimization architectures on the DC link for specific applications. For example, as shown in Figure 4.10, an active energy buffer based on using a switched-capacitor circuit is applied to a current-source photovoltaic inverter. The capacitor, C, is charged in parallel by the front-stage boost converter through the diodes D_1 and D_2 and is discharged through the switch S. A wide variation in the capacitor voltage is allowed, as it is driven by a current source derived from the front-stage boost converter [35].

There are several proposals that utilize the switched-capacitor networks to serve as the energy buffer. A stacked switched-capacitor network that utilizes film capacitors to achieve effective energy densities comparable to electrolytic capacitors is discussed in [36–38]. A multilevel energy buffer and voltage modulator using a switched-capacitor network for reducing the voltage conversion ratio between the input and output is discussed in [39]. To extend the lifetime of the DC-link capacitors, a switchable capacitor network is proposed in [40]. Unnecessary capacitors are disconnected from the circuit, so that the lifetime of the electrolytic capacitors is extended.

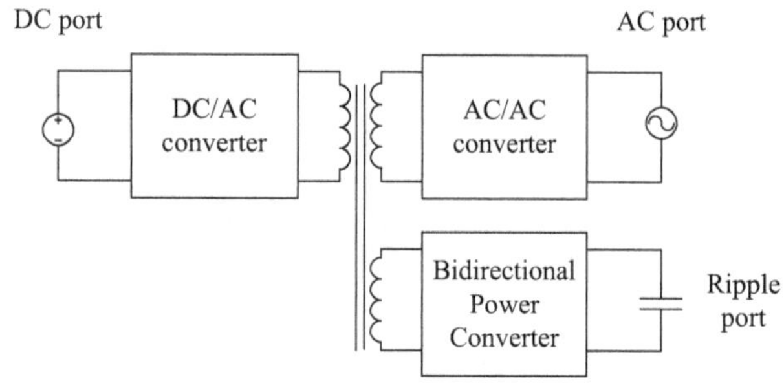

Figure 4.11 Ripple power port

Instead of connecting a bidirectional power converter across the DC link, the concept of using a ripple power port that absorbs the instantaneous power difference between the input and output ports is introduced in [41–44]. It allows designers to decide the capacitor value and voltage. In Figure 4.11, the ripple port is coupled to the system through a transformer.

Some proposals integrate the bidirectional structures into the main power conversion circuits. For example, the flyback transformer has the ripple port or power decoupling circuit connected to the energy storage capacitor for a photovoltaic inverter in [45–51] and LED driver in [52]. In [53], the ripple port is integrated into the inverter with switching network with six switches. In [54], the power decoupling circuit is integrated into the push-pull forward-type inverter. In [55], the decoupling circuit is integrated through the center tap of a transformer.

4.4.2 Ripple cancellation techniques

As shown in Figure 4.12, the ripple voltage on the DC-link capacitor can be cancelled by inserting a voltage source, v_r, in series with the DC-link capacitor. If the magnitude of v_r is the same as the ripple on the DC-link capacitor and the phase of v_r is opposite to the phase of the ripple on the capacitor voltage, the DC-link voltage will have zero ripple. As disclosed in [56] and [57], the added series voltage source can be a linear or switched-mode-type. The attractive feature of this compensation method is that it is unnecessary to redesign the main power conversion stage and controller that have already been used. As shown in [56], the structure can also improve the dynamic response of the system.

4.4.3 Control and modulation techniques

Based on (4.4), the capacitor current, i_C, is equal to the difference between i_A and i_B. Thus, if the two currents can be made the same, $i_C = 0$. Thus, the DC-link capacitor becomes unnecessary. Practically, by reducing i_C, the value of the DC-link capacitor can effectively be reduced. Many control and modulation techniques are based on applying this concept to minimize the DC-link capacitance.

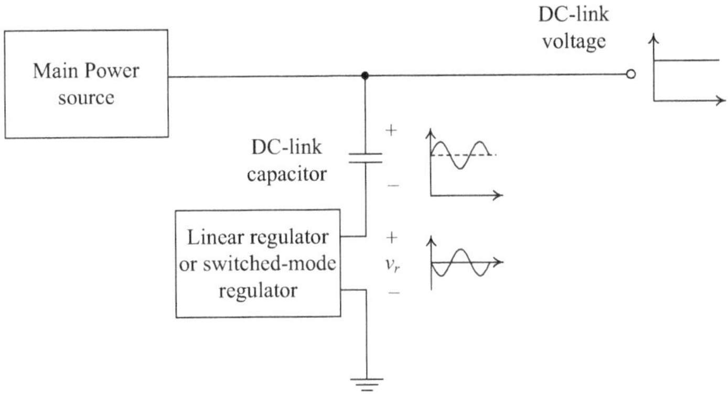

Figure 4.12 Ripple cancellation technique

In [58] and [59], the concept is applied to an AC/DC/AC converter. System A is a three-phase AC/DC converter, while System B is a DC/AC converter for driving an AC motor. A control method is proposed to make the DC-side current output of System A approach the value of the DC-side input current of System B. In [59], the delay in the control method is tackled by introducing a compensation method. In [60], a control method with instantaneous rotor power feedback for a back-to-back PWM converter in a doubly fed induction generator for wind turbine systems to limit the fluctuation range of the DC-link voltage is discussed. A zero-sequence current control between the machine and the AC grid is proposed in [61] to improve the input line current harmonics. In [62], the current injection is obtained from the DC link. In [63], a bidirectional AC/DC power converter and a resonant transfer function to increase gain at the ripple voltage frequency to control the power flow of the converter is presented. In [64], the ripple frequency on the DC-link capacitor current is doubled by using a parallel-connected boost converter to reduce the required value of the DC-link capacitance.

Apart from the control methods, modulation of the switches in the converter can also help to minimize the voltage fluctuation on the DC link. A space vector modulation algorithm to balance the capacitor voltage of a three-level neutral-point-clamped active power filter is proposed in [65]. In [66], the modulation frequency of all power converters connected to the same DC bus is synchronized. The DC-link voltage is stabilized by controlling the phase shifts among the modulation of the converters.

4.4.4 Specialized circuit structures

Apart from using an additional circuit to buffer the power difference between the input and output, some proposals make use of a circuit structure to mitigate the DC-link voltage. Some examples are discussed below.

In [67], three separate AC/DC converters are connected separately to three phases, as shown in Figure 4.13. Their outputs are connected in series. Due to the

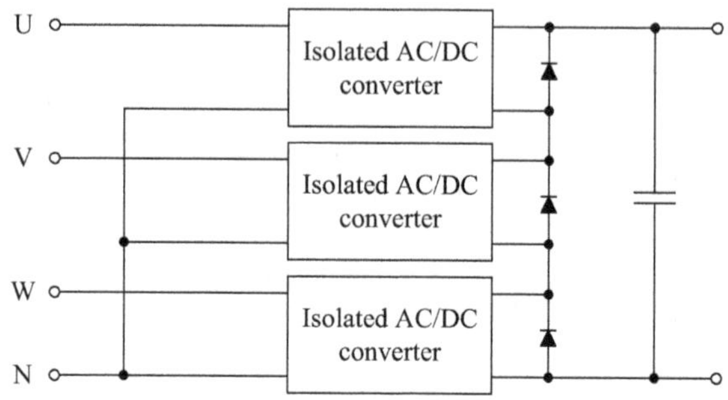

Figure 4.13 Three-phase converter with series-connected output

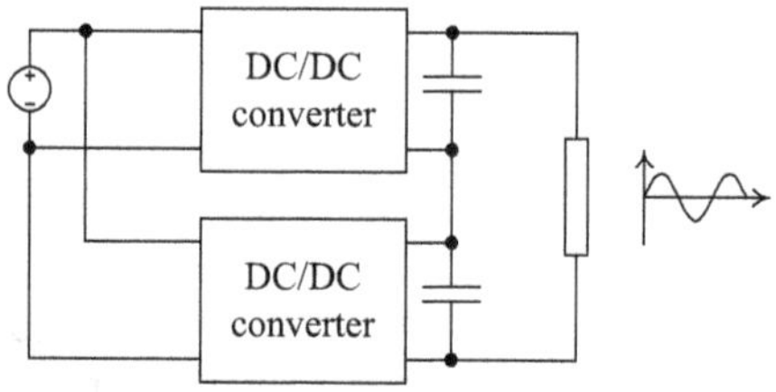

Figure 4.14 Use of differential output of two DC/DC converters

120° phase difference among the three phases, the output can be made to have a very small ripple.

Another example is based on the configuration shown in Figure 4.14, in which the AC output is the differential output of two DC/DC converters. Two boost converters are used in [68], two buck-boost-derived converters are used in [69], and two flyback converters are used in [70]. As discussed in [45], the two output capacitors are used, and the bias voltage induced to these capacitors is modulated so as to reduce the power pulsation on the DC input capacitor and generate sinusoidal voltage at the AC output terminal. The same concept is applicable for an AC/DC converter [71]. A generalized structure on the DC side is discussed in [72]. A stack of capacitors, a set of regulating converters, and power combining converters are configured to provide the buffer for the double-line-frequency on the capacitors.

In [73], a current-fed dual active bridge DC–DC converter is used for a photovoltaic inverter. It employs an advanced phase shift control scheme that allows large ripple voltage on the DC link.

In [74], the DC link that is switched between the output of the direct rectified AC line and DC-link capacitor is proposed. This can effectively reduce the requirement of the DC-link capacitance, but imposes a challenge for designing the input filter.

4.5 Conclusions

Recently, many research investigations have been devoted to improving the reliability of power electronic converter systems by minimizing DC-link capacitance, so that small capacitors of long lifetime can be used to replace electrolytic capacitors. This article gives a review of various techniques, including performance tradeoff, passive approach, and active approach. All of them have their merits and limitations. In particular, the increased circuit complexity might cause efficiency and performance degradation and extra reliability concern. One of the key objectives of this chapter has been to give avenues for further research on a method that can help minimize the use of DC-link capacitance without introducing additional drawbacks.

Acknowledgments

This work was supported by the Research Grants Council of the Hong Kong Special Administrative Region, China, under the Project CityU 112512.

References

[1] L. Gu, X, Ruan, M. Xu, and K. Yao, "Means of eliminating electrolytic capacitor in AC/DC power supplies for LED lightings," *IEEE Transactions on Power Electronics*, vol. 24, no. 5, pp. 1399–1408, 2009.

[2] B. Wang, X. Ruan, K. Yao, and M. Xu, "A method of reducing the peak-to-average ratio of LED current for electrolytic capacitor-less AC–DC drivers," *IEEE Transactions on Power Electronics*, vol. 25, no. 3, pp. 592–601, 2010.

[3] D. Lamar, J. Sebastian, M. Arias, and A. Fernandez, "On the limit of the output capacitor reduction in power-factor correctors by distorting the line input current," *IEEE Transactions on Power Electronics*, vol. 27, no. 3, pp. 1168–1176, 2012.

[4] T. Brekken, N. Bhiwapurkar, M. Rathi, N. Mohan, C. Henze, and L. Moumeh, "Utility-connected power converter for maximizing power transfer from a photovoltaic source while drawing ripple-free current," in *Proceedings of IEEE 33rd Annual Power Electronics Specialists Conference*, 2002, pp. 1518–1522.

[5] A. Kotsopoulos, J. Duarte, and M. Hendrix, "Predictive DC voltage control of single-phase PV inverters with small DC link capacitance," in *Proceedings of IEEE International Symposium on Industrial Electronics*, 2003, pp. 793–797.

[6] S. Khajehoddin, M. Karimi-Ghartemani, P. Jain, and A. Bakhshai, "DC-bus design and control for a single-phase grid-connected renewable converter with a small energy storage component," *IEEE Transactions on Power Electronics*, vol. 28, no. 7, pp. 3245–3254, 2013.

[7] M. Vasiladiotis and A. Rufer, “Dynamic analysis and state feedback voltage control of single-phase active rectifiers with DC-link resonant filters,” *IEEE Transactions on Power Electronics*, vol. 29, no. 10, pp. 5620–5633, 2014.

[8] J. Das, “Passive filters – Potentialities and limitations,” *IEEE Transactions on Industry Applications*, vol. 40, no. 1, pp. 232–241, 2004.

[9] Y. Qin, H. Chung, D. Lin, and S. Hui, “Current source ballast for high power lighting emitting diodes without electrolytic capacitor,” in *Proceedings of 34th Annual Conference on Industrial Electronics*, 2008, pp. 1968–1973.

[10] D. Hamill and P. Krein, “A ‘Zero’ ripple technique applicable to any DC converter,” in *Proceedings of the IEEE 30th Annual Power Electronics Specialists Conference*, 1999, pp. 1165–1171.

[11] S. Hui, S. Li, X. Tao, W. Chen, and W. Ng, “A novel passive offline LED driver with long lifetime,” *IEEE Transactions on Power Electronics*, vol. 25, no. 10, pp. 2665–2672, 2010.

[12] K. K. Sum, “Improved valley-fill passive current shaper,” in *Proceedings of Power Systems World*, 1997, pp. 1–8.

[13] R. Balog and P. Krein, “Automatic tuning of coupled inductor filters,” in *Proceedings of the IEEE 33rd Annual Power Electronics Specialists Conference*, 2002, pp. 591–596.

[14] H. Hu, S. Harb, N. Kutkut, I. Batarseh, and J. Shen, “A review of power decoupling techniques for microinverters with three different decoupling capacitor locations in PV systems,” *IEEE Transactions on Power Electronics*, vol. 28, no. 6, pp. 2711–2726, 2013.

[15] Y. Wang, G. Joos, and H. Jin, “DC-side shunt-active power filter for phase-controlled magnet-load power supplies,” *IEEE Transactions on Power Electronics*, vol. 12, no. 5, pp. 765–771, 1997.

[16] S. Wang, X. Ruan, K. Yao, S. Tan, Y. Yang, and Z. Ye, “A flicker-free electrolytic capacitor-less AC–DC LED driver,” *IEEE Transactions on Power Electronics*, vol. 27, no. 11, pp. 4540–4548, 2012.

[17] S. Dusmez and A. Khaligh, “Generalized technique of compensating low-frequency component of load current with a parallel bidirectional DC/DC converter,” *IEEE Transactions on Power Electronics*, vol. 29, no. 11, pp. 5892–5904, 2014.

[18] T. Larsson and S. Ostlund, “Active DC link filter for two frequency electric locomotives,” in *Proceedings of the International Conference on Electric Railways in a United Europe*, 1995, pp. 97–100.

[19] R. Wang, F. Wang, R. Burgos, D. Boroyevich, K. Rajashekara, and S. Long, “Electrical power system with high-density pulse width modulated (PWM) rectifier,” *US Patent Application 2010/0027304 A1*, February 4, 2010.

[20] R. Wang, F. Wang, D. Boroyevich, *et al.*, “A high power density single-phase PWM rectifier with active ripple energy storage,” *IEEE Transactions on Power Electronics*, vol. 26, no. 5, pp. 1430–1442, 2011.

[21] O. Garcia, M. Martinez-Avial, J. Cobos, J. Uceda, J. Gonzalez, and J. Navas, “Harmonic reducer converter,” *IEEE Transactions on Industrial Electronics*, vol. 50, no. 2, pp. 322–327, 2003.

[22] C. Lee, Y. Chen, L. Chen, and P. Cheng, "Efficiency improvement of a DC/AC converter with the power decoupling capability," in *Proceedings of 27th Annual IEEE Applied Power Electronics Conference and Exposition*, 2012, pp. 1462–1468.

[23] M. Kim, Y. Noh, J. Kim, T. Lee, and C. Won, "A new active power decoupling using bi-directional resonant converter for Flyback-type AC-module system," in *Proceedings of IEEE Vehicle Power and Propulsion Conference*, 2012, pp. 1333–1337.

[24] T. Shimizu, Y. Jin, and G. Kimura, "DC ripple current reduction on a single-phase PWM voltage source rectifier," *IEEE Transactions on Industry Applications*, vol. 36, no. 5, pp. 1419–1429, 2000.

[25] M. Su, P. Pan, X. Long, Y. Sun, and J. Yang, "An active power-decoupling method for single-phase AC–DC converters," *IEEE Transactions on Industrial Informatics*, vol. 10, no. 1, pp. 461–468, 2014.

[26] M. A. Vitorino, R. Wang, M. Beltrao de Rossiter Correa, and D. Boroyevich, "Compensation of DC-link oscillation in single-phase-to-single-phase VSC/CSC and power density," *IEEE Transactions on Industry Applications*, vol. 50, no. 3, pp. 2021–2028, 2014.

[27] Y. Tang, F. Blaabjerg, P. Loh, C. Jin, and P. Wang, "Decoupling of fluctuating power in single-phase systems through a symmetrical half-bridge circuit," *IEEE Transactions on Power Electronics*, vol. 30, no. 4, pp. 1855–1865, 2015.

[28] X. Du, L. Zhou, H. Lu, and H. Tai, "DC link active power filter for three-phase diode rectifier," *IEEE Transactions on Industrial Electronics*, vol. 59, no. 3, pp. 1430–1442, 2012.

[29] T. Shimizu, T. Fujita, G. Kimura, and J. Hirose, "A unity power factor PWM rectifier with DC ripple compensation," *IEEE Transactions on Industrial Electronics*, vol. 44, no. 4, pp. 447–455, 1997.

[30] H. Li, K. Zhang, H. Zhao, S. Fan, and J. Xiong, "Active power decoupling for high-power single-phase PWM rectifiers," *IEEE Transactions on Power Electronics*, vol. 28, no. 3, pp. 1308–1319, 2013.

[31] C. Bush and B. Wang, "Single-phase current source solar inverter with reduced-size DC link," in *Proceedings of IEEE Energy Conversion Congress and Exposition*, 2009, pp. 54–59.

[32] W. Qi, H. Wang, X. Tan, G. Wang, and K. Ngo, "A novel active power decoupling single-phase PWM rectifier topology," in *Proceedings of 29th Annual IEEE Applied Power Electronics Conference and Exposition*, 2014, pp. 89–95.

[33] H. Wang, H. Chung, and W. Liu, "Use of a series voltage compensator for reduction of the DC-link capacitance in a capacitor-supported system," *IEEE Transactions on Power Electronics*, vol. 29, no. 3, pp.1163–1175, 2014.

[34] B. Pierquet and D. Perreault, "A single-phase photovoltaic inverter topology with a series-connected power buffer," *IEEE Transactions on Power Electronics*, vol. 28, no. 10, pp. 4603–4611, 2013.

[35] Y. Ohnuma, K. Orikawa, and J. Itoh, "A single-phase current-source PV inverter with power decoupling capability using an active buffer," *IEEE Transactions on Industry Applications*, vol. 51, no. 1, pp. 531–538, 2015.
[36] M. Chen, K. Afridi, and D. Perreault, "Stacked switched capacitor energy buffer architecture," *IEEE Transactions on Power Electronics*, vol. 28, no. 11, pp. 5183–5195, 2013.
[37] K. Afridi, M. Chen, and D. Perreault, "Enhanced bipolar stacked switched capacitor energy buffers," *IEEE Transactions on Industry Applications*, vol. 50, no. 2, pp. 1141–1149, 2014.
[38] X. Fang, N. Kutkut, J. Shen, and I. Batarseh, "Ultracapacitor shift topologies with high energy utilization and low voltage ripple," in *Proceedings of the 32nd International Telecommunications Energy Conference*, 2010, pp. 1–7.
[39] M. Chen, K. Afridi, and D. Perreault, "A multilevel energy buffer and voltage modulator for grid-interfaced microinverters," *IEEE Transactions on Power Electronics*, vol. 30, no. 3, pp. 1203–1219, 2015.
[40] C. Cojocaru and R. Orr, "Inverter having extended lifetime DC-link capacitor," *US Patent Application 2014/0029308 A1*, January 30, 2014.
[41] Y. Chen and C. Liao, "Three-port Flyback-type single-phase microinverter with active power decoupling circuit," in *Proceedings of the IEEE Energy Conversion Congress and Exposition*, 2011, pp. 501–506.
[42] S. Harb, M. Mirjafari, and R. Balog, "Ripple-port module-integrated inverter for grid-connected PV applications," *IEEE Transactions on Industry Applications*, vol. 49, no. 6, pp. 2692–2698, 2013.
[43] P. Krein, R. Balog, and M. Mirjafari, "Minimum energy and capacitance requirements for single-phase inverters and rectifiers using a ripple port," *IEEE Transactions on Power Electronics*, vol. 27, no. 11, pp.4690–4698, 2012.
[44] P. Krein and R. Balog, "Methods for minimizing double-frequency ripple power in single-phase power conditions," *US Patent 8,004,865*, August 23, 2011.
[45] T. Shimizu, K. Wada, and N. Nakamura, "Flyback-type single-phase utility interactive inverter with power pulsation decoupling on the DC input for an AC photovoltaic module system," *IEEE Transactions on Power Electronics*, vol. 21, no. 5, pp. 1264–1272, 2006.
[46] T. Hirao, T. Shimizu, M. Ishikawa, and K. Yasui, "A modified modulation control of a single-phase inverter with enhanced power decoupling for a photovoltaic AC module," in *Proceedings of the European Conference on Power Electronics and Applications*, 2005, pp. 1–10.
[47] H. Hu, S. Harb, X. Fang, *et al.*, "A three-port flyback for PV microinverter applications with power pulsation decoupling capability," *IEEE Transactions on Power Electronics*, vol. 27, no. 9, pp. 3953–3964, 2012.
[48] H. Hu, S. Harb, N. Kutkut, J. Shen, and I. Batarseh, "A single-stage microinverter without using eletrolytic capacitors," *IEEE Transactions on Power Electronics*, vol. 28, no. 6, pp. 2677–2687, 2013.

[49] S. Kjaer and F. Blaabjerg, "Design optimization of a single phase inverter for photovoltaic applications," in *Proceedings of the IEEE Power Electronics Specialist Conference*, 2003, pp. 1183–1190.

[50] B. Ho and H. Chung, "An integrated inverter with maximum power tracking for grid-connected PV systems," *IEEE Transactions on Power Electronics*, vol. 20, no. 4, pp. 953–962, 2005.

[51] G. Tan, J. Wang, and Y. Ji, "Soft-switching flyback inverter with enhanced power decoupling for photovoltaic applications," *IET Electric Power Applications*, vol. 1, no. 2, pp. 264–274, 2007.

[52] W. Chen and S. Hui, "Elimination of an electrolytic capacitor in AC/DC light-emitting diode (LED) driver with high input power factor and constant output current," *IEEE Transactions on Power Electronics*, vol. 27, no. 3, pp. 1598–1607, 2012.

[53] S. Fan, Y. Xue, and K. Zhang, "A novel active power decoupling method for single-phase photovoltaic or energy storage applications," in *Proceedings of the IEEE Energy Conversion Congress and Exposition*, 2012, pp. 2439–2446.

[54] F. Shinjo, K. Wada, and T. Shimizu, "A single-phase grid-connected inverter with a power decoupling function," in *Proceedings of the IEEE Power Electronics Specialists Conference*, 2007, pp. 1245–1249.

[55] J. Itoh and F. Hayashi, "Ripple current reduction of a fuel cell for a single-phase isolated converter using a DC active filter," *IEEE Transactions on Power Electronics*, vol. 25, no. 3, pp. 550–556, 2010.

[56] H. Chung and W. Yan, "Output compensator for a regulator," *US Patent 8,169,201*, May 1, 2012.

[57] Y. Liu, "Ripple cancellation converter with high power factor," *WO Patent 2013056356A1*.

[58] P. Hammond, "Control method and apparatus to reduce current through DC capacitor link two static converters," *US Patent US 6,762,947 B2*, July 13, 2004.

[59] B. Gu and K. Nam, "A DC-link capacitor minimization method through direct capacitor current control," *IEEE Transactions on Industry Applications*, vol. 42, no. 2, pp. 573–581, 2006.

[60] J. Yao, H. Li, Y. Liao, and Z. Chen, "An improved control strategy of limiting the DC-link voltage fluctuation for a doubly fed induction wind generator," *IEEE Transactions on Power Electronics*, vol. 23, no. 3, pp. 1205–1213, 2008.

[61] H. Yoo and S. Sul, "A novel approach to reduce line harmonic current for a three-phase diode rectifier-fed electrolytic capacitor-less inverter," in *Proceedings of the 24th Annual IEEE Applied Power Electronics Conference and Exposition*, 2009, pp. 1897–1903.

[62] H. Yoo and S. Sul, "A new circuit and control to reduce input harmonic current for a three-phase AC machine drive system having a very small DC-link capacitor," in *Proceedings of the 25th Annual IEEE Applied Power Electronics Conference and Exposition*, 2010, pp. 611–618.

[63] D. Dong, D. Boroyevich, R. Wang, and F. Wang, "Two-stage single phase bi-directional PWM converter with DC link capacitor reduction," *US Patent Application 2012/0257429A1*, October 11, 2012.

[64] J. Ying, Q. Zhang, A. Qiu, T. Liu, X. Guo, and J. Zeng, "DC–DC converter circuits and method for reducing DC bus capacitor current," *US Patent 7,009,852 B2*, March 7, 2006.

[65] H. Zhang, S. Finney, A. Massoud, and B. Williams, "An SVM algorithm to balance the capacitor voltages of the three-level NPC active power filter," *IEEE Transactions on Power Electronics*, vol. 23, no. 6, pp. 2694–2702, 2008.

[66] E. Ganev, W. Warr, and E. Johnson, "Intelligent method for DC bus voltage ripple compensation for power conversion units," *US 7,593,243 B2*, September 22, 2009.

[67] D. Kravitz, "AC to DC power supply having zero frequency harmonic contents in 3-phase power-factor-corrected output ripple," *US Patent 7,839,664 B2*, November 23, 2010.

[68] R. O. Caceres and I. Barbi, "A boost DC–AC converter; analysis, design, and experimentation," *IEEE Transactions on Power Electronics*, vol. 14, no. 1, pp. 134–141, 1999.

[69] N. Vazquez, J. Almazan, J. Alvarez, C. Aguliar, and J. Arau, "Analysis and experimental study of buck, boost, and buck-boost inverters," in *Proceedings of the IEEE Power Electronics Specialists Conference*, vol. 2, 1999, pp. 801–806.

[70] S. B. Kjaer and F. Blaabjerg, "A novel single-stage inverter for AC-module with reduced low-frequency ripple penetration," in *Proceedings of the 10th European Conference on Power Electronics and Applications*, 2003, pp. 2–4.

[71] S. Li, G. Zhu, S. Tan, and S. Hui, "Direct AC/DC rectifier with mitigated low-frequency ripple through waveform control," *IEEE Energy Conversion Congress and Exposition*, 2014, pp. 2691–2697.

[72] S. Lim, D. Otten, and D. Perreault, "Power conversion architecture for grid interface at high switching frequency," in *Proceedings of the 29th Annual IEEE Applied Power Electronics Conference and Exposition*, 2014, pp. 1838–1845.

[73] Y. Shi, L. Liu, H. Li, and Y. Xue, "A single-phase grid-connected PV converter with minimal DC-link capacitor and low-frequency ripple-free maximum power point tracking," in *Proceedings of the IEEE Energy Conversion Congress and Exposition*, 2013, pp. 2385–2390.

[74] B. Bucheru, "DC link voltage chopping method and apparatus," *EP Patent Application 2750274A1*, July 2, 2014.

Chapter 5

Lifetime modeling and prediction of power devices

Mauro Ciappa[1]

Accurate lifetime prediction represents a challenge for the design of complex devices and systems. This work discusses the main techniques for proper data analysis, physical modeling, and efficient calculation.

5.1 Introduction

The need to increase the reliability of high-power devices has been one of the most powerful drivers that have forced engineers to design new products and deploy new materials. New applications (e.g., in the automotive field) have imposed new constraints in terms of space reduction, weight containment, operating temperatures, increasing voltages, and system integration. The steady improvement of the reliability performance of single devices evidenced that traditional reliability assurance programs based simply on the a posteriori assessment of the failure rate were no longer feasible, since they required experimental tests involving many millions of accumulated component-hours to achieve a reasonable statistical significance. Besides the prohibitive costs, the a posteriori assessment strategies mainly developed for mass-produced products like integrated circuits were no longer applicable to the small series that are often encountered in power electronic applications.

Closed-loop built-in reliability programs have been introduced to address the limitations of the traditional approaches. Instead of just assessing statistical values, the reliability performance is actively built into the product already starting from the design and manufacturing phase, and the compliance of the final product with the specifications is continuously monitored. In the case of deviations, devices and processes are redesigned to cope with the initial specs. The main pillars of this strategy are the definitions of the root cause of the observed failure mechanisms, of the relationship among the product specifications and its constituting elements, the variations in the manufacturing process, and the

[1]Integrated Systems Laboratory, Swiss Federal Institute of Technology (ETH), Switzerland

interactions of product materials with the loads and their impact on product reliability.

According to best practices, models for the lifetime of power devices are prepared in parallel (or even prior) to the development phase and are calibrated/validated by targeted experimental procedures. Lifetime models are used during the development phase either to compare different design solutions (or generations) or to confirm that the final product matches the reliability requirements. This step is often integrated with accelerated tests to calibrate and validate the models.

Generally, the lifetime prediction of a system is a bottom-to-top procedure, which starts with the calculation of the reliability of every single component. The lifetime of the whole system is then estimated based on the related reliability block diagram that accounts for possible dependencies and redundancies.

The most popular lifetime prediction models assume that the failure rate of the devices is constant over their operating lifetime and consist of handbooks containing the failure rate models for numerous electronic components (calibrated with historical data), as well as their dependency on a variety of quality and stress (junction temperature, voltage, etc.). The latter are often derived by circuit simulation of the system under representative operating conditions. Nowadays, prediction tools are available, which include an automatic generator of the reliability block diagram, the circuit/thermal simulator, and the database of the failure rate models for the different components. In this case, the calculation is carried out almost automatically once the layout of the circuit is edited.

Due to the relevance of thermo-mechanically activated failure mechanisms, a simple procedure for the prediction of the lifetime of power devices submitted to thermal cycles has been developed based on the law for low-stress fatigue in metals and alloys. This method, which is generally addressed as Coffin–Manson technique, basically starts with a first phase, where the thermal cycles occurring during the operation of the device are counted and binned according to their amplitude and duration. In the second phase, the damage produced by every single cycle is accumulated. The end-of-life of the device is considered to be reached when the accumulated damage exceeds a threshold defined experimentally. Several models have been developed to derive this threshold from the plastic deformation work experienced due to the thermal cycles in bond wires and solder joints.

More recently, lifetime prediction procedures have been introduced that are based on the numerical solutions of the basic physical equations that govern the degradation mechanisms during the aging process of the devices. Normally, due to the lack of accurate experimental data, the exact physical models are simplified and reduced to behavioral models expressed by simple differential equations that are solved either numerically or by finite elements simulators. The main inputs of these calculations are the instantaneous value of the stress parameters (mission profile), which is usually expressed as a time function of the parameter of interest (junction temperature, dissipated power, etc.). In the past [1], a tool has been implemented where different electrical, thermal, and mechanical simulators have been coupled into a single package to predict the lifetime of the IGBTs starting from the mission profile of a locomotive. Since, at present, no comprehensive and coherent simulation tool is available for this scope, this will represent a challenge for the coming years.

Section 5.2 presents a concise review of the main failure mechanisms affecting the power modules with a special focus on IGBT devices. Section 5.3 summarizes the concepts and the best practices for statistically consistent extrapolation of the lifetime from experimental data. Finally, Section 5.4 is devoted to the main lifetime modelling techniques, coverage of the most popular reliability handbooks, and the methods relying on mission profile analysis.

5.2 Failure mechanisms of power modules

The failure mechanisms that affect power device modules can be subdivided into two categories. The first includes the extrinsic mechanisms that result from poorly controlled or poorly designed manufacturing processes. The second category includes the intrinsic failure mechanisms that lead to time-dependent degradation of the performances of the device during its useful lifetime. It is important to notice that in contrast to IC, the lifetime of power modules is usually limited by intrinsic (wear-out) mechanisms, since devices and materials are often operated close to their physical limits. Therefore, one of the main tasks during a prototyping phase is to classify the observed failure mechanisms in order to define appropriate corrective actions for the extrinsic mechanisms and to develop quantitative models for the intrinsic mechanisms, with the scope to design intrinsically reliable devices. The failure mechanisms presented in the following mainly relate to silicon-based IGBT power modules, since at present only poor systematic information is available for devices based on wide bandgap semiconductors.

5.2.1 Package-related mechanisms

Multichip modules for high-power IGBT devices are complex multilayered structures consisting of different materials that have to provide good mechanical stability, good electrical insulation/conduction, and good thermal dissipation properties. The failure mechanisms that are most frequently observed in power modules affect these capabilities and are due to the thermo-mechanical low fatigue stress of some package materials as a consequence of the thermal cycles experienced during operation. The main driving forces of these mechanisms are the mismatch in the coefficient of thermal expansion (CTE) of the different materials, the characteristic length of the layers, and the local temperature swing they are subjected.

5.2.1.1 Bond wire fatigue

Multichip IGBT modules for high-power applications typically include up to 1000 wedge bonds that are connected by ultrasonic bonding either onto aluminum metallization (with a thickness ranging from 3 to 5 μm) or onto strain buffers (e.g., molybdenum plates). Since the majority of them are bonded onto the active area of semiconductor devices (IGBT and freewheeling diodes), they are exposed to almost the full temperature swing imposed by the power dissipation in the silicon and by the ohmic self-heating of the wire itself. Emitter bond wires are usually 300 up to 500 micrometers in diameter. The chemical composition of the wire can differ from manufacturer to manufacturer. However, in all cases, the pure aluminum is hardened by adding some few thousand parts per million of alloying elements such as silicon and magnesium, or nickel for corrosion control. Under normal operating conditions, the current load of a single aluminum bond wire does not exceed 10 A. Thus, the maximum ohmic power dissipation ranges between 100 and 400 mW, depending on the wire diameter. Failures of wire bonds occur predominantly as a result of low cycle fatigue caused either by shear stresses generated between the bond pad and the wire, or by repeated flexure of the wire. Modeling the time-dependent crack propagation within the joint by fracture mechanics is a quite complex issue. Several numerical methods have been attempted, especially by finite element simulations [2] that mainly resulted in quite poor quantitative outcomes.

There is experimental evidence that a crack leading to failure is initiated at the tail of the bond wire and propagates along the internal grain boundaries of the wire material until the bond wire completely lifts off. Due to the more severe exposure to temperature swings in IGBTs, emitter bond wires are more prone to liftoff than are gate connections. The failure of a single wire or of multiple bond wires causes a change either in the contact resistance or in the internal distribution of the current. These effects can be easily monitored by the measurement of V_{CE} [3], which is often used as a degradation monitor. The resulting failure mode strongly depends on the test conditions. In power cycle tests carried out at low voltage, the end-of-life of the device is characterized by the meltdown of the survivors bond wires. On the contrary, during field operation (or high-voltage tests), current crowding effects may occur with consequent triggering of the internal parasitic (latchup).

5.2.1.2 Bond wire heel cracking

Bond wire heel cracking rarely occurs in advanced power modules. However, it can be observed mainly after long endurance tests and especially in cases where the ultrasonic bonding process is not optimized. The failure mechanism is due again to thermomechanical fatigue. In fact, when the wire is subjected to temperature cycles, it expands and contracts undergoing flexure fatigue at the bond heel. In the case of a typical bond wire length of 1 cm and of a temperature swing of 50 °C, the displacement at the top of the loop can be in the 10 μm range producing a change in the bending angle at the heel of about 0.05°. An additional stress is introduced by the fast thermal expansion of the bond wire (e.g., at turn-on) within the silicone gel, which can be considered as a very viscous fluid. Finally, when the driving force of the failure mechanism is ohmic self-heating of the wire, heel cracking can also be observed at the wire terminations bonded on the copper lines of both IGBT chips and freewheeling diodes.

5.2.1.3 Aluminum reconstruction

Although the reconstruction phenomena in the metallization layer of power devices have been reported already in [4], they have been underestimated for long time. Thermal cycles combined with the thermomechanical mismatch between silicon substrate and the overlying aluminum interconnection induce cyclic compressive and tensile stresses in the metallization layer of IGBTs and of freewheeling diodes. Due to the stiffness of the silicon substrate and the high temperature conditions, the softer aluminum metallization layer is strained far beyond its elastic limits. At this point, stress relaxation can occur by diffusion creep, grain boundary sliding, or by plastic deformation through dislocation glide, depending on temperature and stress conditions.

In the case of IGBT devices, the strain rate of the metallization is controlled by the rate of temperature change. Because the typical time constants for thermal transients in IGBTs are in the range of milliseconds, if the devices are operated cyclically at peak junction temperatures above 110 °C, stress relaxation occurs mainly by plastic deformation at the grain boundaries of the aluminum thin film. Depending on the texture of the metallization, this leads either to the extrusion of aluminum grains or to cavitation effects at the grain boundaries, producing hillocks and voids, respectively. The main effect of this mechanism is the reduction of the effective cross-section of aluminum thin film with a consequent increase of the sheet resistance in time. This turns into a steady linear increase of V_{CE} as a function of the number of cycles during power cycling tests. Aluminum reconstruction becomes a reliability hazard in the presence of pre-existing step coverage issues at the emitter contact vias. In this case, thermomechanical and electromigration effects can coalesce, resulting in a complete depletion of the metallization.

5.2.1.4 Solder fatigue and solder delamination

A main failure mechanism of power modules is associated with the thermo-mechanical fatigue of the solder alloys. In this respect, the most critical interface is the solder joint between the ceramic substrate and the base plate, especially in the case of modules using copper base plates [5]. This is due to the fact that this interface combines the worst thermomechanical mismatch of adjacent materials with a quite high temperature swing and large lateral dimensions of the ceramic substrate. The local stress can be reduced by the use of different materials for the base plate such as AlSiC. An additional critical joint is the solder layer between the silicon chip and ceramic substrate. Under low cyclic thermal load, the cracks usually initiate at the fillet of the solder joint and propagate at the interface (chip of substrate) either along the brittle intermetallic phases or precipitates [6]. The crack propagation rate can also be accelerated by the presence in the solder of process-induced voids. Cracks and voids act jointly in worsening the dissipation capability of the package, resulting in the degradation of the thermal resistance with time. The additional temperature increase may have a detrimental effect on other failure mechanisms, in particular bond wire liftoff. The usual indicator of the degradation of the thermal resistance is V_{CE} (as in the case of the bond wire liftoff).

5.2.2 Burn-out failures

Device burnout is a failure mode, which is very frequently observed either as the final act of wear-out, or as a consequence of a robustness issue. Burnout is often associated with a short-circuit condition, where a large current flows through the device (or through a portion of it), while it is supporting the full line voltage. Sustaining a short circuit over a long time interval inevitably leads to thermal runaway and finally to fast destruction of the device. In fact, since IGBTs do not require any *dI/dt* snubbering, the device itself limits the current increase rate. Therefore, after the failure, the current may increase at a rate up to several kiloamps per microsecond, with current peaks in the 100 kA range and decay times down to some few microseconds. In this case, the main part of the stored capacitive energy is released in few hundreds of nanoseconds, reaching a peak power up to 100 MW. The capacitive energy is dissipated by the ohmic components of the circuit, i.e., mainly by the bond wires and by the silicon chip. As a consequence of the adiabatic heating process, the bond wires evaporate by producing a preferential conductive path for arching through the module. The resulting shock wave rapidly propagates through the silicone gel leading to the catastrophic destruction of the device. Advanced power modules have been expressly designed to minimize the secondary effects of an explosion, in particular, fragmentation leading to splinter formation.

There are many systems wherein environmental and wearout-related causes can turn into a short circuit condition. Among these are the operation of the device outside a safe operating area, gate unit malfunction, inhomogeneous current sharing [7], overheating due to the degradation of the thermal resistance, dielectric breakdown, latchup, and cosmic ray radiation. Latchup and cosmic rays are mainly failure mechanisms that are related to the ability of a certain device design to survive stresses out-of-specification. Therefore, strictly speaking, there are mainly robustness-related issues rather than reliability concerns. Latchup (static or dynamic) is a complex phenomenon that leads to the triggering of internal parasitic structures and that results in a sudden collapse of the collector-to-emitter voltage. Once this failure mechanism is activated, the device cannot be longer controlled through the gate. The failure mode associated with latchup is often a catastrophic low-ohmic short circuit of the collector, emitter, and gate.

In high-voltage devices, catastrophic single burn-out events can also be triggered by the interaction of the semiconductor with the neutron component of terrestrial cosmic rays. Neutrons impinging on devices (IGBTs and diodes) biased in the blocking state may produce recoil nuclei that generate an ionization track in the active area. If this happens in a high-field region, carrier multiplication occurs and is followed by a local self-sustaining filamentary discharge with consequent local melting of the semiconductor. Such a robustness-related issue can only be solved either by proper selection (design) of the device or by derating the DC voltage in operation. Due to the almost stochastic nature of the fluence of terrestrial cosmic rays, the associated failure rate can be accurately modeled by an exponential distribution, i.e., with a constant failure rate, λ_{cosmic}. A multiplicative phenomenological model for λ_{cosmic} is currently used [8], which accounts for the operating DC-voltage of the device, temperature, and altitude. All three factors depend exponentially on the respective parameters. It should be noted that λ_{cosmic} exhibits a very strong dependence on the DC voltage such that failure rates on the order of 10,000 FIT (or even more) are not uncommon if the DC voltage is not properly derated.

5.3 Lifetime metrology

In order to estimate the failure rate from experimental data (either from field data or from accelerated lifetime tests), the empiric distribution has to be computed on the base of the observed failure-free-times, t_i. Assuming that n failures with the related failure-free-times $t_1, \ldots, t_n$ have been observed, the first step is to rank these values such that $t(1) \leq t(2) \leq \ldots \leq t(n)$. In the next phase, the empiric distribution is built as a step function, i.e., $F_{emp}(t)$: $t(i) - F_{emp}(t(i)) = i/n$. In the following, the empiric distribution has to be associated with an analytical distribution in order to extract the distribution parameters. This can be done by fitting the empiric distribution by the analytical distribution to be tested according to the maximum likelihood criterion. The goodness of fit can be tested in different ways, i.e., either by using the Kolmogorov–Smirnov test or the Chi-square (χ^2) test [9]. Once the more accurate approximating distribution is identified, the related parameters can be extracted to compute the failure rate, $\lambda(t)$. The usual parameter, which can also be determined from such a plot, is the time required to reach a given percentile of the distribution, as for example the median (t_{50}) that represents the time for getting 50% of the device population failed. Nowadays, this task is performed automatically by dedicated software tools, or by graphic techniques, where the empiric data appear as a straight line in dedicated probability charts associated with each distribution.

The following considerations are focused on the exponential and on the Weibull distributions that are the most frequent lifetime models in use for power devices.

5.3.1 *Lifetime availability*

The lifetime t of a non-repairable device is a random variable, which is defined as the timespan between the initial operation and failure, where the failure condition is clearly identified by a failure criterion. Estimates either of the median (t_{50}), or of the empirical mean ($E(\tau)$), related to a mechanism-specific distribution are measured during lifetime tests (often under accelerated conditions). For engineering purposes, an even more important role is played by the failure rate $\lambda(t)$, which is extracted from the previous distribution, and that is defined as the probability that a device will fail in the next operation hour, provided that it did not fail before. The correct measurement unit for $\lambda(t)$ is the FIT (1 failure in 10^9 operation hours) and not the parts per million, as improperly assumed. This value is very important for the designer, since it univocally expresses the number of failures that can be expected after a given operation time of the devices, and given a specified availability of a repairable system, it is also used to design efficient preventive maintenance strategies.

$E(\tau)$ (or $MTTF$) can also be calculated analytically from the reliability function $R(t)$ as

$$MTTF = E[\tau] = \int_0^\infty R(t)\, dt$$

where $R(t)$ is defined either based on $F(t)$

$$R(t) = 1 - F(t)$$

or on $\lambda(t)$

$$R(t) = \exp\left(-\int_0^t \lambda(x)dx\right) \text{ with } R(0) = 1.$$

For repairable systems, a very important characteristic that is related to the MTTF is the point availability, PA, which is defined as the probability that a system performs its required function under given conditions at a stated instant of time. Indeed, the availability also accounts for factors that are related to the maintainability of a system, as for instance the mean time to repair (MTTR), once the system fails. Exact transient calculations of the availability require the use of stochastic methods. However, in the simple case of the continuous operation of a system, the PA rapidly converges [9] to a steady-state value that is expressed as

$$PA = \frac{MTTF}{MTTF + MTTR}$$

Typical values of PA for reliable systems are in the 0.99999 range.

5.3.2 Exponential distribution

The exponential distribution, $F(t) = 1 - \exp(-\lambda t)$, is typically used if a constant failure rate λ is expected to occur over the whole operating time of the device, i.e., when the stochastic process is memory-less. This means that the failure rate of a device does not depend on how long it has been operated in the past. From a modeling point of view, it reasonably represents the failure regime in the flat region of the bathtub, which is characterized by random failures. This is a realistic model, in the case of mature devices that have been properly screened to eliminate infant failures that have not yet reached the end-of-life region. In power modules, the exponential distribution can be used to model semiconductor-related failure mechanisms only. Under certain circumstances, it can also be used to model robustness-related issues like cosmic rays.

While wearout-related failure mechanisms can be attacked by adequate design rules, random failures are not necessarily related to a given failure mechanism. In fact, they reflect the random character of both the occurrence of physical processes (failure mechanisms) and the variation of manufacturing processes. Random failures play a relevant role in defining the survival probability of (non-reparable) mature systems with a very long operating life, as can be the case for inverters in railway traction applications. For instance, the survival probability after 30 years of operation of a unit consisting of six modules each having a constant failure rate of 100 FIT is close to 0.85. In converse, for a constant failure rate of 400 FIT, the survival probability is close to 0.5 only. This simple, but realistic, example evidences the requirement of an accurate estimate of λ with sufficient statistical significance. For this scope, it is recommended to use a symmetric interval estimate for the constant failure rate, which is defined by a lower (λ_l) and upper (λ_u) limit, in conjunction with a confidence level β (e.g., $\beta = 0.1$, *conf. level* [%] = 100 (1 -2β) = 80%). It can be shown by the use of the Poisson statistics [9] that in the case of k failures observed during the cumulated operating time T, the interval limits are given by

$$\lambda_l = \frac{\chi^2(2k,\beta)}{2T}, \ \lambda_u = \frac{\chi^2(2(k+1), 1-\beta)}{2T}$$

where $\chi^2(x,y)$ is the chi-square distribution with the arguments x and y that can be easily obtained from tabulated values. If no failure is observed during the accumulated operating time T, the interval limits are given by

$$\lambda_l = 0 \text{ and } \lambda_u = \frac{\ln(1/\beta)}{T}$$

When the devices are subjected to wearout, the exponential model with a constant failure rate can be a momentary approximation of a time-dependent failure rate only. Therefore, the use of the mean time to failure, $MTTF = 1/\lambda$, for the calculation of the lifetime of an item can lead to wrong results. For example, an instantaneous failure rate of 10 FIT does not necessarily imply a lifetime of 11,415 years, because the device can fail much earlier due to the intervention of wearout mechanisms, which accelerates with time. Finally, it should be noted that in the case of the exponential distribution, about 63% of the original devices are already failed at $t = MTTF$.

5.3.3 *Weibull distribution*

There is enough experimental evidence that the failure-free times measured and associated with wearout mechanisms can be modeled (at least piecewise) by a Weibull distribution of the form

$$F(t) = 1 - \exp\left(-(\lambda t)^{\beta}\right)$$

where λ is the scale and β is the shape factor, respectively. Therefore, a Weibull distribution exhibiting a failure rate that increases in time is the best-suited model for the wearout (package-related) failure mechanisms observed in power modules. On the contrary, semiconductor-related failure mechanisms are better described through the constant failure rate of an exponential distribution. Furthermore, when repeating a power cycling experiment at a constant ΔT, the observed number of cycles to failure due to thermomechanically activated failure mechanisms (e.g., bond wire liftoff or substrate delamination) is also Weibull distributed.

The related time-dependent failure rate associated with a Weibull distribution is given by

$$\lambda(t) = \beta\lambda(\lambda t)^{\beta-1}$$

and it is monotonically increasing if $\beta > 1$.

In many cases, the experimental data cannot be represented by just a single distribution.

Instead, they require a combination of two or more distributions to be used. This can be the case of the co-existence of two competing failure mechanisms (e.g., bond wire liftoff and substrate delamination during power cycles), which can lead to the termination of the device. If the Weibull parameters associated with both failure mechanisms are λ_1, β_1 and λ_2, β_2, respectively; it can be demonstrated [10] that the failure rate is given by

$$\lambda(t) = \beta_1\lambda_1(\lambda_1 t)^{\beta_1-1} + \beta_2\lambda_2(\lambda_2 t)^{\beta_2-1}$$

For systems using non-repairable devices with a lifetime that is much shorter than the operating time of the systems, the mean time between failures can be estimated by the running average method. Although in some cases the resulting MTBF converges toward a constant value, it does not mean that the failure rate of the devices is also constant in time. In fact, if the failure-free times of the failing devices are distributed after a Weibull distribution, it can be easily demonstrated that the asymptotic limit obtained through the running average algorithm is the mean $E(\tau)$ of the associated Weibull distribution, i.e.,

$$E(\tau) = \frac{\Gamma(1 + 1/\beta)}{\lambda}$$

where Γ is the gamma function.

5.3.4 Redundancy

Although it is very expensive in terms of cost, space, and weight, a possible solution to improve the reliability (availability) of a system is redundancy. Redundancy is used in repairable systems to avoid the interruption of system operation during repair or for short missions of non-repairable units that require high-reliability levels (e.g., spacecraft). In a hot redundancy, all redundant elements are subjected to the same load from the beginning of the system operation. On the contrary, in a cold redundancy, the redundant element is not loaded until the element under operation fails. In this case, for the selection of different elements, an additional switching unit is required. The starting point for the quantification of the improvement due to redundancy is the description of the system in terms of a reliability block diagram, i.e., an event diagram that specifies which elements of the system are necessary for the fulfillment of the required function and which can fail without affecting it. Typical reliability block diagrams are series/parallel structures, where redundant items appear as paralleled elements. The reliability function, R_{system}, of a system, whose reliability block diagram consists of a series structure (no redundancy) of n blocks each associated with a reliability function R_i is

$$R_{system} = \prod_{i=1}^{n} R_i$$

The simplest case of redundancy is the *k-out-of-n* hot redundancy, where k out of n available (identical and with reliability function R) elements are necessary to fulfill the required function of the system. In this case, all elements appear in parallel in the reliability block diagram, and the reliability function R_{system} of the system is calculated as

$$R_{system} = \sum_{i=k}^{n} \binom{n}{i} R^i (1 - R)^{n-i}$$

Besides the analytical approach, the reliability of the most complicated topologies can be easily calculated by Monte Carlo simulation.

It is interesting to note that in a non-repairable system consisting of a single element with a constant failure rate λ (MTTF = 1/ λ), a duplication of the hardware (*1-out-of-2* redundancy) just produces an increase of the *MTTF* by 50% ($MTTF_{1\text{-out-}2}$ = 3/2 λ). Similarly, the gain for a 1-out-of-3 redundancy is less than a factor of two ($MTTF_{1\text{-out-}3}$ = 11/6 λ).

A typical example of redundancy at the device level is the parallel-connected emitter bond wires in power modules. An additional example of redundancy at the system level is press-packaged IGBT modules used in stacks. In case of burn-out, such modules are often designed to fail as low-ohmic short circuits. If the stack is properly designed, the failure of a single device does not impair the blocking capability of the unit. Therefore, it acts as a redundant device. This does not apply to bond-wired power modules, since in this case the main failure mode is an open circuit.

5.4 Lifetime modeling and design of components

The end-of-life period of complex multi-chip modules is often defined by thermomechanical-related failure mechanisms. The lifetime for these wearout mechanisms is normally estimated based on deterministic models that are calibrated by accelerated tests. In a subsequent step, these estimates are extrapolated to account for the real operating conditions of the system.

5.4.1 Lifetime prediction based on mission profiles

The mission profile of a device (or system) is the specific task that must be fulfilled during a stated time and under specified conditions.

One example of a particularly challenging mission profile is power converters used in airborne applications to drive electrical actuators in the close vicinity of the jet engine. In this case, the power devices are asked to survive at the same time under static environment temperatures of about 200 °C during the flight and extreme thermal cycles between –55 °C and 200 °C with a rate of 10 °C/min during landing and take-off. The requested lifetime of the equipment is typically 50,000 hours corresponding to about 500 landing/takeoff cycles [11].

Power devices used in electrical hybrid vehicles are normally specified for a typical lifetime of 10,000–15,000 hours, which corresponds to several millions of power cycles with different durations and amplitudes. In this case, the environment temperature strongly depends on the design of the cooling circuitry of the vehicle. If the power converter shares the same cooling circuitry with the thermal engine, a baseline temperature of 90 °C up to 120 °C can be expected. This leads to peak junction temperatures of the switching devices in the 200 °C range. In vehicles that use a dedicated cooling circuitry for the converter with a typical baseline temperature of 60 °C, the working conditions of the devices are less severe. Car manufacturers often assess the reliability performance of their hybrid vehicles using standardized driving cycles originally intended for the measurement of exhaust emissions and fuel consumption (e.g., NMVEG, Artemis [11]). Reliability-specific standard mission profiles are under development.

The mission profiles of railway systems strongly depend on their main applications. Nevertheless, railway operators usually specify the failure rate of single power devices not to exceed 100 FIT over almost 30 years (i.e., the whole useful life of the system) without preventive maintenance for semiconductors.

5.4.2 Modeling the lifetime of systems with constant failure rate

This procedure applies to large populations of statistically identical and independent items. It assumes that all items (i.e., elementary electronic devices, modules, and electronic subassemblies) are operated in the phase where they exhibit a failure rate that does not depend on time such that the failure flow can be accurately described by a Poisson process. Early failures are preliminarily eliminated by proper screening strategies and the wearout phase is not yet reached. The failure rate of every single device is listed in a database (or handbook) that has been compiled starting from the results of accelerated tests or from field data in the case that environmental and operating conditions are sufficiently known. Besides the list of the reference values, λ_{ref}, of the failure rate for each component, a typical handbook includes prescriptions to account for the environment and product usage conditions (e.g., operating temperature, voltage, vibration, humidity, application, duty cycle), as well as for the quality level, maintenance, design, and maturity of the manufacturing processes. Some recent databases also take into account the effects of overstresses (electrical, mechanical, thermal). Not considered are usually the impact of malfunctions due to software failures, lacking preventive maintenance, and usage outside the specifications.

Failure rate handbooks are issued by several international organizations like the US Department of Defense (MIL-HDBK-217F Notice 2, [12]), International Electrotechnical Commission (IEC TR-62380 [13]), Union Technique de l'Electricité (UTE C 80-810 [14]), FIDES (FIDES DGA-DM/STTC/CO/477A [15], Reliability Information Analysis Center (RIAC HDBK-217Plus [16]), and Telcordia (SR-332 [17]). Unfortunately, not all handbooks report reliability data for active and passive power devices. Furthermore, some traditional handbooks (e.g., MIL-HDBK-217F) make use of partially obsolete data and do not account for the manufacturing processes. This usually turns into quite pessimistic estimates of the lifetime of systems.

In general, all procedures consist of a failure rate prediction at the device level and at the system level. The most simple prediction at the system level (in the absence of redundancy) is the so-called part count method, which assumes that the total failure rate of a system, λ_{system}, is the sum of the failure rates of all components, λ_i

$$\lambda_{system} = \sum_i \lambda_i$$

Devices operated under multiple environmental conditions will have calculations split into the portion of operation time spent in each environment. The part count method is applied, for instance, by the MIL-HDBK-217F and the UTE C 80-810. Some other handbooks, like HDBK-217Plus, multiply the failure rate calculated by the part count method λ_{system} by a correction factor that is computed considering purely system-related issues. These process grading factors consist of a dozen parameters that account for the quality of the devices, observed infant mortality, environmental stress, reliability growth process, etc.

Prediction at the device level does not account individually for the specific failure mechanisms affecting a given device but makes use of simplified multiplicative or multiplicative-additive models. In the simplest multiplicative model, such as those used in MIL-HDBK-217F, the reference failure rate value λ_{ref} for a device that is listed in the handbook is multiplied by parameters that account for the junction temperature (π_T), the application (π_A), the quality (π_Q), and the environment (π_E), yielding

$$\lambda_i = \lambda_{ref,i} \quad \pi_{T,i} \quad \pi_{A,i} \quad \pi_{Q,i} \quad \pi_{E,i}$$

Among these factors, π_T describes the dependency of the failure rate of a device on its operating temperature. More precisely, π_T expresses the acceleration of the degradation mechanisms that occurs if the operating temperature, T_1, of the item under investigation exceeds the temperature T_2, at which the reference failure rate λ_{ref} has been characterized. In the case of a dominant failure mechanism and of a quasi-static temperature profile, π_T is represented by the Arrhenius relationship:

$$\pi_T = \exp\left(-\frac{E_A}{k}\left(\frac{1}{T_2} - \frac{1}{T_1}\right)\right)$$

where E_A is the activation energy of the failure mechanisms (0.3–0.7 eV for silicon devices), and k is the Boltzmann constant ($k = 8.6\ 10^{-5}$ eV/K). In the case of active thermal cycles, more advanced models (e.g., UTE C 80-810) also account for the limited number of cycles, N_f, that the device can withstand as a function of the junction temperature swing, ΔT_j

$$N_f = 10^7 \exp(-0.05\ \Delta T_j)$$

Handbooks like FIDES can also handle more complex application profiles, including combined environmental stresses. The implicit model for the failure rate is a product of a term that includes physical and technological factors ($\lambda_{physical}$), a term representing the quality and technical control of the item (Π_{PM}), and a term that defines the quality and technical control over the development, manufacturing, and usage processes for the product containing the item under investigation ($\Pi_{Process}$):

$$\lambda = \lambda_{physical} \quad \Pi_{PM} \quad \Pi_{Process}$$

The simplicity of this model is misleading. In fact, even for a simple MOS transistor operated in a pulsed mode, the term $\lambda_{physical}$alone includes more that a dozen different parameters. Among these, the acceleration factor due to thermal cycling and the dependency of the failure rate on the humidity, π_{RH}, expressed by a generalized Eyring model

$$\pi_{RH} = \left(\frac{RH_2}{RH_1}\right)^n \exp\left(-\frac{E_A}{k}\left(\frac{1}{T_2} - \frac{1}{T_1}\right)\right)$$

where RH_1, RH_2, T_1, and T_2 are the relative humidity and the temperature at high and low power dissipation conditions, respectively.

For indicative purposes, the MIL-HDBK 217F Notice 2, the UTE C 80-810, and the FIDES handbooks have been used to predict the failure rate of a single chip IGBT (rated for 600 V) in a D2PAK package submitted to a simplified aircraft mission profile. The IGBT is operated at V_{CE} = 270 V (V_{GE} = 15 V) with 50 W dissipation, twice a day, each time for eight hours (flight duration) followed by four hours of parking, during 350 days per year. The device is mounted close to the engine such that its junction temperature is assumed to be 10 °C and 55 °C during the flight and at ground, respectively. The MIL-HDBK 217F predicts a failure rate of 7500 FIT, the UTE C 80-810 of 20 FIT, and the FIDES handbook of 18 FIT. For all three methods, λ_{ref} related to the semiconductor just represents 1 up to 10% of the total failure rate, i.e. 12 FIT, 2 FIT, and 0.3 FIT, respectively. In all cases, the main contribution to the total failure rate is given to extrinsic causes, e.g., to factors related either to the package or to the operating environment. In particular, the main penalty to the value predicted by the MIL-HDBK 217F is due to the environmental factor for airborne inhabited cargo (π_E = 20) and to the application factor (π_A = 8). In the case of the UTE C 80-810 Handbook, almost 99% of the estimated failure rate is due to package-related effects, in particular to thermal cycles. Finally, in the prediction based on the FIDES handbook, $\lambda_{physical}$ accounts for about 65% of the total failure rate.

5.4.3 Modeling the lifetime of systems submitted to low cycle fatigue

In general, the lifetime related to low cycle fatigue of bond wires and solder joints is calculated in terms of the number of thermal cycles to failure, N_f. Once the failure criterion and the mission profile for the item under investigation are defined,

N_f can be transformed into hours and the data can be conveniently processed to obtain the statistical significance of the calculated value.

All calculation methods are based on the principle that every single thermal cycle in the mission profile introduces damage to the joint and that the joint failure occurs, once a given total threshold value of the damage is reached. The usual approaches mainly differ in the way that the damage and the damage accumulation are calculated.

The most popular method for the calculation of the damage produced by a single cycle is the so-called Coffin–Manson procedure, which requires analyzing the mission profile in order to classify all occurring thermal cycles in terms of amplitude and duration. Once the damage of each single cycle is calculated, the total damage is obtained by linear accumulation.

Alternative prediction procedures calculate the damage accumulated by the joint by solving the fundamental equations of thermomechanics. In some academic approaches, the differential equations are solved numerically in three dimensions by finite element simulation tools [2]. This method is very time-consuming; the joints are operated in a very deep plastic regime. In addition, it delivers relatively inaccurate results, since the initial stress conditions and the materials parameters are often completely unknown.

A popular alternative [18] to finite element calculation is the use of a simplified behavioral model consisting of a one-dimensional bimetallic system, where the constitutive equations of the interface material (solder alloy, ultrasonic bond) are solved numerically to obtain the deformation work experienced by the joint.

5.4.3.1 Coffin–Manson-like approach

During thermal cycling of power devices, some materials (solder alloys, bond wires) are repeatedly operated in a very deep plastic regime. This results in failures due to low cycle fatigue.

Under these circumstances, the number of cycles to failure, N_f, for a constant ΔT is calculated starting from the maximum strain experienced by the joint, in a similar way as it is originally done for structural materials according to the Coffin–Manson law.

The derivation of the modified Coffin–Manson relationship is easily illustrated in the case of a simple example of a bimetallic system consisting of a plate of ceramic (alumina) soldered on top of a copper substrate (coefficients of thermal expansion: α_{Cu} and α_{Alu}, respectively). By assuming that the characteristic length of the plate is L and that the multilayer is submitted to a temperature swing ΔT, the total strain, ε_{tot}, in the main direction can be estimated as

$$\varepsilon_{tot} \approx L(\alpha_{Cu} - \alpha_{Alu})\Delta T$$

Since the size of the plates is much larger than the thickness, t_{solder}, of the solder, for a ΔT in the usual range, ε_{tot} has the same order t_{solder} and is much larger than the pure elastic strain

$$\varepsilon_{tot} = \varepsilon_{elastic} + \varepsilon_{plastic} \approx \varepsilon_{plastic}$$

In the case of thermal cycle with amplitude ΔT, the original Coffin–Manson relationship assumes that N_f is expressed as a power law of the total plastic strain experienced by the joint

$$N_f \approx \varepsilon_{plastic}^{-n}$$

In general, N_f is associated with the fth-quantile of the distribution, and n is an exponent that depends on the materials. Accounting for the dependency of ε_{tot} on the temperature yields

$$N_f \approx a(\Delta T)^{-n}$$

where a is a proportionality constant, which is defined experimentally in conjunction with the exponent n.

This model is used to predict N_f in a case where there are different failure mechanisms such as bond wire liftoff, solder delamination, and the bond wire heel cracking. Different failure mechanisms assume different failure criteria. As an example, the failure criterion for delamination is usually defined on the basis of the degradation of the thermal resistance R_{th} of the module (e.g., $\Delta R_{th} = +10\%$). Similarly, the failure criterion for the bond wire lift-off mechanisms is generally assumed to be a given increase of V_{CE} measured.

For the bond wire lift-off, it is common to use the model in its simple form [10], i.e.,

$$N_f = a\left(\Delta T_j\right)^{-n}$$

where ΔT_j is the amplitude of the junction temperature cycle, and the parameters a and n, are usually provided by the module manufacturer in the data sheet.

A similar Coffin–Manson model can be used to predict the number of cycles to failure for the delamination of solder joints at critical interfaces (e.g., DCB-base plate, chip-DCB),

$$N_f = 0,5\left(\frac{L\,\Delta\alpha\,\Delta T_{sub}}{\gamma\,x}\right)^{1/C}$$

where L is the characteristic size of the plate to be considered, $\Delta\alpha$ is the thermomechanical mismatch of both plate materials, Δ_{sub} is the amplitude of the temperature cycle at the interface (e.g., DCB-base plate, chip-DCB), and x and γ are the thickness and the ductility factor of the solder layer, respectively. The exponent C is fitted on experimental data.

Finally, the Schafft model [10] predicts the number of cycles to failure related to heel cracking of the wire bonds, based on

$$N_f = A\left(\frac{r}{\rho_0}\left(\frac{ar\cos(\cos\psi_0(1-\Delta\alpha\,\Delta T))}{\psi_0}-1\right)\right)^n$$

where Δ is the swing of the temperature cycle, A and n are the fit parameters, $\Delta\alpha$ is the thermomechanical mismatch between aluminum and DCB material, r is the radius of the bond wire, ρ_0 is the bending radius at the heel, and ψ_0 is the angle between chip plane and bond wire.

These simple models for N_f depend on Δ Such an approximation is correct, as soon the maximum cycle temperature, T_{max}, is not too close to the melting temperature of the solder alloy. A correction factor is necessary in the case of the solder alloys that exhibit a low melting temperature T_{MP}. An example [10] of such a correction is

$$N_f(\Delta T, T_{\max}) = N_f(\Delta T) \quad \frac{1}{\exp(m(T_{\max} - T_{MP})) + 1}$$

where m is a fitting parameter to be defined experimentally. Alternative expressions are based on the average cycle temperature (T_{aver}) and on an Arrhenius-like term

$$N_f(\Delta T, T_{aver}) = N_f(\Delta T) \quad \exp\left(\frac{C}{k\,T_{aver}}\right)$$

where C is again a fitting factor.

Additional parameters that are relevant for the calculation of N_f are the cycle frequency, cycle dwell-time, and rise- and fall-time of the pulse. In recent years [19], phenomenological models for bond wire lift-off in base plate-less devices have been extended to include the effect of bond wire geometry and load pulse duration. It has to be mentioned that such models include up to eight free parameters that must be extracted by the least square fit of independent experiments. Since they strongly tailored the device under test, such complex models cannot always be applied to the general case.

5.4.3.2 Linear accumulation of the damage (Miner's rule)

The modified Coffin–Manson relationship delivers N_f for a single Δ The application of this method to more complex mission profiles consisting of multiple different cycles requires an additional assumption, which defines how the damage arising from cycles with different amplitudes is accumulated. The usual assumption of prediction methods based on the Coffin–Manson law is the so-called Miner's Rule, i.e., the linear accumulation of the damage.

The simplest implementation of this principle relies on the definition of the fatigue damage function, Q, which expresses the damage accumulated for all cycles, N, in the mission profile with amplitude Δ_1

$$Q(\Delta T_1) = \frac{N(\Delta T_1)}{N_f(\Delta T_1)} = \frac{g(\Delta T_1)}{N_f(\Delta T_1)}$$

where N_f is the number of cycles to failure at ΔT_1. The damage produced by all cycles with arbitrary amplitude ΔT in the mission profile is given by the function $g(\Delta T)$, which delivers the counts of the different cycles. The total damage accumulated during the mission profile is obtained by integration of the function Q over all

ΔT in the mission profile [18], yielding

$$Q(\mathit{after}\ 1\ \mathit{mission\ profile}) = \frac{1}{a}\int_{\Delta T\ \min}^{\Delta T\ \max} \frac{g(\Delta T)}{\Delta T^{-n}}\, d(\Delta T)$$

The ratio $1/Q$ returns the lifetime of the system expressed in terms of the number of mission profiles to failure. Finally, the lifetime of the system expressed in hours is obtained by the multiplication of $1/Q$ by the duration in hours of the mission profile.

The procedure for the extraction of the function $g(\Delta T)$ from the temperature mission profile (e.g., cycle counting) is still an issue under debate. The conventional cycle counting algorithm used in thermomechanics is the *Rainflow method* [20]. Alternative cycle counting procedures have been proposed [18] that are based on physical considerations. It has to be noted that the choice of the correct cycle counting algorithm is very important. In fact, as shown in [18], depending on the assumed criterion, the estimated lifetime may vary by more than a factor of ten. Particular attention has to be paid to filter all spurious cycles introduced either by the limited resolution of the equipment or by the sampling procedure. For the quantification of Q, $g(\Delta T)$ can be used either in its numerical form, or after approximation with a distribution (e.g., Normal, Lognormal, or Weibull) to facilitate analytical calculations.

5.4.3.3 Lifetime prediction from the constitutive equations

In order to overcome the issues related to the arbitrary procedures for cycle counting, alternative prediction procedures have been proposed that are based on the constitutive equations that govern the stress and strain behavior of materials submitted to thermomechanical cycling. A popular model [18] assumes a uniaxial deformation in a bimetallic system, where the total strain, ε_{tot}, is defined as

$$\varepsilon_{tot} = \varepsilon_{elastic} + \varepsilon_{inelastic} = \varepsilon_e + \varepsilon_p + \varepsilon_s + \varepsilon_T$$

where ε_e is the elastic strain, ε_p is the plastic strain, ε_s is the creep strain, and ε_T is the transient creep strain, respectively. Introducing the explicit expression for the strain-stress yields the constitutive model in the form of a differential equation that expresses the strain, ε, as a function of the applied stress, σ, and of the temperature, T:

$$\varepsilon = \frac{\sigma}{E_0 - E_1 T} + C_1\left(\frac{\sigma}{G}\right)^{m_p} + C_2\frac{G}{T}\left[\sinh\left(\alpha\frac{\sigma}{G}\right)\right]^{n_s}\exp\left(-\frac{E_a}{k_B T}\right)t$$
$$+ C_3[1 - \exp(-B\dot{\varepsilon}_s t)]$$

where E_0, E_1, E_a, C_1, C_2, C_3, G, m_p, α, k_B, and B are material parameters and physical constants.

Under the simplified assumption of a one-dimensional bimetallic model, the instantaneous uniaxial strain, $\varepsilon(t,T)$, is calculated from the instantaneous value of the temperature at the solder/joint location and inserted in the constitutive equation.

The instantaneous value of the stress $\sigma(t,T)$ is then calculated by numerical integration of the (discretized) differential equation. Due to the irreversible plastic deformation during a thermal cycle, the representation of the resulting $\sigma(t,T)$ as a function of $\varepsilon(t,T)$ delivers a hysteresis loop. The area included in such a loop represents the deformation energy, W, that is dissipated during the thermal cycle. The total deformation energy, W_{tot}, accumulated during a temperature mission profile is obtained by calculating the total area included in the loops associated with all cycles in the mission profile.

The portion of lifetime consumed during a given mission profile is obtained by calculating the ratio W_{tot}/W_{max}, where W_{max} is the amount of accumulated deformation work at which the device fails. The threshold W_{max} is characteristic for every single power device product and can be calculated from the experimental N_f (ΔT) curves published in the data sheets by the device manufacturers.

Special attention must be paid to the temperature values used for the calculation. In fact, T is the instantaneous value of the temperature at the solder/joint interface and not necessarily the junction temperature $T_j(t)$. However, the instantaneous temperature at the location of interest can be scaled from $T_j(t)$ by means of proper compact thermal models of the package [21].

Alternative procedures based on the Anand viscoplastic model [22] have been attempted to avoid the numerical integration of the constitutive equation. Unfortunately, the related predictions have been shown to be quite inaccurate because of the rough approximation of the transition from the elastic to the plastic regime [23].

Recently, the presented algorithms have been adapted for implementation in mini-computer systems for *in-situ* calculation of the lifetime of power modules in power systems [23]. Starting from the sampled instantaneous junction temperature or dissipated power, this reliability odometer calculates the residual lifetime due to thermo-mechanical failure mechanisms by integration of the constitutive equations.

In spite of its apparent complexity, the model based on the constitutive equations is robust and returns lifetimes that are not strongly dependent on the uncertainty of the assumed parameters. Furthermore, the obtained lifetime does not depend on the algorithms used for cycle counting. More complicated models could be developed, which account for additional mechanisms, such as anisotropic effects and micro-structural changes. However, the excessive number of free parameters associated with these approaches would make the calibration of those models virtually impossible.

5.5 Summary and conclusions

Accurate and robust procedures for lifetime modeling are an important prerequisite for the design of reliable power systems. The basic failure mechanisms of silicon-based power semiconductors have been exposed in conjunction with the major stress factors influencing the failure rate. Failure rate metrology procedures have been discussed for the case of a constant failure rate (exponential) and for a failure rate increasing in time (Weibull). The principles behind the most popular handbook-based prediction models have been shortly presented and the main standards have been benchmarked in the case of a simple device.

Lifetime prediction on the base of mission profiles has been introduced with a special focus on the Coffin–Manson method and on a technique relying on the integration of the constitutive equations of thermomechanics.

Finally, it should be noted that complex multi-dimensional prediction models accounting for all possible stress factors are not necessarily more accurate than simple behavioral models that reproduce the basic evolution of the degradation mechanisms on the base of a few robust calibration parameters.

References

[1] Solomalala, P., Saiz, J., Mermet-Guyennet, M., *et al.* (2007). Virtual reliability assessment of integrated power switches based on multi-domain simulation approach. *Microelectronics Reliability*, *47*(9–11), 1343–1348. http://doi.org/http://dx.doi.org/10.1016/j.microrel.2007.07.006.

[2] Hager, C. Lifetime estimation of aluminium wire bonds based on computational plasticity. *Diss. Technische Wissenschaften ETH Zürich*, Nr. 13763, 2000.

[3] Ciappa, M., Malberti, P., Fichtner, W., Cova, P., Cattani, L., and Fantini, F. (1999). Lifetime extrapolation for IGBT modules under realistic operation conditions. *Microelectronics Reliability*, *39*(6–7), 1131–1136. http://doi.org/http://dx.doi.org/10.1016/S0026-2714(99)00160-2.

[4] Ciappa, M., and Malberti, P. (1996). Plastic-strain of aluminum interconnections during pulsed operation of IGBT multichip modules. *Quality and Reliability Engineering International*, *12*(4), 297–303. http://doi.org/10.1002/(SICI)1099-1638(199607)12:4<297::AID-QRE21>3.0.CO;2-C.

[5] Ciappa, M. (2002). Selected failure mechanisms of modern power modules. *Microelectronics Reliability*, *42*(4–5), 653–667. http://doi.org/http://dx.doi.org/10.1016/S0026-2714(02)00042-2.

[6] Dugal, F., and Ciappa, M. (2014). Study of thermal cycling and temperature aging on PbSnAg die attach solder joints for high power modules. *Microelectronics Reliability*, *54*(9–10), 1856–1861. http://doi.org/http://dx.doi.org/10.1016/j.microrel.2014.08.001.

[7] Castellazzi, A., Ciappa, M., Fichtner, W., Lourdel, G., and Mermet-Guyennet, M. (2006). Compact modelling and analysis of power-sharing unbalances in IGBT-modules used in traction applications. *Microelectronics Reliability*, *46*(9–11), 1754–1759. http://doi.org/http://dx.doi.org/10.1016/j.microrel.2006.07.055.

[8] Kaminski, N. (2004). *Failure Rates of HiPak Modules Due to Cosmic Rays*. Application Note SYA 2042-04. ABB Switzerland Ltd.

[9] Birolini, A. (2014). *Reliability Engineering*. Berlin: Springer.

[10] Ciappa, M. (2001). *Some Reliability Aspects of IGBT Modules for High Power Applications*. Konstanz: Hartung-Gorre.

[11] Ciappa, M. (2005). Lifetime prediction on the base of mission profiles. *Microelectronics Reliability*, *45*(9–11), 1293–1298. http://doi.org/http://dx.doi.org/10.1016/j.microrel.2005.07.060.

[12] MIL-HDBK-217F Notice 2, US Department of Defense. (1995). *Military Handbook, Reliability Prediction of Electronic Equipment.*

[13] *Technical Report: Reliability Data Handbook – Universal Model for Reliability Prediction of Electronics Components, PCBs and Equipment*, International Electrotechnical Commission, IEC TR 62380, (2004).

[14] *Union Technique de l'Electricité et de la Communication, Recueil de données de fiabilité – Modèle universel pour le calcul de la fiabilité prévisionnelle des composants, cartes et équipements électroniques, Août 2005.* UTE-C 80-810, (IEC 62380 TR Edition 1).

[15] FIDES Guide 2010 Edition A, *Reliability Methodology for Electronic Systems*-(DM/STTC/CO/477-A).

[16] RIAC HDBK 217Plus, *Reliability Prediction Models.* Reliability Information Analysis Center. May 2006.

[17] Telcordia SR-332, *Reliability Prediction Procedure for Electronic Equipment*, Issue 2, 2006.

[18] Ciappa, M., Carbognani, F., and Fichtner, W. (2003). Lifetime prediction and design of reliability tests for high-power devices in automotive applications. *IEEE Transactions on Device and Materials Reliability*, *3*(4), 191–196. http://doi.org/10.1109/TDMR.2003.818148.

[19] Scheuermann, U., and Schmidt, R. (2013). Impact of load pulse duration on power cycling lifetime of Al wire bonds. *Microelectronics Reliability*, *53* (9–11), 1687–1691. http://doi.org/http://dx.doi.org/10.1016/j.microrel.2013.06.019.

[20] *Standard Practices for Cycle Counting in Fatigue Analysis*, ASTM E1049 – 85(2011)e1.

[21] Ciappa, M., Fichtner, W., Kojima, T., Yamada, Y., and Nishibe, Y. (2005). Extraction of accurate thermal compact models for fast electro-thermal simulation of IGBT modules in hybrid electric vehicles. *Microelectronics Reliability*, *45*(9–11), 1694–1699. http://doi.org/http://dx.doi.org/10.1016/j.microrel.2005.07.083.

[22] Anand, L. (1985). Constitutive equations for hot-working of metals. *International Journal of Plasticity*, *1*(3), 213–231. http://doi.org/10.1016/0749-6419(85)90004-X.

[23] Ciappa, M., and Blascovich, A. (2015). Reliability odometer for real-time and in situ lifetime measurement of power devices. *Microelectronics Reliability*, *55*(9–10), 1351–1356.

Chapter 6
Reliability of power electronics packaging

Simon S. Ang[1] *and H. Alan Mantooth*[1]

6.1 Introduction

The increasing performance requirements for power electronic systems require multiple power semiconductor die on a single power substrate in a half-bridge or full-bridge configuration housed in a single power electronic module. These power electronic modules not only demand improved electrical performance in terms of circuit element parasitic mitigation but also improved thermo-mechanical performance. Wide bandgap power semiconductors offer improved performance in terms of faster electrical switching as well as high-temperature operation. Current packaging research focuses on optimizing the thermo-mechanical performance of power electronic module packages to minimize thermo-mechanical strains and stresses as well as to enhance reliability while providing the improved electrical performance offered by the wide bandgap power semiconductors. New and novel nano-composites and metallurgical materials are being investigated for applications in power electronic module packaging to optimize heat transfer efficiency and improve temperature distributions to minimize the thermo-mechanical stresses that drive conventional power device packaging failure modes.

These power packages and power electronic modules are subjected to large temperature cycling and power cycling requirements compared to conventional small-signal integrated circuit packages due to their high power density requirement. Moreover, these power packages and modules need to endure high operating temperatures. As such, the reliability testing requirements for these power electronic packages and modules are different from those for the small-signal integrated circuit packages. In this chapter, the basic reliability concepts are first reviewed, followed by a description of the reliability testing for power packages and power electronic modules. This is then followed by the common failure modes for the power packages and power electronic modules. This chapter concludes with the specific reliability concepts for the high-temperature power packages and power electronic modules.

[1]Department of Electrical Engineering and Computer Science, University of Arkansas, USA

6.2 Reliability concepts for power electronic packaging

Some high-performance power packages and modules require a 30-year operation life, with a 100 failures in time (FIT) requirement [1]. It is well known that FIT increases with junction temperatures greater than 120 °C. The current technology trend in power packages and power electronic modules towards higher junction temperatures of the power semiconductor devices, approaching and even surpassing 200 °C, indicates that the FIT figure of merit would increase. As such, more emphasis should be placed on the thermo-mechanical reliability of the power packages and power electronic modules.

In lifetime data or a Weibull analysis, a statistical distribution of lifetime data from a representative sample of power packages is generated. These parameterized statistical data sets are used to estimate the reliability, probability of failure at a specific time, mean lifetime, and failure rate. The probability of failure at a given time is the probability that a device will fail at a particular time and is the reciprocal of the reliability. Mean lifetime is the average time that the power packages in the population are expected to operate before failure and is often expressed as "mean time to failure" (MTTF) or "mean time before failure" (MTBF). Failure rate is the number of failures per unit time that can be expected to occur for the power packages. Statistically, it is the rate of change of the cumulative failure probability divided by the probability that the power device will not already be failed at time t. The MTBF and failure rates are related, as shown below.

$$\text{MTBF} = \frac{\text{Total device operating hours}}{\text{Total number of failures}} \tag{6.1}$$

Failure rate is the reciprocal of the MTBF.

$$\alpha = \frac{1}{\text{MTBF}} \tag{6.2}$$

An exponential distribution function is usually used to describe random electrical failures,

$$F(t) = 1 - e^{-\alpha t} \tag{6.3}$$

where the failure rate α can be defined as

$$\alpha = \frac{f}{n \cdot t} \tag{6.4}$$

where f is the number of failures, n is the number of devices, and t is the test time.

For reliability testing, it is common practice to increase the test temperature to which the power semiconductor devices are subjected to accelerate the occurrence of failure mechanisms. Suppose a device is stressed at a high junction temperature, T_{j2}, for time t_2. To estimate the same level of stress applied at a lower junction temperature, T_{j1}, for time t_1, we can define an acceleration factor as the ratio of the time at T_{j1} to T_{j2}. Since the acceleration factor is a dimensionless ratio, it can also be expressed as the ratio of failure rates that are inversely proportional to stress times. The Arrhenius equation is often used to estimate failure rates. The acceleration factor is then defined as

$$a_f = e^{\left\{E_a\left(T_{j2}-T_{j1}/(T_{j2}\cdot T_{j1}\right)/k\right\}} \tag{6.5}$$

where E_a is the activation energy, k is Boltzmann's constant (8.6×10^{-5} eV/K), T_{j1} is the operating junction temperature, and T_{j2} is the junction temperature during the accelerated test. The activation energy, E_a, is found experimentally and depends on the predominant failure mechanism that occurred during the accelerated testing.

Thermo-mechanical failures dominate the failure rates of power packages and power electronic modules. The Manson–Coffin [2] relationship relates the number of temperature cycles to failure of a solder joint and is given as

$$N_f = \frac{K_M}{(\Delta T)^a} \tag{6.6}$$

where N_f is the number of cycles to failure, K_M is a material dependent constant, ΔT is the difference in temperature during cycling, and a is an experimentally determined constant. As can be seen from (6.6), the time to failure depends on the characteristics of the material and inversely on the temperature excursion, ΔT. The number of cycles to failure with two different temperature excursions, ΔT_1 and ΔT_2, is

$$\frac{N_1}{N_2} = \left(\frac{\Delta T_1}{\Delta T_2}\right)^a \tag{6.7}$$

However, the Manson–Coffin relationship may not adequately estimate the thermal fatigue life of solders since it was developed for temperatures below 50% of the melting point of the solder. In many cases, some solders operate at higher than 50% homologous temperatures in power packages. Modified Manson–Coffin relationships will need to be developed for these cases.

6.3 Reliability testing for power electronics packaging

Reliability testing for power electronics packages and modules can be classified as either environmental or endurance testing. These tests are usually subjected to international standards such as JEDEC [3] and MIL-Std. 883 [4]. Environmental testing includes thermal shock (Mil-Std 883 method 1011.9, IEC68-2-14 [5]), temperature cycling (MIL-STD-883 method 1010 Condition C, IEC68-2-14, JEDS22-A-104D [6]), vibration (IEC68-2-6), lead integrity (IEC-2-21), solderability (IEC-2-20), autoclave (JESD22-A110AB), and mounting torque. Endurance testing includes power cycling (JEDS22-A-122), high-temperature storage (IEC68-2-2), low-temperature storage (IEC68-2-1), moisture resistance (IEC-2-3), high-temperature reverse bias, high-temperature gate bias, and power cycling (JESD22-A122). Many of these tests are further described below. Figure 6.1 shows some of the reliability test apparatus used for these measurements.

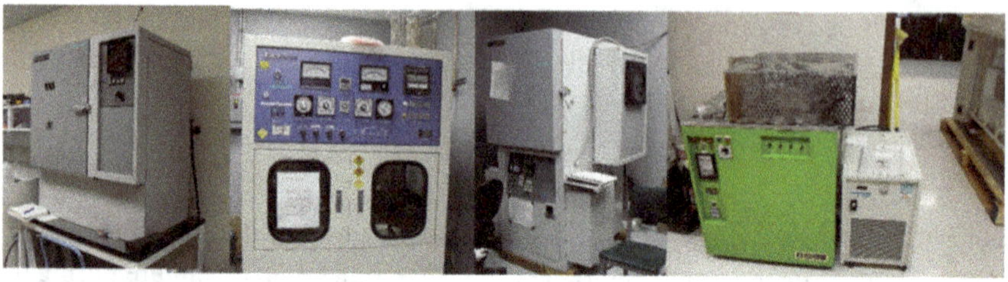

Figure 6.1 Reliability test apparatus at the High Density Electronics Center, University of Arkansas

6.3.1 Thermal shock testing

Liquid-to-liquid thermal shock testing is commonly used to exert repetitive thermal stress to power electronic packages and modules. Thermal shock is used to predict the ability of the power electronic modules to withstand thermally-induced size changes and to evaluate package hermetic integrity, crack resistance, and changes in electrical properties. Thermal shock standards can be found in Mil-Std 883 Method 1011.9. Thermal shocks are performed at temperatures of 100 °C soak for 5 minutes and 0 °C soak at 5 minutes for 5 cycles. This test is usually performed in a thermal shock chamber with the power packages or modules placed in a basket of inert fluorinated fluid.

6.3.2 Temperature cycling

Temperature cycle testing is conducted to examine the change in electrical characteristics and physical structure of the power packages or modules to alternate exposures at extremes of high and low temperatures. The differences in the coefficients of thermal expansion (CTEs) of the different layers in the power packages or modules can create cracking and delamination of packages and internal structures, and changes in electrical characteristics resulting from thermo-mechanical damage. According to MIL-STD-883 Method 1010 Condition C, the power packages or module must be placed in a temperature cycling test chamber so that there is no substantial obstruction to the flow of circulating air across each device under test. These power packages or modules are then cycled between temperature extremes, usually from −40 °C to +125 °C for the required number of cycles. The dwell time at each extreme temperature shall be greater than 10 minutes, and the devices under test must reach the temperature in less than 15 minutes. The transfer time from hot to cold or from cold to hot must not exceed one minute. There are various methods of predicting lifetimes from thermal tests [7]. The Manson–Coffin relations can be used to predict lifetimes from temperature cycling tests. The acceleration coefficient, α_T, can be obtained as shown below:

$$\alpha_T = N1/N2 = (\Delta T_1/\Delta T_2)^K \tag{6.8}$$

where $N1$ and $N2$ are the failed cycle counts, ΔT_1 and ΔT_2 are the operating and test temperature range, and K is an experimentally determined factor. The package or module lifetime in years is then given by

$$(N \times \alpha_T)/365 \tag{6.9}$$

where N is the number of days to failure.

6.3.3 Power cycling test

During a power cycling test, the power device or module under test is mounted on a heat sink similar to a typical power electronics application. The power device or module is heated up electrically by conduction losses (i.e., I^2R losses). When the maximum target junction temperature in the power device or module is achieved, the active heating of the device is stopped, and the power device or module cools down to a specified temperature, thus completing the power cycling test. As such, considerable junction temperature gradients are generated inside the power device or module. Figure 6.2 shows the temperature evolution during the power cycling test. The heat sink temperature is usually used to control the power cycling test. When the upper limit of T_h is reached, the active heating of the device is stopped and cooling commences immediately. As such, the device junction temperature decreases. When the lower limit T_L is reached, the load current is turned on to initiate active heating again and the cycle is repeated. The characteristic parameter for power cycling tests is defined as:

$$\Delta T_j = T_h - T_L. \tag{6.10}$$

The medium temperature for the power cycling test is defined as:

$$T_m = T_L + \frac{T_h - T_L}{2}. \tag{6.11}$$

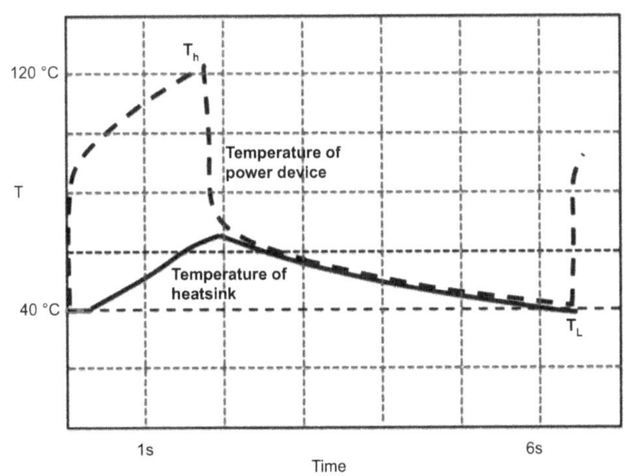

Figure 6.2 Power cycling plot for a power package or module

The thermo-mechanical stress caused by the repetitive heating/cooling on/off cycles can lead to failure cracks in the power package or module. Usually, 6000 cycles for a ΔT_j of 100 °C and 3000 cycles for a ΔT_j of 125 °C are required for large power packages or modules according to JESD22-A122. A usual on/off cycle is 5 minutes of on time and 5 minutes of off time. A long power cycle time usually creates a higher stress for the devices. Differences in CTEs of different layers of materials in the power package or module during the temperature excursion create thermo-mechanical stress at the interfaces between these layers. From Figure 6.2, it can be seen that the power module's case temperature changes comparatively little, while large temperature excursion occurs at the junction of the power semiconductor devices. When this stress is repeated over the power cycling test, cracks may occur in the solder attachment between the devices and power substrate. When these cracks reach the bottom of the power chip, they cause an increase in thermal resistance and a subsequent increase in junction temperature of the power semiconductor devices. This thermo-mechanical stress often leads to the fatigue of materials and interconnections. Due to the interdependency of many failure modes, failure analysis of failed power packages or modules is challenging. Another challenge in power cycling testing is the measurement of T_j due to its transient nature; thus, a relatively fast measurement technique must be employed.

6.3.4 Autoclave

The autoclave or pressure cooker test is performed to evaluate the moisture resistance of non-hermetic power packages or modules with no electrical bias applied. It is performed at 100% room humidity and 121 °C under a pressure of 15 psig or 103.4 kPa according to JEDS22-A102C. This test is destructive to the power device or module and is mainly used for comparative evaluations, such as lot acceptance or process monitoring, since accelerating factors relating to the test conditions have not been well established. Moreover, the autoclave test may produce failures that are not typical of power packages or modules. As such, the autoclave test is usually used to qualify a power package or module only.

6.3.5 Gate dielectric reliability test

For power packages or modules that contain power devices with gate oxides, this test is usually performed at 150 °C junction temperature in static operation at or near the maximum-rated gate oxide breakdown voltages according to JEDS22-A108B. For SiC power MOSFETs, it is also advisable to perform the time-dependent dielectric breakdown (TDDB) test to assess the reliability of their gate dielectrics at high junction temperatures. TDDB is a charge injection process characterized by a build-up stage followed by a runaway stage. Charges are trapped in the gate dielectric, increasing in number with time, as current flows. When the electric field exceeds the dielectric breakdown voltage threshold in the weakest point of the gate dielectric, the runaway stage commences. Large currents can conduct through these weakest gate dielectric points to heat up the dielectric, which causes a further increase in the current flow. This positive feedback mechanism eventually results in electrical runaway, followed by thermal runaway, which destroys the gate dielectric oxide. TDDB testing is usually performed at a constant electric field or voltage close to the breakdown voltage, and the test temperature is generally 225 °C or higher for SiC power MOSFETs due to their higher junction temperatures capability. The failure criterion is generally a gate dielectric current of a few microamps (μA) according to Fowler-Nordheim breakdown curves generated for the gate dielectric of the device under test. Moreover, the lifetime for the larger SiC power MOSFETs is expected to be less than the lifetime for the smaller ones.

6.3.6 Highly accelerated stress test

Highly accelerated stress test (HAST) is performed to test the moisture resistance of non-hermetic packages or modules in high-humidity operating environments. A low bias is applied to result in a small current being drawn during the test. The combination of pressure, humidity, temperature, and operating bias accelerates the moisture penetration into the packages or modules. Corrosion may occur in the presence of contaminants such as chlorine and phosphorous when moisture reacts with the aluminum conductors to eventually cause catastrophic electrical failure. According to JESD22-A110B, the HAST stress conditions are 85% humidity at 18.6 psig and 130 °C or 85% humidity at 3 psig and 110 °C.

6.3.7 High-temperature storage life test

The high-temperature storage life (HSTL) test is a stabilization test used to determine the effects of elevated temperature storage without external electrical stresses. It is also used to assess the long-term reliability of die attaches and wire bonds by promoting intermetallic growth. According to JESD22-A103B, the power packages or modules are placed in an air-circulated chamber heated to 150–175 °C for a specific length of time.

6.3.8 Burn-in test

Burn-in tests are used to screen out infant mortality failures by correlating their lifespan at room temperature with their lifespan at elevated temperature. The packages and modules are usually placed in an oven maintained at the desired test temperature by applying a high voltage to the devices under test. High temperature accelerates gate oxide defects (such as pin holes, uneven layer growth, etc.), ionic contamination, and bulk defects. Burn-in tests are usually performed for 24–72 hours for electronic assemblies such as modules and 4–8 hours for components. Burn-in tests for power semiconductor devices and modules should be carefully monitored to avoid thermal runaway that can destroy test sockets and the test apparatus. As such, burn-in tests for power semiconductor devices and modules are expensive and not cost-effective. With burn-in, the initial FIT is expected to be improved.

6.3.9 Other tests

The lead integrity test is used to determine the integrity of leads according to JESD22-B105B. A solderability test is used to determine the ability of all external package or module leads and terminations to be wetted by solder. This solderability test is usually performed via a dip-and-inspect procedure. Packages or modules are aged by exposure to steam for a period of 8 hours, and then flux is applied before they are dip-heated to a temperature of 215 °C or 245 °C for 5 seconds according to JESD22-B102C.

6.4 Power semiconductor package or module reliability

Due to the high-current operating conditions, the multiple interfaces between materials of different coefficient of thermal expansion (CTE) create tremendous reliability challenges for a power semiconductor package or module. Figure 6.3 shows the structural view of a power electronic module. Power semiconductor devices are attached to the power substrate, and then the power substrate is attached to a base plate that is usually made of copper. Wire bonds are used to provide electrical interconnections, and they are attached from the power semiconductor devices to the power substrate as well as to the output connectors. The module is then encapsulated to provide mechanical and environmental protection. Significant thermo-mechanical stresses are created within the power package or module, and these stresses lead to potential reliability problems. For a power package or module, bond wire fatigue and solder fatigue are usually the major failure mechanisms. These thermo-mechanical fatigue–induced failures are mainly due to CTE mismatches and temperature excursion during operation.

6.4.1 Solder joint reliability

Soft solders such as tin-silver, indium or tin-lead alloys are frequently used in power electronic modules. Of the above soft solders, the fatigue of the tin-lead solder has been studied extensively. When a direct bond copper (DBC) substrate is soldered using a Sn-3.2Ag-0.8Cu lead-free solder alloy, a Cu_5Sn_6 intermetallic phase can be observed at the copper/solder interface [8]. Plastic inhomogeneity

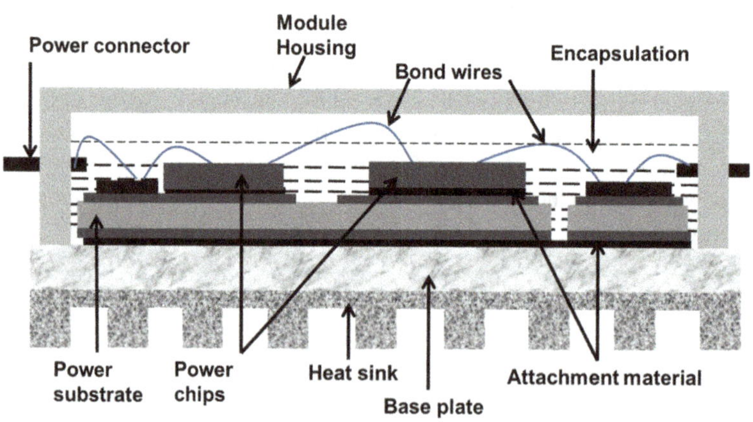

Figure 6.3 Structural view of a power electronic module

results in stress concentrations at or near the interface between the solder and the intermetallic. Hence, fatigue cracks are usually found within the intermetallic layer next to the DBC ceramic substrate. Due to the larger CTE mismatch, fatigue cracks are usually found near the intermetallic layer immediately next to the ceramic substrate. As such, cracks initiate at the vicinity of the solder joint border with maximum shear stress [9].

For low ΔT (<100 °C) temperature cycling, the number of cycles to failure of the solder joint between a DBC ceramic substrate and base plate depends weakly on the temperature excursion [10]. However, a typical ΔT of 165 °C (for −40 °C to +125 °C) is called for in a typical temperature cycling test. As such, these temperature cycling tests should not be extrapolated to practical power electronic operating conditions. Moreover, the solder joint failure mechanism during this type of accelerated testing is not representative of the lower operating temperatures usually encountered in actual field applications [10]. The number of cycles to the failure of large solder joints due to thermo-mechanical fatigue is given by the modified Manson–Coffin relationship in the form of:

$$N_f = \frac{1}{2}\left(\frac{\Delta\alpha\Delta TL}{\gamma x}\right)^{c^{-1}} \tag{6.12}$$

where L is the lateral size of the solder joint, $\Delta\alpha$ is the CTE mismatch between the upper and the lower joint materials, ΔT is the temperature swing, c is the fatigue exponent, x is the thickness of the solder joint, and γ is the ductility factor of the solder. Equation (6.12) implies that the solder joint lifetime is improved by using a smaller solder joint area, by having a smaller CTE mismatch, and by increasing the solder thickness.

6.4.2 Bond wire reliability

Bond wire fatigue mechanical issues include bond wire lift-off and bond wire heel cracking. The shear stresses between bond pad and bond wire and repeated flexure of the bond wire contribute to the bond wire fatigue. In a typical power electronics module, there can be hundreds of bond wires making the electrical interconnection from the power devices to the power substrate, from the power substrate to the external connectors, and from the power devices to the external connectors, as well as between power devices, as shown in Figure 6.4.

As shown in Figure 6.4, many of these bond wires are bonded onto the active areas of the power semiconductor devices. As such, these bond wires experience the temperature swing caused by power dissipation in the power devices as well as by the self-heating of the wire itself due to ohmic heating. Aluminum bond wires are currently being used, but copper bond wires are becoming common in some applications. Aluminum bond wire is usually hardened by alloying with a few thousand parts per million of magnesium or silicon to retard the corrosion mechanism in pure aluminum bond wire. The fusing current for a 10 mm long, 225 μm diameter aluminum bond wire loop in air is about 10 A due to ohmic heating [11]. Failure of a wire bond is usually caused by repeated flexure of the

Figure 6.4 A power electronic module with bond wires

bond wire or by shear stresses generated between the bond wire and its termination. In addition to the lower cost for the copper bond wires, slower intermetallic growth in copper bond terminations enhances reliability for high-power packages and modules [12]. Figure 6.5 shows two 12-mil aluminum bond wires providing the electrical connections between two conducting plates typical of a power electronic module. As can be seen, the bond loop heights as well as the bond angles are typically quite large.

(a) **Bond wire lift-off**

Bond wire lift-off usually occurs at the aluminum bond wire terminations bonded onto silicon power devices due to large temperature excursions. However, bond wire lift-off seldom occurs at the bond wire terminations bonded onto copper traces or pads. This is because there are fewer temperature excursions between the bond wires and the copper pads on the power substrate compared to those bond wire terminations on the power semiconductor die. Cracks leading to bond wire lift-off are usually initiated at the tail of the bond wire and propagate until the bond wire eventually lifts off. A molybdenum/aluminum bond pad can provide some strain relief to mitigate thermo-mechanical fatigue by distributing the CTE mismatch of aluminum and silicon across a thick layer [10]. The number of thermal cycles-to-failure, N_f, can be modeled using a simple bimetallic approach to approximate the thermo-mechanical stresses arising at the interface between an aluminum bond wire joint and silicon under a temperature excursion of ΔT:

$$\varepsilon_t = L(\alpha_{Al} - \alpha_{Si})\Delta T \tag{6.13}$$

where α_{Al} and α_{Si} are the CTEs of aluminum and silicon, respectively, and L is the aluminum bond wire joint length. The large thermo-mechanical mismatch

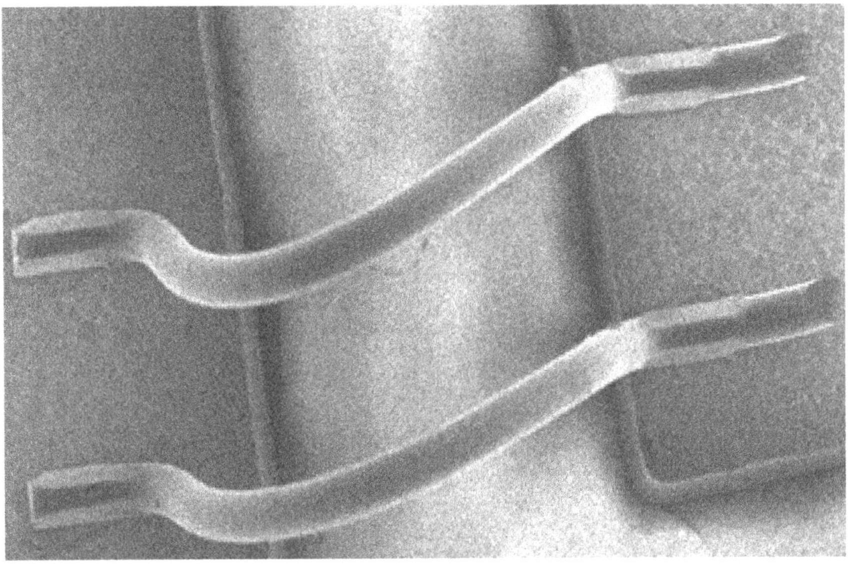

Figure 6.5 *Two 12-mil aluminum bond wires illustrating the bond heights and bond angles*

means that the joint is deeply in the plastic regime. As such, only plastic strain exists in the joint. Hence, the number of cycles-to-failure, N_f, can be expressed by the Manson–Coffin relation as:

$$N_f = a(\Delta T)^{-n} \tag{6.14}$$

The coefficients a and n can be experimentally measured based either on thermal or power cycling of power devices or modules. The simple model above has been shown to accurately predict the bond wire fatigue for temperatures less than 120 °C. For higher temperature ranges, alternative models that take into account the deviation from the power law are necessary.

(b) **Bond wire heel cracking**

Bond wire heel cracking, due to thermo-mechanical effects, is observed after long endurance tests. When the bond wires are subjected to temperature cycling, they expand and contract, and flexure fatigue of the bond wires occurs. The wire bond loop height is an important factor as it determines the angle at the heel of the bond wire. Package or module encapsulation also plays a role in this bond wire heel cracking. Heel cracking is usually slower than the lift-off failure mechanism. Schafft analytically predicted the number, N_f, of thermal cycles to heel cracking due to bending stress based on the power law [13]:

$$N_f = A\varepsilon_f^n \tag{6.15}$$

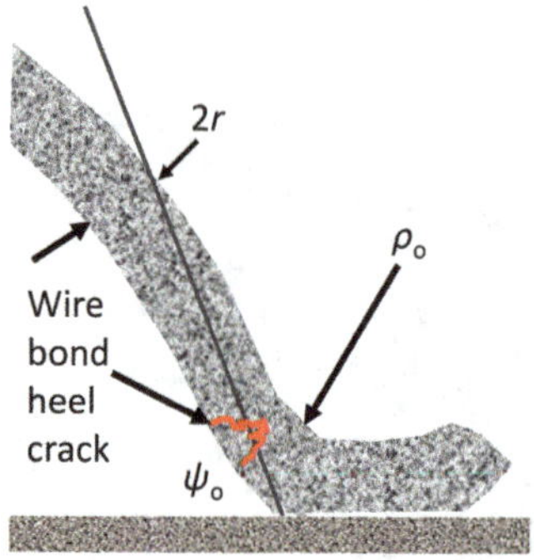

Figure 6.6 Bond wire parameter definitions

where A and n are constants for the bond wire material and the wire strain ε_f is computed according to [10,13]:

$$\varepsilon_f = \frac{r}{\rho_o}\left(\frac{\cos^{-1}((\cos\varphi_o))(1-\Delta T\Delta\alpha)}{\varphi_o} - 1\right) \tag{6.16}$$

where $\Delta\alpha$ is the mismatch in the CTE of aluminum bond wire and silicon, and φ_o, ρ_o, and r are geometrical parameters, as shown in Figure 6.6.

6.5 Reliability of high-temperature power electronic modules

Conventional power packaging technologies have many concerns with operating at high temperatures, including electrical parasitic contributions, inter-metallic formations, and long-term reliability depending on bonding surfaces. At high temperatures above the traditional operating temperatures of semiconductor electronics (i.e., above 125–150 °C), some failure modes manifest themselves more readily. For example, polymeric materials utilized for voltage isolation and encapsulation degrade more readily at high temperatures, presenting a major reliability problem for power semiconductor modules. The need for high power density, high operating voltage, and high temperature for power semiconductor devices in the power electronics industry presents additional reliability challenges for power semiconductor devices and packaging technologies [14]. Wide bandgap power semiconductor devices, such as silicon carbide (SiC) and gallium nitride (GaN) devices, offer the opportunity to circumvent the limits of device physics for silicon. SiC and GaN power devices are commercially available, but their packaging technologies limit high-temperature applications. Material properties are of great importance in high-temperature power electronic modules. Tradeoffs between material properties, such as thermal, electrical, and mechanical properties and the reliability, module size, and the total cost, must be carefully made when designing high-temperature power electronic modules. The key material reliability issues to be considered are related to the power substrate, power die attachment, and encapsulation.

6.5.1 Power substrate

The power substrate provides the electrical interconnections both within and outside of the power electronic module. It serves as the interconnection layer for various power semiconductor devices within the power module. As such, the desired power substrate should have a CTE matched with the power semiconductor devices as well as the base plate. The desired properties of the power substrate are high thermal conductivity, large fracture toughness, and high flexural strength. Power substrates are normally made of copper or aluminum plates sandwiched around a ceramic material ranging from alumina (Al_2O_3) to aluminum nitride (AlN) and silicon nitride (Si_3N_4), depending on the power electronic application requirements. Depending on the electrical conducting materials, these power substrates are known as direct bond copper (DBC) and direct bond aluminum (DBA) substrates. While the bonding mechanisms for DBC and DBA substrates are very similar, the DBA substrates yield a better thermal cycling reliability compared to the DBC substrates [15]. Another high reliability power substrate that is suitable for high-temperature applications is the active metal brazed (AMB) power substrate. AMB has a Si_3N_4 ceramic layer that has a very high tensile strength (800 MPa), compared to those of AlN (270 MPa) and Al_2O_3 (400 MPa) to avoid ceramic fracture. For high-temperature cycling, a thinner metallization layer demonstrates better results [16].

Common failure mechanisms for DBC substrates are delamination of the metal traces from the ceramic and cracking of the ceramic layer underneath the metal traces. Grain boundary restructuring and reorientation cause the surface roughness of the DBA (both AlN and Si_3N_4) to increase after each thermal cycle. Copper layer delamination from the Al_2O_3 layer is observed especially when a large copper trace in the DBC substrate was exposed without being encapsulated, as shown in Figure 6.7. This is due to the CTE mismatch and thermal stress between the Al_2O_3 (7.1 ppm/°C) and the copper layers (17 ppm/°C).

The theoretical strain between the copper and Al_2O_3 layers can be calculated using

$$\varepsilon_x = (\alpha_{Cu} - \alpha_{Al2O3})\Delta T \tag{6.17}$$

For typical wide temperature operation from 250 °C down to −55 °C, or a ΔT of 305 °C, the strain is 3.02×10^{-3}. The Young's modulus of copper is about 120 GPa; this yields a normal stress σ_x of 362.4 MPa. Since the force on the copper layer is equal to the product of thermal stress and area, it is directly proportional to the area of the copper plate. As such, smaller copper traces are less prone to delamination failure. Moreover, large encapsulated copper traces exert force from their thermal stress onto the encapsulation instead of directly onto the copper traces. The copper area without encapsulation should be minimized to reduce stress acting on a large copper trace.

Figure 6.7 Delamination of copper conducting plane from the Al_2O_3 ceramic substrate in DBC

6.5.2 High-temperature die attach reliability

For power modules to continuously operate reliably at high temperatures, the die attach material should be thermo-mechanically stable within 50% of the homologous temperature, which is defined as the maximum junction temperature of the power semiconductor die on the power substrate to the melting temperature of the die attached material. There are few high-temperature lead-free solder alloys available for die attachment at high junction temperatures that comply with the international standard for the Restriction of Hazardous Substrates Directive (RoHS). Furthermore, many module fabrication processes require the need to consider the temperature hierarchy of different solder alloys. All these considerations limit the options with the common solder alloy attachment techniques for these high-temperature power electronic modules. Several alternative die attachment techniques, such as nano-silver sintering, transient liquid phase (TLP) diffusion bonding [17–19], and reactive bonding, have been investigated for high-temperature power electronic modules.

For solder die attach joints, phase transition or dislocation movement help to release their residual stress. At high stress levels, voids or cracks would form, and stress would be released by crack propagation. In SAC305 (96.5% tin, 0.5% copper and 3% silver) solder joints, cracks with lengths of 10–20 μm are usually detected after temperature cycling [20]. These solder joints exhibit sphere or rod-like textural evidence of intermetallic growth that initiated the growth of micro-cracks [21]. Voids form at the interface between Ag_3Sn intermetallic particles and the Sn matrix phase. These voids propagate through the Sn matrix phase, resulting in crack propagation and the growth of micro-cracks.

Nano-silver sintering involves the low-temperature sintering of nano-silver paste at about 220–280 °C, performed under pressure or without pressure. Typically, these nano-silver joints have a 40 MPa bonding strength with no voids due to solidus inter-diffusion [22]. Furthermore, the microporous structure in these nano-silver joints relieves the thermo-mechanical stress. During thermal cycling, stress is released in these nano-silver joints because of plastic deformation. Moreover, micro-cavities nucleate and grow at the grain boundaries, causing the bonding strength to decrease after temperature cycling. As the number of thermal cycles increases, a de-sintering process breaks the larger grains into several smaller grains. This de-sintering process changes the microstructure and further loosens the sintered silver from the grain boundaries [22].

The TLP die attachment process combines traditional liquid phase soldering with a diffusion process to yield a die attachment alloy that can withstand a higher temperature than its processing temperature. TLP bonding attachment materials comprise a low-melting-point tin (Sn) or indium (In) -based metal or alloy inserted between two high melting point base metals such as gold (Au), copper (Cu), nickel (Ni), or silver (Ag). The low melting point interlayer between the two base metals is to be joined based on inter-diffusion and isothermal solidification at the bonding temperature. The presence of the liquid phase during bonding increases the integrity of the bond. Once solidification is complete, the bond is able to withstand operation at temperatures above the bonding temperature. One of the key process parameters for the TLP process is the weight percent ratio of the low melting metal or alloy to the base metal with a high melting temperature. Other process parameters such as bonding reaction temperature and time, bond surface preparations, and applied external pressure can affect the quality and reliability of the TLP bonds. In void-free Au-In TLP joints, the formation of inter-metallic phases of AuIn and $AuIn_2$ is responsible for their joint strength and reliability [17]. Similarly, in Cu-Sn TLP joints, the formation of a uniform layer of a $Cu_{41}Sn_{11}$ intermetallic phase, a two-phase microstructure comprising $Cu_{41}Sn_{11}$ phase and a dispersion of ductile (Cu) particles, and a uniform Cu solid solution, is responsible for their strength and reliability. The $Cu_{41}Sn_{11}$ phase is strong (300 MPa) but brittle (5 MPa). Improvements in both strength and toughness can be obtained through the conversion of the $Cu_{41}Sn_{11}$ phase to the $Cu_{41}Sn_{11}$ + (Cu) microstructure. Conversion of the microstructure to (Cu) would increase the toughness, but decrease their un-notched strength at higher temperatures [23].

6.5.3 Die top surface electrical interconnection

Aluminum bond wires are the most widely used top-side electrical interconnections for the power semiconductor devices in power electronic modules. New electrical interconnection solutions are required for high-temperature power electronic modules to improve both enhanced thermal management and reduced parasitic inductance for the power electronic modules. Aluminum ribbon bonding, as a replacement of many paralleled bond wires, offers a higher current capability, a lower parasitic inductance at higher frequencies, as well as lower and more consistent bond loop heights than the conventional bond wires [24]. Copper wire bonding is another potential top-side electrical interconnection for high-temperature power electronic modules. In addition to their higher current-carrying capability over the conventional aluminum bond wires, copper wires offer a high pull strength and a high bond foot shear strength with improved reliability [25]. However, the requirements of power semiconductor die surface metallization and the need for a high ultrasonic bonding power and force present some technological challenges. Planar packaging offers another way to realize the die top surface interconnection [26–28]. This die top electrical interconnection was realized either through a flexible circuit board or a patterned DBC with direct solder ball contact as well as using a patterned metallization layer. For high-temperature wire bond reliability, it is essential that the nickel-plated bond surfaces on the power substrate should be phosphorous-free.

6.5.4 Encapsulation

Encapsulation is needed to protect power electronics modules from environmental and mechanical hazards. This encapsulation also functions as a dielectric material and ensures that the power electronics modules operate at a high voltage potential without electrically breaking down. Thus, the adhesion between the encapsulation, the power substrate, and the power semiconductor die is critical for the long-term reliability of these power electronic modules. While a cleaned interface helps to improve the adhesion, some silicone gels offer primers to strengthen the adhesion between the encapsulation and the substrate [29]. Most of the commercially available silicone encapsulations are limited to less than 250 °C. For high-voltage applications, a high dielectric breakdown strength passivation can be used to relieve high electrical field intensification areas in the power electronics modules. Polyamide imide [30], polyimide, and parylene HT dielectrics [31] have been used for this application. Filled silicone elastomers are used for high-temperature electronic encapsulation applications. Silicone elastomers are silicone-based polymers that have been vulcanized. Silicone elastomers have a high concentration of chemical crosslinks that make them less susceptible to thermal degradation. Inorganic fillers are added to increase mechanical hardness and chemical stability. Increasing the filler content prevents environmental erosion of the silicone's surface and ensures its long-term ability to insulate electronic components. Isothermal testing is a common test technique for silicone. In low power electronics, typical evaluation temperatures are from 150 °C (1000 hours) to 200 °C (48 hours). In high power electronics, typical evaluation temperatures range from 250 °C to 300 °C. Typically, a 50% drop in mechanical, thermal, or electrical properties is deemed a material failure.

When polyimide (PI) dielectric is aged between 250 °C and 360 °C, it is found that while thermal degradation occurs within the polyimide material, oxidative degradation occurs predominantly near the surface layers [32]. Thus, thin polyimide film exhibits a thermo-oxidative degradation mechanism up to the glass transition temperature [32]. The initial reaction of the PI with the oxygen from atmospheric air leads to oxidation of the outer layer of the polyimide, which subsequently leads to the formation of both volatile and nonvolatile byproducts, causing transverse micro-cracks in the rigid oxidized layer [33]. The cracks in the oxidized layer allow oxygen to penetrate deeper into the polyimide bulk, further accelerating the degradation mechanism. The time to failure of the polyimide dielectric is thickness-dependent because of the surface degradation mechanism, and the results show that thickness is the most appropriate aging marker. The initial deposited polyimide film thickness must be >14.5 μm to withstand 600 V of breakdown voltage at 300 °C [32].

Moisture influences the long-term durability of the polymeric encapsulation. High moisture content could decrease the ultimate fatigue and brittle strength of the molding compound. Low fracture toughness of the molding compound decreases its effectiveness to resist the initiation and propagation of fatigue and brittle cracks [34]. Hygrothermal stresses and hygroscopic stresses caused by the swelling and warpage of the polymeric material increase the probability of package cracking [35]. Cracking of the encapsulation, which leads to loss of protection from environmental factors, is a common failure mode of encapsulations. Degradation of dielectric properties, such as dielectric constant, dissipation factor, and dielectric breakdown strength, is the result of the change of material properties at high operation temperatures.

6.6 Summary

In this chapter, basic reliability concepts for power electronic packaging and modules are presented. Standard reliability testing, such as temperature cycling, power cycling, high-temperature gate bias, is described. Due to the typical large operation temperature excursions for power packages and modules, thermo-mechanical stresses become important for many failure mechanisms. Manson–Coffin and Arrhenius relationships are useful in describing these failure mechanisms. At the time of writing this chapter, new standards are critically needed to standardize the testing and reliability of these power packages and power electronic modules operating at high temperatures, defined as temperatures higher than 175 °C.

References

[1] Harris, P.G., and Chaggar, K.S., The role of intermetallic compounds in lead-free soldering, *Soldering and Surface Mount Technology*, 1998;10(3):38–52.

[2] National Institute of Standard and Technology. *Engineering Statistics Handbook*. Available from: http://www.itl.nist.gov/div898/handbook/apr/section1/apr153.htm (accessed: 2 February 2025).

[3] JEDEC Solid State Technology Association. Available from: http://www.jedec.org/ (accessed: 2 February 2025).

[4] MIL-STD-883 – Test method standard for microcircuits.

[5] webstore.iec.ch/p.../info_iec60068-2-14%7Bed5.0%7Den_d.img.pdf.

[6] JEDEC Solid State Technology Association. JESD22-A104D (Revision of JESD22-A104C, May 2005). 2009 March; Available from: www.jedec.org/sites/default/files/docs/22a104d.pdf (accessed: March 2015).

[7] Groothuis, S., W. Schroen, and M. Murtuza, Computer aided stress modeling for optimizing plastic package reliability, *IEEE International Reliability Physics Symposium*, 1985; pp. 184–191.

[8] Siewert, T.A., Madeni, J.C., and Liu, S., Formation and growth of intermetallics at the interface between lead-free solders and copper substrates, *Proceedings of APEX*, 1994; pp. 1–7.

[9] Rodriguez, M., Shammas, N., Plumpton, N., Newcombe, D., and Crees, D., Static and dynamic finite element modeling of thermal fatigue effects in IGBT modules, *Microelectronics Reliability*, 2000;40:455–463.

[10] Ciappa, M., Selected failure mechanisms of modern power modules, *Microelectronics Reliability*, 2002;42:653–667.

[11] http://heraeus-contactmaterials.com/media/webmedia_local/media/downloads/documentsbw/brochure/HERAEUS_BondingWire_Brochure_2012.pdf.

[12] Hamidi, A., Beck, N., Thomas, K., and Herr, E., Reliability and lifetime evaluation of different wire bonding technologies for high power IGBT modules, *Microelectronics Reliability*, 1999;39:1153–1158.

[13] Schafft, H., Testing and fabrication of wire bonds electrical connections – A comprehensive survey, *National Bureau of Standards, Technical Note*, 1972; 726:106–109.

[14] Ang, S.S., Rowden, B.L., Balda, J.C., and Mantooth, H.A., Packaging of high-temperature power semiconductor modules, *Electrochemical Society Transactions*, 2009;27(1):909–912.

[15] Knoll, H., Weidenauer, W., Ingram, P., Bennemann, S., Brand, S., and Petzold, M., Ceramic substrates with aluminum metallization for power application, *Proceedings of the Electronic System-Integration Technology Conference (ESTC)*, 2010; pp. 1–5.

[16] Dupont, L., Lefebvre, S., Khatir, S., and Bontemps, S., Evaluation of substrate technologies under high temperature cycling, *Proceedings of the 4th International Conference in Integrated Power Systems (CIPS)*, 2006; pp. 1–6.

[17] Mustain, H.A., Brown, W.D., and Ang, S.S., Transient liquid phase die attach for high-temperature silicon carbide devices, *IEEE Transaction on Component, Packaging, and Manufacturing Technologies*, 2010;33(3):563–570.

[18] Yoon, S.W., Glover, M., Shiozaki, K., and Mantooth, H.A., Highly reliable double-sided bonding used in double-sided cooling for high temperature power electronics, *Proceedings of IMAPS High Temperature Electronics Conference (HiTEC)*, 2012; pp. 1–6.

[19] Yoon, S.W., Glover, M.D., Mantooth, H.A., and Shiozaki, K., Reliable and repeatable bonding technology for high temperature automotive power modules for electrified vehicles, *Journal of Micromechanics and Microengineering*, 2013;23(1):15–17.

[20] Jiang, L., Thermo-mechanical reliability of sintered-silver joint versus lead-free solder for attaching large-area devices, *Master thesis*, Virginia Polytechnic Institute and State University, Blacksburg, VA, 2010.

[21] Lang, F.Q., Hayashi, Y., Nakagawa, H., Aoyagi, M., and Ohashi, H., Joint reliability of double-side packaged SiC power devices to a DBC substrate with high temperature solders, *Proceedings of the 10th Electronics Packaging Technology Conference*, 2008; pp. 897–902.

[22] Bai, G., Low-temperature sintering of nanoscale silver paste for semiconductor device interconnection, *PhD dissertation*, Virginia Polytechnic Institute and State University, Blacksburg, VA, 2005.

[23] Bosco, N.S., and Zok, F.W., Strength of joints produced by transient liquid phase bonding in the Cu–Sn system, *Acta Materialia*, 2005; 53:2019–2027.

[24] Ong, B., Helmy, M., and Chuah, S., Heavy al ribbon interconnect: An alternative solution for hybrid power packaging, *Proceedings of the 37th International Symposium on Microelectronics (IMAPS)* 2004; Long Beach, California, pp. 14–18.

[25] Guth, K., Siepe., D, Görlich, J., *et al.*, New assembly and interconnects beyond sintering methods, *Proceedings of the PCIM Europe*, 2010; Nuremberg, Germany.

[26] Stockmeier, T., Beckedahl, P., Gobl, C., and Malzer, T., SKiN: Double side sintering technology for new packages, *Proceedings of the IEEE 23rd International Symposium on Power Semiconductor Devices and ICs (ISPSD)*, 2011; pp. 324–327.

[27] Zhang, H., Ang, S.S., Mantooth, H.A., and Krishnamurthy, S., A high temperature, double-sided cooling SiC power electronic module, *IEEE Energy Conversion Congress and Exposition (ECCE)*, 2013; pp. 2877–2883.

[28] Ozmat, B., Korman, C.S., McConnelee, P., Kheraluwala, M., Delgado, E., and Fillion, R., A new power module packaging technology for enhanced thermal performance, *Proceedings of the Seventh Intersociety Conference on Thermal and Thermomechanical Phenomena in Electronic Systems*, 2000; pp. 287–296.

[29] Rhodes, K., Riegler, B., Thomaier, R., and Sarria, H., *Silicone adhesives and primers on low surface energy plastics and high strength metals for medical devices*, Nusil Technology. Available from: http://nusil.com/en/news/post?id=2875ac5f-45d9-44a4-9b5b-3b7b0c3994d8&mediaId=4282395d-d524-418d-9456-c51e71aa1b87.

[30] Zhou, J., Ang, S.S., Mantooth, H.A., and Balda, J.C., A nano-composite polyamide imide passivation for 10 kV power electronics modules, *Proceedings of the 2012 IEEE Energy Conversion Congress and Exposition*, 2012, pp. 4262–4266.

[31] Kumar, R., Molin, D., Young, L., and Ke, F., New high temperature polymer thin coating for power electronics, *Proceedings of the Nineteenth Annual IEEE Applied Power Electronics Conference and Exposition (APEC '04)*, 2004; pp. 1247–1249.

[32] Khazaka, R., Locatelli, M.L., Diaham, S., and Bidan, P., Endurance of thin insulation polyimide films for high-temperature power module applications, *IEEE Transactions on Components, Packaging and Manufacturing Technology*, 2013;3(5):811–817.

[33] Tsotsis, T.K., Keller, S., Bardis, J., and Bish, J., Preliminary evolution of the use of elevated pressure to accelerate thermo-oxidative aging in composite, *Polymer Degradation Stability*, 1999;64(2):207–212.

[34] Deshpande, A., Study and characterization of plastic encapsulated packages for MEMS, *Thesis*, Worcester Polytechnic Institute, Worcester, MA, 2005.

[35] Wong, E.H., Rajoo, R., Koh, S.W., and Lim, T.B., The mechanics and impact of hygroscopic swelling of polymeric materials in electronic packaging, *Transactions of the ASME* 2002;124:122–126.

Chapter 7

Modelling for the lifetime prediction of power semiconductor modules

Ivana F. Kovačević-Badstuebner[1], Johann W. Kolar[1] and Uwe Schilling[2]

Reliability engineering has emerged as a relatively new branch of Power Electronics (PE) supporting the fast progress towards advanced Power Electronic Converter Systems (PECS) with significantly improved reliability ratings. PECS operate under increasingly severe temperature profiles, i.e., fast temperature cycling between extreme temperature levels. Accordingly, the reliability requirements for power semiconductor modules as fundamental components of PECS are significantly increased. Power module manufacturers have been working on new power module designs and packaging technologies in order to increase endurance and prolong the lifetime of power modules in the future, and subsequently enable high performance of the PECS also concerning reliability [1]. In the future, the reliability aspects have to be included into novel multi-domain optimization tools that will further improve the design of PECS. The first step towards this goal is to allow the integration of lifetime models of the system components into the design process.

The reliability of power modules represents a highly interdisciplinary topic as it requires a deeper knowledge in different areas: (1) mechanical design and thermal capabilities of power modules, (2) physics of failure (PoF) explained by material science, and (3) power electronics as its application field. The prediction techniques and sophisticated models for lifetime estimation of power modules have been studied and received more and more attention. Major investigations have been carried out by the manufacturers of power modules as they possess detailed product data, the expertise and resources necessary for developing and verifying lifetime models of power modules. Most of these models are empirical lifetime models used for the characterization of power cycling capabilities of power modules, e.g., the lifetime models presented in [2,3]. They are practically based only on experience and statistical analysis of the large databases gained in long-running accelerated

[1]Power Electronic Systems Laboratory, Swiss Federal Institute of Technology (ETH) Zurich, Switzerland
[2]SEMIKRON Elektronik GmbH & Co. KG, Sigmundstrasse, Nuremberg, Germany

cycling tests. The existing empirical models are often used in practice by PE engineers as the only means for an end-of-life (EOL) estimation of power modules and, further then for the prediction of PECS reliability under specified mission profiles. However, these lifetime models must be carefully applied, as normally, an extrapolation is conducted when applying the models gained from accelerated testing to field conditions. The validity of this extrapolation is sometimes doubtful.

Accelerated power and temperature cycling tests are used to emulate the actual operation conditions in the field with the aim to collect relevant data for building the lifetime models of power modules. A lifetime model can be defined as a tool for the EOL estimation of a power module exposed to cyclic thermal loading. By carefully controlling test conditions, it can be assumed that failures are only due to wear-out effects. Interconnection failures are reported to be the most common causes of failure of power modules operated in these cycling tests. Usually, the critical parts of standard power modules in power and/or temperature cycling experiments are three interconnections: wire bonds, the chip solder joint, and the substrate-baseplate solder layer. In the study presented in [4], it was shown that the bond wire lift-off failure mode and solder fatigue have different physical behaviour and thus have to be separately analysed and treated differently. In the sense of the empirical lifetime modelling approach, this requires developing a specific model for each failure mode separately. This, in turn, implies performing the separation of failure modes in the first place and then an extensive number of power cycling experiments that typically take considerable time.

The main drawback of the empirical models is that they are purely statistical and do not directly describe and assess the deformation mechanisms of power modules invoked under complex thermal loadings. Therefore, with new concepts of power module packages, and ever-increasing lifetime expectations for power semiconductor devices operating under very challenging environmental conditions, many research groups have started investigating physics-based lifetime models [5–9]. The physical modelling of failure includes the analysis and modelling of actual failure mechanisms, i.e., the stress and strain development under thermal loading. The calibration of physics-based lifetime models typically requires only several power cycling tests. The physics-based lifetime models provide a better insight in the physical process leading to a failure, and furthermore, they can be integrated in to multi-domain modelling tools for the virtual design of PECS. In addition, they can contribute to the validation of the extrapolation of existing empirical models. As a result, the PoF approach has emerged as a new methodology in power electronics that can potentially improve the lifetime estimation and enable the reliability engineering to be integrated into the development and research cycles of the overall design process of PECS. However, today, physics-based lifetime modelling is still missing in common engineering practice with respect to the design for reliability in PE.

This chapter reviews the basic ideas and difficulties with the physics-based lifetime modelling of power semiconductor modules, demonstrating the capabilities of a developed physics-based lifetime model for solder interconnections. The chapter is organized as follows. In Section 7.1, accelerated power and temperature cycling tests are briefly described. The power (active) and temperature (passive) cycling tests are used by manufacturers to verify the required functionality of the power module during the total field lifetime. Power cycling tests are important for the lifetime calculation of power modules operating in the actual applications. Section 7.2 presents the standard power module package and its intrinsic failure modes. The main focus is put on the most-stressed parts of the power module assembly that define the dominant failure mechanisms occurring during the power cycling experiments. An overview of existing empirical and physics-based lifetime models for power modules is given in Section 7.3. These lifetime models represent the state-of-the-art of the lifetime modelling of power modules. The necessary steps for further investigations and improvements are presented, which should improve the design for reliability of PECS. In Section 7.4, the theoretical background of the physics-based lifetime modelling of solder joints is given. Subsequently, in Section 7.5, a proposed physics-based lifetime model for the power module's solder joints is introduced. Finally, the theoretical considerations are verified based on a set of power cycling test data provided by SEMIKRON Elektronik GmbH & Co. KG [10].

7.1 Accelerated cycling tests

PECS, like those used in cars, locomotives, and airplanes, require long-ranging lifetime guarantees of five up to 30 years [11]. Collecting the system information for the reliability assessment under real application conditions even over a mid-term period of several years is not feasible since the technology changes very fast and the manufacturers of power modules must provide certain warranties for new products. Therefore, accelerated power cycling (PC) and temperature cycling (TC) tests are performed and well established among the manufacturers of power modules as the means for a lifetime estimation of products. PC and TC tests are designed to accelerate the failure modes occurring in the real applications due to thermo-mechanically induced wear-out. The IEC International Standards for semiconductor devices such as IEC 60747-34 and IEC 60747-9 [12,13] define endurance and reliability test setups; however, the exact procedures for PC and TC tests, e.g., setting the level of load current and the heating-cooling times, are mostly product-oriented, and, hence, defined independently by manufacturers [14].

TC tests are performed using temperature-controlled chambers to evaluate the influence of ambient temperature variations on power modules and are mainly aimed at the lifetime testing of large area solder interconnections such as the solder joint between the baseplate and substrate of the power module [15]. In the PC tests, a power module is actively heated by switching on and off the current, resulting in defined heating and cooling phases. Accordingly, the semiconductor chip represents the source of heat dissipation and, hence, the interconnections in close vicinity to the chip, i.e., bond wires and chip solder, are more stressed and likely to fail. PC tests are commonly used by manufacturers to trigger wear-out mechanisms of power modules in use in a shorter time period and gain useful data for the lifetime modelling.

7.2 Dominant failure mechanisms

Power modules represent main functional elements of power electronic converters and typically contain several semiconductor devices, e.g. MOSFETs, IGBTs, and diodes. The package of power modules has a very important function as an interface between the semiconductor devices and the circuit application.

Standard power modules are designed in a multi-layer structure, as shown in Figure 7.1. The heat dissipated inside a power semiconductor chip, illustrated by the power flow P_V in Figure 7.1, is conducted through the multi-layer structure into a heatsink, and is then transferred by convection to the ambient. The chip (silicon layer) is soldered to a direct-bonded-copper (DBC) ceramic substrate and then, at the bottom side, the substrate is soldered to a metal baseplate. The baseplate and heatsink are interfaced via a layer of thermal grease to achieve a better thermal contact. The layers silicon(Si)-copper(Cu)-ceramic(AlN, Al_2O_3), characterized by different thermal properties, have great influence on the heating and cooling rates of the module and determine its overall thermal performance. The thermo-mechanical stress appearing within a power module can be directly related to the temperature gradients formed across the layers and the difference between their thermal properties, such as the coefficients of thermal expansion (CTE). Consequently, fatigue failures of power modules are directly related to the power modules assembly structure and packaging technology.

There are different failure mechanisms of power modules, such as bond wire fatigue, reconstruction of chip metallization, solder fatigue, fatigue crack propagation within the ceramic substrate, corrosion of aluminium bond wires, and burn-out failures, which can occur within the power module when it operates in the field and fatally affect its functionality [16]. Particularly, the failures can be a result of wear-out, a poorly designed and optimized manufacturing process, some external sources, e.g., cosmic rays, a usage out of specifications. The reliability of a power module can then be defined as a probability that the power module performs its required function for a specified time interval under certain operating conditions.

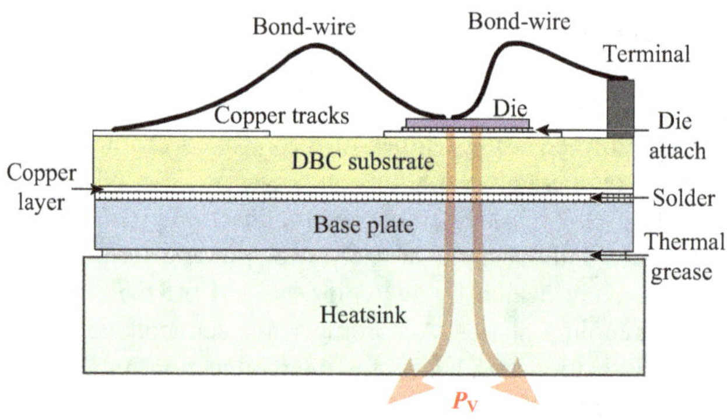

Figure 7.1 Multi-layer structure of a standard power module design

Typically, the failure rate over time is described by the so-called bathtub-curve, consisting of early life failures, random failures, and EOL failures. Lifetime models, e.g., described in [2,3,17], serve to describe EOL failures due to wear-out, i.e., aging of material. They are used to gain a better understanding of the useful functional period of a component operating in the field. Moreover, lifetime models play an important role in the thermal and electrical design of a PECS as they describe the power and temperature cycling capabilities of power modules, the influence of the cycling parameters on the number of cycles to failure, N_f, which in turn determines the EOL of the power module under the assumed conditions. Accordingly, the lifetime modelling represents a valuable step for a general reliability analysis of power modules.

The root cause of wear-out is the thermo-mechanical stress that is invoked in the materials of the layers of the power module structure operating under intrinsic cyclic temperature profiles. It was shown that during accelerated power/temperature cycling tests, the inner interconnections of the power module, such as bond wires, the chip solder layer, and the substrate-baseplate solder joint (see Figure 7.1), determine the lifetime of the whole assembly. In practice, it is hard to distinguish the dominant failure mode as there is a close interaction between different failure mechanisms. For example, solder fatigue results in a higher thermal resistance of the power module $R_{th(j-s)}$ (= $\Delta T_{j-s}/P_V$), where ΔT_{j-s} is the temperature difference measured between the chip and the heatsink) that leads to higher junction temperatures T_j and thus, to higher stress levels at the place of the bond wires, which results eventually in a lift-off of bond wires. Similarly, bond wire lift-off causes uneven current distribution that produces higher power losses and/or higher T_j and, hence, more thermal stress at the solder interconnections.

Carefully specified accelerated cycling tests are performed to decouple different failure modes, and failure criteria must be defined in order to acquire useful data for the generation of lifetime models. Typically, the bond-wire lift-off failure mechanism is detected by an increase of the forward voltage drop of 5% to 20%. An indicator of solder fatigue is an increase of 20% to 50% of the thermal resistance, $R_{th(j-s)}$, of the power module measured between the chip and the heatsink, which is caused by crack propagation within the solder interconnecting layer [14,18].

Driven by the automotive industry and improvements in traction systems, the main requirement for new generations of power modules is to allow an operation at higher temperature levels, e.g. above 150 °C. The reliability of power module's interconnections is a factor that limits the increase of the allowed junction (chip) temperature. Accordingly, the idea for novel design concepts of power modules has been to eliminate and/or reinforce the most-stressed interconnections, i.e., wire bonds and solder interconnecting layers. The first solder-free IGBT power module, the SEMIKRON SKiM module, became commercially available in 2008 and has demonstrated significantly enhanced power cycling capability, as reported in [19]. In this novel power module concept, the baseplate is removed and the DBC substrate is pressed onto the heatsink, i.e., pressure is induced by a pressure part on top which is screwed to the heatsink. Additionally, a silver sintering technology is applied for die-attach, and spring contacts are used instead of soldering for the interconnections between the substrate and the terminals for the load current, the gate control circuit, and auxiliary contacts such as the auxiliary emitter and temperature sensor. The main advantage of this novel concept is the capability to operate at much higher temperatures due to the significantly higher melting temperature of the silver sinter layer in comparison to the melting temperature of solder alloys.

7.3 Lifetime modelling

The power modules of PECS are rated for high switching current and blocking voltage levels. The power losses generated under the field/test operation conditions lead to high temperature variations, which in turn produce thermo-mechanical stresses within the power modules that gradually degrade their functionality and lead to wear-out. As the first step of the lifetime modelling, the power losses of the PECS must be calculated for a characteristic load profile by means of electrical circuit simulations and translated via thermal modelling into corresponding temperature profiles, to which the power module's layers are exposed. Specifically, thermal modelling and simulation is performed to gain knowledge and information about the temperature distribution within the power module's structure, since the direct temperature measurements of the power module's layers are practically not possible. As the electrical properties of switching devices, i.e., power modules, are temperature-dependent, the thermo-electric modelling might be necessary for the accurate determination of the actual temperature behaviour. Accordingly, thermal (and/or thermo-electric) modelling can be seen as a constitutive part of the lifetime prediction approach.

The main principle of the lifetime modelling is based on the correlation of the number of cycles to failure, N_f, directly gained in the PC and/or TC tests to the corresponding temperature profiles. This further allows a quantitative evaluation of the response of the power module to the temperature changes in the application field. Lifetime modelling of power modules therefore strongly depends on the cycling test results. Two lifetime modelling approaches can be distinguished: empirical lifetime models and physics-based lifetime models.

This section presents the main steps of the lifetime modelling procedure for power semiconductor modules. First, the principles of thermal modelling of power modules are described, and then, the existing empirical and physics-based lifetime models are briefly summarized, pointing out the main differences of the physics-based modelling approach in comparison to the well-established empirical lifetime models.

7.3.1 Thermal modelling

Thermal problems in power electronics can be solved in a simplified way by means of one-dimensional thermal networks, defined by the so-called Cauer and Foster models [20]. These thermal networks are built of thermal circuit elements, i.e., thermal resistances R_{th} and thermal capacitances C_{th}. Similar to an electric circuit, C_{th} are included to model the transient response, while only R_{th} can be used to model the steady-state thermal behaviour. The Cauer model has a physical interpretation as the internal nodes of the Cauer network can be directly associated with the layers of the power module, which is not the case for the Foster model. In the Foster model, only the input node with respect to the power source corresponds to an actual physical point of the power module. Both thermal models return the same transient behaviour of this input node, i.e., the same overall thermal system impedance, Z_{th}. In particular, the same thermal behaviour is obtained for different sets of R_{th} and C_{th} that define the Cauer model and the Foster model, respectively. The main advantage of the Foster model is that it provides a straightforward analytical expression for Z_{th}, which can be used to numerically calculate the thermal response of the overall power module structure to an application given mission profile. Under the real operation conditions, the power module's layers experience different temperature distributions, both in the lateral and vertical directions. The layers below the chip typically exhibit a similar temperature pattern as the chip (silicon), and their maximum temperature is lower than the maximum temperature of the chip. Having non-constant temperature over the surface area, the main question that arises with the physics-based lifetime modelling is which temperature value should be applied in the corresponding lifetime models, i.e., the average, the maximum, or the minimum temperature. In particular, when modelling the thermo-mechanical stress inside a power module's interconnecting layer, its lateral temperature gradients have to be taken into account. Accordingly, three-dimensional (3D) thermal solvers allowing finer discretization of the power module's structure have to be employed. This allows a more comprehensive and detailed calculation of the temperature evolution, which is required for the physics-based lifetime modelling.

Thermal solvers are typically based on a finite difference method (FDM), and they use a 3D network of equivalent thermal resistances, R_{th}, and capacitances, C_{th}, for the modelling of the power flow inside the power module package. 3D thermal modelling provides both lateral and vertical temperature gradients of the power module's layers. This lateral and vertical temperature distribution is used for a comprehensive thermal characterization of the power module package. On the other hand, thermal characterization based on measurements provides only the so-called virtual junction (chip) temperature T_{vj}, since the inner points of the power module's structure are not accessible for direct temperature measurements without alterations of the structure. Accordingly, thermal modelling represents a necessary step for a valid lifetime estimation process, which specifically provides information about the temperature distribution in the observed solder layer for lifetime modelling of the solder intersections.

What distinguishes power modules from other physical systems is the ability to accommodate very fast temperature changes. The fast temperature transients make the temperature measurements non-trivial and thus standard thermocouples cannot be applied. The virtual junction temperature, T_{vj}, is therefore typically derived from measurements of the temperature-dependent electrical parameters of the power semiconductor devices, e.g., the on-state saturation voltage $V_{ce}(T)$ between the collector and emitter contacts at low (sense) current levels. In [21], the physical meaning of this single temperature value T_{vj} is related to the temperature distribution across the active surface of the chip. Specifically, it was shown that T_{vj} corresponds to the area-weighted average temperature over the active chip area.

7.3.2 Empirical lifetime models

Empirical models are deduced from experience and large databases of PC test results collected over many years for different module technologies; they express lifetime in terms of the number of cycles to failure, N_f. Specifically, empirical models describe the N_f-dependence on the parameters of PC tests, such as maximum, mean, or minimum temperature, cycle frequency, heating and cooling times, load current, and furthermore the power module's properties such as blocking voltage class, and the geometry of bond wires. PC tests are conducted to investigate the influence of PC test parameters on the N_f-results. It was observed that the dominant failure mechanisms under PC experiments are the failures of bond wires and solder interconnecting layers, i.e., the EOL of power modules is typically limited by the lifetime of these interconnections.

The first well-established empirical model, which resulted from the so-called LESIT project [2], did not distinguish between different failure mechanisms, i.e., a single analytical model was used to describe the N_f-lifetime of power modules. The main reason was the difficulty to set the PC experimental conditions to control dominant failure modes in order to make the database of PC N_f-results correspond to a single failure mechanism, i.e., either wire-bond lift-off or solder failure. These two failure mechanisms cannot be explained by the same dependencies on PC test parameters, as the natures of these failures are different. This implies that a separation of failure modes has to be performed in the first place, and then two analytical models have to be derived from PC experiments. Advanced interconnection technologies that became available in last couple of years have allowed making a clear distinction between these two failure modes under PC test conditions [22], and hence, enable power module manufacturers to develop different empirical models for each failure mode, which was not possible for the early lifetime models.

7.3.2.1 Examples of empirical lifetime models

As power modules are exposed to periodic thermo-mechanical stress, the well-known Coffin–Manson law can be used to relate the number of cycles to failure, N_f, and the temperature change ΔT (= $T_{max} - T_{min}$) reached in the applied power cycles,

$$N_f = a \cdot (\Delta T)^{-n}, \tag{7.1}$$

where the model constants a and n are the empirical parameters depending on the module design. The experience gained in PC experiments indicated that the other PC parameters besides ΔT have a strong influence on the number of cycles to failure. In the 1990s, the LESIT project [2] was conducted for standard power modules with Al_2O_3 ceramic substrate and copper baseplate from different manufacturers. As a result, the influence of the average junction temperature, $T_{j,m}$, was included into the Coffin–Manson N_f-model via the Arrhenius approach as

$$N_f = a \cdot (\Delta T_j)^{-n} \cdot e^{\frac{E_a}{k_b \cdot T_{j,m}}}, \tag{7.2}$$

where ΔT_j is the peak-to-peak variation of the junction temperature cycle ($T_{j,\,max} - T_{j,\,min}$), $T_{j,m}$ is the mean temperature value $(T_{j,\,max} + T_{j,\,min})/2$, and k_b is the Boltzmann's constant; a, n, and E_a are the model parameters that are determined by fitting (7.2) to the experimental PC data. E_a represents the activation energy that characterizes the deformation process. This analytical lifetime model corresponds to linear functions in a log-log, i.e. $\log(N_f) - \log(\Delta T_j)$, plane for different average junction temperatures, $T_{j,m}$.

As a next step, the N_f-lifetime dependency on the other PC parameters, such as the heating time and the power density, was examined.

ABB (Hamidi *et al.* [23]) conducted PC tests to characterize the influence of the heating time t_{on} on the thermal fatigue of high-voltage (HV) IGBT power modules. Two dominant failure mechanisms were taken into account: bond wires lift-off and baseplate solder failure. It was shown that longer t_{on} has a severe impact on the lifetime of power modules and represents an important ageing accelerator.

INFINEON (Bayerer *et al.* [3]) proposed a lifetime model,

$$N_f = K \cdot (\Delta T_j)^{\beta_1} \cdot e^{\frac{\beta_2}{T_j + 273}} \cdot t_{on}^{\beta_3} \cdot I^{\beta_4} \cdot V^{\beta_5} \cdot D^{\beta_6}, \tag{7.3}$$

which includes, in addition to (7.2), the influence of the heating time t_{on}, the current per wire bond stitch I, the diameter of bond wires D, and the chip voltage class V (defined by blocking voltage of chip) on the power-cycling capabilities. The impact of these parameters is assumed to have power law nature and, thus, the dependency of N_f on t_{on}, I, V, and D can be described by a linear log-log characteristic. This empirical lifetime model, referred to in the literature as CIPS2008 model, is based on a purely statistical analysis of a large number of PC test results from different module technologies. Additionally, the CIPS2008 lifetime model does not cover the lifetime limitation due to substrate-baseplate solder failures and it is not applicable for the power modules used in traction applications, as the assumed relation between the blocking voltage and chip thickness does not apply to modules for traction application [3]. The second limitation is that the model parameters cannot be set independently during the PC experiment, e.g., the heating time and a certain maximum temperature are correlated by the nature of power module and PC tests. The authors suggested therefore that the model should be carefully used and only in the range where the PC tests were performed.

Similarly, SEMIKRON (Scheuermann *et al.* [17]) has recently proposed a new lifetime model for the advanced power modules with sintered chips, e.g., SKiM modules. These modules exhibit significantly improved lifetime as the classical solder process for the die attach is replaced by Ag diffusion sintering technology, and the Al wire bond geometry is optimized. In this way, the bond wire lift-off and heel cracking were the only observed failure modes, so that the developed lifetime model corresponds only to the failure mechanisms due to thermo-mechanical stress of bond wires. The proposed lifetime model also includes the impact of the aspect ratio of Al bond wire (*ar*) as significantly higher lifetime was observed by increasing the height of the bond wire loops,

$$N_f = A \cdot (\Delta T_j)^{\alpha} \cdot ar^{\beta_1 \cdot \Delta T_j + \beta_0} \cdot \left(\frac{C + t_{on}^{\gamma}}{C + 1} \right) \cdot e^{\frac{E_a}{k_b \cdot T_{jm}}} \cdot f_{Diode}, \tag{7.4}$$

where A is a general scaling factor, ΔT_j is the junction temperature swing given in Kelvin, t_{on} is the load puls duration, T_{jm} is the absolute medium junction temperature in Kelvin, k_b is the Boltzmann's constant, E_a is an activation energy in eV, and f_{Diode} is a de-rating factor applied for the test on free-wheeling diodes; β_0 are β_1 are the model coefficients determined together with the other model parameters A, α, C, γ, E_a, and f_{Diode}, using a least square fitting procedure. The parameterization of this analytical model was based on a set of 97 PC tests, gained in the experiments that took approximately five years. According to the authors, more PC tests are still required to gain better understanding of the influence of the model parameters on the lifetime. Furthermore, the Ag-diffusion sintered modules and the standard modules with baseplate were compared in [4,22] that allowed distinguishing and analysing dominant failure mechanisms under different PC test conditions. The comparison has shown that: (a) wire bond lifetime is less affected by the medium junction temperature than the lifetime of the chip solder, which is reflected by lower activation energy, E_a; (b) the chip solder degradation is dominant for high temperatures; and (c) for an intermediate temperature range both failure mechanisms, i.e., solder failure and bond wire lift-off occur and lead gradually to the module's EOL. This referenced study represents an approach for separating failure modes that allowed building empirical lifetime models for each failure mode alone, with the aim to develop more accurate lifetime prediction methods.

7.3.3 Physics-based lifetime models

Physical modelling requires failure and deformation mechanisms to be known so that the stress and strain development within the power module assembly is modelled and directly correlated to the number of cycles to failure. Physical modelling represents the basis for the PoF analysis. PoF provides a deeper physical description of the observed failure mechanism, and hence, represents a promising alternative to empirical lifetime models.

Direct measurements of stress and strain in electronic packages demand the usage of high resolution measuring methods, e.g. infrared and scanning electron microscopy. The other way to determine the stress and strain deformations within a power module is by means of a stress–strain simulation via computational mechanics, e.g. finite element analysis (FEA). However, the physics-based modelling by FEA requires detailed knowledge of the material and geometry properties of the power module assembly, which is often not available in the power module datasheets and only accessible by the manufacturers. An alternative to FEA for the stress–strain modelling is based on a numerical approach for calculating the stress–strain response under a given temperature profile and using a parameterization procedure based on the PC experimental data, as it is described in the following. The existing physics-based lifetime models of power modules published in the literature are briefly summarized in this section with the aim of highlighting the state-of-the-art for the PoF analysis in power electronics with regard to power modules as reliability-critical components of PECS.

7.3.3.1 Model 1—ETHZ-PES lifetime model

One of the first physics-based models in power electronics [5,24,25], referred to as the ETHZ-PES lifetime model in this book, is dedicated to the planar solder joints within power semiconductor modules. It is based on a numerical algorithm for calculating the stress–strain evolution under a cyclic thermal load applied to the solder interconnecting layer and the thermo-mechanical model described in [26]. The applied numerical algorithm was originally developed for the solder joints of surface mount devices (SMD). The basic idea is that the solder response to a cyclic thermal load can be described by means of a hysteresis loop [27]. The number of cycles to failure N_f can then be calculated by Morrow's type of fatigue law as,

$$N_f = W_{crit} \cdot (\Delta w_{hys})^{-n}, \tag{7.5}$$

where the constant n ($n > 0$) depends on the solder type and varies from less than 1 up to 2.2 [25], W_{crit} is the energy leading to the failure, and Δw_{hys} represents the accumulated deformation energy per cycle, i.e., inelastic strain energy density calculated as the area enclosed by the hysteresis loop. As a result, the stress–strain solder response, i.e., hysteresis behaviour, can be employed as a tool for lifetime estimation of solder joints in electronic devices [28]. The material and geometry properties of the power module have to be known in order to calculate Δw_{hys}. Similarly to (7.5), the plastic strain range per cycle $\Delta\varepsilon_{hys}$ can also be used to calculate N_f using a Coffin–Manson's type of fatigue law,

$$N_f = \varepsilon_{crit} \cdot (\Delta\varepsilon_{hys})^{-n_1}, \tag{7.6}$$

which can be applied for many metals subjected to repetitive plastic deformation. The hysteresis solder behaviour is calculated from the constitutive equations for the specific type of eutectic solder alloy describing the elastic and (visco)plastic strain deformation of the solder joint under the applied stress. The material-dependent constants of the constitutive solder equations for various solder alloys are available in the literature. The main problem, as reported by Clech [29], is that constitutive solder equations were fitted to data sets of various sources corresponding to different solder joints so that the material parameters found in the literature show a significant scatter. Specifically, the solder deformation is typically measured on solder joints of SMD electronic assemblies, e.g., ball-grid arrays (BGA), flip-chip (FC), and ceramic chip carriers (CCC). The geometry of the solder joint has an influence on the stress distribution and, thus, the solder deformation is expected to differ for bulk solder specimen and large area planar solder joints of power modules from SMD solder joints. Due to a lack of available data, the material parameters corresponding to SMD electronic assemblies found in the literature were applied for the lifetime modelling based on the ETHZ-PES model in [5,24,25]. It should be emphasized that there is a strong requirement to verify the employed constitutive model for the solder interconnections in power modules in order to be able to correctly and accurately calculate the lifetime expectations.

Besides the material parameters, the constitutive solder equations include other parameters depending on the geometry of the assembly such as the effective assembly stiffness K and the imposed strain per degree temperature change D_1. These parameters have to be determined for the specific power module structure.

In [5,24], a parameterization procedure that takes a set of the experimental PC tests as input was suggested for the calculation of the unknown parameters K and D_1.

In [25], the authors used an FE method simulation of the power module's structure to find K and D_1 based on the approach suggested by Darveaux [30]. They proved a good agreement between the PC experiments and the developed physics-based lifetime model. The exponent n in (7.5) was the only parameter fitted to the experimental data.

Furthermore, using this energy-based model, the influence of different arbitrary temperature profiles on the lifetime of a power module can be estimated by means of W_{crit}, e.g., as illustrated in [24]. However, for accurate lifetime estimation, it is necessary to extract the actual temperature profile of the solder layer under investigation for the given operational conditions. Therefore, in comparison to empirical models, which are mostly based on the virtual junction temperature value, the described energy-based model requires the knowledge of local maximum solder temperature values. This potentially leads to computationally expensive calculations and modelling techiques in practice, e.g., 3D thermal modelling performed to gain the solder temperature from the measured junction temperature. Accordingly, mission profile evaluation based on this physics-based lifetime model has to be further analysed and investigated in the future, including also the extraction of stress–strain hysteresis loops from complex stress–strain curves that can be defined as response of the solder material to arbitrary temperature profiles.

The theory of the ETH-PES lifetime model is described in more detail in Section 7.4, and the verification of this proposed approach by means of results of PC tests is presented in Section 7.5.

7.3.3.2 Model 2—O. Schilling *et al.*

A second physics-based lifetime modelling approach was presented by O. Schilling *et al.* in [7]. This lifetime model is dedicated for the power module structure with the EOL determined by the failure of Al bond wires. The developed lifetime model is also based on a Morrow's type of fatigue law,

$$N_f = c_1 \cdot (\Delta w_{hys})^{c_2}, c_2 < 0. \tag{7.7}$$

The integrated deformation energy per cycle, Δw_{hys}, within Al bond wires was calculated from the stress–strain hysteresis loop using a 2D FEM simulation. Al bond wires are described by tensile strength, Young's modulus and yield strength in the 2D FEM simulation, while the solder layers are described by the viscoplastic ANAND model implemented in the commercial ANSYS FEM software tool [31]. The authors assumed $c_2 = -1.83$ taken from previous investigations presented in [32] that addressed the failure of power modules due to heel cracking of Al bond wires. However, the authors in [32] used the Coffin–Manson type of fatigue law based on the inelastic strain range defined by (7.6).

As the constant c_1 has no universal value and depends on geometry, the proposed approach returns only normalized N_f values. The lifetime modelling approach was compared with the normalized N_f vs. ΔT_j curve gained from empirical PC data assuming the power law $N_f(\Delta T_j)$ dependency. The authors stated that the 2D FEM simulation data and the empirical data agreed well within the uncertainty of measurements.

7.3.3.3 Model 3—Steinhorst *et al.*

A third physics-based modelling idea was recently presented by Steinhorst *et al.* in [6]. The proposed model calculates the crack initiation and propagation in the solder layers due to fatigue, on the basis of the Darveaux's energy-based model [30],

$$N_0 = K_1 \cdot (\Delta W)^{K_2}, \tag{7.8}$$

$$\frac{da}{dN} = K_3 \cdot (\Delta W)^{K_4}, \tag{7.9}$$

where N_0 is the number of cycles to crack initiation calculated based on the plastic energy density, ΔW, integrated during one temperature cycle, a is the crack length, and da/dN is the crack propagation. $K_j, j = 1\ldots4$ are the model parameters that have to be fitted to the experimental curves. The academic FEM package SPC-Pm2Ad [33] is used to simulate the stress–strain hysteresis loops of the solder layers. From the integrated deformation energy, the crack length can be obtained. Simultaneously with the crack development, a thermal simulation of the power module is performed, and an increase of more than 20% of the thermal resistance, R_{th}, is used as the failure criterion. The model is still in the development phase. The parameters of the model are arbitrarily selected, and also more comprehensive constitutive solder equations have to be implemented. The physics of crack propagation under thermal loading is complex in nature, as the crack propagation changes the structure of the solder layer. Additionally, the change of solder material properties with temperature should be also taken into account. The Darveaux's energy-based model given by (7.8) and (7.9) was also used in [34] to investigate the crack behaviour within the baseplate solder layer under different TC tests.

7.3.3.4 Model 4—Déplanque *et al.*

Another lifetime prediction method for the solder joints in power modules was presented by Deplanque *et al.* in [35,36]. It is based on the crack propagation within the assembly of a chip soldered on a copper substrate. Damage of the solder joint was characterized by three methods: scanning acoustic microscopy (SAM), measurements of the thermal resistance, and an FE method used to predict the crack initiation and propagation using the Paris law,

$$N_0 = C_1 \cdot (\varepsilon_{\mathrm{acc,\ int}})^{C_2}, \tag{7.10}$$

$$\frac{\mathrm{d}a}{\mathrm{d}N} = C_3 \cdot (\varepsilon_{\mathrm{acc,\ int}})^{C_4}, \tag{7.11}$$

where $\varepsilon_{\mathrm{acc,\ int}}$ is the mean value of the integrated accumulated creep strain along the lines coinciding with the direction of the crack propagation, N_0 is the number of cycles until crack initiation, $\frac{\mathrm{d}a}{\mathrm{d}N}$ is the crack propagation rate, and the constant parameters C_i, $i = 1\ldots4$, are the material-dependent coefficients that are determined by means of FEM simulations. The authors calculated the values of C_i, $i = 1\ldots4$, for two examined solder alloys, SnPb and SnAgCu (SAC305). The crack length L after N cycles can then be calculated as

$$L = \frac{\mathrm{d}a}{\mathrm{d}N} \cdot (N - N_0). \tag{7.12}$$

It was shown in [35] that the damage surface can be used as a damage indicator instead of using the crack length, i.e., the equivalent accumulated creep strain integrated along the damage surface of the solder layer using the commercial FEM tool ANSYS. The implemented solder model includes both the primary and secondary creep, which makes the model more accurate. As the authors emphasized, the developed lifetime model is not fully tested, and more investigations are required in order to be able to apply it universally for every planar solder joint.

Based on (7.10) and (7.11), Newcombe *et al.* suggested in [37] a PoF approach for the substrate-baseplate solder interconnection, i.e., the solder layer between the DBC substrate and the baseplate,

$$\frac{\mathrm{d}L}{\mathrm{d}N} = C_3 \cdot (\Delta\varepsilon_{\mathrm{p}})^{C_4}, \tag{7.13}$$

where L is the crack length, and $\Delta\varepsilon_{\mathrm{p}}$ is the accumulated plastic strain per cycle. The coefficients C_3 and C_4 were taken from [36] for SnPb solder alloy. Using this lifetime model, an increase in lifetime for higher solder thickness was observed. A similar work was performed in [9], calculating C_3 and C_4 coefficients by FEA computer simulation for SnAg solder.

Furthermore, in [37], the main idea was to show that a PoF method using (7.13) integrated in a design optimization tool has a tremendous advantage over the standard Coffin–Manson approach. The parameters of a Coffin–Manson model (7.1) always correspond to a specific power module structure that has to be fully characterized in advance. However, the physics-based lifetime model defined by (7.13) enables an efficient lifetime prediction when some of the design paremeters are altered, e.g., the thickness of the solder layer.

Based on Model 4, a PoF approach for reliability prediction of IGBT modules, including four failure modes/locations—chip solder joint (die attach), substrate solder joint, the busbar solder joint and Al wire bonds—was presented in [38]. Two physics-based lifetime models were used: the model (7.13) was used for the lifetime estimation of solder interconnecting layers, and the $N_f(\Delta\varepsilon_p)$ relation given by (7.6) was employed for the lifetime prediction of the bond wires. The advantages of such physics-based reliability design tools were pointed out by the authors.

7.3.3.5 Model 5—Yang *et al.*

The damage-based crack propagation model in the time domain described in [8,39] represents a very comprehensive physics-based model for wire bonds in power electronics modules. It takes into account both the damage accumulation and the damage removal processes. It was shown that both mechanisms have an important impact during thermo-mechanical cycling. Specifically, the measurements of the bond shear force showed that the bonds subjected to temperature cycles with higher maximum value, T_{max}, had slower wear-out rates even though they were exposed to larger temperature swings, ΔT. This behaviour was explained by the diffusion-driven damage removal mechanism and it was stated that neglecting this mechanism can lead to wrong lifetime predictions. Similarly to the ETHZ-PES lifetime model, the main idea of the damage-based crack propagation model is to avoid extensive accelerated testing, and additonally, to enable a more accurate lifetime estimation for arbitrary mission profiles.

The proposed damage-based lifetime model was evaluated against several TC tests only and the parameterization procedure was not explained in a comprehensive way. Additionally, a detailed verification that would allow for the usage of this lifetime estimation approach in typical PC test applications is needed in order to prove its functionality and accuracy.

The authors also summarized in [8] the state-of-the-art lifetime prediction models for wire bonds analysing their limitations and necessary investigations that have to be performed to gain better understanding of the real deformation mechanisms occurring in the bond wires under thermal loading, e.g. grain coarsening and softening.

7.3.4 Lifetime prediction based on PC lifetime models

The lifetime prediction of power modules using a PC test lifetime model, i.e., N_f-based model, is performed based on Miner's rule for damage accumulation [26]. According to Minor's rule, each temperature swing, ΔT_k, contributes independently to the decrease of the power module's lifetime with an extent defined by the corresponding number of cycles to failure, $N_f|_{\Delta T_k}$, $Q(\Delta T) = \frac{N|_{\Delta T_k}}{N_f|_{\Delta T_k}}$, where $Q(\Delta T_k)$ is the total damage produced by N cycles at the temperature swing of ΔT_k. EOL is reached when Q equals unity, $Q = 1$. The lifetime of a power module operating under n different temperature conditions, $k = 1\ldots n$, is then defined by superposition of all individual damages $Q(\Delta T_k)$, and the EOL is reached for $\sum_{k=1}^{n} Q(\Delta T_k) = 1$.

In a real operating environment, power modules are exposed to rather non-uniform temperature variations, and thus, a proper definition of a temperature cycle within a given mission profile is required for the N_f-based lifetime modelling. Typically, the Rainflow algorithm is used to extract the temperature cycles from an arbitrary temperature profile. The Rainflow-counting algorithm was originally developed to define cycles as closed stress/strain hysteresis loops within a complex stress–strain curve in order to asses the fatigue life of material exposed to complex loading [40]. If it is assumed that the junction temperature is proportional to the stress in a given design, the Rainflow algorithm can be applied to count and measure temperature cycles contained within an arbitrary temperature profile. Generally, a cycle-counting algorithm cannot easily describe the impact of time-dependent effects and non-linear damage accumulation on the wear-out behaviour of power module interconnections. Thus, the transformation of a mission-profile into a sequence of temperature cycles of certain duration remains a problem that introduces some uncertainties into the lifetime prediction especially for highly irregular temperature profiles such as those experienced in automotive applications.

7.4 Physics-based lifetime estimation of solder joints within power semiconductor modules

The physics-based lifetime modelling of the solder joints used in power semiconductor modules is based on the stress-analysis of the module's assembly. The stress levels are used as indicators for the activation of different failure modes. Firstly, the nature of the stresses has to be examined, i.e., how stress appears and acts within materials, and secondly, the calculated or measured stress levels have to be correlated with the observed failures. The wear-out failures typically result from the thermally induced stresses, which in turn originate from the mismatch in thermal expansion of the adjoining layers. Specifically, mechanical and physical properties of the interconnecting interfaces determine the nature of stresses that the power module is able to survive. Furthermore, the time parameter also has to be included as a relevant variable in the stress analysis as the damage mechanisms within power modules are time-, geometry- and material-dependent.

The deformation occurring in a solder interconnecting layer can be numerically described by the constitutive solder equations. These constitutive equations are based on two state variables represented by two physical measures: stress and strain.

The thermo-mechanical deformation is often described by the equivalent von-Misses stress σ^e and strain ε^e in the 3D space, using three principle planes [41]. The equivalent *von-Misses* stress and strain can be expressed in terms of the actual normal (σ, ε) and shear (τ, γ) stress and strain components that are developed when the solder interconnecting layer is exposed to a cyclic thermal load. For the planar solder joints within power modules, the shear stress and strain components are important due to the geometry and material properties of the power module's assembly. The effective shear stress τ and strain γ can be derived from the equivalent *von-Misses* stress σ^e and strain ε^e as [41]

$$\tau^* = \frac{1}{\sqrt{3}} \cdot \sigma^e, \tag{7.14}$$

$$\gamma^* = \sqrt{3} \cdot \varepsilon^e. \tag{7.15}$$

In the literature, the consitutive solder equations are either given in the form of shear or tensile solder data. The transformations given by (7.14) and (7.15) are then used to convert the normal stress τ and strain ε components into the shear components τ and γ and vice versa [42].

Two different deformation mechanisms of solder material can be distinguished: fatigue and creep/stress relaxation. Fatigue occurs in solder due to the constant accumulation of damage when the solder material is exposed to cyclic thermo-mechanical loading. However, at sufficiently high stresses and temperatures, the solder material shows the tendency to move or deform permanently to relieve stress. These phenomena are referred to as creep and stress relaxation. Compared to fatigue, i.e., time-independent deformation caused usually by the initiation and propagation of cracks, creep, and stress relaxation are types of time- and rate-dependent damage mechanisms and are usually caused by the formation and growth of voids within the solder material [43]. The term "creep" is more specifically used for straining under a constant stress, while stress relaxation corresponds to the deformation under a constant strain.

At the temperatures developed under PC and TC test conditions or in the field defined by the mission profile, both the creep and fatigue damaging processes can be activated in the power module. Instead of pure creep or pure fatigue, the regime of creep-fatigue interaction is more often present [43]. It is very hard to say which of these two mechanisms will predominate and finally cause the failure of the weakest solder interconnecting layer.

The next sections (Sections 7.4.1–7.4.4) address the theory of the stress and strain deformations of solder material exposed to cyclic thermal loading. The constitutive solder equations relevant for the physics-based lifetime models of solder joints in power semiconductor modules are summarized.

7.4.1 *Stress–strain (hysteresis) solder behaviour*

Due to the difference of thermo-mechanical properties of the adjacent components connected by solder, shear force acts across the solder interconnecting layer and produce simultaneous bending and stretching of the whole assembly. Analysing deformation mechanisms of the solder joint layer between ceramic chip-carriers (CCC) and printed wiring boards in the Leadless CCC assemblies, Hall found that the shear stress–strain response of the solder joint to periodical temperature cycling has the shape of a hysteresis loop [27]. The accumulation of damage in the solder material is reflected by the area enclosed by the hysteresis loop. A hysteresis loop describes the complex stress–strain history of the solder layer exposed to repetitive cyclic loading, and the shape of the hysteresis can be explained by the behavioural model of solder material. Furthermore, it was observed that the points on the hysteresis curve corresponding to the same temperatures form a set of parallel lines called the stress reduction lines.

A resulting stress–strain loop that is not closed means that the stress–strain response may change with successive load cycles. In the simulation, the hysteresis loop usually tends to stabilize after several load cycles. With respect to the thermo-mechanical deformations occurring in power modules, a load cycle corresponds to a temperature cycle that is either generated under the accelerated cycling tests or in the application field. The parameters that determine the shape of a stabilized hysteresis loop are presented schematically in Figure 7.2. The minimum and maximum shear stress and strain values, τ_{min},τ_{max}, γ_{min} and γ_{max}, correspond to the extreme values of the temperature cycle, T_{min} and T_{max}.

At different stress levels and temperatures, the solder material experiences different physical deformations that gradually contribute to damage accumulation and finally to the failure of the solder interconnection. As a result, the stress–strain plot can be employed as a tool for lifetime estimation of solder joints in electronic devices, which was also described by the proposed thermo-mechanical model in [26]. Solder constitutive equations numerically describe time-independent elastic and plastic deformations, i.e., fatigue, and time-dependent plasticity (viscoplasticity), i.e., creep. Specifically, for each temperature change, the total increase of strain can be described by three components: $\gamma_{elastic}$, $\gamma_{plastic}$, and γ_{creep}. All three components are incorporated in a general equation for total strain,

$$\gamma_{TOT} = \gamma_{eleastic} + \gamma_{plastic} + \gamma_{creep}. \tag{7.16}$$

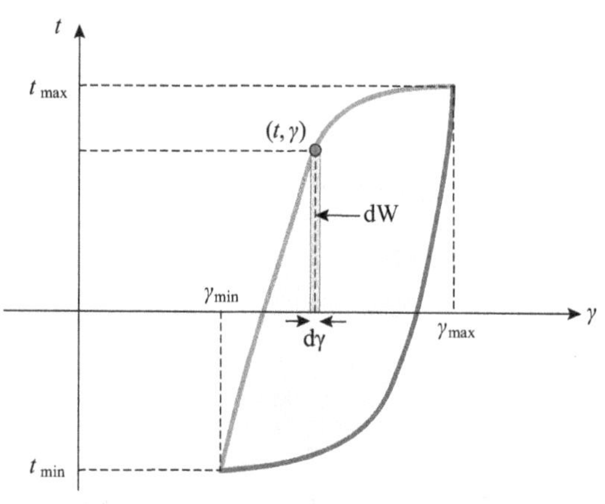

Figure 7.2 The response of solder to a periodic temperature cycle between T_{min} and T_{max} temperature values: a stabilized hysteresis loop. The minimum and maximum shear stress and strain values, τ_{min}, τ_{max}, γ_{min}, and γ_{max}, are defined by T_{min} and T_{max} and the material and geometrical properties of solder.

Furthermore, the strain components can be defined as a function of stress τ, temperature T, and time t. As elasticity and plasticity are time- and rate- independent, the elastic and plastic strain components, γ_{elastic} and γ_{plastic}, are only functions of stress and temperature. However, creep represents the time-dependent plasticity, and it is a function of all three variables. The elastic, plastic, and creep solder behaviour are derived based on the methods of classical mechanics and physics. The corresponding constitutive solder equations are summarized in the following subsection.

7.4.2 Constitutive solder equations

The elastic behaviour is described by Hooke's law,

$$\gamma_{\text{eleastic}} = \frac{\tau}{G(T)}, \tag{7.17}$$

where $G(T)$ is the shear modulus. The temperature sensitivity of the shear modulus is given by the shear modulus constants G_0 and G_1,

$$G(T) = G_0 - G_1 \cdot (T - 273\text{K}), \tag{7.18}$$

which depend only on the type of material.

Several models for time- and rate-independent plasticity of eutectic solder alloy were published in the literature [44–46]. The commonly used model was proposed by Darveaux [45,47],

$$\gamma_{\text{plastic}} = C_{\text{p}} \cdot \left(\frac{\tau}{G}\right)^{m_{\text{p}}}, \tag{7.19}$$

where C_{p} and m_{p} are the material-dependent parameters.

Typically, creep becomes relevant at temperatures of about $0.3T_{\text{M}}$ for pure metals and about $0.4T_{\text{M}}$ for alloys and most ceramics, where T_{M} is the melting point of material. Three typical stages of creep can be identified: primary, secondary, and steady-state creep. Primary creep is a transient regime of deformation in which the strain rate decreases over time until it reaches the minimal value defined by the steady-state creep rate. The tertiary creep is an unstable regime with very fast increasing strain-rate, which precedes the final rapture of the material. The steady state creep has been mainly investigated by engineers in the past, as it was assumed to be the dominant creep over time. The majority of strain-rate equations in the literature correspond therefore to the steady-state regime. For most of the existing physics-based lifetime models of solder interconnections in electronic devices, often only the steady state is taken into account. Moreover, the creep mechanisms are mainly investigated for the steady-state creep regime. Under these investigations of steady-state creep, different atomistic processes responsible for creep development are analysed, such as the gliding and climbing motions of dislocations in the crystal structure, and the diffusion of atoms (or vacancies) through the lattice of a crystal or along the grain boundaries. These two mechanisms refer to the dislocation-controlled and diffusion-controlled creep mechanisms, respectively.

They become dominant at different stress and temperature levels. Specifically, the dislocation creep occurs and becomes dominant for higher levels of applied stress, while the diffusion-controlled creep appears at lower stress levels and higher temperatures, which is roughly at about $0.3T_M$ for metals. A schematic representation of a deformation mechanism map can be used to identify the dominant damage mechanism [43].

Different types of models based on, e.g., a power law approach, a hyperbolic sine (sinh) law, a two-cell model, or an obstacle-controlled model, as summarized in [42], have been used for modelling the steady-state creep behaviour. These equations can be directly applied if all equation parameters are given for the specific solder type of interest. Here, the Darveaux's constitutive solder model is presented as a comprehensive model that includes both the equations for the primary and the steady state creep:

$$\frac{\mathrm{d}\gamma_s}{\mathrm{d}t} = C_1 \cdot \frac{G(T)}{T} \cdot \left[\sinh\left(\alpha \frac{\tau}{G(T)}\right)\right]^n \cdot e^{-\frac{Q}{k \cdot T}}, \tag{7.20a}$$

$$\gamma_{\text{prim}} = \gamma_T \cdot \left(1 - e^{-B \cdot t \cdot \frac{\mathrm{d}\gamma_s}{\mathrm{d}t}}\right), \tag{7.20b}$$

$$\gamma_{\text{tot}} = \gamma_{\text{prim}} + \frac{\mathrm{d}\gamma_s}{\mathrm{d}t} \cdot t, \tag{7.20c}$$

where $G(T)$ is the temperature-dependent shear modulus (7.18), γ_s is steady-state creep strain component, γ_{prim} is the strain component due to primary creep, α, B, Q, γ_T, C_1, G_0 and G_1 are material constants dependent on solder type, and τ is the value of shear stress in MPa. Additionally, the material-dependent model constants for several solder types can be found in the literature.

Another stress–strain relation describing the thermo-mechanical solder behaviour is defined by the isothermal stress reduction lines,

$$\gamma + \frac{\tau}{K} = D_1 \cdot (T - T_0), \tag{7.21}$$

where T_0 is a fixed temperature reference, K is the effective assembly stiffness, and D_1 is the imposed strain per degree temperature change. The parameters K and D_1 depend both on solder type and on the geometry of the power module. In the special cases when the slope of the stress reduction lines, K, tends to a very high or very low value, i.e. $K \to \infty$ or $K \to 0$, it reduces to the stress relaxation line and pure creep line, respectively. This relationship is derived from a simplified spring model of an assembly of two materials connected by solder [48].

The constitutive solder equations (7.17)–(7.21) are used to model the hysteresis behaviour of the solder material exposed to an arbitrary temperature profile. Clech derived the algorithm for calculating the stress–strain response of the solder joints exposed to temperature changes. As a numerical approach, the Clech's algorithm can be implemented as a computer application that further allows building a computer-based design-for-reliability tool [49]. Clech's algorithm is therefore introduced in the next subsection.

7.4.3 Clech's algorithm

Clech's idea was to find a way to accurately simulate the response of the solder joints of SMDs exposed to cyclic thermal loading. Clech's algorithm can be explained by Figure 7.3.

Knowing the stress–strain state (γ_1, τ_1) at time *t*, the stress–strain state (γ_2, τ_2) at time $t + \Delta t$ can be calculated using the constitutive solder equations. It is assumed that:

(a) Δt is a relatively small time step ($\Delta t \to \mathrm{d}t$).
(b) The stress is built up instantaneously for the temperature increase from $T = T_1$ to $T_2 = T_1 + \mathrm{d}T$. For this instantaneous stress change, it is assumed that dominant strain components are time-independent elastic and plastic components $\mathrm{d}\gamma_{\mathrm{elastic}}$ and $\mathrm{d}\gamma_{\mathrm{plastic}}$.
(c) At temperature T_2, the change of stress follows the T_2-stress reduction line building the total strain component, which is in turn mainly defined by the creep component, $\mathrm{d}\gamma_{\mathrm{creep}}$.

Clech's algorithm allows the numerical calculation of the stress–strain hysteresis loops of solder interconnecting layers exposed to a given temperature profile (mission profile) with an arbitrary shape. Accordingly, Clech's algorithm represents the computational core of a computer-aided-design (CAD) lifetime estimation tool.

7.4.4 Energy-based lifetime modelling

Physical lifetime modelling of solder joints in chip packages is based on models that can be classified in four groups: stress-based, strain-based, damage-based, and energy-based models [50]. All these models require the stress and strain data to be known in order to predict service life. In comparison to the other methods,

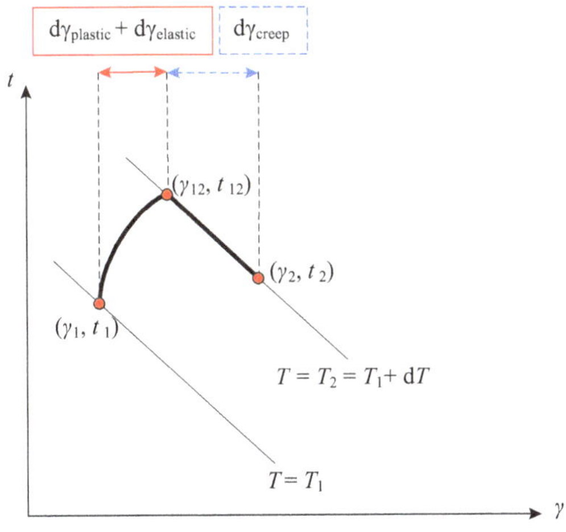

Figure 7.3 Calculation of stress–strain (hysteresis) response according to Clech's algorithm

energy-based models are seen to be the most convenient, as they have the ability to capture test conditions with more accuracy.

The energy-based lifetime models use the information of the energy enclosed by the stress–strain hysteresis loop to predict the EOL of an electronic device. Most of the energy-based models are derived based on studies that investigated the lifetime of solder joints in high-density electronics packages mostly used in telecommunication and consumer electronics, such as surface-mount packages, e.g., chip size packages (CSP), chip carrier (CC), and BGA. The developed lifetime models account for the geometry of the solder joint and, thus, cannot be straightforwardly applied to an arbitrary type of solder joint.

Energy-based modelling is based on the assumption that the EOL of a device is determined by the total deformation energy accumulated within the solder joint during the operation of a device. A device fails when the deformation work reaches the critical value, W_{tot}. Specifically, the energy-based models correlate the hysteresis energy density, i.e., the deformation energy per cycle, and the number of cycles to failure, N_f,

$$N_f = C \cdot (\Delta w_{hys})^{-n}, \tag{7.22}$$

where C and n are the model constants that often depend on the material and geometrical properties of the chip package, and Δw_{hys} is the accumulated energy per cycle or inelastic strain energy density. The value of Δw_{hys} is calculated by integrating the stress along strain in the strain range $(\gamma_{min}, \gamma_{max})$ defined by the hysteresis loop, see Figure 7.2,

$$\Delta w_{hys} = \oint_{HysLoop} \tau d\gamma. \tag{7.23}$$

The energy-based lifetime modelling using Clech's algorithm can then be summarized by the following steps:

1. the stress–strain response of a solder joint is generated by means of Clech's algorithm,
2. the parameters C and n are derived from experiments,
3. the parameterized energy-based N_f-model is applied to calculate EOL for different temperature profiles.

The other type of energy-based models involves the crack initiation and propagation inside the solder layer, based on the Darveaux's energy-based model (7.8)–(7.9) that links the number of cycles to the crack initiation, N_0, and the crack propagation rate to the accumulated energy density per cycle. The EOL is then defined by a critical crack length.

7.5 Example of physics-based lifetime modelling for solder joints

The ETHZ-PES lifetime model for the solder joints inside of power semiconductor modules described in Section 7.3.3 was tested using experimental PC results, conducted by SEMIKRON Elektronik GmbH & Co. KG [10]. The aim was to verify the proposed lifetime prediction approach for its application to PC test conditions. Nine samples of standard baseplate modules in half-bridge configuration, SKM200GB12T4 [22], were used for the verification. The specifications of the corresponding PC experiments are given in Tables 7.2 and 7.4. In the SKM200GB12T4 module, the chips are soldered to the DBC substrate using Sn3.5Ag solder. Under all PC tests, the chip solder fatigue was the dominant failure mechanism that justifies the selection of the employed set of PC cycles for the model verification.

The modelling steps using the proposed algorithm are schematically presented by a flowchart in Figure 7.4 and described in the following. Additionally, the solder material- and geometry-dependent parameters used in the model are specified in Table 7.1.

7.5.1 Thermal simulation

As a first step, 3D thermal simulations of the power module were conducted in order to calculate the temperature evolution in the module during a power cycle of each experimental PC test. In order to perform a physical simulation of the

Table 7.1 Material- and geometry-dependent Sn3.5Ag solder parameters

Time-independent elastic strain (7.17) [47]	
G_0 [MPa]	19,310
G_1 [MPa/K]	68.9
Time-independent plastic strain (7.19) [47]	
C_p	$2 \cdot 10^{11}$
m_p	4.
Steady-state creep strain (7.20)(a) [47]	
C_1 [K/s/MPa]	0.454
α	1500
n	5.5
Q [eV]	0.5
Primary creep strain (7.20)(b) [47]	
γ_T	0.086
B	147
Range of the unknown geometry-dependent parameters (7.21)	
K [MPa]	$(5 \cdot 10^2, 10^4)$
D_1 [1/K]	$(10^{-4}, 10^{-3})$

Table 7.2 Specifications of PC tests used for the parameterization

PC test	T_{min} [°C]	T_{max} [°C]	ΔT [°C]	t_{on}, t_{off} [s]	P_{loss} [W]	N_f [cycles]
PC_1	40	155	115	50, 31	925.9	31,332
PC_2	80	148	68	0.95, 4	900	220,279
PC_3	78	148	70	5, 1	480	168,390

solder aging in which both creep and time-independent elasticity and plasticity are considered, it is necessary to know not only the maximum and minimum temperature of the junction but also the total time dependence of the solder temperature $T(t)$.

In the experimental setup, the devices were mounted on a water-cooled heatsink. During the heating phase of a power cycle, the water flow through the heatsink was stopped in order to obtain a sharp temperature increase. By contrast, the water flow was turned on during the cool-down time of each power cycle, so that the lower temperature limit was reached more quickly. Because of this design of the experiment, a simulation is not straightforward: Foster models cannot be used, since two different models describe the system with and without water flow, and it is impossible to switch the models during simulation (since there is no way of determining the state of charge of the equivalent capacitors). At the same time, a full-fledged thermo-hydrodynamic simulation of the complete system, including the water flow, is very time-consuming, especially because multiple power cycles have to be simulated, before the system reaches thermal equilibrium. Therefore, a physical approach has been chosen based on a simplified 3D-Cauer-based model (see Section 7.3.1). Experience has shown that the simulation of the junction temperature evolution of a power module on a heatsink can be well approximated if a solid heatsink is used and a (non-physical) transition layer with negligible thermal capacitance is inserted beneath it. The thermal resistivity of this transition layer is then adapted such that the measured total thermal resistance from junction to ambient is reproduced. This approach can be applied to both systems with and without water flow, resulting in two different values for the thermal resistivity of the transition layer. Specifically, in an electric equivalent circuit of the thermal system of the power module, the transition layer is replaced by two sets of thermal resistors. The switches, which are placed in front of the resistors, connect the corresponding set of the resistors to the rest of the electric equivalent circuit. These switches are then operated in accordance with the heating and cooling cycles.

Using this approach, a 3D thermal simulation model was built using the program MAKENET [51], as illustrated in [21], which generates an electrical equivalent circuit of resistors and capacitors from the 3D geometry that can consequently be simulated with PSPICE. The simulation results of the junction temperature change during a power cycle are found to be in good accordance with the temperature monitoring during the PC test. In order to reach even higher accuracy for the simulation of the solder aging, the power cycle is then rescaled linearly in order to describe the measured temperature change exactly. The solder temperature is read from these rescaled simulation results. The maximum temperature points of the solder layer under the chip centre obtained from the thermal modelling are used further as the input for the lifetime modelling.

7.5.2 Stress–strain modeling

The second step was to calculate the stress–strain response to the simulated temperature profiles according to the procedure described in Section 7.4.3. The required inputs of the algorithm for calculating hysteresis solder behaviour are the constitutive solder equations (7.17)–(7.20), the geometry-dependent parameters that define the stress-reduction lines (7.21), and the exponent parameter n of the energy-based N_f-model, cf. (7.22).

The complete constitutive equations including creep and plastic time-independent deformations of Sn3.5Ag solder alloy have not been published. A majority of publications concentrates on the ultimate strength and steady-state creep [29]; however, it was shown in [45] that Sn3.5Ag undergoes more primary creep and, hence, the primary creep should not be neglected for this type of solder material. Dareveaux's solder constitutive relations were found to provide the most comprehensive solder model of Sn3.5Ag solder alloy available, and, thus, they were implemented in the proposed lifetime model [47], cf. Table 7.1.

The parameters K and D_l in (7.21) define the stress-reduction lines, and both depend on the solder type and the module geometry. These model parameters are determined in a parameterization procedure implemented as MATLAB optimization routine using the N_f results from three PC experiments, PC_{1-3}. The properties of the tests PC_{1-3} are defined in Table 7.2, and the corresponding temperature profiles (TP) gained from the thermal modelling decribed above are shown in Figure 7.5.

The optimal parameters K, D_l, and n should minimize the difference of three deformation energies, $W_{\mathrm{crit}i}$, $i = 1, 2, 3$, calculated by (7.5) for the corresponding simulated temperature profiles, T_i,

$$W_{\mathrm{crit}i} = N_{\mathrm{f}i} \cdot (\Delta w_{\mathrm{hys}i})^n, i = 1, 2, 3. \tag{7.24}$$

The free parameters of the optimization algorithm are K, D_l, and the exponent n. The values of other model parameters are the solder constants taken from the literature, cf. Table 7.1. The error r of parameterization is defined as the maximum ratio between the calculated critical energies, $r = \max(W_{\mathrm{crit}i}/W_{\mathrm{crit}j})$, $i, j = 1, 2, 3$, $i \neq j$. In an ideal case, r would be unity, i.e., $W_{\mathrm{crit1}} = W_{\mathrm{crit2}} = W_{\mathrm{crit3}}$. In practice, there is an optimal r-value higher than unity. Accordingly, the estimated number of cycles to failure, $N_{\mathrm{f,estim}}$, can be defined as a N_f-range, (N_{fmin}, N_{favg}, N_{fmax}), which corresponds to the minimum $W_{\mathrm{crit}min}$, average $W_{\mathrm{crit}avg}$, and maximum $W_{\mathrm{crit}max}$ of the calculated critical energies for three PC tests resulting from the optimization procedure (7.5).

$$W_{\mathrm{critFunc}} = \mathrm{Func}(W_{\mathrm{crit1}}, W_{\mathrm{crit2}}, W_{\mathrm{crit3}}) \tag{7.25a}$$

$$N_{\mathrm{fFunc}} = \frac{W_{\mathrm{critFunc}}}{(\Delta w_{\mathrm{hys}})^n}, \mathrm{Func} = avg, max, min. \tag{7.25b}$$

The error, r, of the parameterization for the PC tests specified in Table 7.2 is $r =$ 1.714. Namely, for the PC_1 test, the actual N_{f1} value is closer to the minimum of the estimated N_f-range, for the PC_2 test the actual N_{f2} value coincides with the maximum, and for the PC_3 test the actual N_{f3} coincides with the minimum of the estimated N_f-range, as shown in Table 7.3.

Table 7.3 *N_f prediction for $PC_{1,2,3}$ tests*

PC test	N_{fmin}, N_{favg}, N_{fmax} **(7.5)**	$N_{f,estim}$	**Rel. error [%]**	$\frac{dT}{dt}$ **(at T_{max})**
1	31,302, 38,762, 53,654	(N_{fmin}, N_{favg})	$(-0.09, 23.7)$	$0.66K/s$
2	128,514, 159,142, 220,279	(N_{favg}, N_{fmax})	$(27.7, 0)$	$10.67K/s$
3	168,390, 208,521, 288,627	(N_{fmin}, N_{favg})	$(0, 23.8)$	$0.37K/s$

Table 7.4 *Specifications of PC tests used for the model verification*

PC test	T_{min} **[°C]**	T_{max} **[°C]**	ΔT **[°C]**	$t_{on}, t_{off}[s]$	P_{loss} **[W]**	N_f
A	40	155	174	64.5, 48.7	921.9	28,780
B	79	146	67	0.95, 4	912	248,710
C	80	150	70	1.2, 4	920	234,632
D	77	150	73	2.9, 0.9	944	149,125
E	39	148	109	13.6, 10.3	1122	38,441
X	40	176	136	2, 3	1569.4	21,956

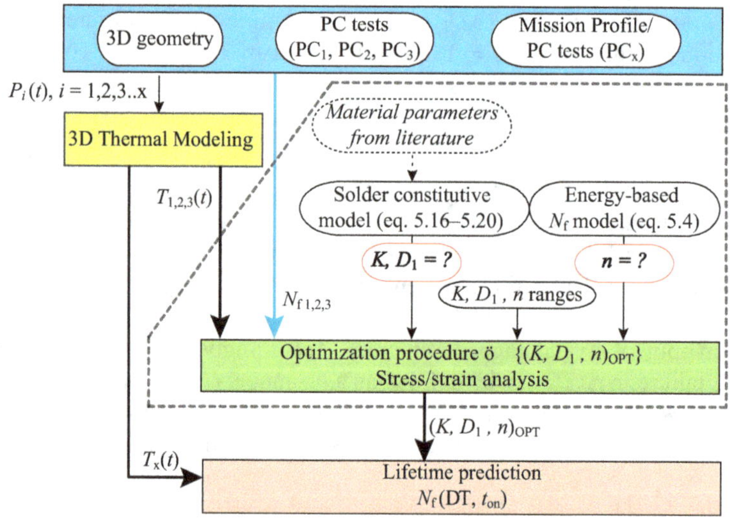

Figure 7.4 *Flowchart describing the main modelling steps of the ETHZ-PES physics-based lifetime model for the solder joints of power semiconductor modules. The block marked by a dashed line includes the procedure for finding the unknown material and geometry-dependent solder parameters.*

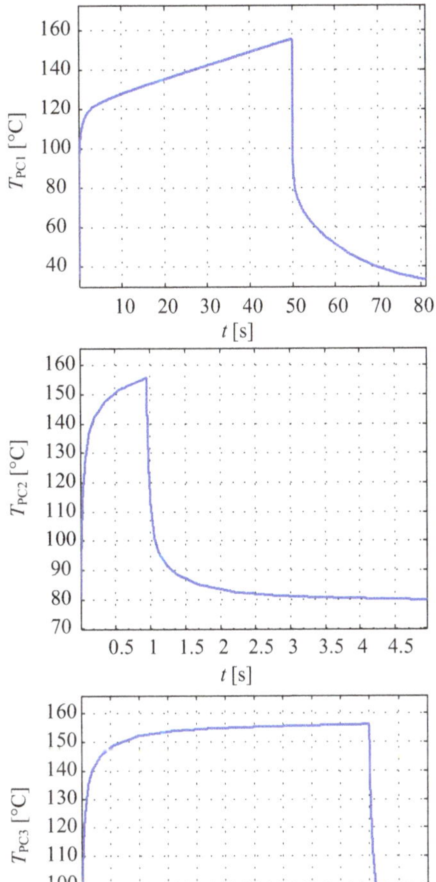

Figure 7.5 Temperature profiles T_{PC1}, T_{PC2}, and T_{PC3} developed under PC_1, PC_2, *and* PC_3 *tests, respectively*

The exponent, n, is set to 2.2 as the best fit to the experimental data. The range of the parameters as input of the optimization procedure is selected such that the stress–strain behaviour of solder alloy is correctly modeled, i.e., creep is higher at high maximum temperatures and low temperature rates, e.g., for the PC_1 and PC_3 tests, and time-independent plasticity is dominant for fast temperature changes, e.g., the PC_2 test. Specifically, the order of 10^3 and of 10^{-4} is selected for K and D_1, respectively, as specified in Table 7.1, which returns a good agreement with the simulated stress–strain response of the solder joint to the expected physical behaviour of solder, see Section 7.5.6. It was shown in [25] that the unknown parameters K and D_1 are correlated, i.e., a higher D_1 corresponds to a lower value of K and vice versa. An optimal parameter set is calculated to be $(K, D_1) = (1403.03, 9.9 \cdot 10^{-4})$.

7.5.3 Stress–strain analysis

The N_f prediction is performed based on the analysis of the simulated stress–strain curves. The parameters of interest are the temperature swing amplitude, ΔT, the maximum temperature level, T_{max}, and the temperature rate, $dT/dt|_{T_{max}}$. The values of ΔT, T_{max}, and $dT/dt|_{T_{max}}$ for PC_{1-3} are given in Tables 7.2 and 7.3.

The temperature profiles generated in PC_i experiments can be analysed to gain a better understanding of the expected stress–strain response. For example, both the PC_3 test and the PC_2 test have a lower temperature amplitude of 70K, but the PC_3 test is characterized by slower temperature change at the maximum temperature level in comparison to the PC_2 test. Accordingly, the chip solder layer experiences more creep deformation for the PC_3 test and, thus, shorter lifetime. However, the PC_1 experiment is characterized by a higher temperature swing of 115K and a longer heating time, which explains the resulting shorter lifetime.

The share of creep strain component in the total strain deformation can be used as a quantitative measure of creep deformation. Specifically, by observing the stress–strain deformation under a temperature cycle, it can be concluded that the temperature evolution at T_{max} is relatively slow for PC_3, and thus it can be expected that the creep strain component has a similar or higher share in the total strain deformation than the elastic/plastic time-independent strain component. After the simulation, with the optimal parameter set, the creep component at T_{max} was calculated for the PC_3 test to be approx. 50%. Similarly, for the PC_1 experiment, the creep component reaches 40% of the total strain deformation, while for the PC_2 test, the time-independent elastic-plastic strain component is dominant, i.e., more than $\approx$ 80% of the total strain.

Due to longer heating times, the stationary state is reached for the PC_1 and PC_3 experiments, and not reached for the PC_2 test with a shorter t_{on}. As an example, for the described stress–strain analysis, the hysteresis loops for the tests PC_1, PC_2 and PC_3, are presented in Figure 7.6, where also the fast temperature change occurring in power modules under typical power cycling test conditions and the strain components for test PC_1 are depicted.

7.5.4 Model verification

By observing the properties of the calculated stress–strain curves at the simulated temperature profiles T_{1-3}, the N_f-lifetime is estimated within the calculated N_f-range. For PC tests, which resemble the tests PC_1 or PC_3, N_f is predicted in the lower N_f-range, (N_{fmin}, N_{favg}); and for PC tests which are similar to the PC_2 test, the N_f predictions are located in the upper part of the N_f-range, (N_{favg}, N_{fmax}).

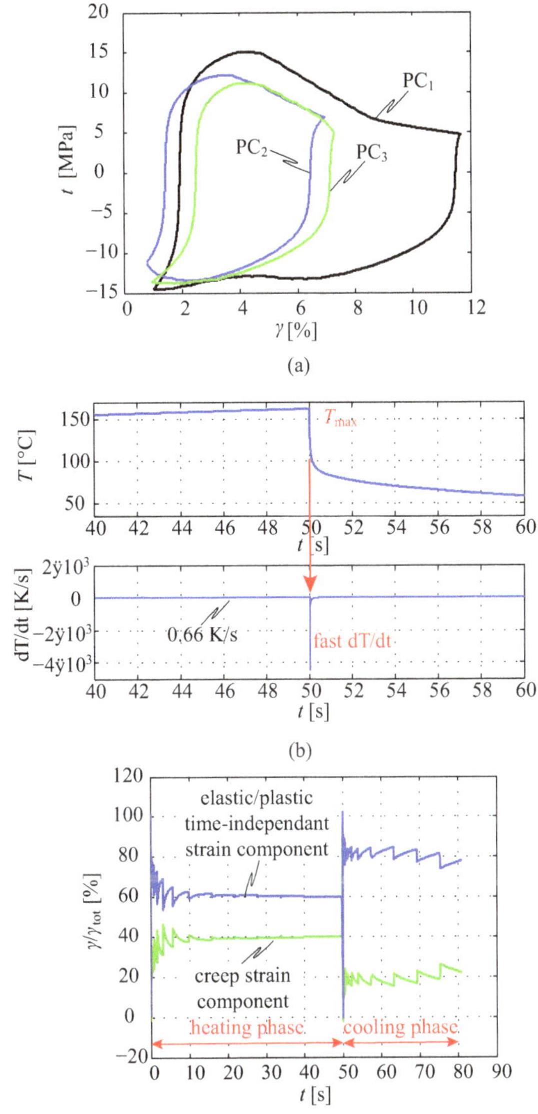

Figure 7.6 *Stress–strain analysis: (a) stress–strain response, i.e., hysteresis loops, under PC_1, PC_2 and PC_3 tests; (b) zoomed part of the temperature profile T_{PC1} of the PC_1 test at the transition between the heating and cooling phases illustrating a fast temperature change; and (c) strain components developed for the PC_1 test*

Table 7.5 N_f*prediction for* PC_{A-X} *tests*

PC test	N_{fmin}, N_{favg}, N_{fmax} (7.5)	$N_{f,estim}$	Rel. error [%]	$\frac{dT}{dt}$ (at T_{max})
A	23,216, 28,749, 39,793	(N_{fmin}, N_{favg})	(19.3, −0.1)	$0.7K/s$
B	140,949 174,541, 241,594	(N_{favg}, N_{fmax})	(−29.9, −2.8)	$10.4K/s$
C	122,054, 151,142 209,205	(N_{favg}, N_{fmax})	(−35.6, −10.8)	$10.1K/s$
D	164,966, 204,281, 282,758	(N_{fmin}, N_{favg})	(10.6, 36.9)	$1.2K/s$
E	31,393, 38,875, 53,809	(N_{fmin}, N_{favg})	(−18.3, 1.1)	$0.9K/s$
X	21,853, 27,061, 37,457	(N_{fmin}, N_{favg})	(−0.5, 23.2)	$5.2K/s$

Following the proposed procedure, the prediction of the number of cycles to failure for other power cycle experiments, PC_A, PC_B, PC_C, PC_D, PC_E, PC_X, specified in Table 7.4, was performed. The estimated N_f-ranges and relative errors are shown in Table 7.5.

The temperature profiles of the power cycles PC_B and PC_C are very close to the temperature profile of PC_2, so that N_f is evaluated as the maximum of the calculated N_f-range with relative errors of less than 11%.

Tests PC_D and PC_3 exhibit similar ΔT and T_{max}. Analysing the stress–strain curves, it was observed that the creep strain component has a similar share in the total strain for both PC tests, i.e., tending to 50%. Accordingly, the estimated N_f value is set in the lower N_f-range, as for test PC_3.

Tests PC_A and PC_E exhibit higher temperature amplitudes, similar to the PC_1 test, the stationary state is reached in test PC_A and nearly reached in test PC_E. The share of the creep strain component is approximately 40%, as for test PC_1, so that the estimation of the lifetime belongs to the lower N_f-range, according to the parameterization of the PC_1 test.

Test PC_X represents a more specific case. It exhibits a shorter heating time but a very high maximum temperature of 176 °C and a high temperature amplitude of 134 °C. The creep component was calculated to be less than 40% of total strain, as for test PC_2. As the high temperature amplitude and high maximum temperature are shown to have a direct influence on the lifetime and the creep strain component is a derived factor, the decision-making is performed based on the parameterization of test PC_1, which is characterized by a high ΔT.

The relative estimation errors are summarized in Table 7.5. A maximum estimation error of less than 24% indicates a good agreement between the lifetime prediction based on the ETHZ-PES model and the PC N_f-test results.

7.5.5 Lifetime curves extraction

The presented lifetime modelling enables extracting the lifetime curves of the SKM200GB12T4 power module. A lifetime curve represents the N_f dependency on the temperature swing amplitude ΔT for a defined heating time t_{on}, i.e., $N_f(\Delta T)|_{t_{on}}$.

Temperature profiles for different values of ΔT are calculated by keeping the average temperature, the heating time, and the cooling time constant, and scaling the simulated temperature profile of the power module to obtain the desired

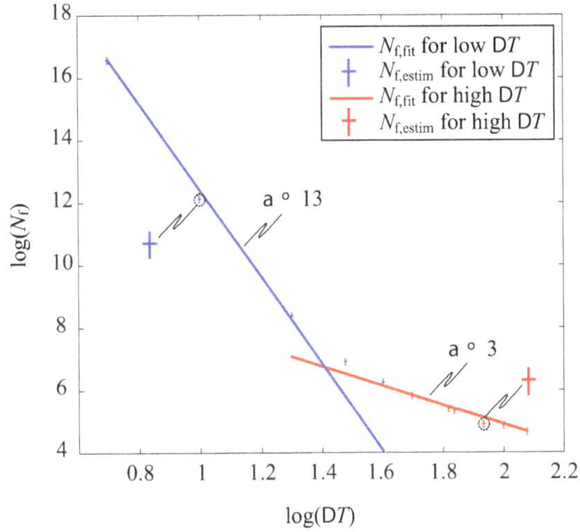

Figure 7.7 Lifetime curves $N_f(\Delta T)$ *extracted for the average temperature* T_{avg}*, heating time* t_{on}*, and cooling time* t_{off} *of test* PC_2*; the cross symbols indicate the estimated* N_f *values with an estimation error of less than 25% in the low and high* ΔT *ranges; the full lines represent the linear asymptotic curves fitted to the estimated* N_f *values with the calculated slopes of* $\alpha = 13$ *and* $\alpha = 3$ *for lower and higher values of* ΔT*, respectively*

temperature change. This corresponds to power cycles that differ only in the amount of dissipated power and the cooling medium temperature. Using the calculated temperature profiles for lower and higher temperature swings, the lifetime prediction, i.e., N_f, in both low and high stress regimes can be calculated by applying the proposed procedure described by the flowchart in Figure 7.4. The $N_f(\Delta T)$ curve corresponding to the medium temperature T_{avg}, heating time t_{on}, and cooling time t_{off} of test PC_2 is shown in Figure 7.7.

As can be seen in Figure 7.7, the calculated N_f values show two asymptotic linear dependencies in a log-log plot, i.e., $N_f \propto \Delta T^{-\alpha}$, for low and high temperature swings, with exponents of $\alpha \approx 3$ and $\alpha \approx 13$, respectively. The lifetime curves extracted for the tests PC_1 and PC_3 show similar behaviour, i.e., the exponent α varies around 3 for higher ΔT and around 13 for lower values of ΔT. A high exponent of $\alpha = 13$ indicates that the common extrapolation of the power cycling results towards lower temperatures is much too conservative, at least for solder failures. This can be also ascribed to the constitutive model used, i.e., Darveaux's solder model. The parameters of Darveaux's model for Sn3.5Ag solder were derived from tensile and shear loading experiments employed in the temperature range between 25 °C and 135 °C [47]. Accordingly, it must be considered that Darveaux's solder model must be further evaluated and verified for lower and/or higher temperatures for accurate lifetime prediction in a wide temperature range.

7.5.6 Model accuracy and parameter sensitivity

It should be pointed out that the scattering of cycles to failure by $\approx 20\%$ around the mean value is common for PC tests even if the tests are conducted under as far as possible equal conditions. Specifically, the PC tests PC_2, PC_B, and PC_C can be considered as nearly same; however, a deviation of the number of cycles to failure between these PC tests can be observed. Similar to tests PC_D and PC_3, even though the power module under test PC_3 is expected to fail faster than under test PC_D due to longer heating time [17], a shorter lifetime was observed for test PC_D. Accordingly, an estimation error in the range of about 20% can be characterized as a good N_f prediction.

As previously mentioned, a reliable constitutive model for the solder alloy employed in the power module has to be implemented first in order to correctly model the lifetime assuming the wear-out of the solder layer as the dominant failure mechanism. The elastic deformation is described by the Young's modulus, and the strain-hardening power law is typically used to describe time-independent plastic strain deformation, but there is no unique relation describing the creep deformation mechanism of solder. In order to simplify creep modelling, the steady-state (secondary) creep is commonly the only creep mechanism considered in the thermo-mechanical modelling of solder joints in microelectronics packaging. Different types of creep models, e.g., a power law model, a hyperbolic sine (sinh) law, a two-cell model, and an obstacle-controlled model [42], have been developed to describe the steady-state creep behaviour. However, it was shown that primary creep cannot be neglected for lead-free solder alloys [52]. The parameters of the constitutive relations are determined by curve fitting using data gained in monotonic and creep test experiments of the solder material under investigation [53]. Therefore, the model parameters depend on solder material and solder joint configuration. To collect data for building an accurate constitutive model, the solder joint used in the experiment setup should closely resemble the real solder joint design to be modeled [54]. These types of experiments actually are missing for the solder joints of power modules. The data of Sn3.5Ag solder alloy used in the tested SKM200GB12T4 modules are not well covered in the literature. Darveaux's model for Sn3.5Ag solder was implemented in the ETHZ-PES lifetime model as all model parameters of the solder constitutive model can be found in the literature, including primary and secondary creep strain deformation and elastic/plastic time-independent deformation [47].

The sensitivity of N_f estimation to the change of K, D_1, and n, which are fitted to the experimental PC test data in the parameterization procedure, is performed to gain a better insight into the physical meaning of the unknown parameters. The parameter K represents the effective assembly stiffness, and D_1 is the imposed strain per degree temperature change, which reflects the geometry and the difference of the CTEs between the layers connected by soldering. As can be observed by increasing the stiffness K and/or decreasing D_1 by an order of magnitude, the creep strain component becomes smaller than the elastic/plastic

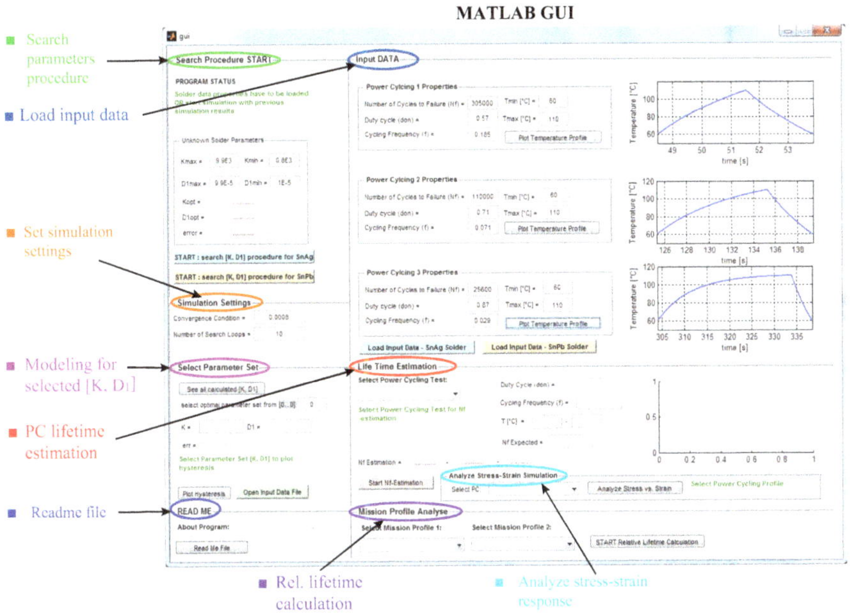

Figure 7.8 GUI of the MATLAB software tool for lifetime estimation of solder layers of power modules based on the ETHZ-PES model

time-independent strain component for all PC temperature profiles and vice versa; by decreasing the stiffness K and/or increasing D_1 by an order of magnitude, the creep component becomes dominant over the elastic/plastic time-independent strain component even for the PC_2 test, characterized by a fast temperature change and a short heating time. This behaviour agrees to the isothermal stress reduction curve, which tends to a stress relaxation line and a pure creep line, for two extremes, $K = \infty$ and $K = 0$, respectively [48]. Even though a similar error of parameterization can be achieved by values of K and D_1 of different magnitude, the stress–strain curves obtained with these parameters do not describe the expected physical solder behaviour.

By finding an optimal parameter set (K, D_1) in the same parameters ranges for different values of n, it was observed that the relative error of the N_f estimation is significantly higher for lower values of n, which is best visible for the test PC_X, e.g., the relative error is about 64% for $n = 1$. Accordingly, the best estimation is achieved for a value of $n = 2.2$, which is also the result of the optimization procedure with three variables (K, D_1, n).

By changing the value of the parameters (K, D_1) by 10% around their nominal (calculated) values, the error of the parameterization r is changing up to 1.76, which does not have a significant influence on the lifetime prediction, i.e., less than 10%. Higher D_1 shifts the hysteresis loop in the area of higher strain, and higher K increases the hysteresis loop area, which implies higher damage.

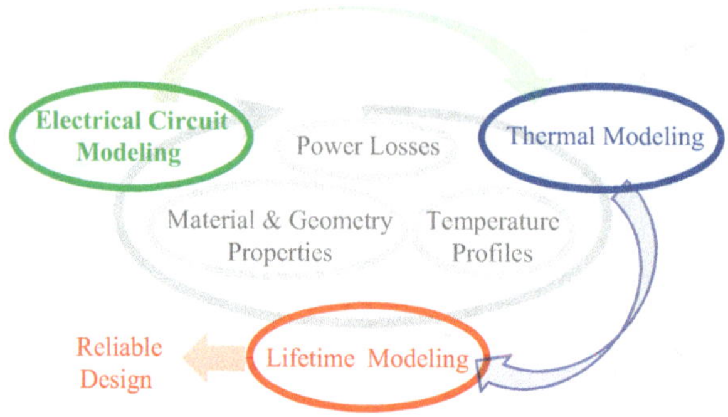

Figure 7.9 Concept for a virtual prototyping platform for reliability analysis of power modules and/or PECS

7.5.7 Lifetime estimation tool

The ETHZ-PES lifetime model is implemented in a MATLAB software tool, which has been partially developed in the course of the ECPE research project "Reliability and Lifetime Modeling and Simulation of Power Modules and Power Electronic Building Blocks" [55]. Through an easy-to-use GUI (Figure 7.8), users can specify the temperature profiles of power cycling tests, run a parameterization for the defined range of model parameters, analyse the calculated stress–strain response, and estimate the relative lifetime of power module for two arbitrary temperature profiles. The constitutive solder models for different solder types found in the literature [42,45,56] have also been implemented. This software tool still needs further improvements, but nevertheless represents a useful basis for developing a universal physics-based lifetime modelling tool for power modules.

7.6 Conclusions

PoF lifetime modelling approaches can be seen as a new methodology in power electronics that potentially can improve the lifetime estimation and enable reliability engineering to be integrated into the development and research cycles of the overall design process of power electronics systems. As illustrated in Figure 7.9, the coupling of different domains within a virtual prototyping platform, e.g., electrical circuit simulations, thermal modelling, and lifetime prediction, will allow egineers to develop reliable PECSs in a time- and cost-efficient way in future. The developed MATLAB software with the implemented ETH-PES lifetime model can be seen as a first step towards a virtual prototyping platform.

Acknowledgements

The authors would like to thank Prof. Dr. Uwe Scheuermann and Dr. Ralf Schmidt from SEMIKRON for providing experimental data, helpful suggestions, and inspiring discussions based on their long-term experience in lifetime modelling. Furthermore, the authors are very much indebted to the European Center for Power Electronics (ECPE) for financing the research project "Reliability and Lifetime Modeling and Simulation of Power Modules and Power Electronic Building Blocks" [55], which significantly contributed to the development of the presented approach.

References

[1] J. Lutz, "Packaging and reliability of power modules," in: *Proc of 8th Int. Conf. on Integrated Power Systems (CIPS)*, 2014, pp. 17–24.

[2] M. Held, P. Jacob, G. Nicoletti, P. Scacco, and M.-H. Poech, "Fast power cycling test of IGBT modules in traction application," *International Journal of Electronics*, vol. 86, no. 10, pp. 1193–1204, 1999.

[3] R. Bayerer, T. Herrmann, T. Licht, J. Lutz, and M. Feller, "Model for power cycling lifetime of IGBT modules – various factors influencing lifetime," in: *Proc of 5th Int. Conf. on Integrated Power Systems (CIPS)*, 2008, pp. 37–42.

[4] R. Schmidt, F. Zeyss, and U. Scheuermann, "Impact of absolute junction temperature on power cycling lifetime," in: *Proc. of 15th European Conf. on Power Electronics and Applications (EPE)*, 2013, pp. 1–10.

[5] I. Kovacevic, U. Drofenik, and J. W. Kolar, "New physical model for lifetime estimation of power modules," in: *Proc. of the Int. Power Electronics Conf (ECCE Asia)*, 2010, pp. 2106–2114.

[6] P. Steinhorst, T. Poller, and J. Lutz, "Approach of a physically based lifetime model for solder layers in power modules," *Microelectronics Reliability*, vol. 53, no. 8–10, pp. 1199–1202, 2013.

[7] O. Schilling, M. Schaefer, K. Mainka, M. Thoben, and F. Sauerland, "Power cycling testing and FE modelling focussed on Al wire bond fatigue in high power IGBT modules," *Microelectronics Reliability*, vol. 52, no. 9–10, pp. 2347–2352, 2012.

[8] L. Yang, P. A. Agyakwa, and C. M. Johnson, "Physics-of-failure lifetime prediction models for wire bond interconnects in power electronic modules," *IEEE Transactions on Device and Materials Reliability*, vol. 13, no. 1, pp. 9–17, 2013.

[9] H. Lu, T. Tilford, and D. Newcombe, "Lifetime prediction for power electronics module substrate mount-down solder interconnect," in: *Proc. of Int. Symp. on High Density Packaging and Microsystem Integration (HDP)*, 2007, pp. 1–10.

[10] Application Manual Power Semiconductors [Online]. 2015. Available: https://assets.danfoss.com/documents/latest/456901/AB509142199100en-000101.pdf.

[11] H. Wang, M. Liserre, F. Blaabjerg, *et al.*, "Transitioning to physics-of-failure as a reliability driver in power electronics," *IEEE Journal of Emerging and Selected Topics in Power Electronics*, vol. 2, no. 1, pp. 97–114, 2014.

[12] Semiconductor devices-Mechanical and climatic test methods-Part 34: Power cycling (IEC 60747-34), Int. Electrotechnical Commission (IEC) Std.

[13] Semiconductor devices-Discrete devices-Part 9: Insulated-gate bipolar transistors (IGBTs) (IEC 60747-9), Int. Electrotechnical Commission (IEC) Std.
[14] J. Lutz, H. Schlangenotto, U. Scheuermann, and R. DeDoncker, *Semiconductor Power Devices Physics, Characteristics, Reliability*. Berlin: Springer; 2011.
[15] T. Herrmann, M. Feller, J. Lutz, R. Bayerer, and T. Licht, "Power cycling induced failure mechanisms in solder layers," in: *Proc. of the European Conf. on Power Electronics and Applications*, 2007, pp. 1–7.
[16] M. Ciappa, "Selected failure mechanisms of modern power modules," *Microelectronics Reliability*, vol. 42, no. 4, pp. 653–667, 2002.
[17] U. Scheuermann and R. Schmidt, "A new lifetime model for advanced power modules with sintered chips and optimized Al wire bonds," in: *Proc. of Int. Exhibition and Conf. for Power Electronics, Intelligent Motion, Renewable Energy and Energy Management (PCIM)*, 2013, pp. 810–817.
[18] H. Huang and P. A. Mawby, "A lifetime estimation technique for voltage source inverters," *IEEE Transactions on Power Electronics*, vol. 28, no. 8, pp. 4113–4119, 2013.
[19] U. Scheuermann and P. Beckedahl, "The road to the next generation power module – 100% solder free design," in: *Proc. of 5th Int. Conf. on Integrated Power Systems (CIPS)*, 2008, pp. 111–120.
[20] U. Drofenik and J. Kolar, "Teaching thermal design of power electronic systems with web-based interactive educational software," in: *Proc. of 18th Annual IEEE Applied Power Electronics Conf. and Exposition (APEC)*, vol. 2, 2003, pp. 1029–1036.
[21] R. Schmidt and U. Scheuermann, "Using the chip as a temperature sensor - the influence of steep lateral temperature gradients on the V_{ce} (T)-measurement," in: *Proc. of 13th European Conf. on Power Electronics and Applications (EPE)*, 2009, pp. 1–9.
[22] U. Scheuermann and R. Schmidt, "Impact of solder fatigue on module lifetime in power cycling tests," in: *Proc. of 14th European Conf. on Power Electronics and Applications (EPE)*, 2011, pp. 1–10.
[23] A. Hamidi, A. Stuck, N. Beck, and R. Zehringer, "Time dependent thermal fatigue of HV-IGBT-modules," in: *Proc. of 27th Kolloquium Halbleiter-Leistungsbauelemente und Materialgüte von Silizium*, Freiburg/Breisgau, 1998, pp. 26–27.
[24] U. Drofenik, I. Kovacevic, R. Schmidt, and J. W. Kolar, "Multi-domain simulation of transient junction temperatures and resulting stress–strain behavior of power switches for long term mission profiles," in: *Proc. of the 11th IEEE Workshop on Control and Modeling for Power Electronics (COMPEL)*, 2008, pp. 1–7.
[25] G. Riedel, R. Schmidt, C. Liu, H. Beyer, and I. Alapera, "Reliability of large area solder joints within IGBT modules: numerical modeling and experimental results," in: *Proc. of 7th Int. Conf. on Integrated Power Systems (CIPS)*, 2012, pp. 288–298.

[26] M. Ciappa, "Lifetime modeling and prediction of power devices," in: *Proc. of 5th Int. Conf. on Integrated Power Systems (CIPS)*, 2008, pp. 1–9.

[27] P. M. Hall, "Forces, moments, and displacements during thermal chamber cycling of leadless ceramic carriers soldered to printed boards," *IEEE Transactions on Components, Hybrids, Manufacturing Technology*, vol. 7, no. 4, pp. 314–327, 1984.

[28] R. Darveaux, "Effect of assembly stiffness and solder properties on thermal cycle acceleration factors," in: *Proc. of 11th Int. Workshop on Thermal Investigations of ICs and Systems (THERMINC)*, 2005, pp. 192–203.

[29] J.-P. Clech, *Lead-Free Electronics: iNEMI Projects Lead to Successful Manufacturing*. New York: Wiley-IEEE Press, 2007.

[30] R. Darveaux, "Effect of simulation methodology on solder joint crack growth correlation," in: *Proc. of 50th Electronic Components and Technology Conf.*, 2000, pp. 1048–1058.

[31] G. Z. Wang, K. Becker, J. Wilde, and Z. N. Cheng, "Applying ANAND model to represent the viscoplastic deformation behavior of solder alloys," *Journal of Electronic Packaging*, vol. 123(3), pp. 247–253, 1998.

[32] S. Ramminger, N. Seliger, and G. Wachutka, "Reliability model for Al wire bonds subjected to heel crack failures," *Microelectronics Reliability*, vol. 40, no. 8–10, pp. 1521–1525, 2000.

[33] A. Meyer, *Programmer's Manual for Adaptive Finite Element Code SPC-PM 2Ad*, Preprint SFB393/01-18 TU Chemnitz, 2001.

[34] T.-Y. Hung, C.-J. Huanga, C.-C. Leed, C.-C. Wange, K.-C. Lue, and K.-N. Chiang, "Investigation of solder crack behavior and fatigue life of the power module on different thermal cycling period," *Microelectronic Engineering*, vol. 107, pp. 125–129, 2013.

[35] S. Déplanque, "Lifetime prediction for solder die-attach in power applications by means of primary and secondary creep," *Ph.D. dissertation*, The Brandenburg University of Technology Cottbus-Senftenberg, Cottbus, 2007.

[36] S. Déplanque, W. Nuchter, B. Wunderle, R. Schacht, and B. Michel, "Lifetime prediction of SnPb and SnAgCu solder joints of chips on copper substrate based on crack propagation FE-analysis," in: *Proc. of 7th Int. Conf. on Thermal, Mechanical and Multiphysics Simulation and Experiments in Micro-Electronics and Micro-Systems (EuroSime)*, 2006, pp. 1 – 8.

[37] D. Newcombe and C. Bailey, "Rapid solutions for application specific IGBT module design," in: *Proc. of Int. Exhibition and Conf. for Power Electronics, Intelligent Motion, Renewable Energy and Energy Management (PCIM)*, 2007.

[38] H. Lu, C. Bailey, and C. Yin, "Design for reliability of power electronic modules," *Microelectronics Reliability*, vol. 49, pp. 1250–1255, 2009.

[39] L. Yang, P. A. Agyakwa, and C. M. Johnson, "A time-domain physics-of-failure model for the lifetime prediction of wire bond interconnects," *Microelectronics Reliability*, vol. 51, no. 9–11, pp. 1882–1886, 2011.

[40] S. D. Downing and D. F. Socie, "Simple rainflow counting algorithms," *International Journal of Fatigue*, vol. 4, no. 1, pp. 31–40, 1982.

[41] J. H. L. Pang, *Lead Free Solder*. Berlin: Springer, 2012.

[42] J.-P. Clech, "An obstacle-controlled creep model for Sn–Pb and Sn-based-lead-free solders," in: *Proc. of SMTA Int. Conference (SMTAI)*, 2004.

[43] D. Rubesa, *Lifetime Prediction and Constitutive Modeling for Creep-Fatigue Interaction*. Berlin: Gebrueder Borntraeger, 1991.

[44] W. Ramberg and W. R. Osgood, *Description of Stress–Strain Curves by Three Parameters*, National Advisory Committee for Aeronautics, Washington DC, Tech. Rep., 1943.

[45] R. Darveaux and K. Banerji, "Constitutive relations for tin-based solder joints," *IEEE Transactions on Components, Hybrids, and Manufacturing Technology*, vol. 15, no. 6, pp. 1013–1024, 1992.

[46] S. Knecht and L. R. Fox, "Constitutive relation and creep-fatigue life model for eutectic tin-lead solder," *IEEE Transactions on Components, Hybrids, and Manufacturing Technology*, vol. 13, no. 2, pp. 424–433, 1990.

[47] R. Darveaux, K. Banerji, A. Mawer, and G. Dody, Reliability of plastic ball grid array assemblies (Chapter 13), in: J. H. Lau, (ed.), Ball Grid Array Technology. New York: McGraw-Hill, 1995.

[48] C. H. Raeder, L. E. Felton, R. W. Messier, and L. F. Coffin, "Thermo-mechanical stress–strain hysteresis of Sn–Bi eutectic solder alloy," in: *Proc. of 17th IEEE/CPMT Int. Electronics Manufacturing Technology Symp.*, 1995, pp. 263–268.

[49] J.-P. Clech, "Solder reliability solutions: a PC-based design-for-reliability tool," *Soldering & Surface Mount Technology*, vol. 9, pp. 45–54, 1997.

[50] W. Lee, L. Nguyen, and G. Selvaduray, "Solder joint fatigue models: review and applicability to chip scale packages," *Microelectronics Reliability*, vol. 40, no. 2, pp. 231–244, 2000.

[51] U. Scheuermann and J. Lutz, "High voltage power module with extended reliability," in: *Proc. of 8th European Conference on Power Electronics and Applications (EPE)*, 1999.

[52] D. Shirley, Transient and steady-state creep in Sn–Ag–Cu lead free solder alloys: experiments and Modeling, *Ph.D. dissertation*, University of Toronto, Toronto, 2009.

[53] K. Mysore, G. Subbarayan, V. Gupta, and R. Zhang, "Constitutive and aging behavior of Sn3.0Ag0.5Cu solder alloy," *IEEE Transactions on Electronics Packaging Manufacturing*, vol. 32, no. 4, pp. 221–232, 2009.

[54] H. Yang, P. Deane, P. Magill, and K. Murty, "Creep deformation of 96.5Sn–3.5Ag solder joints in a flip chip package," in: *Proc. of Electronic Components and Technology Conf.*, 1996, pp. 1136–1142.

[55] The European Center for Power Electronics (ECPE). [Online]. 2015. Available: http://www.ecpe.org/home/.

[56] S. Wiese and K.-J. Wolter, "Microstructure and creep behaviour of eutectic SnAg and SnAgCu solders," *Microelectronics Reliability*, vol. 44, pp. 1923–1931, 2004.

Chapter 8

Active thermal control for improved reliability of power electronics systems

Shihao Xia[1], Ke Ma[1], Zian Qin[2] and Dao Zhou[3]

As previously discussed, the thermal stress of power electronics components is intimately associated with some significant wear-out failure mechanisms observed in converter systems. In many emerging applications of power electronics, including renewable energy, electric vehicles, and power transmission, the operating conditions of the converter are often complicated, resulting in adverse thermal behaviors in the power electronics components with various time constants. This can lead to a significant risk of reliability issues. This chapter introduces the active control methods that are designed to relieve the critical thermal stress in power electronics components. As one of the most vulnerable and expensive components, power semiconductor devices will be the primary focus, with case studies of a few selected converter applications. It can be concluded that the thermal stress, especially the thermal cycling in the power semiconductor devices, can be significantly relieved by proper controls, modulation strategies, and the activation of energy storage systems. Further studies on this topic are still underway, with promising results.

8.1 Introduction

8.1.1 Thermal stress and reliability of power electronics

The failure mechanisms of power electronics are intricate and influenced by a multitude of factors [1–5]. It has been revealed that thermal cycling, such as temperature fluctuations inside or outside the components, represents a significant and pervasive failure mechanism in power electronics systems. The temperature fluctuations of different materials with mismatched coefficients of thermal expansion (CTEs) may result in a disconnection of contacting areas after a certain number of cycles, thereby leading to the wear-out of the devices. A significant number of manufacturers of power electronics devices, including power semiconductors and

[1]Department of Electrical Engineering, Shanghai Jiao Tong University, China
[2]Department of Electrical Sustainable Energy, Delft University of Technology, Netherlands
[3]Department of Energy, Aalborg University, Denmark

capacitors, have developed their own reliability models. These models are typically based on accelerated or aging tests and are capable of evaluating the lifetime information of components according to certain thermal behaviors [6–9]. An illustrative example can be found in Figure 8.1, which depicts the results of an important project conducted in the 1990s. This project, known as LESIT, aimed to correlate the numerical relationship between the loads on the power device insulated-gate bipolar transistor (IGBT) and the number of cycles to fail. As illustrated in Figure 8.1, an increase in both the amplitude of temperature fluctuation, ΔTj, and the mean temperature level, Tm, has been observed to result in a reduction in the lifetime of power devices. This relationship has also been explained by some analytical models, as summarized in [7], and it has also been corroborated by numerous recent lifetime tests of the power devices [10–13].

In an energy conversion system, the power or current flowing through the converter is typically set according to the available electrical or mechanical power level. Consequently, in the event that the available power inputted to the converter system is not constant, as is the case in renewable energy or motor drive applications, the intricate and variable mission profiles will be directly reflected by the loading variation in the power electronics components. This, in turn, will result in complicated thermal cycling, which can rapidly lead to the wear-out of the components. An example can be observed in the context of wind power conversion systems, as shown in Figure 8.2. In this figure, the primary disturbances that contribute to thermal cycling in power electronics devices are organized according to their dominant time constants of variation. As illustrated in Figure 8.2, the period of thermal cycling in power devices for wind power applications can be influenced by a multitude of factors, spanning from microseconds (AC grid alternating) to years (ambient temperature or wind-speed variation) [14].

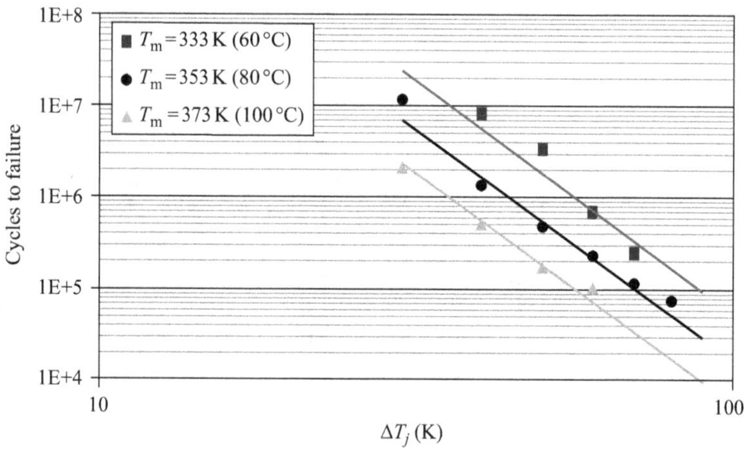

Figure 8.1 Number of cycles to failure of IGBT vs. temperature fluctuation amplitude

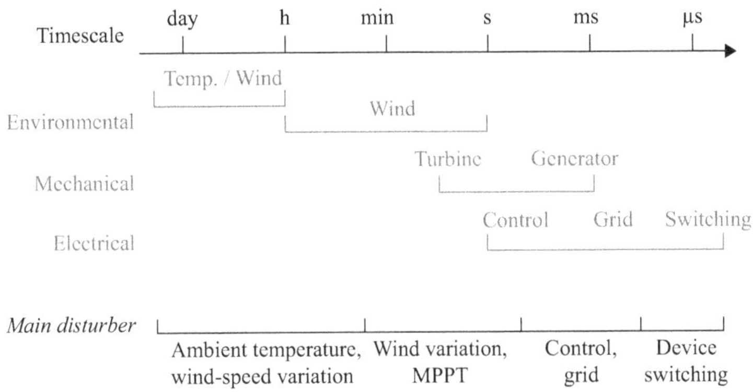

Figure 8.2 Typical time constants of different impact factors to the thermal behaviors of power device. MPPT: maximum power point tracking

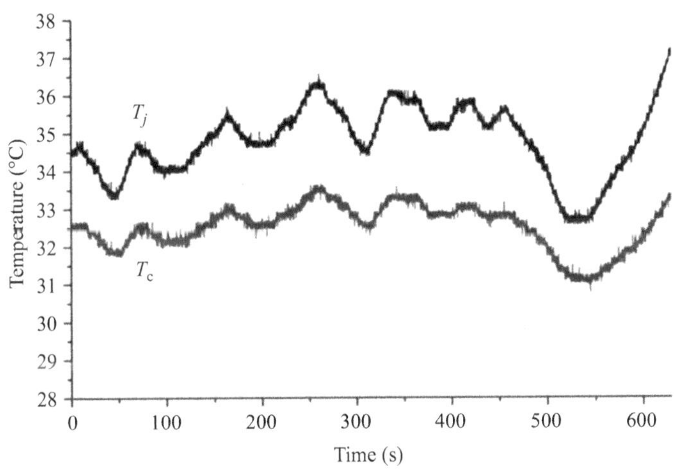

Figure 8.3 Experimental results of medium-term thermal behaviors within 10 min (with temperature sampling rate at 10 Hz)

Besides the wide spread of the time constants, the thermal cycling behavior of power devices exhibits significant variability depending on the device location and the nature of the disturbances. Figures 8.3 and 8.4 illustrate the experimental-based temperature measurement of an opened power module using an infrared camera. The figures demonstrate the temperature of a power device in a 10-kW three-phase photovoltaic (PV) inverter under real-time operation. More details about this study can be found in [15]. Two distinct testing points are utilized for measurement: the chip temperature, Tj, and the base plate temperature, Tc. The thermal profiles of the measured Tj and Tc within 10 min are illustrated in Figure 8.3. It can be observed that the converted power undergoes a corresponding change in accordance with the

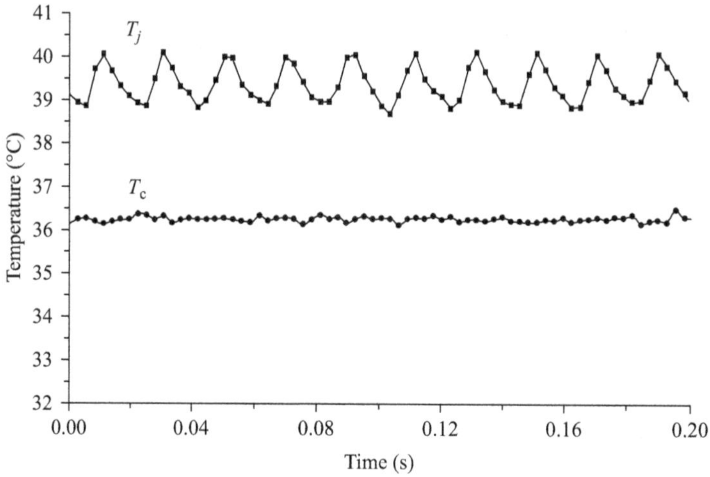

Figure 8.4 Experimental results of short-term thermal behaviors within 0.2 s (with temperature sampling rate at 350 Hz)

variation of solar irradiance within the same amount of time. It can be seen that Tj is higher than Tc with a larger cycling amplitude, and they fluctuate slowly in accordance with the power variation of the converter.

The thermal behaviors of the Tj and Tc are illustrated in Figure 8.4, which depicts a much shorter time period of 0.2 s. In this figure, the solar irradiance remains constant, and the current loading of the converter is maintained at a rated output power of 10 kW. It can be observed that a distinct type of thermal cycling can be identified, which differs significantly from the one illustrated in Figure 8.3. The Tj exhibits a markedly faster fluctuation with a smaller but constant amplitude at a frequency of 50 Hz, which is disrupted by the alternating current synchronized with the frequency of the power grid.

It can be concluded that the thermal cycling of power devices is a significant factor influencing the reliability of the converter. The thermal cycles of the converter can be influenced by a number of factors related to the mission profile of the entire energy conversion system. Consequently, the thermal profiles within power devices may exhibit disparate behaviors and time constants, depending on the device's location and the disturbances affecting the converter loads.

8.1.2 Concept of active thermal control for improved reliability

The root cause of a component's failure can be attributed to a discrepancy between the stress and strength levels to which it is subjected. For a power semiconductor device, the stress can be represented as the loading level or thermal cycling, and the strength can be represented as the inherent ability of the device to withstand the thermal cycles. As shown in Figure 8.5, the actual stress level of a power device is significantly

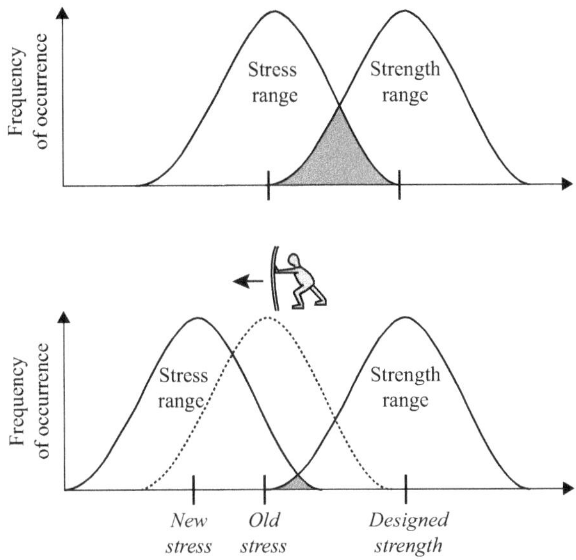

Figure 8.5 Concept of active thermal control for improved reliability of power electronics by reducing the stress

influenced by the user's actions or operational environments. Consequently, it is essential to distribute the stress levels across a broader range rather than concentrating them at a single point. Furthermore, the strength of components should exhibit a certain range of distribution due to inherent manufacturing differences. Consequently, a point of intersection between the stress and strength ranges emerges, indicating a shorter lifespan or heightened probability of failure than anticipated.

One effective method for reducing the crossing area between the stress and strength, or for extending the lifetime/reliability of components, is to reduce the overall stress range to a lower level, as shown in Figure 8.5. For a converter system, this improvement relieves the thermal cycling of the components either by reducing the fluctuation amplitude or by reducing the mean level of temperature [16]. Furthermore, the strength of the converter does not need to be changed, which eliminates the additional cost associated with enhancing the converter design or components.

As previously stated, the thermal cycling of a converter is caused by a multitude of disturbances with various time constants. Conversely, the loading of power electronics components can be easily modified by altering the operating mode of the converter. Consequently, there are potential avenues for actively controlling the thermal cycling of components in a manner that optimizes reliability.

Due to high loss dissipation density and intense thermal cycling, power semiconductor devices are identified as the most vulnerable components in converter systems, representing a significant portion of the overall cost [3–5]. Accordingly, this chapter will examine the application of active thermal control methods to reduce stress levels in power semiconductor devices, which have been selected as a case study.

8.2 Modulation strategies achieving better thermal loading

8.2.1 Impacts of modulation strategies on thermal stress

The modulation strategies employed in power electronics have the potential to influence not only the harmonics of a converter's output voltage and current but also the current flow among the power devices themselves. Specific modulation techniques have been devised to enhance the utilization of the DC-link voltage, minimize harmonics, or redistribute losses among devices [17,18]. As modulation strategies have the capacity to change the thermal loading of power devices, they can be employed to improve the loss distribution in power converters, thereby improving reliability performance. Basically, the modulation strategies can impact the thermal loading of devices in two key ways: through switching loss and through conduction loss of power devices.

It has been demonstrated that the switching loss can be significantly reduced by implementing discontinuous Pulse Width Modulation (PWM) or reducing the carrier frequency [19–25]. The disadvantages introduced concurrently comprise elevated current ripples resulting from the diminished switching count. Moreover, since the maximum length of the discontinuous interval has to be limited in one switching cycle, the effect of switching loss reduction is thus limited. To further lower the switching loss, the carrier frequency can be reduced. Again, the higher current ripple will be created by a lower switching frequency; but, if it is only used in a short dynamic process, the temporary degradation of the power quality may still be acceptable [26].

In some cases, conduction loss can be regulated in order to optimize the thermal loading, particularly in circuit topologies that have switching redundancy (e.g., three-level neutral point clamped topology [25] or nine-switch converters [27].) By adding a special common mode offset to the PWM references, the input and output of the converters will not change considerably, but the conduction loss can be redistributed among the power devices, and therefore it can be used to reduce the thermal loading of the power device [27,28]. More details based on this case study are given below to illustrate the impact of the modulation strategies on the thermal performance of the power converters under different operating conditions.

8.2.2 Modulation under normal conditions

8.2.2.1 Thermal improvement by switching loss reduction

The reduction of switching loss through the utilization of discontinuous PWM (DPWM) in B6 converters (Figure 8.6) represents a well-established strategy, as illustrated in Figure 8.7. The principle of DPWM is to clamp the appropriate voltage/current reference of the converter output to the upper or lower rail of the carrier within a specified interval. This enables the corresponding power device to maintain its status without switching (on or off), thereby reducing the switching loss within that interval. The reduction in IGBT loss at the peak value consequently results in a reduction in both the average value and the variation of the junction temperature. It is also noteworthy that the impact on the conduction loss of the device may be modified

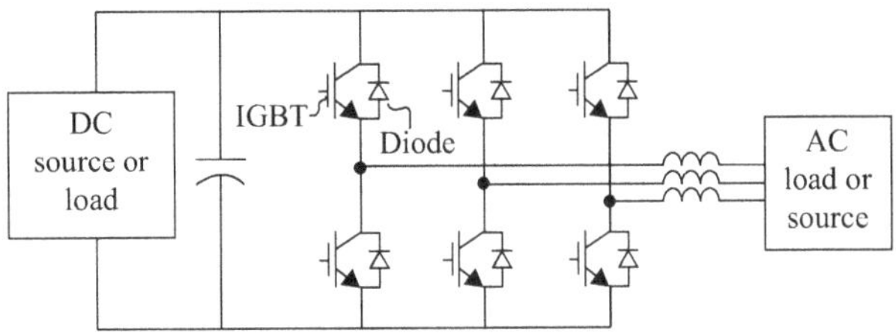

Figure 8.6 The topology of the B6 converter

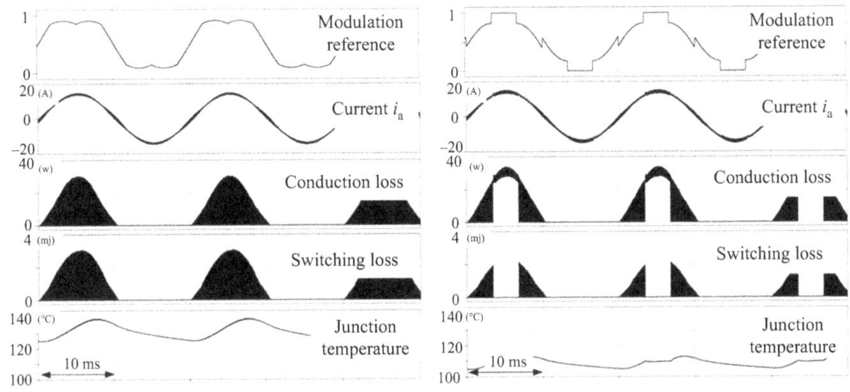

Figure 8.7 Power loss and junction temperature of the IGBT in a B6 converter with (a) SPWM and (b) DPWM (switching loss:conduction loss = 3.56:1)

by the use of DPWM, as illustrated in Figure 8.7(b). The conduction loss of the IGBT in the discontinuous segment is, in fact, increased by the use of DPWM, as the device is kept in an on state. Consequently, the prerequisite for attaining enhanced thermal performance through DPWM is that the augmented conduction loss must be less than the diminished switching loss. In contrast, the application of DPWM may potentially result in an elevated temperature of the IGBT.

The utilization of DPWM is not only limited to two-level converters, but multilevel converters can also take advantage of it. A case study on a three-level neutral point clamped (NPC) converter can be found in [25], where different modulation methods in terms of optimal zero sequence injection (Opt-ZSSPWM) [29], conventional 60° discontinuous PWM (CONV-60° DPWM) [30], and alternative 60° discontinuous PWM (ALT-60° DPWM) [30] are applied to the converter. It has been reported in [25] that with CONV-60° DPWM and ALT-60° DPWM, the device with the highest temperature (Cl. Diode1) can be relieved significantly.

The modulation strategy can also be used to achieve soft-switching in DC/AC inverters, including single phase, three-phase, half bridge, and full bridge [31]. The switch current is controlled to be bidirectional in every switching cycle; thus, zero-voltage switching can be realized. However, the application is limited to low power due to the large d*i*/d*t*.

8.2.2.2 Thermal improvement by conduction loss reduction

The conduction loss can also be modified by modulation methods to improve the thermal performance, and a typical case can be found for the nine-switch converter [27], as seen in Figure 8.8(a). The nine-switch converter is a switch-reduced version of the twelve-switch back-to-back converter, and if the application criteria are followed, the nine-switch converter can get an even lower loss than the 12-switch converter [21]. Since the input and output current are combined in the same leg, the nine-switch converter has unevenly distributed loss among the three switches of each leg, where the upper and lower switches normally have lower losses than the middle one, as shown in Figure 8.8(b). A solution to relieve this uneven loss distribution and reduce the hotspot temperature is by increasing the distance between the upper and lower references to shift the conduction loss of the middle switch (hottest one) to the upper and lower ones [22]. Meanwhile, the 120°-DPWM can achieve the maximum distance between the upper and lower references, as indicated in Figure 8.9(b). The conduction loss is then more equally shared among the three switches, as illustrated in Figure 8.10(a). In addition, by introducing the 120°-DPWM, the switching loss is naturally influenced; but in the case study, the effect of the switching loss is mitigated significantly by choosing the specific power devices to achieve a low switching and conduction loss ratio. Nevertheless, the overall loss reduction of the middle switch is about 30% when using 120°-DPWM. The pictures captured by the infrared camera of the nine-switch converter with sinusoidal pulse width modulation (SPWM) and 120°-DPWM are compared in Figure 8.10(b). Good matching with the loss distribution (Figure 8.10(a)) can be observed, and the hotspot temperature is reduced significantly by using 120°-DPWM.

8.2.3 Modulations under fault conditions

The ride-through ability during the grid faults is required more and more for grid-tied converters, especially in renewable energy applications with large power capacity [32]. It has been reported in [32,33] that the thermal loading of the 3L-NPC converter shown in Figure 8.11(a) could be rather unbalanced among the power devices, and undesired

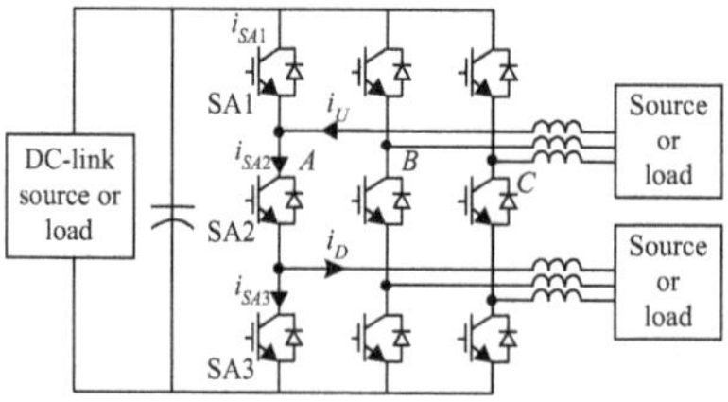

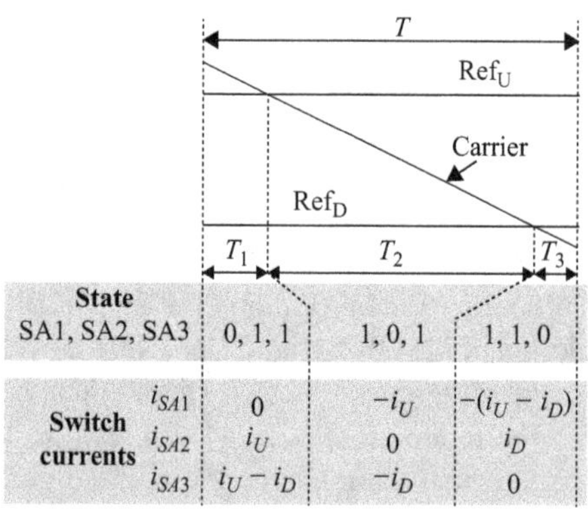

Figure 8.8 Nine-switch power converter system and its switch currents: (a) circuit topology and (b) instantaneous currents in a half-carrier period for Phase A

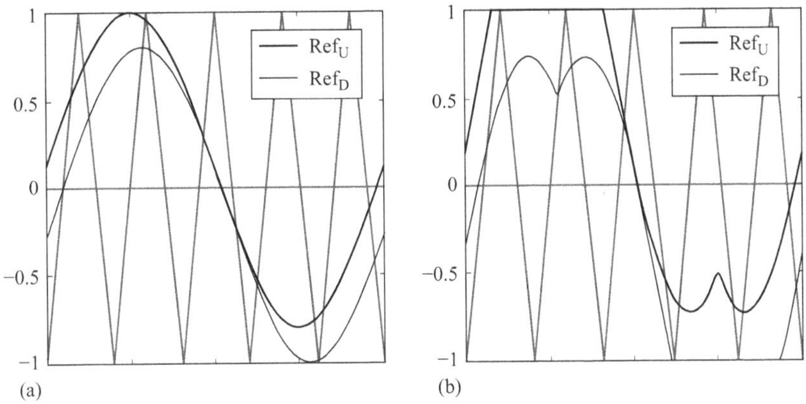

Figure 8.9 Modulation references of the nine-switch converter in Figure 8.8(a): (a) SPWM and (b) DPWM

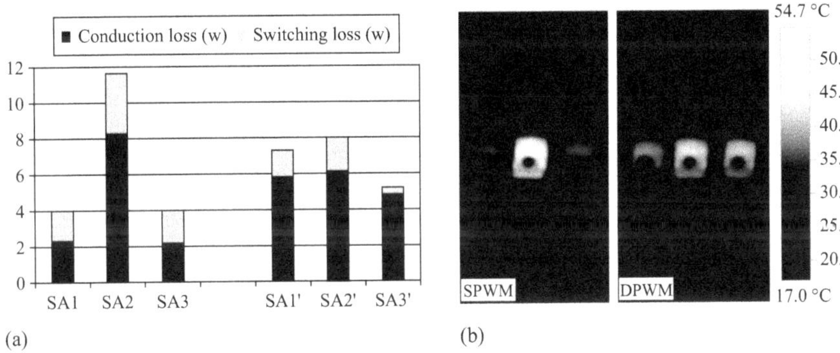

Figure 8.10 Loss distribution and temperature of the nine-switch converter with SPWM and DPWM: (a) loss distribution (SPWM: SA1, SA2, SA3; DPWM: SA1', SA2', SA3'); (b) thermal picture (from left to right: SA3, SA2, SA1)

overheating of the inner switch, as well as the clamped diode, will be present during the low-voltage ride-through (LVRT) operation of the 3L-NPC converter.

An examination of the space vector diagram (SVD) of the 3L-NPC converter, as illustrated in Figure 8.11(b), reveals that all of the state vectors within the inner hexagon of the SVD exhibit switching redundancies. This diagram is typically employed for the space vector modulation (SVM) of the three-phase converter. The aforementioned switching redundancies afford the requisite control flexibility to modify the current paths flowing in the power devices. This is particularly pertinent when the reference vector is situated in the inner hexagon of Figure 8.11, in conjunction with the LVRT operation of a 3L-NPC converter.

Consequently, a series of optimized modulation sequences is proposed in [32], and one of the SVM sequences within one switching cycle is illustrated in Figure 8.12.

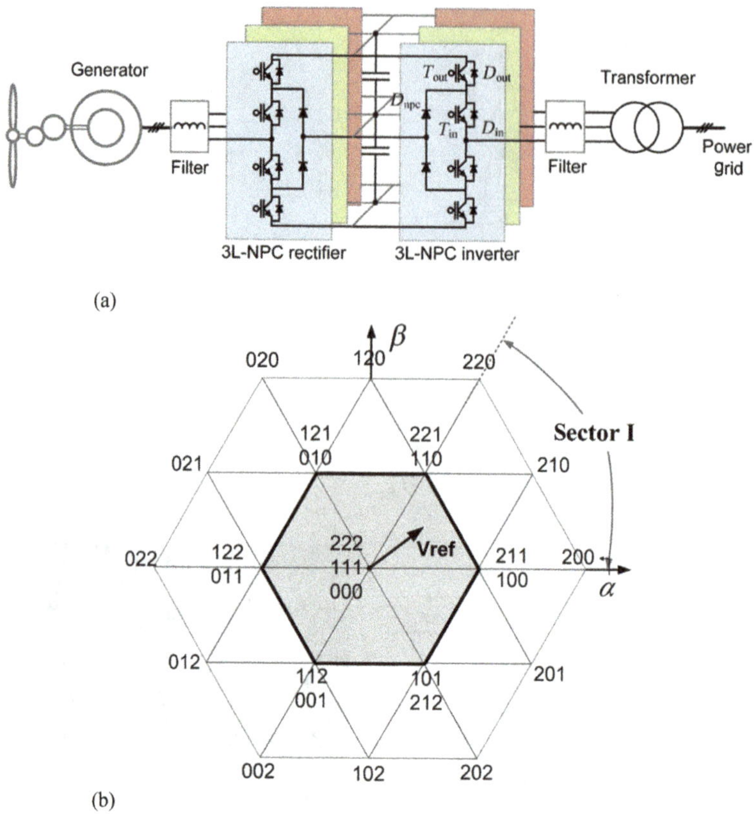

Figure 8.11 (a) 3L-NPC converter used in a wind turbine. (b) Space vector diagram for the 3L-NPC converter.

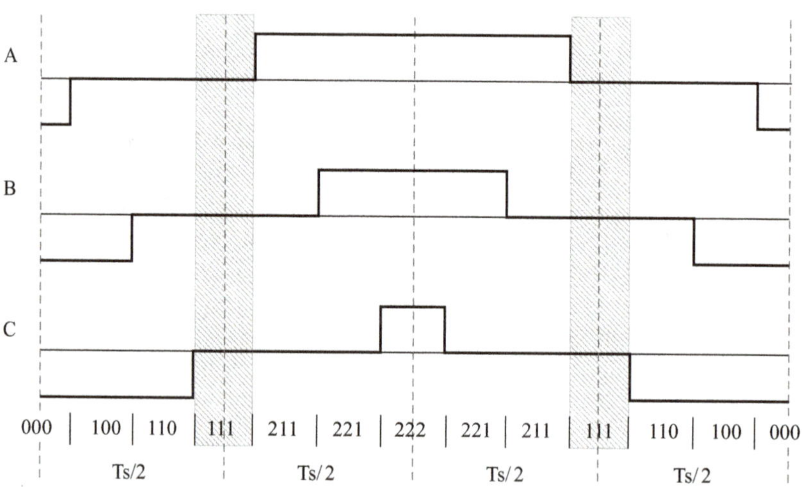

Figure 8.12 Optimal modulation sequence for relieved device temperature under LVRT

It can be observed that the state vector 111, which connects the three-phase output of the converter to the neutral point of the DC bus, has been eliminated. This indicates that the time required for the converter to be in a state where the current is flowing through the clamped diode and inner switch of the 3L-NPC will be reduced. The simulation results of the device temperature in the 3L-NPC converter under a severe LVRT are shown in Figure 8.13. It is evident that the implementation of the optimized modulation sequence results in a more balanced thermal loading across the devices, with a notable reduction in the thermal loading of the most stressed devices during LVRT.

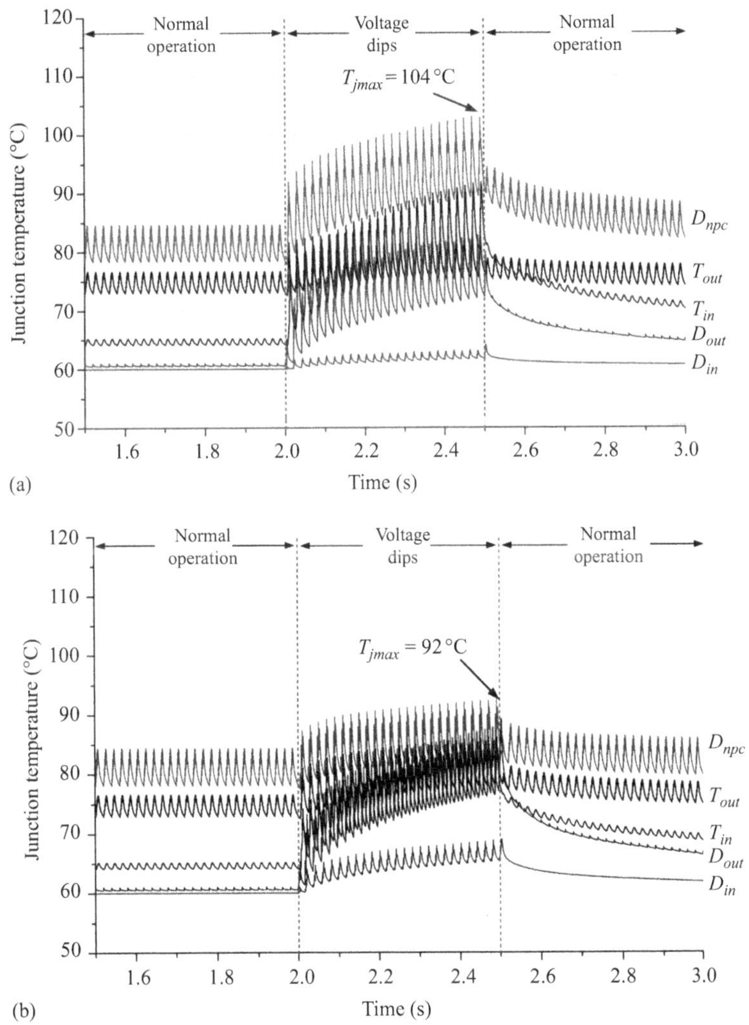

Figure 8.13 Device temperature of 3L-NPC under LVRT of 500 ms [32]. (a) With normal modulation and (b) with optimal modulation.

8.3 Reactive power control achieving better thermal cycling

8.3.1 Impacts of reactive power

The reactive power delivered by a converter is typically not constrained by the available mechanical/electrical power input to the converter system. However, it can significantly impact the loading of components, making it an ideal candidate for achieving active thermal control and enhancing the reliability of the converter [34–36]. A straightforward illustration can be observed in Figure 8.14, where a

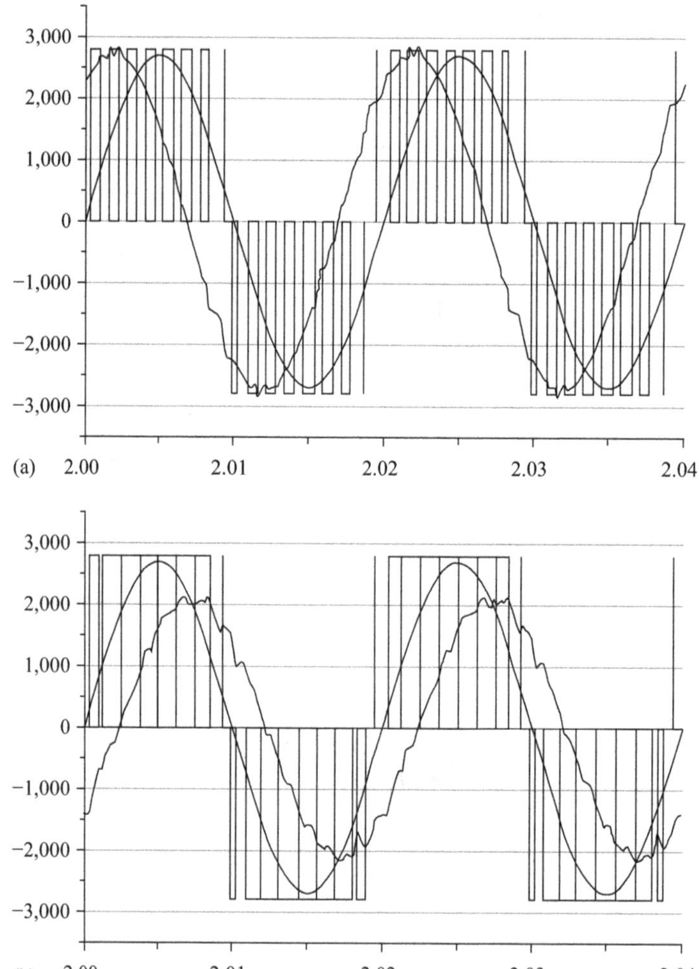

Figure 8.14 Output current and voltage of a grid-connected 3L-NPC converter. (a) Maximum unexecuted reactive power. (b) Maximum over-executed reactive power.

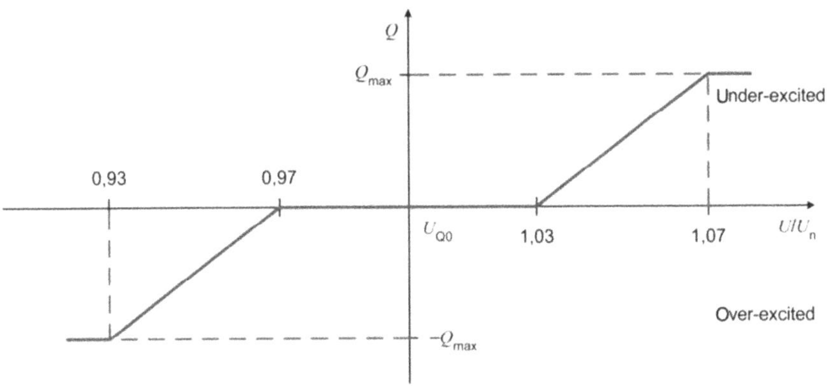

Figure 8.15 Reactive power requirement specified in German grid codes [37]

3L-NPC converter is operating under maximum over-excited (OE) as well as under-excited (UE) reactive power, respectively, with rated active power output. It can be observed that the reactive power has the effect of modifying both the phase angle between the output voltage and current of the converter and the current amplitude flowing in the power devices. These effects are related to the loading level and thermal distribution among the power devices.

However, the reactive power delivered to the power grid is closely related to its stability and voltage level, particularly in the case of "weak" grids with limited power capacity [37,38]. Consequently, an increasing number of grid codes have established rigorous limitations on the quantities of reactive power that can be injected by the converter. Figure 8.15 shows an example of the German grid code [37], where both the boundaries for the maximum under-excited (UE) and over-excited (OE) reactive power that can be injected into the grid are defined with relation to the level of the voltage at the terminals of the power generation unit. These grid codes will significantly restrict the thermal control ability through the use of reactive power delivered by a converter.

This section introduces several active thermal control methods that can overcome the limitations of grid codes by utilizing reactive power. A double-fed induction generator (DFIG)–based wind turbine system as well as a full-scale parallel converter–based wind turbine system will be used as case studies.

8.3.2 Case study on the DFIG-based wind turbine system

A typical DFIG-based wind turbine system is shown in Figure 8.16. Due to the doubly fed mechanism of the DFIG wind turbine system, the produced power from the wind energy can be transferred from both the stator side and rotor side of the induction generator. Similarly, the reactive power can be supported from either the stator side of the DFIG or the GSC in back-to-back power converters [39,40].

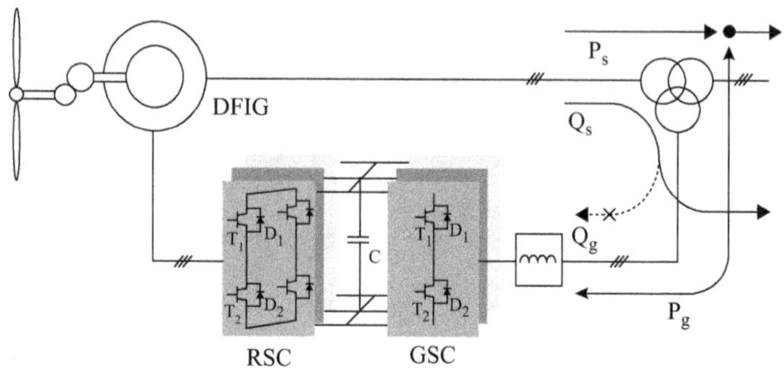

Figure 8.16 Compensation scheme of the reactive power in the DFIG wind turbine system

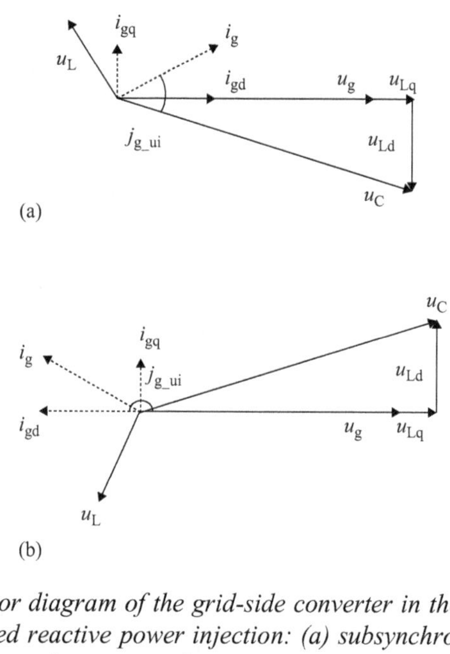

Figure 8.17 Phasor diagram of the grid-side converter in the case of the over-excited reactive power injection: (a) subsynchronous mode; (b) supersynchronous mode

Because both the rotor-side converter (RSC) and the GSC have the ability to control the reactive power, the reactive power can be circulated within the DFIG system. As shown in Figure 8.16, the reactive power delivered to the power grid will not be changed in the case that the reactive power of the back-to-back power converters is controlled in the opposite direction. However, some realistic limitations exist by using the control scheme. For the convenience of explanation, the phasor diagram for the grid-side converter (GSC) in Figure 8.16 is shown in

Figure 8.17. Due to the opposite polarity of the active power through the GSC between the subsynchronous mode and the supersynchronous mode, the direction of the active power is shown in Figure 8.17(a) and (b), respectively. In the case of the UE reactive power, the phasor diagram can be obtained by rotating the q-axis current 180 degrees.

The analytical formula for the converter output voltage, U_c, is expressed as follows:

$$U_c = \sqrt{(U_g + i_{gq}X_g)^2 + (i_{gd}X_g)^2} \leq \frac{U_{dc}}{\sqrt{3}} \tag{8.1}$$

where U_g and U_{dc} denote the rated peak phase grid voltage and dc-link voltage respectively; X_g denotes filter reactance at 50 Hz; and i_{gd} and i_{gq} denote the grid-side converter peak current in the d-axis and q-axis, respectively.

It is evident that regardless of the operation modes, the amplitude of the converter voltage is increased in the case that the OE reactive power is introduced. Therefore, one constraint is the OE current should not lead to over modulation.

The second restriction lies in the capacity of the power device,

$$\sqrt{{i_{gd}}^2 + {i_{gq}}^2} \leq I_m \tag{8.2}$$

where I_m denotes the peak current of the power module, which should not be exceeded by increasing reactive current.

The third limitation is that the capacity of the induction generator Q_s must be taken into account [41]:

$$\frac{3}{2}U_g i_{gq} \leq Q_s \tag{8.3}$$

Similar limitations can be seen from the RSC, as has been addressed in [42].

By considering the above-mentioned limitations, the possible ranges of the reactive power in the GSC and the RSC are summarized in Figure 8.18, where the current amplitude and the power factor angle of the power converter, which are two indicators for power device loading, are shown.

The sudden fluctuations in wind speed that occur during gusts will be reflected by the converter's adverse thermal cycling, as demonstrated in [42]. Therefore, it is feasible to regulate the junction temperature oscillation in response to wind gusts through an appropriate thermal-oriented reactive power control in a DFIG system, as illustrated in Figure 8.19.

A typical 1-year return period wind gust is defined in IEC [43]. The thermal cycling of the back-to-back power converters with and without thermal-oriented reactive power control is illustrated in Figure 8.20(a) and (b), respectively. As

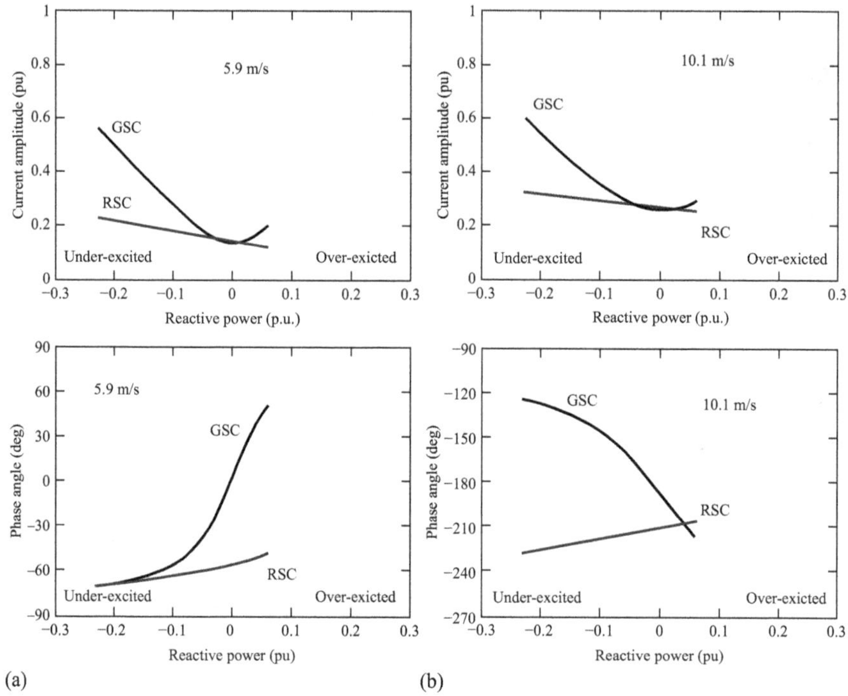

Figure 8.18 Influence of reactive power circulation on the back-to-back power converters in a DFIG system: (a) subsynchronous mode at a wind speed of 5.9 m/s; (b) supersynchronous mode at a wind speed of 10.1 m/s

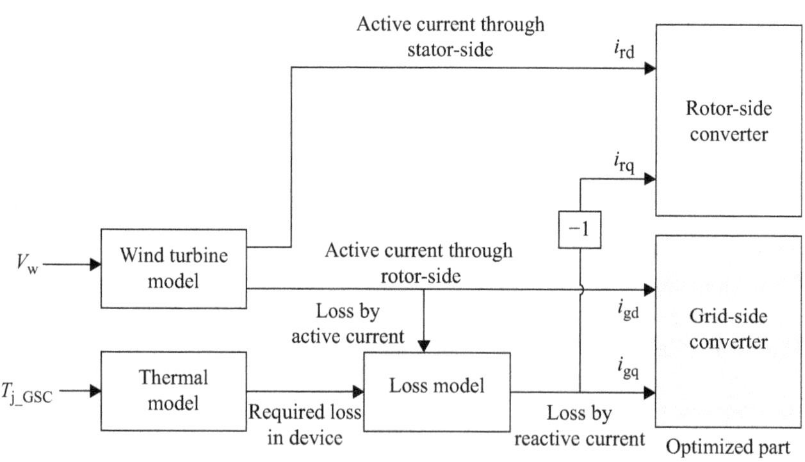

Figure 8.19 Thermal-oriented control diagram of the back-to-back power converters during wind gusts. V_W: wind speed. T_{j_GSC}: junction temperature of the grid-side converter

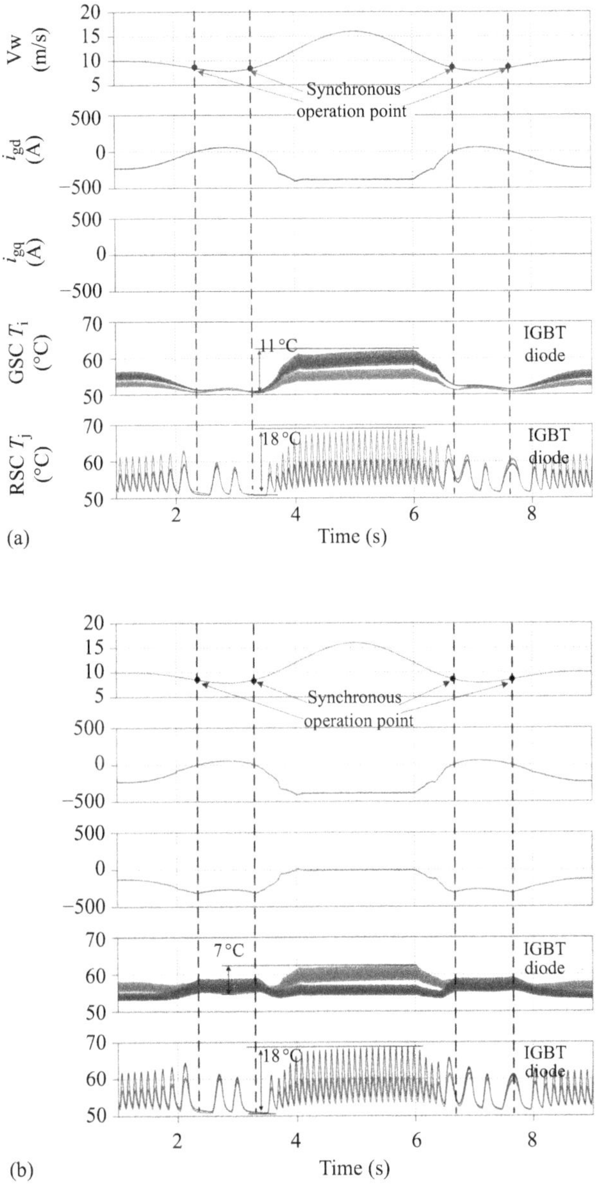

Figure 8.20 Thermal cycling of the back-to-back power converters during wind gusts in a DFIG system: (a) without thermal-oriented reactive power control; (b) with thermal-oriented reactive power control

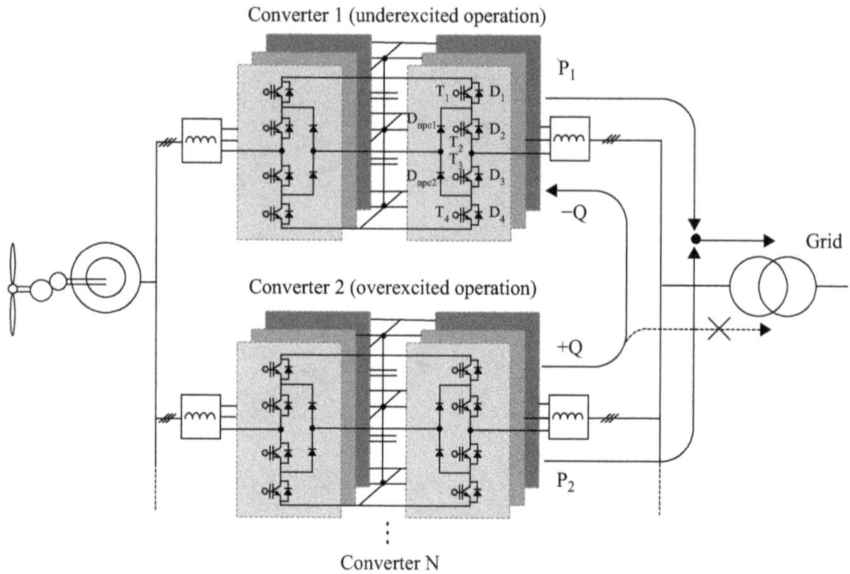

Figure 8.21 Reactive power circulating among parallel converters in a wind turbine system

illustrated in Figure 8.20(a), the thermal stress is minimized at a synchronous wind speed due to the absence of active power flows in the GSC. Conversely, it is maximized in the RSC due to the low frequency of the rotor current. As illustrated in Figure 8.20(b), the maximum junction temperature fluctuation in the GSC is observed to decrease from 11 °C to 7 °C as a result of the incorporation of supplementary reactive power. Conversely, the maximum junction temperature fluctuation in the RSC remains unaltered at 18 °C.

8.3.3 Study case of paralleled converters

It is also possible to control the temperature of the device by utilizing the reactive current in a full-scale converter–based wind turbine system, as illustrated in Figure 8.21. Typically, the converter in this wind turbine system needs to handle 3 times more power than the wind turbine system shown in Figure 8.16, while maintaining the same capacity. As a result, paralleling converter units are frequently employed in full-scale converter–based wind turbines. This configuration allows for the temperature of power devices to be controlled by circulating the reactive power within the converter system, as detailed in [44]. The control results under a wind gust are illustrated in Figure 8.22. It can be seen that the temperature fluctuations in the most stressed device can be significantly reduced from 32 °C in Figure 8.22(a) to 12 °C in Figure 8.22(b).

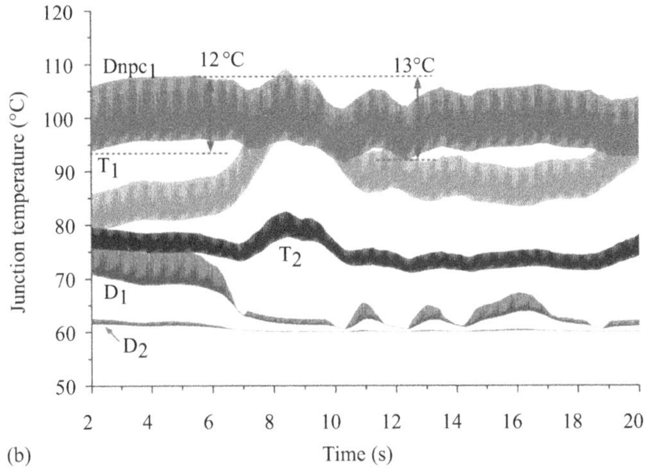

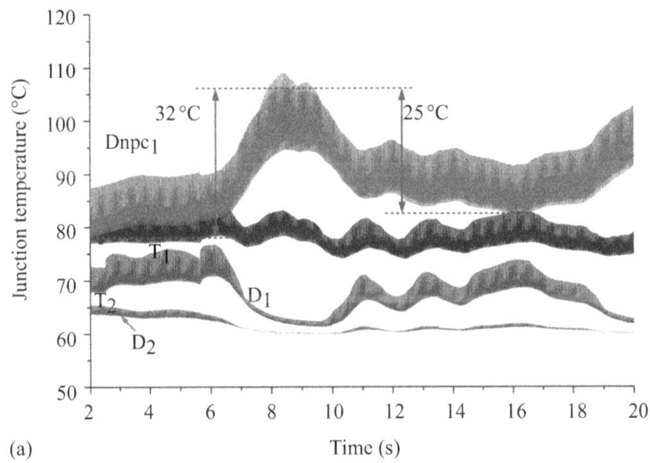

Figure 8.22 Device temperature of a 3L-NPC converter under wind gust. (a) Wihout reactive power control and (b) with reactive power control.

8.4 Thermal control strategies utilizing active power

8.4.1 Impacts of active power on the thermal stress

The idea to reduce the thermal excursions by regulating the reactive power has been illustrated in Section 8.3. However, one shortcoming is that the reactive power can only heat up the power devices while cooling down the power devices is impossible. Thus, the efficiency is always reduced. Another drawback is that the reactive power needs to be circulated among the grid-tied power converters or DFIG generators since the total reactive power injected into the utility grid is not able to be freely regulated. In this section, another type of active thermal control method that utilizes the active power from an energy storage system (ESS) is going to be introduced.

The active power exerts a considerable influence on the current amplitude, thereby affecting the loss and thermal loading of power devices. In contrast to the reactive power, the active power in the converter typically exhibits a fixed correlation with wind speed, with the objective of optimizing energy generation. This relationship is inherently uncontrollable unless an alternative source, such as an ESS, is integrated into the converter. An ESS can absorb or release power when the wind power is beyond or below the average value; therefore, the power delivered to the grid via the converter can be smoothed, leading to smaller thermal excursions.

The ESS for a large-scale wind turbine is still quite expensive; however, power fluctuations caused by the wind-speed variation are not only harmful to the power converter but also unfriendly to the utility grid. Actually, it increases the risk of grid instability and requires more backup generation capability for power balancing of the grid. As a result, several grid-tied energy storage projects have been started [45]. Moreover, energy storage systems for wind turbines have been already commercially available [46]. If the ESS is situated on the DC link of the wind power converter, as illustrated in Figure 8.23, the thermal excursions in the grid-side converter as well as the power fluctuations getting into the grid can be reduced. The primary considerations are the selection of an appropriate energy storage device, the design of the ESS, and the

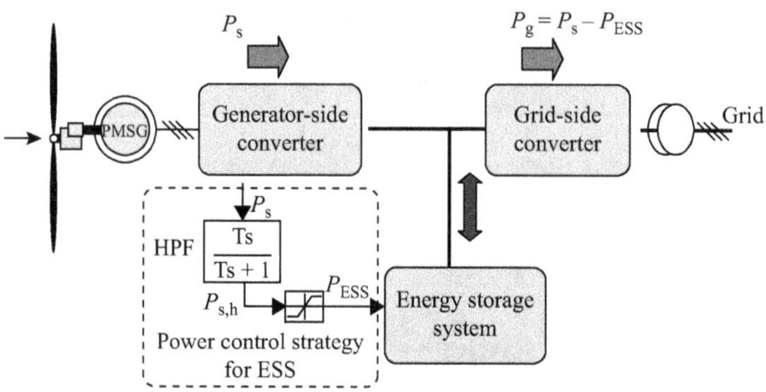

Figure 8.23 Energy storage system (ESS) integrated in a back-to-back wind power converter

potential for thermal performance enhancement through the use of the ESS. A case study provides further insight into this topic.

8.4.2 Energy storage in large-scale wind power converters

The ESS, which comprises a bidirectional power converter and an energy storage device (ESD), can be connected to the DC link of the wind power converter. The design of the power control strategy for the ESS can be accomplished through a straightforward and readily comprehensible approach, as illustrated in Figure 8.23 [47]. Initially, the power generated by the wind turbine is quantified. Subsequently, a high-pass filter is employed to extract the short-term power fluctuations, which are then utilized as the reference charging power for the ESS. This results in a more stable power flow in the grid-side converter. Additionally, the control of the wind power converter is decoupled from the ESS, with the generator-side converter employed for maximum power point tracking (MPPT) and the grid-side converter utilized for DC-link voltage stabilization through grid current regulation.

According to the spectrum in Figure 8.24 [48], the wind-speed variations can be categorized into long-term and short-term ones. The period of the former is in the hundreds of hours, while the latter is on the scale of seconds or minutes. Considering the volume, weight, and cost of the required energy storage, it is more feasible to mitigate short-term power fluctuations by using the energy storage. The short-term ESD candidates are listed in Table 8.1 [49–52]. In detail, the high-speed flywheel is superior in power density, cost of power, deep discharging, and environmental impact. But in order to reduce the power loss caused by friction, the applied magnetic-bearing system is rather complex and it makes the maintenance difficult as well. The Li-ion batteries are developing very fast due to their wide application in mobile electronics products and the cost has been

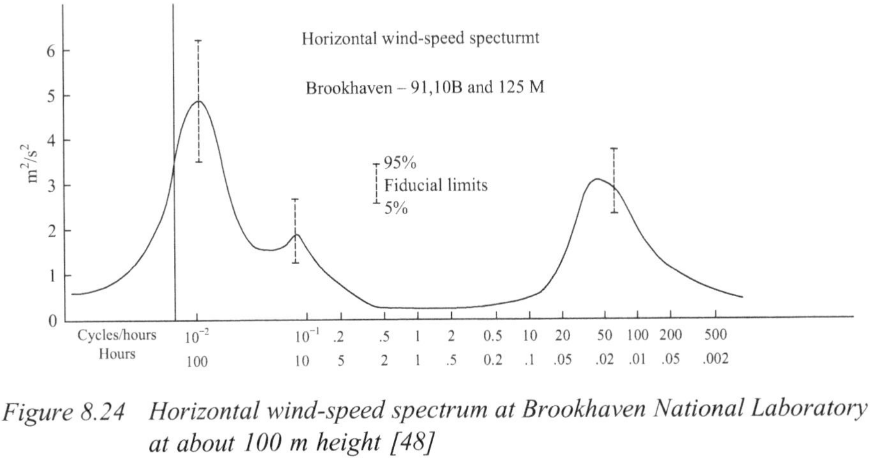

Figure 8.24 Horizontal wind-speed spectrum at Brookhaven National Laboratory at about 100 m height [48]

Table 8.1 Comparison ofshort-term energy storage technologies [49–52]

		Li-ion batteries	Super capacitors	High-speed flywheels
Energy density (Wh/kg)		+++	+	+
Power density (W/kg)		+	+++	+++
Self-discharge		-	----	----
Safety		++	+++	++
Complexity		--	-	---
Stressor impact to lifetime	Temperature	--	-	-
	Deep discharging	-	No	No
Maintenance		-	-	--
Number of deep charge/discharge cycles		+	++	+++
Current Minimum Cost for energy [$/kWh]		-	--	--
Current Minimum Cost for power [$/kW]		--	-	-

reduced significantly. So far, technical limitations, such as low power density, temperature, and deep discharging sensitivity still exist, but they continue to be improved. Another candidate is the supercapacitor (double-layer capacitor), which can achieve high-power density, deep discharging, wide operation temperature range, and high cycle life. These advantages make supercapacitors a good choice for high-power applications. Although the cost is still high for larger-scale applications, it is getting lower and lower due to the developing of the technology.

A case study regarding energy storage sizing has been done in [53], where a supercapacitor bank is chosen to be the ESD because of its excellent high-power density. The cost of the ESS, including the bidirectional converter, is first fixed, and then the relationship between the power and energy of the ESS can be illustrated by the curves in Figure 8.25. The thermal performance of the wind power converter shown in Figure 8.23 is evaluated when operating with ESSs at different sizes and different power and energy ratios. The deformation energy of the wind power converter is obtained by using the stress–strain lifetime model [54], and it is shown in Figure 8.26. The power and energy ratios, which can lead to the lowest deformation energy, are labeled in Figure 8.26.

As seen in the figure, although the cost of the ESS increases significantly, the energy is more or less the same while the power increases a lot. It is thus concluded that the power is much more critical than the energy of the ESS from a reliability improvement point of view. The power and temperature profiles of the wind power converter operating with ESS are illustrated in Figure 8.27. As illustrated, the power is smoothed by employing the ESS, thereby reducing the thermal excursion.

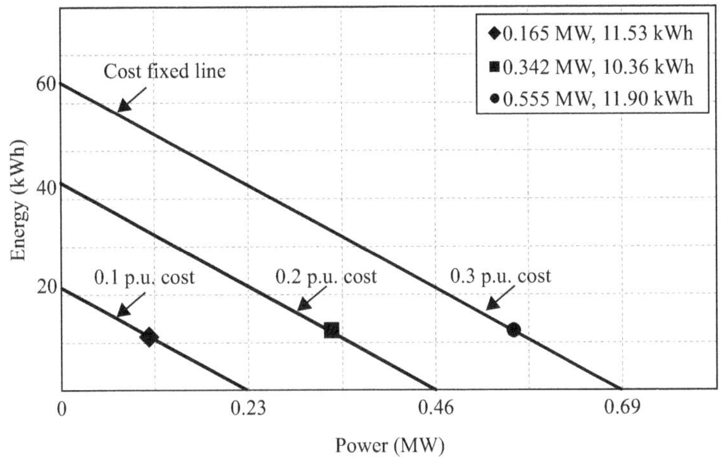

Figure 8.25 Energy as a function of the power of the energy storage system when the cost is fixed

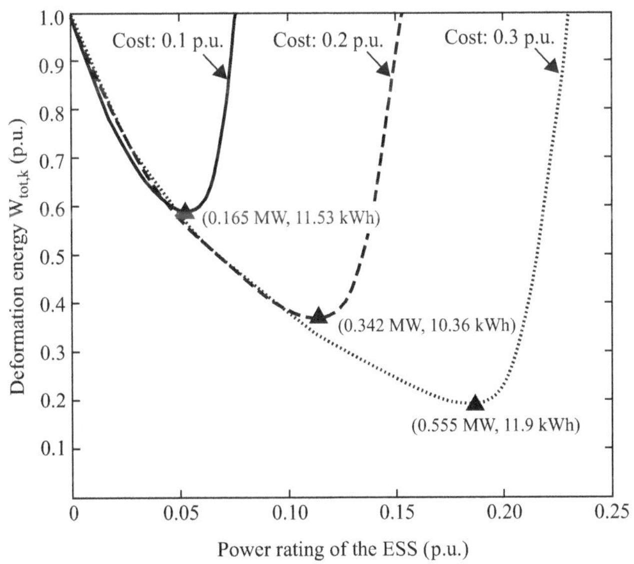

Figure 8.26 Deformation energy as a function of the power rating of the energy storage system with various fixed cost

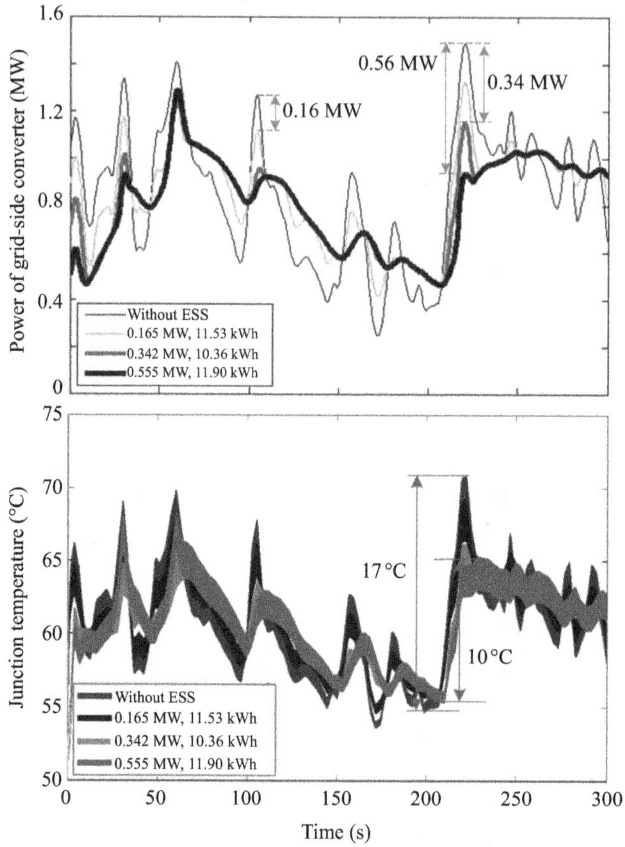

Figure 8.27 Power and junction temperature of the grid-side converter using energy storage

As the ESS capacity increases, the temperature becomes smoother, and the reliability is enhanced.

8.5 Current and DC voltage control achieving optimal thermal stress redistribution under abnormal conditions for a motor drive

The thermal stress under abnormal conditions will dramatically rise and be distributed unevenly in short-term timescale, which is also of great importance to the reliability of power electronics components and systems. In this section, a conventional abnormal condition, the stall condition of an electric machine drive inverter shown in Figure 8.28, is selected as a study case. The stall condition will result in a significant rise and unbalanced distribution of junction temperature because the electric machine will be switched to the operation mode with DC load

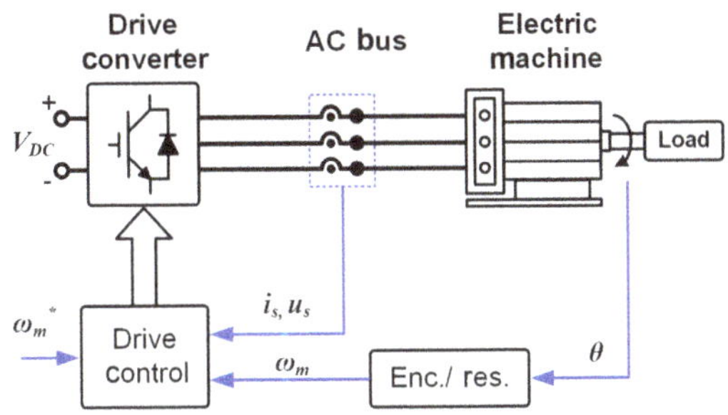

Figure 8.28 A typical electric machine drive system

(where the fundamental frequency is 0 Hz) in this case. Active thermal control methods under this condition will be discussed in this section.

8.5.1 Impacts of current and DC voltage on the thermal stress

According to the reference [55], the power losses of the power semiconductor devices are composed of conduction loss and switching loss, whose independent variables are different. Thus, in order to optimize the total loss, the modeling processes of conduction loss and switching loss need to be unified.

First, when neglecting the switching behavior and applying the switching cycle average model, the conduction loss of the power semiconductor devices can be constructed as (8.4), with the independent variable as stator current i_s and duty ratio d.

$$p_{\mathrm{cond@}T_H/T_L} = V_{\mathrm{cond@}T_H/T_L}(i_s) \cdot i_s \cdot d = f_{\mathrm{cond@}T_H/T_L}(i_s, d) \tag{8.4}$$

where $V_{\mathrm{cond@}T_H/T_L}$ regards the conduction voltage of the power semiconductor device at a certain reference junction temperature T_H or T_L, i_s is the stator load current, and d is the duty ratio of the target device.

By replacing the duty ratio d with stator voltage u_s and DC voltage U_{DC} according to the inverter modulation model, the conduction loss under the stall condition can be remodeled as (8.5) based on SPWM, neglecting nonlinear factors.

$$p_{\mathrm{cond@}T_H/T_L} = V_{\mathrm{cond@}T_H/T_L}(i_s) \cdot i_s \cdot d = V_{\mathrm{cond@}T_H/T_L}(i_s) \cdot i_s \cdot \left(\frac{u_s}{U_{DC}} + 50\%\right) \tag{8.5}$$

where u_s is the stator voltage, which can be further described as $u_s = R_s \cdot i_s$ according to the mathematical model of the electric machine (PMSM for example and neglecting the mechanical dynamics, which is not focused on the power loss model part). Thus, the conduction loss under the stall condition can be remodeled as (8.6), whose independent variables are stator current i_s and DC voltage U_{dc}.

$$p_{\mathrm{cond@}T_H/T_L} = V_{\mathrm{cond@}T_H/T_L}(i_s) \cdot i_s \cdot \left(\frac{i_s \cdot R_s}{U_{DC}} + 50\%\right) = f_{\mathrm{cond@}T_H/T_L}(i_s, U_{DC}) \tag{8.6}$$

Second, the switching loss of the power semiconductor device under the stall condition can be reconstructed as given in (8.7), with the independent variable as stator current i_s and DC voltage U_{dc}:

$$p_{sw@T_H/T_L} = f_{sw} \cdot \left(\frac{U_{DC}}{U_{ref}}\right)^{K_V} \cdot \left(E_{onref@T_H/T_L}(i_s) + E_{offref@T_H/T_L}(i_s)\right)$$
$$= f_{sw} \cdot f_{sw@T_H/T_L}(i_s, U_{DC}) \tag{8.7}$$

where f_{sw} is the switching frequency, E_{onref} and E_{offref} are the turn-on loss and turn-off loss under the reference DC voltage U_{ref}, which can be obtained from the datasheet or experimental measurements, K_v is the coefficient of DC voltage correction.

To sum up, when the equivalent stator resistance R_s and the switching frequency f_{sw} are set to be const or time-invariant, the total power loss $p_{total@T_j}$ of the power semiconductor device under the stall condition can be described as (8.8). The accuracy of the proposed model can be improved through more detailed datasheets or other parameter measurements of the target devices [56].

$$p_{total@T_j} = f_{cond@T_j}(i_s, U_{DC}) + f_{sw@T_j}(i_s, U_{DC}) = f_{total@T_j}(i_s, U_{DC}) \tag{8.8}$$

8.5.2 Adaptive DC voltage under abnormal condition

In order to analyze the effect of DC voltage control, a 40-kW permanent magnet synchronous motor (PMSM) is chosen as the electric machine for further study. The key parameters of the PMSM and experimental setup are given in [57].

According to the selected IGBT, PMSM, and (8.8), the relationship of stator current, DC voltage, and power loss of the IGBT under the stall condition can be constructed as shown in Figure 8.29. It can be figured out in Figure 8.29 that there is a DC voltage U_{DC_lim} (green line) corresponding to the modulation guaranteed bound (95% depth), based on the nonrotating electric machine electromagnetic model. Furthermore, there is an optimal DC voltage U_{DC_opt} (red line) corresponding to the optimal distribution between conduction loss and switching loss of IGBT under the stall condition (i.e., minimum total power loss) at each current level.

By projecting the lines in Figure 8.29 to the i_s-U_{DC} axis plane, the relationship $f_{DC\_opt}(i_s)$ of the optimal DC voltage U_{DC_opt} and stator current i_s (red lines), and the relationship $f_{DC\_lim}(i_s)$ of the U_{DC_lim} and i_s (green line), are illustrated in Figure 8.30. Thus, from the active thermal control perspective, once the stall condition is detected or predicted, under the guarantee of the modulation depth with U_{DC_lim}, the DC voltage can be regulated to the optimal DC voltage (ODV) U_{DC_opt} in order to realize the minimum total power loss. Moreover, it can be observed from Figure 8.30 that the optimal DC voltage U_{DC_opt} can accommodate a wide range of operating switching frequencies (9–15 kHz). Generally, the $f_{DC\_opt}(i_s)$ can be expressed as (8.9), using a fitting method. The accuracy of the fitting method can

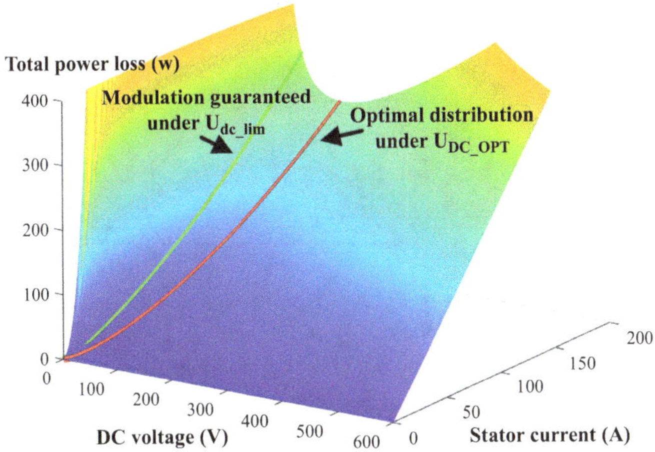

Figure 8.29 *Relationship of stator current, DC voltage, and total power loss of the power semiconductor devices under the stall condition for the drive system in Figure 8.28*

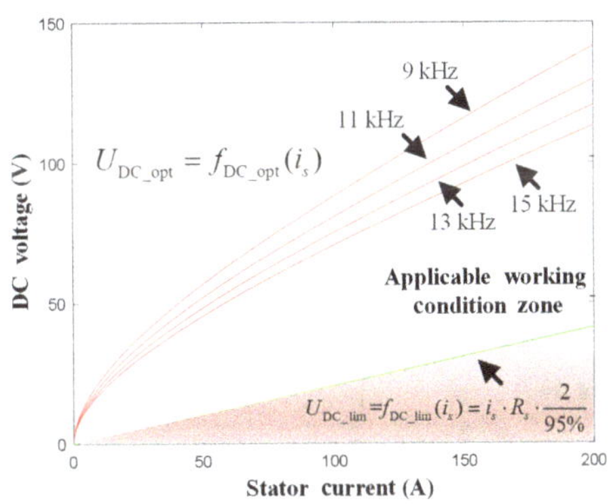

Figure 8.30 *Relationship between stator current and optimal DC voltage (ODV) under stall conditions with different switching frequencies (9–15 kHz)*

be enhanced by improving the resolution of the original data.

$$f_{DC\_opt}(i_s) = a_1 i_s^{a_2} + a_3 i_s \tag{8.9}$$

Besides, the DC voltage control has to be performed in a smooth transition by a speed-based-coefficient (S-B-C) digital filter algorithm, which can be seen as a first-order low-pass filter.

To sum up, the complete control diagram of the proposed ADVC can be constructed, which is illustrated in Figure 8.31. To verify the feasibility of the

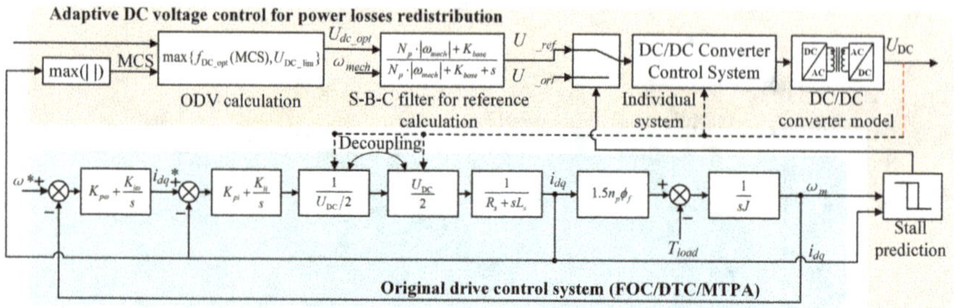

Figure 8.31 Control diagram of the proposed adaptive DC voltage control (ADVC) based on the original electric machine drive system. FOC: field-oriented control. DTC: direct-torque control. MTPA: maximum torque per ampere

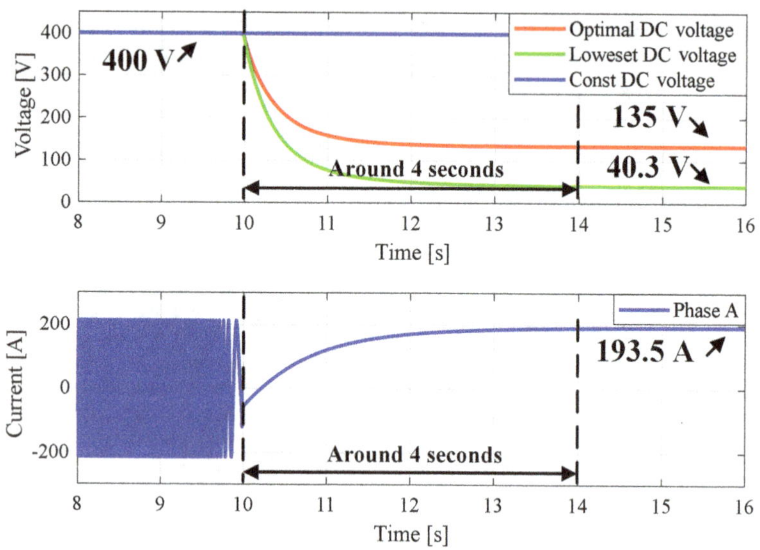

Figure 8.32 DC bus voltage (red line: ODV; yellow line: lowest DC voltage control; blue line: static DC voltage) and stator current

proposed ADVC, a study case is given in [57], as well as the mission profile. The DC voltage and the stator current are shown in Figure 8.32. With the ADVC, the DC bus voltage of the drive converter is controlled at 135 V, corresponding to the load current of 193.5 A, as shown in Figure 8.30. While with the lowest DC voltage control (modulation depth guaranteed), the DC bus voltage of the drive converter is controlled at 40.3 V. With the proposed speed-based-coefficient (S-B-C) filter algorithm for DC voltage reference calculation, it can be found that the DC voltage can perform self-adaptive to the stator current in terms of the

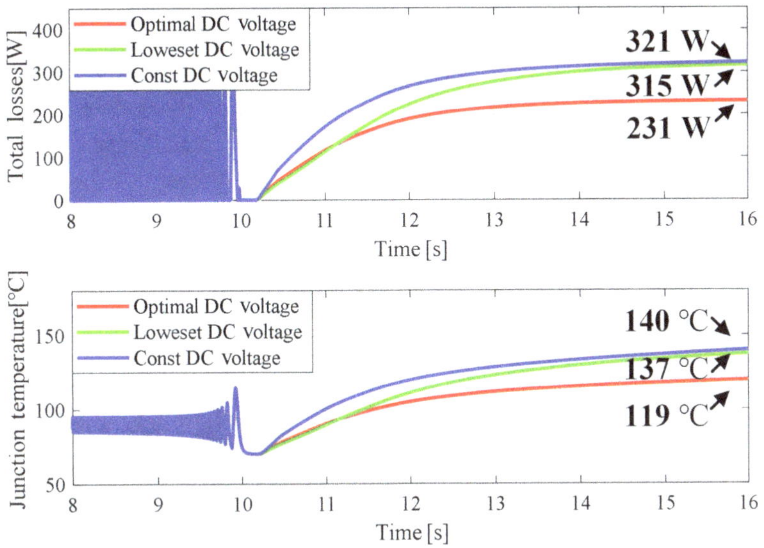

Figure 8.33 Total power loss and junction temperature of the upper IGBT in Phase A (red line: ODV; yellow line: lowest DC voltage control; blue line: static DC voltage)

transition time (around 4 s) and adjustment speed, corresponding to the rotor speed and original control system.

The total power loss and junction temperature response of the upper IGBT in Phase A are shown in Figure 8.33. Even applying the lowest DC voltage, the total power loss is barely reduced (from 321 W to 315 W, 2%), and the thermal concentration in the IGBT is not relieved (from 140 °C to 137 °C). However, when applying the proposed ADVC, the thermal stress is significantly relieved. The total power loss can be reduced from 321 W to 231 W (28%), and the junction temperature is reduced, from 140 °C to 119 °C.

8.5.3 Optimal phase angle control under abnormal condition of motor drive

In order to relieve the current stress on the power semiconductor devices of the electric machine drive inverter under the stall condition, a reasonable stall electrical phase angle should also be designed, instead of controlling the current amplitude in the price of the stability of the mechanical system [58].

The maximum current stress, which can significantly decide on the reliability of the target drive module in a short-term timescale, is shown in Figure 8.34(a), while the corresponding phase angles of stator current and rotor position are shown in Figure 8.34(c). Within one fundamental cycle of the current, it can be found that the maximum current stress of power semiconductor devices reaches a minimum at a specific electrical phase angle PA_{min}, which can be expressed as (8.10):

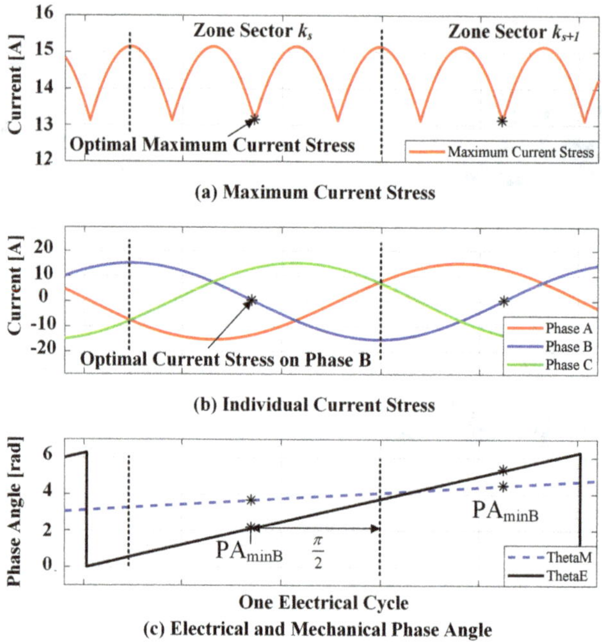

Figure 8.34 Maximum and individual current stress with corresponding phase angle (dotted line: stall condition detection boundary; star point: optimal stall operation condition in case of Phase B optimization). (a) Maximum current stress, (b) individual current stress, and (c) electrical and mechanical phase angle.

$$\begin{cases} PA\frac{\pi}{3}_{idq\,\min} \\ \text{where } k = 0, 1, 2, 3, 4, 5, 6\ \theta_{idq} = \arctan\frac{i_d}{i_q} \end{cases} \tag{8.10}$$

where i_d, i_q are the stator currents in dq axis, and θ_{idq} is the phase angle offset caused by stator current in d-axis.

In order to realize the optimal distribution, current stress should be redistributed among the three phases according to the thermal characteristics of the drive converter. Figure 8.34(b) shows the individual current stress on each phase, while Figure 8.34(c) shows the corresponding electrical phase angle. Considering the minimum current stress on specific phases, PA_{min} in (8.10) can be classified according to selected phase k, which can be expressed as (8.11):

$$\begin{cases} PA_{mink} = \theta_k + \theta_{idq} \text{ or } \theta_k + \theta_{idq} + \pi \\ \text{where } \theta_{k=\text{A,B,C}} = 0, \frac{2\pi}{3}, -\frac{2\pi}{3};\ \theta_{idq} = \arctan\frac{i_d}{i_q} \end{cases} \tag{8.11}$$

where θ_k represents the variable phase angle const for the selected phase k.

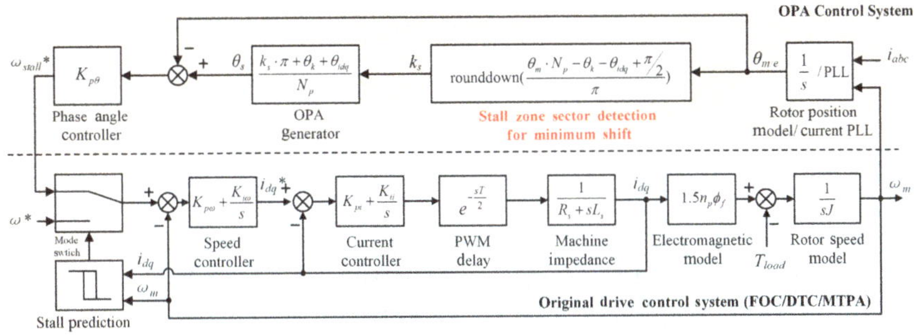

Figure 8.35 Control diagram of the proposed OPA control and the original drive control system. FOC: field-oriented control; DTC: direct-torque control; and MTPA: maximum torque per ampere.

According to the analysis above, the optimal current distribution can be realized by controlling specific electrical phase angles, which are difficult to regulate directly. However, the electrical phase angle and mechanical phase angle are linearly dependent under stall conditions (ω_e=0), so that the control target can be simplified to the mechanical phase angle.

Figure 8.35 shows the block diagram of the proposed mechanical optimal phase angle (OPA) control system, based on the original drive control system, which is designed as follows:

Firstly, the optimal phase angle reference of the rotor position should be generated. To guarantee an acceptable mechanical/electrical phase angle shift, the optimal phase angle of the stator current is separated into different stall zone sectors k_s (dotted line boundary). In each stall zone, there exist only one optimal phase angle θ_s (star point position), as shown in Figure 8.34(c). The stall zone and optimal phase angle can be calculated by (8.12).

$$\begin{cases} k_s = \text{rounddown}\left(\dfrac{\theta_m \cdot N_p - \theta_k - \theta_{idq} + \dfrac{\pi}{2}}{\pi}\right) \\ \theta_s = \dfrac{k_s \cdot \pi + \theta_k + \theta_{idq}}{N_p} \\ \text{where } \theta_{k=\text{A,B,C}} = 0, \dfrac{2\pi}{3}, -\dfrac{2\pi}{3}, \theta_{idq} = \arctan\dfrac{i_d}{i_q} \end{cases} \tag{8.12}$$

Secondly, the speed reference under the stall condition should be regenerated. The difference between the current mechanical phase angle and the generated optimal mechanical phase angle is calculated. Instead of the original speed reference, the speed reference under the stall condition will be generated through a proportional controller.

Finally, the speed reference mode should be switched from the original signal to the generated one. In order to pre-switch the control mode of the drive

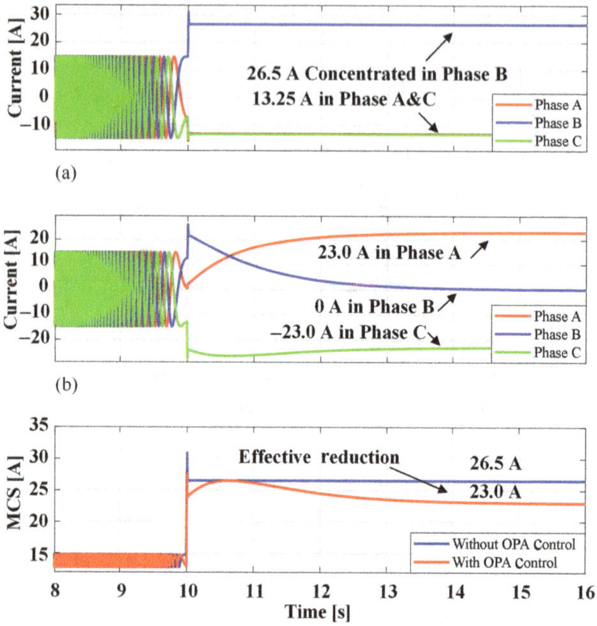

Figure 8.36 (a)Current stress with original control (without OPA control) and (b) current stress with OPA control in simulation validation.

system, the coming stall condition must be predicted. The prediction can be realized through speed and electromagnetic torque detection, whose threshold should be designed according to the detailed operation capacity and device selection.

To verify the feasibility of the proposed OPA control method, a study case is provided in [58]. The electric machine is controlled in field-oriented-control (FOC) mode. A mission profile is constructed, consisting of acceleration, deceleration, and stall conditions, which can be found in [58] with more details.

The current stress on the power semiconductor devices is compared in Figure 8.36. With OPA control, the maximum current stress is reduced from 26.5 A to 23 A, within less than 4 s adjustment. As the thermal coupling effect of Phase B is designed the strongest, the current stress of Phase B is redistributed to the minimum as well as the thermal stress shown in Figure 8.37.

8.5.4 Enabling technology for experimental validation of active thermal control under abnormal conditions

Abnormal conditions, such as the stall condition, will result in dramatically raised thermal stress in short-term timescales. For safety and repeatable experiments, it is difficult to conduct experiments with actual grid or electric machines. Recently,

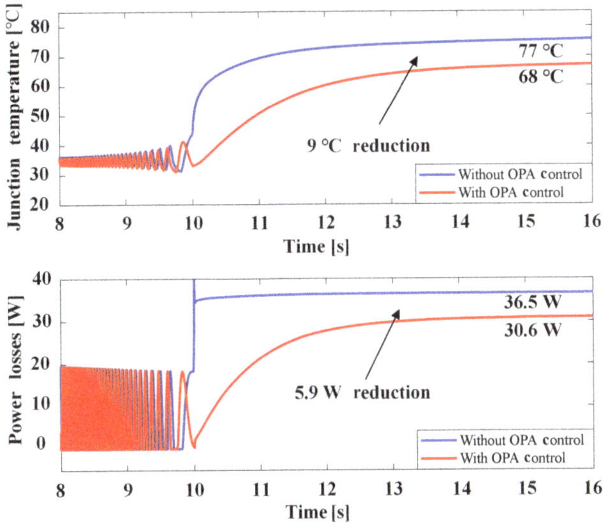

Figure 8.37 Thermal stress with original control (without OPA control) and thermal stress with OPA control in simulation validation

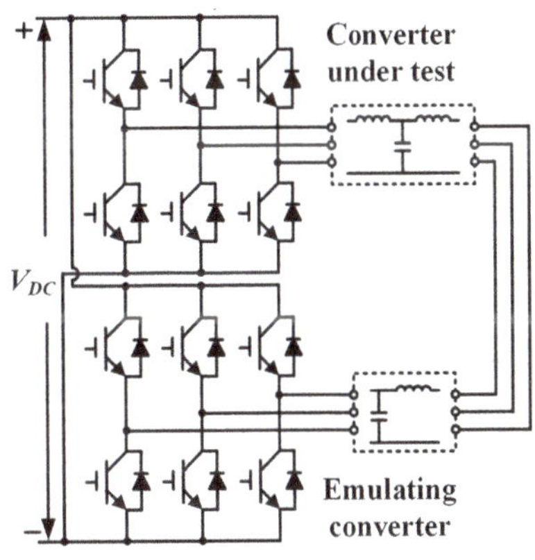

Figure 8.38 Typical circuit configuration for mission-profile-emulation-based testing system

mission profile emulation has become a promising testing tool aimed at improving the power electronics converters for their high flexibility and efficiency.

Figure 8.38 shows a typical circuit configuration for a mission-profile-emulation-based testing system. Two converters, namely the converter under test and the emulating converter, are connected at both AC and DC terminals. Different types of coupling filters can be used for different testing scenarios, such as *L* filters for drive converters and *LC*/*LCL* filters for grid-tied converters.

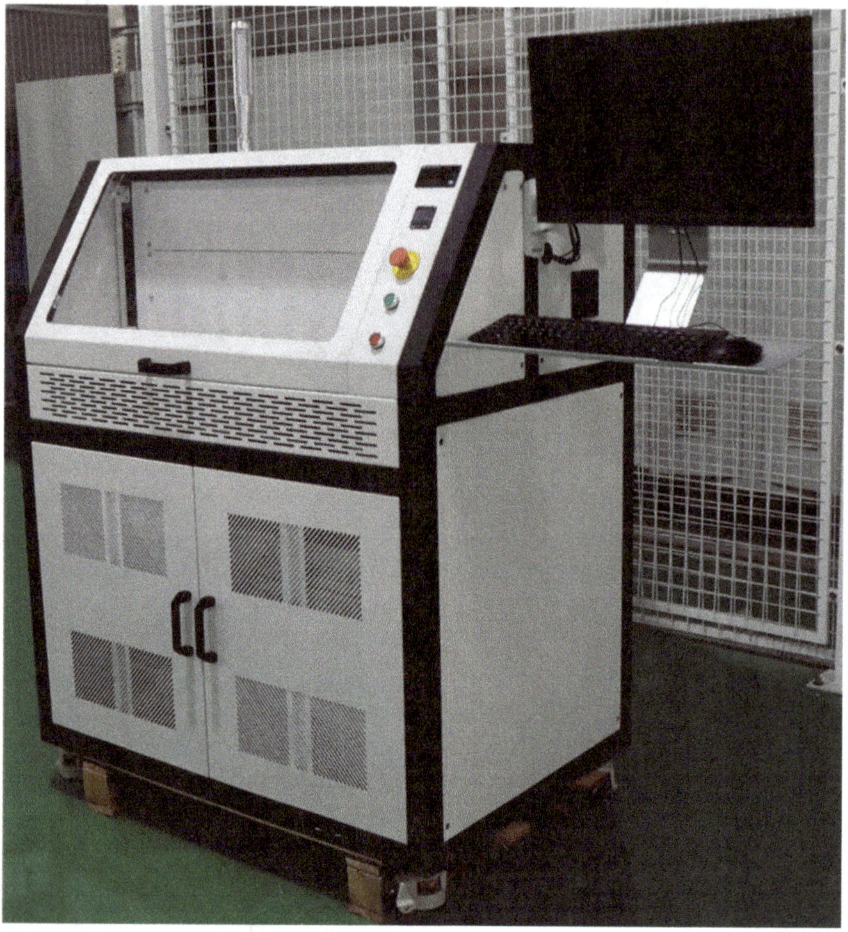

Figure 8.39 Experimental setup

A mission-profile-emulation-based testing platform was established for both drive converters and grid-tied converters as presented in Figure 8.39, as well as the graphic user interface provided in Figure 8.40. Electrothermal parameters, such as case temperature and on-voltage, can be monitored online, which provides the possibility to comprehensively test the power electronics converters and the corresponding control algorithms under actual mission profiles.

Two typical abnormal conditions created by the platform are provided in Figures 8.41 and 8.42, including low-voltage ride-through for grid-tied converters and stall condition for drive converters. With the mission-profile-emulation-based testing system, abnormal conditions can be flexibly programmed and safely recreated. Thus, both the reliability of the power electronics converters and the corresponding advanced control algorithms can be detailed validated.

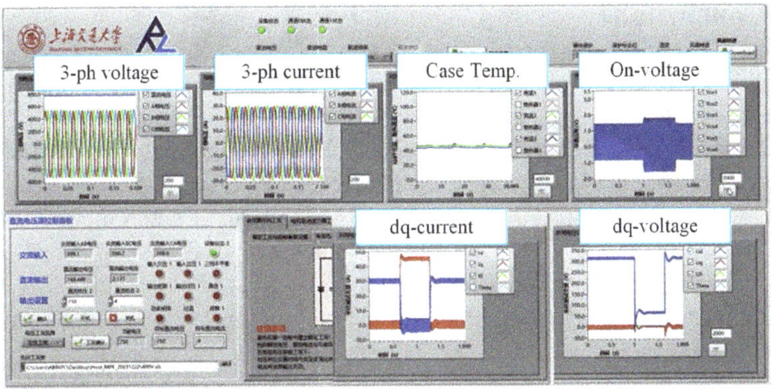

Figure 8.40 Graphic user interface for the control of mission profile emulator

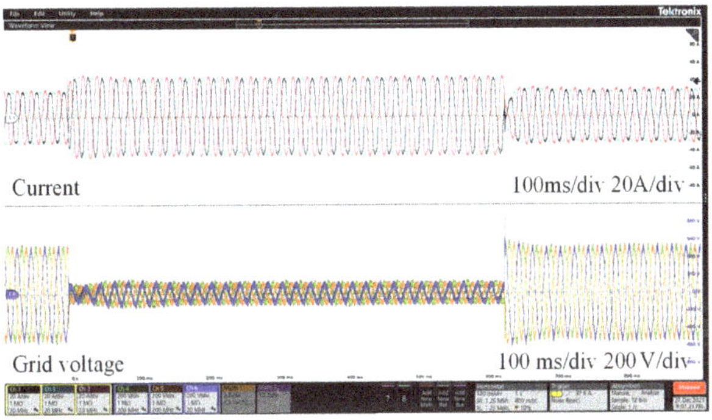

Figure 8.41 Low-voltage ride-through by mission-profile-emulation-based testing system for grid-side operation

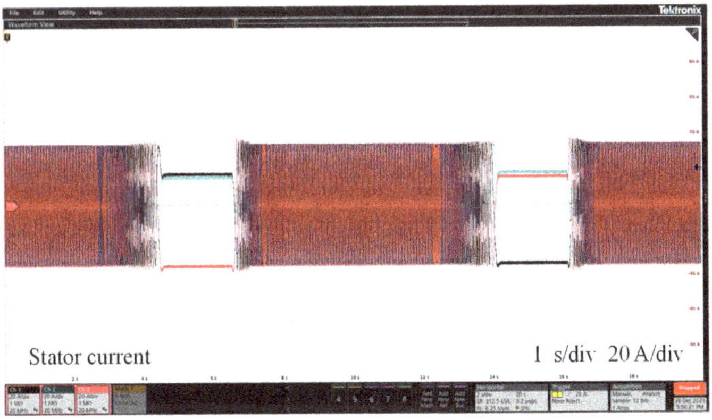

Figure 8.42 Stall condition of drive converter by mission-profile-emulation-based testing system for electric machine drive operation

8.6 Conclusions

The thermal stress, especially the thermal cycling and dramatical junction temperature rise under abnormal conditions in the power semiconductor devices for power converters, can be significantly reduced by active thermal controls. Different ways, such as modulation strategies to reduce power losses, current/voltage controls to redistribute thermal stress, power controls to achieve better thermal cycling, and the activation of energy storage systems to achieve smaller thermal excursions, can all achieve optimal thermal performances, as well as better reliability of the power converters. The corresponding studies on this topic are still underway with promising opportunities.

References

[1] S. Faulstich, P. Lyding, B. Hahn, and P. Tavner "Reliability of offshore turbines – Identifying the risk by onshore experience," in Proc. of European Offshore Wind, Stockholm, 2009.

[2] B. Hahn, M. Durstewitz, and K. Rohrig "Reliability of wind turbines – Experience of 15 years with 1500 WTs", *Wind Energy*, Springer, Berlin, 2007.

[3] E. Wolfgang, L. Amigues, N. Seliger and G. Lugert, "Building-in reliability into power electronics systems". *World Electron. Packag. Syst. Integr.*, pp. 246–252, 2005.

[4] D. Hirschmann, D. Tissen, S. Schroder, and R. W. De Doncker, "Inverter design for hybrid electrical vehicles considering mission profiles," in IEEE Conference on Vehicle Power and Propulsion, vols. 7–9, pp. 1–6, Sep. 2005.

[5] S. Yang, A. T. Bryant, P. A. Mawby, D. Xiang, L. Ran, and P. Tavner, "An industry-based survey of reliability in power electronic converters," *IEEE Trans. Ind. Appl.*, vol. 47, no. 3, pp. 1441–1451, 2011.

[6] J. Due, S. Munk-Nielsen, and R. Nielsen, "Lifetime investigation of high power IGBT modules", in Proc. of EPE'2011, Birmingham, UK, pp. 1–8, 2011.

[7] C. Busca, R. Teodorescu, F. Blaabjerg, *et al.*, "An overview of the reliability prediction related aspects of high power IGBTs in wind power applications," *Microelectron Reliab*, vol. 51, nos. 9–11, pp. 1903–1907, 2011.

[8] N. Kaminski and A. Kopta, "Failure rates of HiPak modules due to cosmic rays," ABB Application Note 5SYA 2042-04, Mar. 2011.

[9] E. Wolfgang, "Examples for failures in power electronics systems," presented at *ECPE Tutorial on Reliability of Power Electronic Systems*, Nuremberg, Germany, April 2007.

[10] A. Wintrich, U. Nicolai, and T. Reimann, "Semikron application manual," ISLE Verlag 2011, Nuremberg, Germany pp. 128, 2011.

[11] J. Berner, "Load-cycling capability of HiPak IGBT modules," ABB Application Note 5SYA 2043-02, 2012.

[12] U. Scheuermann, "Reliability challenges of automotive power electronics," *Microelectron Reliab*, vol. 49, nos. 9–11, pp. 1319–1325, 2009.

[13] U. Scheuermann and R. Schmidt, "A new lifetime model for advanced power modules with sintered chips and optimized Al wire bonds," in Proc. of PCIM'2013, pp. 810–813, 2013.

[14] K. Ma, M. Liserre, F. Blaabjerg, and T. Kerekes, "Thermal loading and lifetime estimation for power device considering mission profiles in wind power converter," *IEEE Trans. Power Electron.*, vol. 30, no. 2, pp. 590–602, 2015.

[15] K. Ma and F. Blaabjerg, "Transient modelling of loss and thermal dynamics in power semiconductor devices," in Proc. of ECCE'2014, pp. 5495–5501, Sep 2014.

[16] K. Ma, "Chapter 26 - Thermal loading control in power converters," in *Control of Power Electronic Converters and Systems*, F. Blaabjerg, Ed., Academic Press, New York, pp. 403–423, 2018.

[17] D. G. Holmes and L. A. Thomas. *Pulse Width Modulation for Power Converters: Principles and Practice*, vol. 18. Wiley, New York, 2003.

[18] M. H. Ahmet, J. K. Russel, and L. A. Thomas, "Carrier-based PWM-VSI overmodulation strategies: Analysis, comparison, and design," *IEEE Trans. Power Electron.*, vol. 13, no. 4, pp. 674–689, 1998.

[19] A. M. Hava, R. J. Kerkman, and T. A. Lipo, "A high performance generalized discontinuous PWM algorithm," in Proc. of APEC'1997, vol. 2, pp. 886–894, 1997.

[20] L. Dalessandro, S. D. Round, U. Drofenik, and J. W. Kolar, "Discontinuous space-vector modulation for three-level PWM rectifiers," *IEEE Trans. Power Electron.*, vol. 23, no. 2, pp.530–542, 2008.

[21] E. Demirkutlu and A. M. Hava, "A scalar resonant-filter-bank based output-voltage control method and a scalar minimum-switching-loss discontinuous PWM method for the four-leg-inverter-based three-phase four-wire power supply," *IEEE Trans. Ind. Appl.*, vol. 45, no. 3, pp. 982–991, 2009.

[22] Y. Wu, M. A. Shafi, A. M. Knight, and R. A. McMahon, "Comparison of the effects of continuous and discontinuous PWM schemes on power losses of voltage-sourced inverters for induction motor drives," *IEEE Trans. Power Electron.*, vol. 26, no. 1, pp. 182–191, 2011.

[23] Z. Qin, M. Liserre, and F. Blaabjerg, "Thermal analysis of multi-MW two-level generator side converters with reduced common-mode-voltage modulation methods for wind turbines," in Proc. of PEDG'2013, pp. 1–7, Jul. 2013.

[24] Z. Zhang, O. C. Thomsen, and M. A. E. Andersen, "Discontinuous PWM modulation strategy with circuit-level decoupling concept of three-level neutral-point-clamped (NPC) inverter," *IEEE Trans. Ind. Electron.*, vol. 60, no. 5, pp. 1897–1906, 2013.

[25] A. Isidoril, F. M. Rossi, F. Blaabjerg, and K. Ma, "Thermal loading and reliability of 10-MW multilevel wind power converter at different wind roughness classes," *IEEE Trans. Ind. Appl.*, vol. 50, no. 1, pp. 484–494, 2014.

[26] V. Blasko, R. Lukaszewski, and R. Sladky, "On line thermal model and thermal management strategy of a three-phase voltage source inverter," in Proc. of IEEE-IAS Annual Meeting, pp. 1423–1431, 1999.

[27] Z. Qin, P. C. Loh, and F. Blaabjerg, "Application criteria for nine-switch power conversion systems with improved thermal performance," *IEEE Trans. Power Electron.*, vol. 30, no. 8, pp. 4608–4620, 2015,

[28] Z. Qin, P. C. Loh, and F. Blaabjerg, "Power loss benchmark of nine-switch converters in three-phase online-UPS application," in Proc. of ECCE'2014, pp. 1180–1187, Sep. 2014.

[29] H. Wang, R. Zhao, Y. Deng, and X. He, "Novel carrier-based PWM methods for multilevel inverter," in Proc. of IECON'2003, vol. 3, pp. 2777–2782, 2003.

[30] T. Bruckner and D. Holmes, "Optimal pulse-width modulation for three-level inverters," *IEEE Trans. Power Electron.*, vol. 20, no. 1, pp. 82–89, 2005.

[31] Q. Zhang, H. Hu, D. Zhang, X. Fang, Z. J. Shen, and I. Bartarseh, "A controlled-type ZVS technique without auxiliary components for the low power DC/AC inverter," *IEEE Trans. Power Electron.*, vol. 28, no. 7, pp. 3287–3296, 2013.

[32] K. Ma, and F. Blaabjerg, "Modulation methods for neutral-point-clamped wind power converter achieving loss and thermal redistribution under low-voltage ride-through", *IEEE Trans. Ind. Electron.*, vol. 61, no. 2, pp. 835–845, 2014.

[33] K. Ma, and F. Blaabjerg, "Thermal optimized modulation method of three-level NPC inverter for 10 MW wind turbines under low voltage ride through," *IET J. Power Electron.*, vol. 5, no. 6, pp. 920–927, 2012.

[34] Z. Chen, J. M. Guerrero, and F. Blaabjerg, "A review of the state of the art of power electronics for wind turbines," *IEEE Trans. Power Electron.*, vol. 24, no. 8, pp. 1859–1875, 2009.

[35] F. Blaabjerg, Z. Chen, and S. B. Kjaer, "Power electronics as efficient interface in dispersed power generation systems," *IEEE Trans. Power Electron.*, vol. 19, no. 5, pp. 1184–1194, 2004.

[36] M. Liserre, R. Cardenas, M. Molinas, and J. Rodriguez, "Overview of multi-MW wind turbines and wind parks," *IEEE Trans. Ind. Electron.*, vol. 58, no. 4, pp. 1081–1095, 2011.

[37] Generators connected to the low-voltage distribution network: VDE-AR-N 4105: 2018-11.

[38] M. Tsili, and S. Papathanassiou, "A review of grid code technical requirements for wind farms," *IET Renew. Power Gener.*, vol. 3, no. 3, pp. 308–332, 2009.

[39] A. Camacho, M. Castilla, J. Miret, R. Guzman, and A. Borrell, "Reactive power control for distributed generation power plants to comply with voltage limits during grid faults," *IEEE Trans. Power Electron.*, vol. 29, no. 11, pp. 6224–6234, 2014.

[40] S. Engelhardt, I. Erlich, C. Feltes, J. Kretschmann, and F. Shewarega, "Reactive power capability of wind turbines based on doubly fed induction generators," *IEEE Trans. Energy Convers.*, vol. 26, no. 1, pp. 364–372, 2011.

[41] C. Liu, F. Blaabjerg, W. Chen, and D. Xu, "Stator current harmonic control with resonant controller for doubly fed induction generator," *IEEE Trans. Power Electron.*, vol. 27, no. 7, pp. 3207–3220, 2012.

[42] D. Zhou, F. Blaabjerg, M. Lau, and M. Tonnes, "Thermal behavior optimization in multi-MW wind power converter by reactive power circulation," *IEEE Trans. Ind. Appl.*, vol. 50, no. 1, pp. 433–440, 2014.

[43] *Wind Energy Generation Systems, Part 1: Design Requirements, IEC 61400-1*, 4th edn, International Electro-technical Commission, Geneva, 2019.

[44] K. Ma, M. Liserre, and F. Blaabjerg, "Reactive power influence on the thermal cycling of multi-MW wind power inverter," *IEEE Trans. Ind. Appl.*, vol. 49, no. 2, pp. 922–930, 2013.

[45] J. Eyer and G. Corey, "Energy storage for the electricity grid: Benefits and market potential assessment guide – A study for the DOE energy storage systems program," *Sandia National Laboratories, Sandia Report*, SAND2010-0815, Feb. 2010.

[46] Energy Storage Systems for Wind Turbines https://ampowr.com/energy-storage-systems-wind-turbines. [Online: accessed 30-Jul-2024].

[47] Z. Qin, M. Liserre, F. Blaabjerg, and H. Wang, "Energy storage system by means of improved thermal performance of a 3 MW grid side wind power converter," in Proc. of IECON'2013, pp. 736–742, 2013.

[48] I. V. der Hoven, "Power spectrum of horizontal wind speed in the frequency range from 0.0007 to 900 cycles per hour," *J Meteorol*, vol. 14, pp. 160–164, 1957.

[49] D. I. Stroe, A. I. Stan, R. Diosi, *et al.*, "Short term energy storage for grid support in wind power applications," in Proc. of OPTIM'2012, pp. 1012–1021, 2012.

[50] How does a Supercapacitor Work? https://batteryuniversity.com/article/bu-209-how-does-a-supercapacitor-work [Online: accessed 30-Jul-2024].

[51] X. Liu, T. Zhao, H. Deng, P. Wang, J. Liu, and F. Blaabjerg, "Microgrid energy management with energy storage systems: A review," *CSEE J. Power Energy Syst.*, vol. 9, no. 2, pp. 483–504, 2023.

[52] K. Yoshimoto, T. Nanahara, and G. Koshimizu, "New control method for regulating state-of-charge of a battery in hybrid wind power/battery energy storage system," in Proc. of PSCE'2006, pp.1244–1251, 2006.

[53] Z. Qin, M. Liserre, F. Blaabjerg, and P. C. Loh, "Reliability-oriented energy storage sizing in wind power systems," in Proc. of IPEC-ECCE-ASIA'2014, pp. 857–862, 2014.

[54] I. F. Kovačević, U. Drofenik, and J. W. Kolar, "New physical model for lifetime estimation of power modules," in Proc. of IPEC'2010, pp. 2106–2114, 2010.

[55] K. Ma, A. S. Bahman, S. M. Beczkowski, and F. Blaabjerg, "Complete loss and thermal model of power semiconductors Including device rating information," *IEEE Trans. Power Electron.*, vol. 30, no. 5, pp. 2556–2569, 2015.

[56] K. Ma, J. Lin, and Y. Zhu, "Statistical characterization for loss distributions of power semiconductor devices", *IEEE Trans. Power Electron.*, vol. 36, no. 7, pp. 7384–7388, 2021.

[57] A. Wang, Y. Qi, and K. Ma, "Adaptive DC voltage control for optimal junction temperature redistribution under stall condition of electric machine drive inverter," *IEEE Trans. Power Electron.*, vol. 38, no. 4, pp. 4229–4234, 2023.

[58] Y. Qi, K. Ma, and S. Xia, "Active thermal control with optimal phase angle under stall condition of machine drive inverter," *IEEE Trans. Power Electron.*, vol. 37, no. 9, pp. 10128–10132, 2022.

Chapter 9

Power module lifetime test and state monitoring

Stig Munk-Nielsen[1], Pramod Ghimire[2], Ionut Trintis[3], Bjørn Rannestad[4] and Paul Thøgersen[5]

9.1 Overview of power cycling methods

Designing power modules (PM) for a lifetime represents the power cycling (PC) capability of that device. The purpose of PC is to heat the PM specifically with a higher average temperature (T_{avg}) at certain temperature swing (ΔT), which are also identified as key stressors in semiconductors [1,2]. The temperature is mainly controlled by the power dissipation in the device. During PC, temperature is usually kept higher than the field operation to obtain an accelerated end of life [3]. PC is a well-established test method and also formulated in IEC 60749-34, which is caused by a pulsating DC current. With turn-on time in seconds, it is also possible to determine the thermal impedance of the test device. The method is widely used and many companies offer mobile test systems up to higher current ratings. Normally in PC, many modules can be connected in series and parallel, where each string can be tested, to obtain the population of test results necessary for a statistical analysis.

The failure mechanisms of power modules are not dominated by semiconductor die failure; this is implied since the temperature of the junction defines the stress level. Common power cycling failures are caused by interconnections such as bond wires, die metallization, and soldering [4]. Here, it is noticed that the semiconductor junction is not the root cause of failure but aluminum reconstruction, heel cracks, and bond wire lift-off are seen at relatively low ΔT [5]. At higher ΔT, the solder cracks can be provoked within the accelerated test. It is possible to model the lifetime of the bond wire, metallization, and soldering based on physical material properties. Physics of failure models are useful for optimizing the material fatigue lifetime but this requires detailed information of interconnected materials. Such information is often not available to the application engineer. In short, the

[1]AAU Energy, Aalborg University, Denmark
[2]Siemens Gamesa Renewable Energy A/S, Esbjerg, Denmark
[3]Siemens Gamesa Renewable Energy, Brande, Denmark
[4]KK Wind Solutions, Ikast, Denmark
[5]Department of Energy Technology, Aalborg University, Denmark

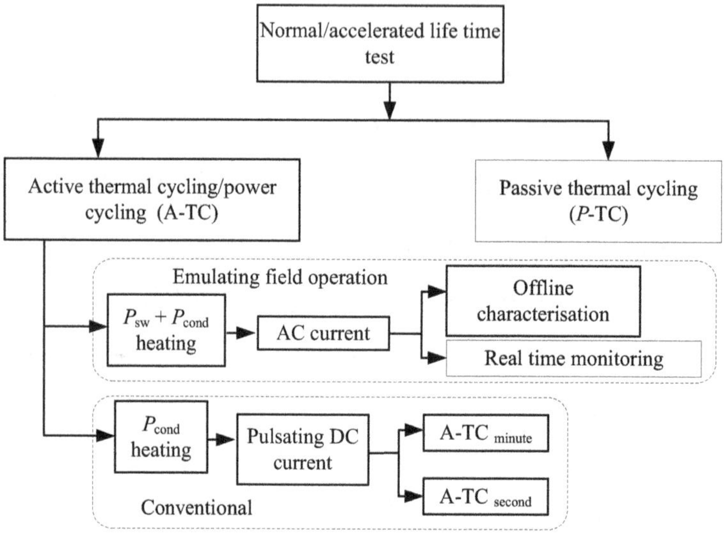

Figure 9.1 Classification of power cycling test methods

metallization failure mechanisms are due to the coefficient of thermal expansion (CTE) mismatch between the aluminum metallization and the semiconductor die [4]. The bond wire heel crack and bond wire lift-off are caused by metal fatigue due to cyclic expansion and contraction of bond wires. The soldering paste between semiconductors metallization and heat sink structures may have voids that increase local temperature.

In field operation, the power modules often will be part of the AC circuit and the current will be sinusoidal. PC emulating filed operation can be built using full-scale power converters, as classified in Figure 9.1 [6–9]. However, one has to be careful not to trigger nonrealistic failures. PC method with switching and conduction losses is described in this chapter.

9.2 AC current power cycling

9.2.1 Introduction

Modern PC methods are emulating field operating points for power electronic devices under test (DUT) as close as possible to real-life applications. The basic test setup required to apply a realistic stress condition on PM was introduced in [8], and it is represented in Figure 9.2. The DUT is an IGBT half-bridge, connected to the same DC-bus with a control (CTRL) IGBT half-bridge. The output/input of the two half-bridges is connected to a load inductor. A DC power supply is required to keep the DC-link capacitors charged and supply the power losses in the system. A cooling system is required to control the baseplate temperature of the DUT, while another cooling system should control the CTRL temperature as close as possible to

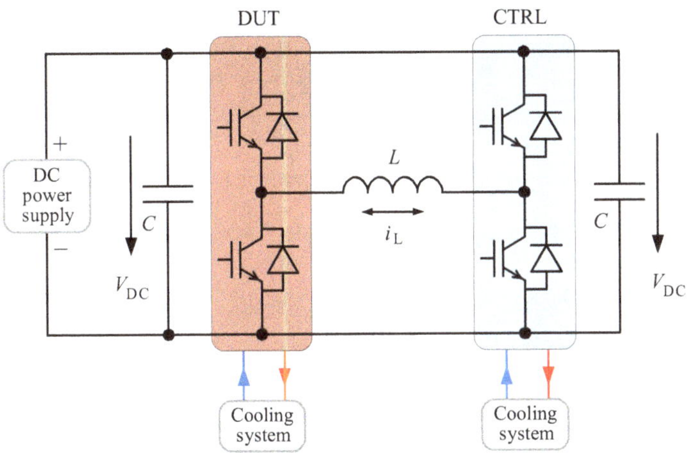

Figure 9.2 Basic AC power cycling test setup

the ambient temperature. It is also desired to use the CTRL side to a higher rating, to avoid aging of this part of the system.

To circulate power between the two converter legs, an open loop voltage must be produced by one leg (e.g., DUT). The other leg (e.g., CTRL) is controlled in a closed loop to regulate the current to a given reference. The phase and sign of the inductor current can be controlled to achieve four-quadrant operation. The frequency of the voltage and current, the switching frequency, and the DC-link voltage are operating parameters that can be adjusted to achieve the desired temperature profile and stress condition. The basic concept of the test setup can be applied to other converter topologies as well [7,9]. The building blocks can be H-bridges or three-phase bridges as well as multilevel converter topologies.

9.2.2 Stressors in AC power cycling

The stressors that have the highest contribution to the aging of power electronic devices are ΔT, T_{avg}, temperature gradient (dT/dt), voltage, humidity, pollution, and vibrations [10]. The humidity, pollution, and vibration stress factors are independent of the PC topology and are usually created by external equipment.

9.2.2.1 Junction temperature swing (ΔT)

The main advantage of such an accelerated lifetime test setup is that the power loss is created by both switching and conduction loss in both transistors and diodes. Depending on the chosen stress condition, the focus can be on IGBTs or diodes, as the sign of the current can be chosen as well as the power factor. To create an accelerated test condition, there are more degrees of freedom to generate certain ΔT. The reference AC frequency of the current and voltage can be adjusted, resulting in different ΔT. In Figure 9.3, an example of such variation is shown for the following conditions: $T_{cooling} = 80°C$, $f_{sw} = 2.5$ kHz, $V_{DC} = 1100$ V, $i_{ref} = 500$ A, PF = 1, inverting

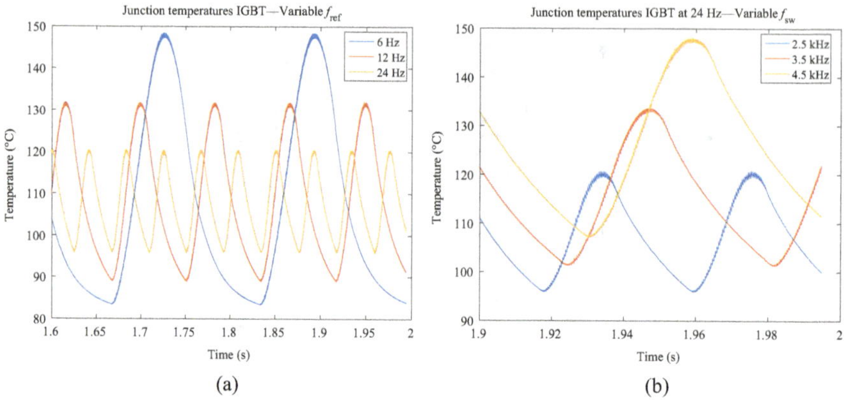

Figure 9.3 Temperature swings at (a) adjusted fundamental frequency and (b) adjusted switching frequency

operation with respect to the DUT. The switching frequency can be adjusted to increase the switching losses. Temperature swings of such test conditions are shown, where the fundamental frequency is varied as constant switching frequency, see Figure 9.3(a), and the switching frequency is adjusted at constant fundamental frequency, see Figure 9.3(b).

9.2.2.2 Average junction temperature (T_{avg})

The T_{avg} in an AC power cycling system is adjusted using a combination of both baseplate temperature and the electrical operating point ($V_{DC}, V_{AC}, i_{AC}, f_{ref}, f_{sw}$). The baseplate temperature is achieved using cooling system, controlling the coolant temperature in the case of liquid cooling [11], or based on the heat sink temperature in the case of air cooling [7].

9.2.2.3 Junction temperature gradient (d*T*/d*t*)

The AC frequency will determine the period of the $\mathrm{d}T/\mathrm{d}t$, and therefore it will impact the temperature gradient. For a given thermal interface from junction to cooling of DUT, the switching and conduction loss for the AC current being circulated will define the $\mathrm{d}T/\mathrm{d}t$. Taking an example from Figure 9.3, the variation of f_{sw} for the same AC frequency and the same thermal impedance will dictate $\mathrm{d}T/\mathrm{d}t$ being: 1.5K/ms at 2.5 kHz, 2 K/ms at 3.5 kHz, and 2.5 K/ms at 4.5 kHz. The increase of the loss and therefore, the $d\mathrm{T}/dt$ can also be realized by modifying the rated parameters of the gate drive.

9.2.2.4 Voltage

The blocking voltage that the power devices are commutating during operation imposes thermal and electrical stress. The thermal stress is due to the switching power loss, which increases with the blocking voltage capability. The electrical stress is due to the cosmic radiation. Depending on the DC-link voltage and prototype altitude, the probability of failure due to cosmic rays can be estimated [12]. Dynamically changing the DC-link voltage during operation can optimize efficiency and reliability [13]. The cosmic radiation failure can be realized using the system from Figure 9.2, having an operating point with a low-temperature swing and high DC-link voltage.

9.3 Wear-out status of power modules

Wear-out is a process of material deterioration due to both electrical and environmental loadings, i.e., interconnection materials of power modules. Wear-out failure is defined as an intrinsic failure in [14]. Usually, the lifetime distributions of semiconductor devices are described using the famous bathtub curve Figure 9.4, which is a superposition of three failure rates such as

- extrinsic failure: attributed to unintended defects and production errors which occur predominantly in the beginning of life,
- intrinsic failures: attributed to wear-out failures,
- random failures: attributed to extrinsic failures but random in nature.

PMs are stressed from electrical, thermal, and mechanical cycling in both normal and abnormal operations. Generally, most of the failure mechanisms are put into the same category even though they behave differently at load. Here, online monitoring at the component- and system-levels assists in understanding the development of degradation and also aids in identifying the root cause of failure mechanisms. In regards to an IGBT, mainly three electrical and thermal parameters are found sensitive to wear: on-state collector-emitter voltage ($v_{ce,on}$), gate-emitter voltage (V_{ge}) characteristics, and thermal resistance (R_{th}) [1,4,15]. These parameters are also identified as a possible source of known failure criteria. Failures occur mainly due to a lack of robustness, bad quality production, lack of understanding of failure modes, lack of protection against catastrophic failures, etc. This section describes the $v_{ce,on}$ measurement method as a potential method to identify the wear of the IGBT when in realistic operation. Initially, $v_{ce,on}$ method is

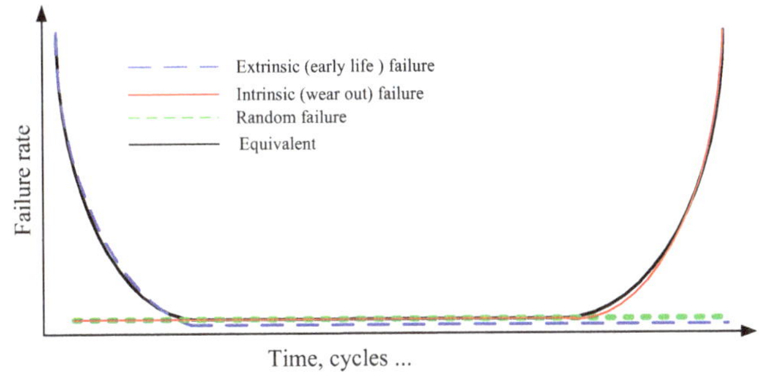

Figure 9.4 Representation of bathtub curve

introduced. The example of voltage measurement in an IGBT and a free-wheeling diode (FWD) is presented using the online method. Intrinsic failures and failure mechanisms are briefly included. Finally, current and cooling temperature measuring methods are shortly explained.

9.3.1 On-state voltage measurement method

Normally, $v_{ce,on}$ depends on intrinsic parameters: device structure, materials, and doping level, and extrinsic parameters: V_{ge}, collector current (I_c), and T_{vj}. Static characteristics of IGBT are determined by $v_{ce,on}$, I_c, and T_{vj} at a given V_{ge}. Similarly, for a FWD V_{FD}, I_{FD}, the threshold voltage (V_{to}), and T_{vj} are used. The $v_{ce,on}$ is influenced by both electrical and thermal effects at all times, but not necessarily at the same level as outlined in Figure 9.5. The former could be due to the degradation of interconnects (wire bond, solder, metallization) or semiconductor chips, whereas the latter is primarily due to thermal resistance.

Several methods are proposed to measure $v_{ce,on}$ when the converter is active and is in the off-state [16]. However, most of them are limited to a laboratory test only. A method [17,18] is suitable that uses a similar technique known as desaturation protection in modern IGBT gate drivers. The $v_{ce,on}$ is sensed to detect when the transistor current enters the saturation region.

Primarily, online monitoring of $v_{ce,on}$ topology should fulfill the following criteria when the converter is active.

- Voltage-blocking capability: The voltage-blocking components should withstand a module-rated collector-emitter voltage level.
- Voltage insulation: The minimum physical (clearance and creepage) distance requirement as defined on standard IEC60950-1, UL60950-1.
- Self-protection capability: Protect low voltage/current components in measurement circuit from high dI/dt and dV/dt that originated during switching of the device.
- Isolate circuit during fault (avoid fusing): Self-islanding capability if fault occurs in the measurement unit.

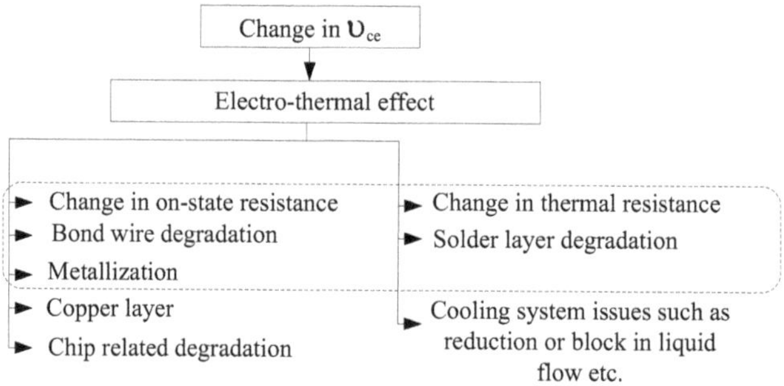

Figure 9.5 Collector-emitter voltage (v_{ce}) influencing the degradations in a power module [6]

- Rugged on temperature fluctuation: No/minimal error in the voltage measurement due to a change in the ambient temperature.
- Less offset voltage: Minimal offset in the measurement from amplifiers to improve the accuracy of the measurement.
- Low impedance: Measurement circuit should not influence the gate switching performance such as oscillation in the gate signal.

9.3.1.1 On-state voltage measurement circuit

An additional measurement circuit is required on a gate driver, as shown in Figure 9.6 [17,18]. As the υ_{ce} swings between kV and mV during transistor

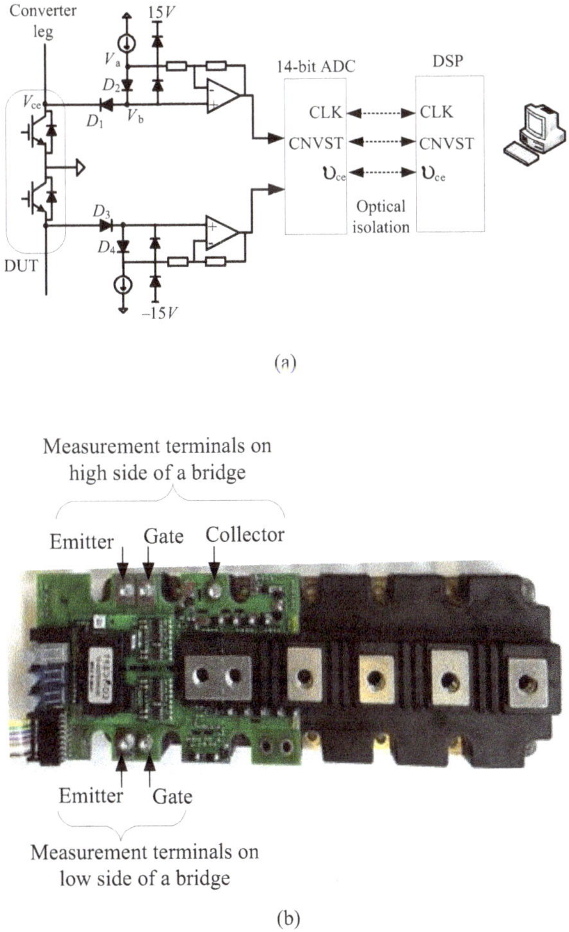

Figure 9.6. On-state collector-emitter $\upsilon_{ce,on}$ monitoring (a) a measurement circuit and (b) an image of a gate driver including a prototype measurement circuitry

switching, maintaining the accuracy in the measurement is critical. In this circuit, two fast-switching diode pairs (D_1/D_2 and D_3/D_4), as shown in Figure 9.6(a), are used to block the off-state high voltage and measure the forward voltage drops on high/low side (HS/LS) of DUT. These BY203 diodes have a high breakdown voltage of 2 kV, 300 nS reverse recovery time, and zero forward voltage thermal coefficients at certain current levels.

With HS IGBT on, D_1 and D_2 are forward biased with a 10 – mA current source, they are thermally coupled externally and display similar forward voltage temperature coefficients [17]. The on-state v_{ce} is measured using a bipolar 14-bit ADC with resolution of 0.61 mV. Equation (9.1) is used to derive the absolute $v_{ce,on}$

$$v_{ce,on} = V_b - V_{D_2} = V_b - (V_a - V_b) = 2V_b - V_a, \tag{9.1}$$

Figure 9.6(b) shows a prototype gate driver including a voltage measurement circuit to demonstrate the application as well as an upgraded version of an existing system.

9.3.1.2 Implementation of $v_{ce,on}$ measurement in a converter

An H-bridge test converter is depicted in Figure 9.7, where the proposed measurement circuit is used to measure $v_{ce,on}$ in DUT. The converter is operated at electrical parameter as outlined on Table 9.1.

In Figure 9.7, $I_{DUT,H}$, $D_{DUT,H}$, $I_{DUT,L}$, and $D_{DUT,L}$ are HS IGBT, HS FWD, LS IGBT, and LS FWD of DUT. V_{DUT} is open loop voltage reference at DUT, i_L is load current, and L_1, L_2, and L_3 are load inductors. The voltage waveforms of the collector-emitter $v_{ce,on}$ and gate-emitter V_{ge} are shown in Figure 9.8.

9.3.2 *Current measurement*

Digital signal processor (*TMS*320*F*2812) interface is used to implement control in a converter. A 12-bit analog to digital converter (ADC) channel in the DSP is utilized to sample a signal converted by LEM LF 1005-*S* current transducer for the measurement, the control, and the over-current protection, as shown in Figure 9.9(b). An interface circuit is built to protect the DSP, which should be operated in between 0 and 3.3 V. Here, the interface circuit adds a voltage shift of 1.5 V and limits the higher voltage to less than 3.3 V during positive and negative peaks of the load current.

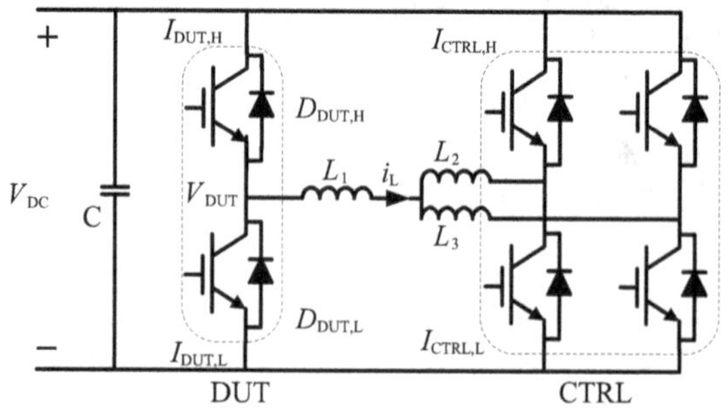

Figure 9.7 Power converter topology

In Figure 9.9(a), R_m is measuring resistance and V_{meas} is output voltage for a corresponding current. The current signal has a 1.5 V DC offset and current to voltage ratio of 0.00135 $\frac{V}{A}$. Hence, maximum ±1000 A will yield the measurement signal in between 2.85 V and 0.15 V. The over-current protection is also built based

Table 9.1 Converter operating parameters

Symbol	Meaning	Online parameters
V_{DC}	DC-link voltage	1000 V
V_{DUT}	Forward voltage reference	253 V_{rms}
i_L	Load current	922 A_{peak}
θ	Phase (V_{DUT} to i_L)	2.7 Rad
F_{out}	Fundamental frequency	6 Hz
F_{SW}	Switching frequency	2.5 kHz
C	DC-link capacitance	4 mF
L	Inductor	380 μH
$T_{cooling}$	Cooling temperature	80 ± 0.5 °C

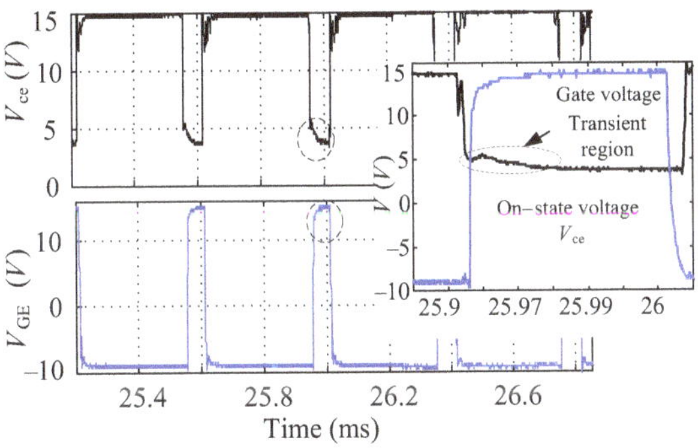

Figure 9.8 Collector-emitter $v_{ce,on}$ and gate-emitter V_{ge} waveforms

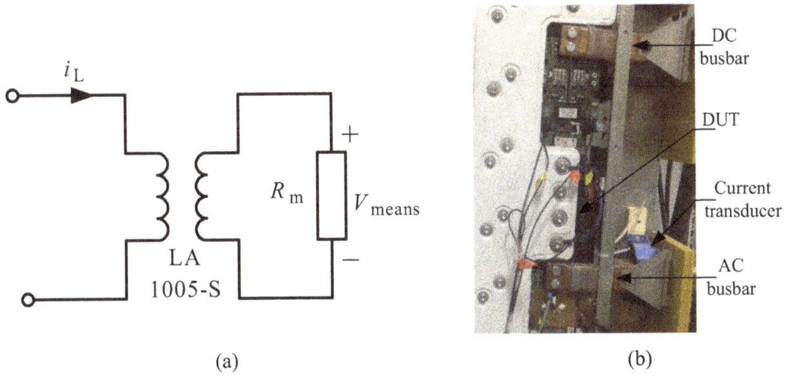

Figure 9.9 Current measurement using (a) a layout of the current measurement technique and (b) an image of the current measurement in the setup

on these measurement signals. An offset correction technique is included in the control/calibration sequence in the measurement. The offset by the transducer and the circuit is measured during converter stops/before starting and compensated during operation if there is any.

9.3.3 Cooling temperature measurement

The baseplate temperature of power modules is maintained by using Danfoss Shower Power® with water-mixed glycol liquid on a pressurized cooling pipe. A heavy-duty Pt100 thermister, shown in Figure 9.10(a), is used as a temperature transducer for liquid temperature logging as well as for control. The thermister is inserted into a cooling pipe as shown in Figure 9.10(b).

The output of the transducer is fed into a National Instrument (NI) data acquisition (DAQ) system (USB-6215). The output is used for the control as well as for logging in the normal operation of the converter. The cooling temperature variations for four different tests on PM1, PM2, PM3, and PM4 are shown in Figure 9.11.

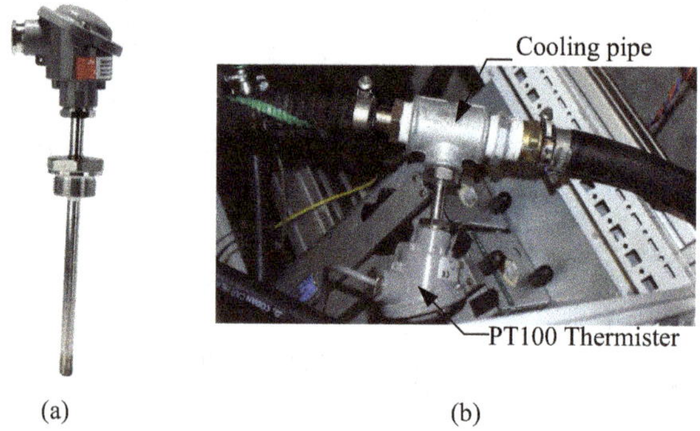

Figure 9.10 Cooling temperature measurement using (a) a PT100 thermister and (b) an image of temperature transducer on a cooling pipe

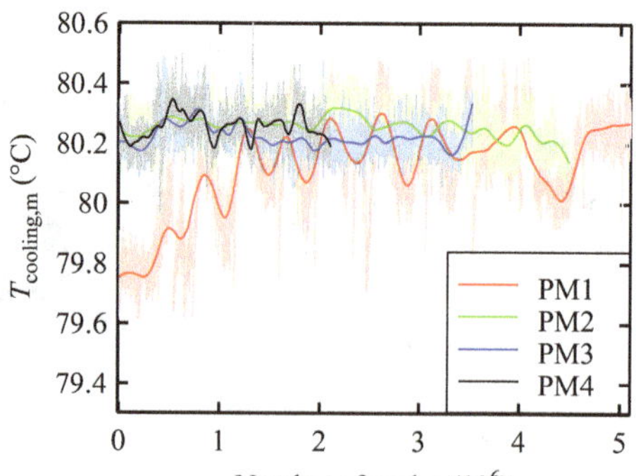

Figure 9.11 Liquid cooling temperature for four tested modules

9.4 Voltage evolution in IGBT and diode

The voltage evolution can be investigated online (i.e., when converter is on-state) and offline (i.e., when converter is off-state) [18]. In offline measurement, the converter operates for a certain number of cycles and stops a normal operation to measure the $v_{ce,on}$ at a DC current. On the other hand, in online measurement, $v_{ce,on}$ is measured without interrupting the converter operation [6,17]. In normal working conditions, the operating points of devices are changed with the load. Figure 9.12 (a) shows current loading and $v_{ce,on}$ measurement for one fundamental cycle. However, one current level can be chosen for a wear-out investigation, as shown in Figure 9.12(b), where 900 A is chosen. However, the V_{ge} should be at the same level to compare between the cycles.

The $v_{ce,on}$ is higher on the falling side of the current because of the rise in the chip temperature. The voltage evolution on the rising and the falling side of the AC load is depicted in Figure 9.13(a) for an $I_{DUT,H}$. Similarly, voltage evolution in $D_{DUT,L}$ is shown in Figure 9.13(b). In Figures 9.13 and 9.14, PM1, PM2, PM3, and PM4 are referred for different power modules which are loaded for different number of cycles [6]. In this test, diodes are more stressed than IGBT in DUT because of the longer conduction time [6]. Between two sides, the $D_{DUT,L}$ shows faster degradation. Furthermore, a step rise in voltage is noticed after $4.5 \times e^6$ cycles showing the effect of bond wire lift-off, as shown in Figure 9.14(a). After this event, the evolution in the rate of degradation is accelerated and the module fails catastrophically with a nearly 7% rise in V_{FD} before failure. The smaller variation in voltage measurement is attributed to cooling temperature variation during the measurement.

In high-power modules, the rise in on-state resistance is attributed to the degradation of electrical interconnections from terminal to chip [5], whereas the rise in thermal resistance is predominantly attributed to degradation in the solder layer and the thermal system [19–21]. In the presented method, the origin of the

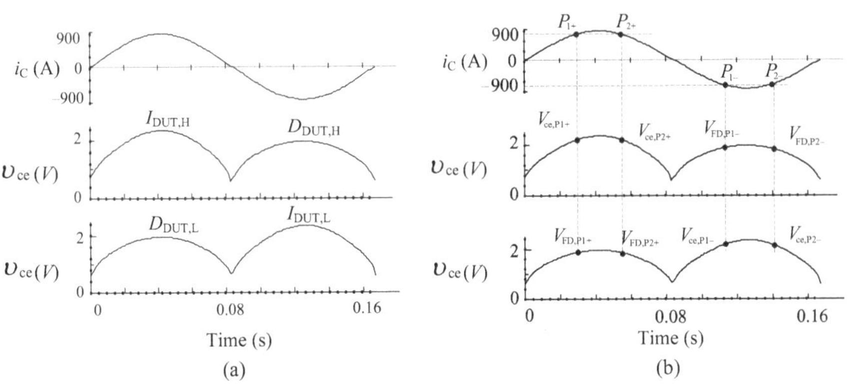

Figure 9.12 Real-time monitoring of on-state collector-emitter voltage v_{ce} and load current i_C (a) for one cycle (b) observation points sampling at 900 A

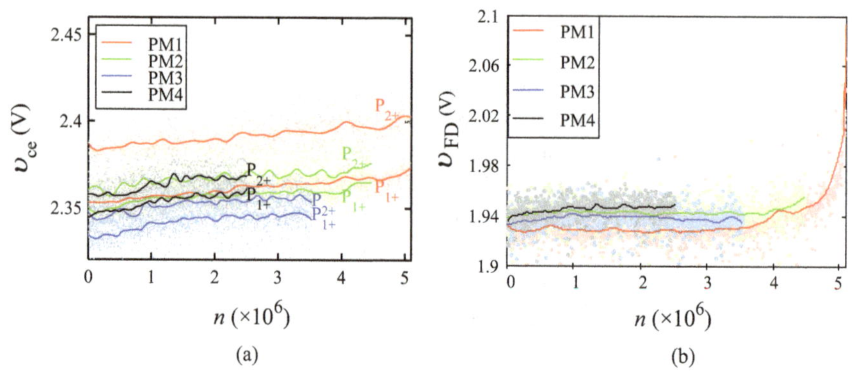

Figure 9.13 Evolution of on-state collector-emitter voltage $v_{ce,on}$ on (a) rising and falling side for high side IGBT and (b) low side diode on a half-bridge module

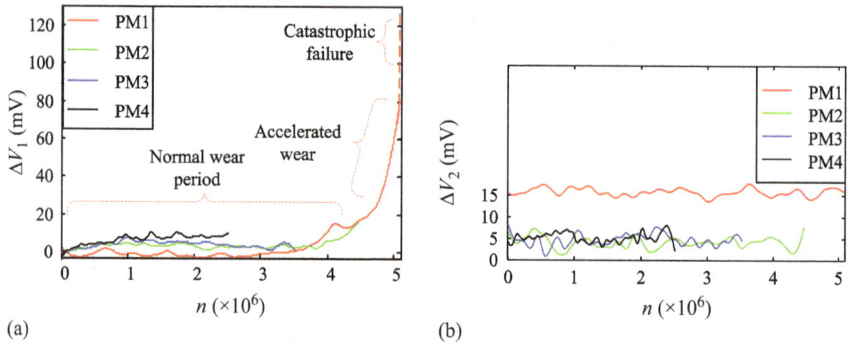

Figure 9.14 Real-time monitoring of (a) ΔV_1 and (b) ΔV_2 for LS diode at 900 A

degradation source can be identified. Equations (9.2)–(9.4) describe data evaluation theory to separate failures in bond wires and solder layers. ΔV_1 shows the rise in voltage from electrical degradation, whereas ΔV_2 is attributed to a rise in R_{th} [6], as shown in Figure 9.14(b). In the given result, the variation in 2–4 mV resulted from $T_{cooling}$ and error in the measurement.

$$V_{P_{ref}} = \frac{1}{N}\sum_{i=1}^{100} V_{P_1}(n_i), \tag{9.2}$$

$$\Delta V_1(n) = V_{P_1}(n) - V_{P_{ref}}, \tag{9.3}$$

$$\Delta V_2(n) = V_{P_2}(n) - V_{P_1}(n), \tag{9.4}$$

where n is the number of cycles.

9.4.1 Application of $v_{ce,on}$ monitoring

Online monitoring shows real-time development of degradation in components [5,6], which offers appropriate action control to be taken, such as derating of the system, operational management or preventive maintenance scheduling, replacement of weak components, etc. The real-time monitoring has three basic functions:

- Detection: state of health or condition monitoring.
- Diagnosis: find root cause of fault.
- Prognosis: estimate remaining useful life for operational management or preventive maintenance.

The proposed measurement approach is suitable for state of health monitoring of power IGBT modules with smaller time resolution. It can be implemented for wind turbines, aeronautics, marine, industrial applications, etc.

With the implementation of online monitoring, wear-out mechanisms can also be interpreted using a Weibull approach together with an understanding of physical analysis. In this approach, failure criteria are defined based on physical analysis and an appropriate statistical model [14]. In this test, a 7% rise in v_{FD} in an LS diode leads to catastrophic failure of power modules. A control action shall be taken after detecting the certain percentage rise in voltage when the wear-out failure begins. For an IGBT, from 10% up to a 20% rise in voltage are defined as failure criteria in the literature [16]. Figure 9.15 shows that the PM1 was in operation for 19hrs when the device was loaded at the same electrical load even after the wear-out mechanism accelerated and before the catastrophic failure of the device occurred.

The monitoring is beneficial to reduce operation and maintenance costs and protect the system from catastrophic damage such as explosion or even fire in a converter. For example, in onshore wind turbine applications, the survey-based result shows that power electronic converters contributed to a 13% failure rate as

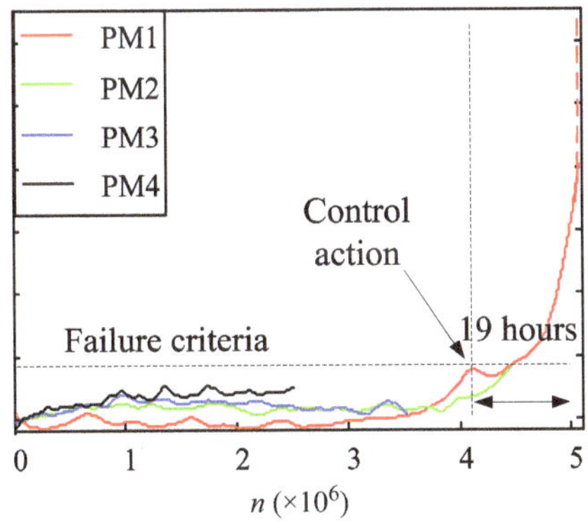

Figure 9.15 A general scheme to generate control signal for a preventive maintenance application

well as an 18% downtime [22]. Here, power modules and capacitors are found as the most fragile components. In addition, for offshore wind farms, currently, the operation and maintenance cost includes 18–23% of the total cost of energy. Hence, these data show the essence of real-time measurement at a component level [23].

9.4.2 Degradation and failure mechanisms

The electrical properties of power modules are governed by the quality of used materials, packaging technology, and their degradation under the load. It is difficult to assess degradation and failure mechanisms after the failure of a device. Degradation assessments are investigated using micro-sectioning, FIB, SEM, and a four-point probing approach [5,24]. The latter method is a non-damaging technique in contrast to previous methods. The four-point probing method is useful for individual electrical measurements on semiconductor chips and interconnections: bond wires, solders, and metallization in power modules [24]. However, this requires opening and the removal of gel before conducting the sample. Microscopy-based micro-analysis failure mapping allows the identification of the contribution to the degradation of the power module from each interconnection, which is not equal for all layers.

Known failure mechanisms are identified mainly due to degradation in interconnections and materials used in IGBTs and diodes. The failures occur due to electrical, thermal, and mechanical stress in interconnections of a device. A major degradation is observed in wire bond and metallization due to wear-out.

9.4.2.1 Bond wire degradation

Heavy Al wire bonds have been dominating the interconnection technology in high-power modules [25]. Bond wire lift-off and heel-cracking are two major fatigues in Al wires. However, those degradations highly depend on the load applied and a geometry of the module. The degradation process is accelerated by thermal, power, or mechanical cycling processes. In all processes, the stress in a certain region may exceed the elastic limit, thereby creating plastic deformations. Over time, cracks are initiated at natural locations such as voids or at the interface heel. Due to the strength of the crystalline grains cracks are expected to propagate at grain boundaries [25,26]. The result is that cracks are often observed to have initiated near the interface and have propagated inwards on the wire where the grain diameter increases. The reason for this can be explained by the Hall–Petch relation [27]:

$$\sigma_y = \sigma_0 + k_y d^{\frac{-1}{2}} \tag{9.5}$$

where σ_y is the yield strength, k_y is the dislocation locking term, d is the grain diameter, and σ_0 is the yield strength for dislocation along slip planes [26].

Figure 9.16 shows the bond wire/IGBT interface for sample PM4. In this figure, the bond wire is almost lifted off the chip and the fracture is close to the wire/chip interface which is attributed to a large granular structure [5]. In addition, partly delamination of the bond wire will change the local current distribution as well as

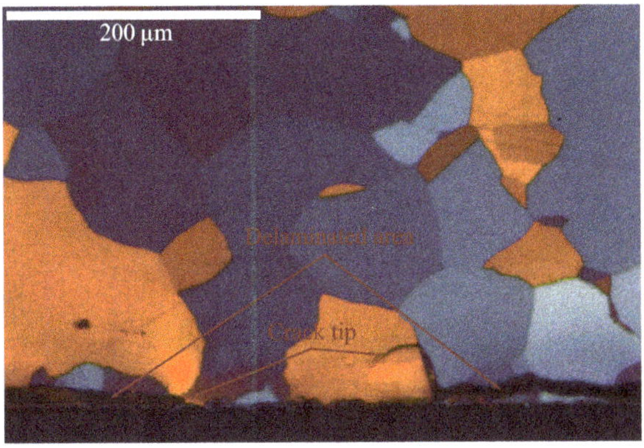

Figure 9.16 Bond wire/IGBT interface of PM4 sample [5]

the effective resistance of the given interface. The effective resistance of the wire is increased with loading cycles in mΩ [5]. In addition, the current distribution depends on length, because the shortest wires carry higher current. In wear-out tests, central wires are degraded more than those attached to the edges of the chip in the tested module. Fracture development goes through the area in the wire which corresponds to a gradual interface between the refinement and deformation regions [25,26]. Based on the medium T_m and amplitude ΔT of the temperature field experienced during loading, the HS and LS of PMs are damaged differently.

9.4.2.2 Metallization degradation

The reconstruction observed in the metallization is a well-known degradation process originated by thermal cycling [4]. Reconstruction occurs mainly due to the unmatched CTE between the Al metallization and the Si chip. The stiffness of the Si chip is large compared to Al; hence, the induced stress in the Al can exceed the plastic limit. The plastic deformation, such as sliding of grain boundaries and dislocations, occurs, making a reconstruction. Reconstruction affects the forward voltage of the section if the metallization is losing contact with channels, becoming thinner, or causing delamination or lift-off.

The metallization mostly changes in the topmost 1–3 μm Al layer. Sometimes the reconstruction propagates down to the surface of the semiconductor chip through the entire metallization layer. An increased degree of degradation is observed at the center of the chip compared to the edges in the test mentioned in Section 9.3.1. Figure 9.17 shows SEM images of metallization of IGBT and diode for fresh and cycled (PM3) [6]. The IGBT shows no sign of degradation, whereas metallization reconstruction occurs in the LS diode.

9.4.2.3 Solder layer degradation

Solder is used to attach a Si chip onto the DCB and the DCB onto the baseplate. Solder layers experience significant temperature variation during converter

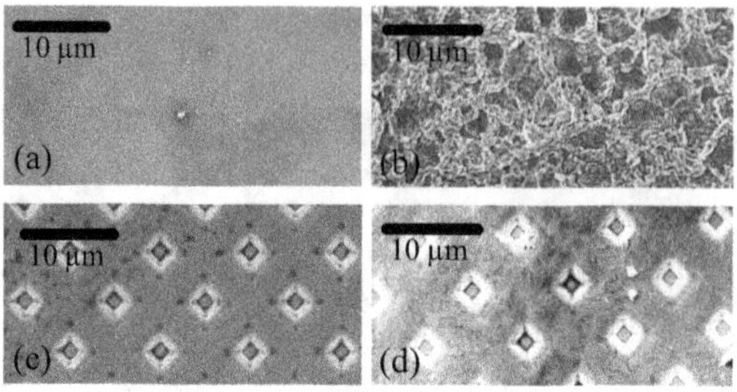

Figure 9.17 SEM images of the metallization surface of (a) new LS diode, (b) PM3 LS diode, (c) New LS IGBT, and (d) PM3 LS IGBT

operation, which dissipates from the chip to the baseplate. The stress induced during temperature fluctuations causes crack formation in solders. Mainly, cracks are initiated from voids and impurities [4].

9.4.3 Post-mortem investigation

Post-failure investigation often risks losing valuable information in the final stage of the device's lifetime. To assess the degradation and failure mechanism process, PMs can be investigated by stopping during functional life after reaching a certain degradation level. The online monitoring approach gives the freedom to select the degradation level and stop it for an investigation purpose. However, careful handling is necessary during the investigation. Basically, the post-mortem investigation process includes [5]:

- The PM disassembly where the module housing, silicone gel could be removed and the module could be separated into smaller sections.
- Four-point-probing could be used to assess the electrical degradation on each individual layer from terminal to chip.
- The metallization of Al could be observed using scanning electronic microscopy (SEM) and focused ion beam(FIB) process.
- The micro-sectioning process could be initiated to investigate the degradation on individual bond wires, chips, etc. This process includes cutting out and polishing the subject, which should be carried out at the end of the investigation.

9.5 Chip temperature estimation

9.5.1 Introduction

Operating characteristics of PM are limited by temperature limitations on semiconductor devices and their packaging. Based on a load mission profile and thermal limitations, the converter is designed to operate in a safe virtual junction temperature (T_{vj}), which is a key parameter for the design. However, due to unpredictable changes in load and faults, it is essential to measure or estimate junction temperature while the system is running. Now, the reliability of power electronic devices are key concern, which largely depends on temperature fluctuations [4,28]. Additionally, real-time measurement of chip temperature can also be used for fast overload control, failure analysis of the device, lifetime analysis, etc.

Semiconductors are widely used in commercial, industrial, military, and automotive applications. Especially in automotive and wind power applications, high-temperature oscillation occurs because of the intermittent nature of source and load [29,30]. In addition, low-frequency operation elevates the T_{avg} and ΔT, which could make a serious impact on the device performance [31].

This section presents a brief overview of chip temperature estimation methods, and a case study of the possibility to implement v_{ce}–load current method for high-power application. The presented results are preliminary outcomes obtained in the ongoing research activity. However, more research such as sensitivity analysis and different operating points needs to be considered before going into an application. The measurements are evaluated using direct IR thermography in spatial and time distribution in an open module.

9.5.2 Overview of junction temperature estimation methods

Modern devices have a built-in NTC thermistor on top of the baseplate, which is located far hence is not able to detect a chip temperature. Thus, two major approaches, such as analytical (electro-thermal models of devices) and physical (measurement), are being used for chip temperature estimation [32]. In the first method, an accurate 3D structure finite element model (FEM) could give a close temperature field, but it requires post-processing and often takes longer computational time [32]. In the second method, direct and indirect measurements can be done. In a direct method, a temperature sensor can be integrated directly onto a chip, such as a thermocouple or fiber optics with/without a temperature sensor. Similarly, the temperature field can be measured by an IR thermal camera. However, these methods require modification on a device packaging as well as pose functional limitations to implement in a converter. In an indirect method, several temperature-sensitive electrical parameters (TSEPs) can be used, which is more applicable to the converter. Hence, innovative measurement methods have been introduced lately. Mainly, dynamic and static characteristics of a PM can be utilized to find temperature dependencies with a function of gate voltage, I_C, and forward voltage drop [33]. In dynamic characteristics, IGBT turn-on, turn-off, and rise time are used, where gate resistance and parasitic capacitances are the dominating parameters.

In static characteristics, $v_{ce,on}$ is used in both low-current and high-current methods. In low-current method, small current runs through the device when the converter is offline. Low current has a negligible contribution to heat up the device and also underestimate the parasitics during measurement. However, it requires a modification in converter control, halting normal operation for a short period of time during the measurement. On the other hand, high-current method introduces self-heating, if the calibration is not conducted in a short time. in addition, $v_{ce,on}$

changes due to the temperature gradient in series interconnection. Nonetheless, this method is regarded as a suitable and cheaper technique in real-time operation, though maintaining accuracy and implementing in converter operation are still challenging. A key benefit is that this method does not require any functional or structural modifications in the converter. However, the measurement circuitry has to fulfill certain requirements to estimate a temperature average in space ($T_{\text{avg,space}}$) in the chip area [34].

- Sensitivity: 1–4 mV/°C,
- Measurement accuracy: 1–4 mV,
- Measurement resolution: 0.61 mV,
- Calibration time: As short as possible, in the case of the P3 module, it is good for less than 300 uS,
- Same temperature field across module during calibration,
- Aging compensation: typically 20 mV due to normal aging in P3 module, varies with packaging and rating of PM.
- Failure compensation: mainly due to bond wire lift-off, typically 5 mV to 7 mV in P3, it varies with the packaging.

9.5.3 $v_{\text{ce,on}}$–load current method

The method includes two major steps in estimating a chip temperature. First, a calibration of $v_{\text{ce,on}}(T)$ is required at different currents and temperatures. Generally, the manufacturer provides characteristic curves for 25 °C, 125 °C and 150 °C. Especially, for multi-chip modules, it does not refer to any specific chip and the given characteristic is only a typical value for the module. Due to the requirement of higher accuracy in the measurement, the calibration of $v_{\text{ce,on}}$-T on each test sample is necessary. This will also eliminate the discrepancies in product variation. Nearly, a 20 − mV variation is observed between the modules for chips located in the same place by authors [6] in a 1.7 kV/1000 A module. A second step includes a real-time $v_{\text{ce,on}}$ measurement, when the converter is active, but with a higher accuracy, as specified in Section 9.3.1.

9.5.3.1 Calibration of $v_{\text{ce,on}} - T$

Calibration is a process to obtain static characteristics $I-v_{\text{ce,on}}(T)$ of individual components of PM. Parallel chips are calibrated at the same time. As outlined in the previous section, calibration time is an essential requirement to avoid self-heating. Hence, two approaches are developed to calibrate the module at load current in a converter. Table 9.2 outlines switching sequences and current waveforms in both methods for the converter, as shown in Figure 9.7.

- Current ramp method
- Current plateau method

In the current ramp-up method, the current rises through the device within 300 μS, as shown in the figure in the upper half of Table 9.2. Here, $v_{\text{ce,on}}(T)$ is measured for a wide range of I_c and T at constant DC-link voltage. Whereas in the current plateau method, the current ramps up very fast and is maintained at a steady current, as shown in the lower half of Table 9.2. Thereafter, $v_{\text{ce,on}}(T)$ is measured

Table 9.2 The $\upsilon_{ce}(T)$ callibration strategies

1. Current ramp-up method		
υ_{ce} **and** I_c	**Active components**	**Waveform**
υ_{ce} at $I_{DUT,H}$, I_c	$I_{DUT,H}$ is ON, $I_{CTRL,\ L}$ is ON (current ramp) $I_{CTRL,L}$ is OFF	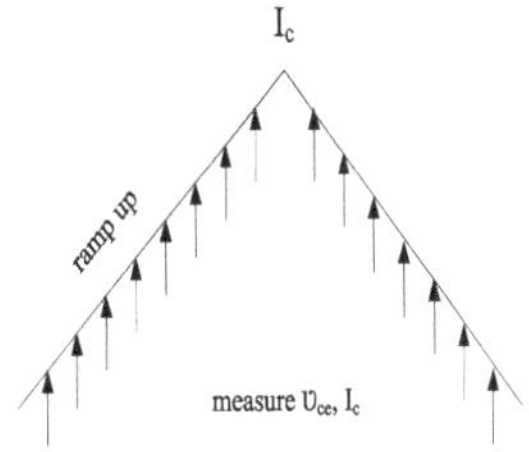
υ_{FD} at $D_{DUT,H}$, I_c	$I_{CTRL,H}$ is ON, $I_{DUT,L}$ is ON (Current ramp) $I_{DUT,L}$ is OFF	
υ_{ce} at $I_{DUT,L}$, I_c	$I_{DUT,L}$ is ON, $I_{CTRL,H}$ is ON (Current ramp) $I_{CTRL,H}$ is OFF	
υ_{FD} at $D_{DUT,L}$, I_c	$I_{CTRL,L}$ is ON, $I_{DUT,H}$ is ON (Current ramp) $I_{DUT,H}$ is OFF $I_{CTRL,L}$ is OFF	
2. Current plateau method		
υ_{ce} at $I_{DUT,H}$, I_c	$I_{DUT,H}$ is ON, $I_{CTRL,L}$ is ON (Current ramp) $I_{CTRL,L}$ is OFF(Plateau)	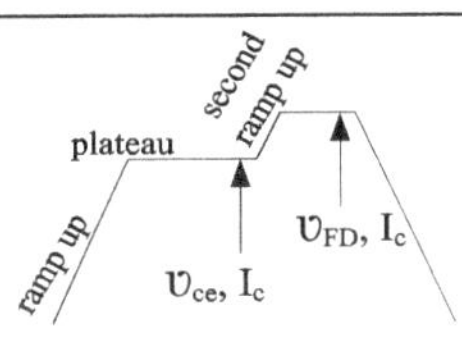
υ_{FD} at $D_{DUT,L}$, I_c	$I_{CTRL,L}$ is ON, (Second current ramp) $I_{DUT,H}$ is OFF (Plateau)	
υ_{ce} at $I_{DUT,L}$, I_c	$I_{DUT,L}$ is ON, $I_{CTRL,H}$ is ON (Current ramp) $I_{CTRL,H}$ is OFF(Plateau)	
υ_{FD} at $D_{DUT,H}$, I_c	$I_{CTRL,H}$ is ON, (Second current ramp) $I_{DUT,L}$ is OFF (Plateau)	

for a wide range of I_c and T. The latter method is good for calibrating at smaller current steps in comparison to the former. In both methods, the temperature field is monitored continuously and verified uniform temperature field on the surface using thermography. However, during calibration at higher temperatures, the IR measurement on a surface is also affected by a convectional flow of air in an open module, as shown in Figure 9.18.

Figure 9.19 shows $I-\upsilon_{ce}(T)$ characteristics for $I_{DUT,H}$ (Figure 9.19(a)) and $D_{DUT,H}$ (Figure 9.19(b)) at 5 °C and 5 A steps for 25 °C to 125 °C.

Figure 9.20 exhibits $\upsilon_{ce,on}$ dependencies on temperature at both NTC (Figure 9.20(a)) and PTC (Figure 9.20(b)) for $I_{DUT,H}$. Similarly, Figure 9.21 exhibits $\upsilon_{ce,on}$ dependencies on temperature at both NTC (Figure 9.21(a)) and PTC (Figure9.21(b)) for $D_{DUT,H}$.

From the calibration, as shown in Figures 9.20 and 9.21, the current dependencies temperature calibration factor (K-factor) ($°C/mv$) is formulated [35]. Here,

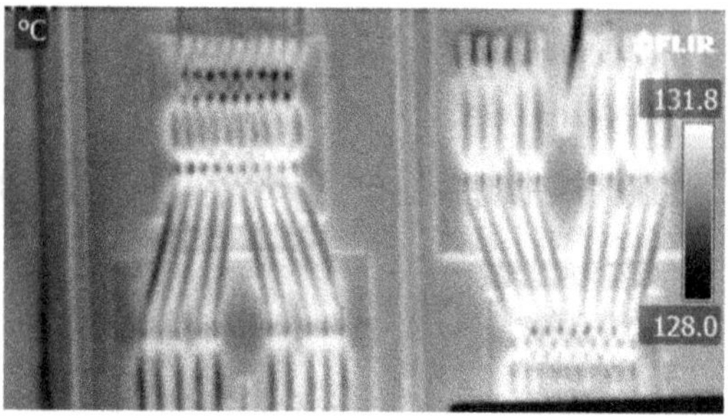

Figure 9.18 IR image of a single section of module during calibration at 125 °C

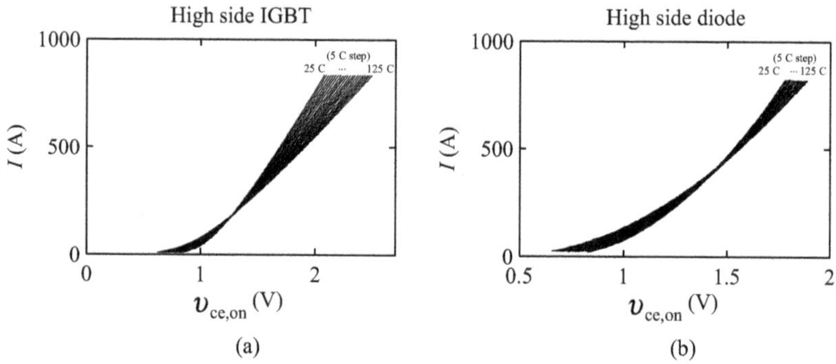

Figure 9.19 Calibrated IV curve from 25 °C *to* 125 °C *for (a) a high side IGBT and (b) high side diode*

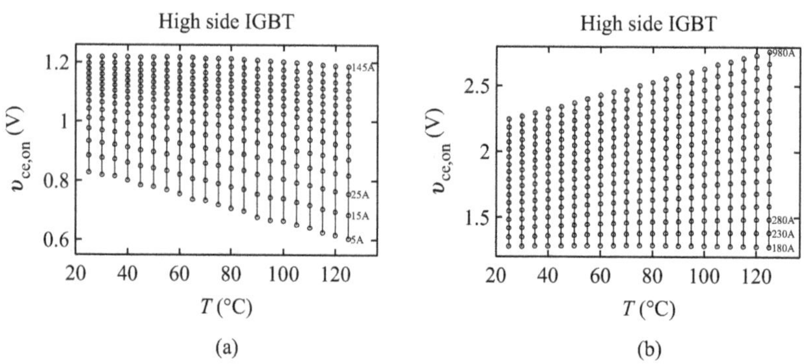

Figure 9.20 Voltage and temperature dependency at (a) NTC and (b) PTC for high-side IGBT

The temperature sensitivity shall be updated using recalibration after a certain degradation level, which varies on the load applied. Figures 9.22 and 9.23 show the K-factor for $I_{\mathrm{DUT,H}}$ and $D_{\mathrm{DUT,H}}$.

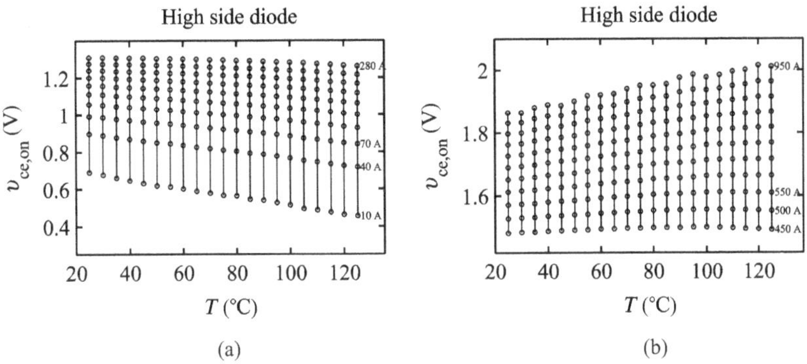

Figure 9.21 Voltage and temperature dependency at (a) NTC and (b) PTC for high-side diode

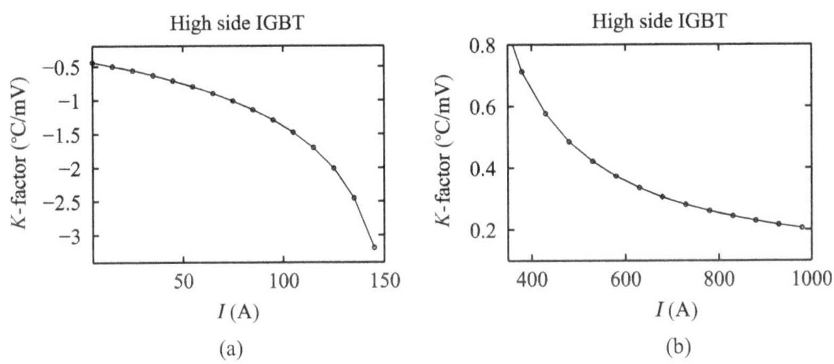

Figure 9.22 Current and temperature dependency at (a) NTC and (b) PTC for high-side IGBT

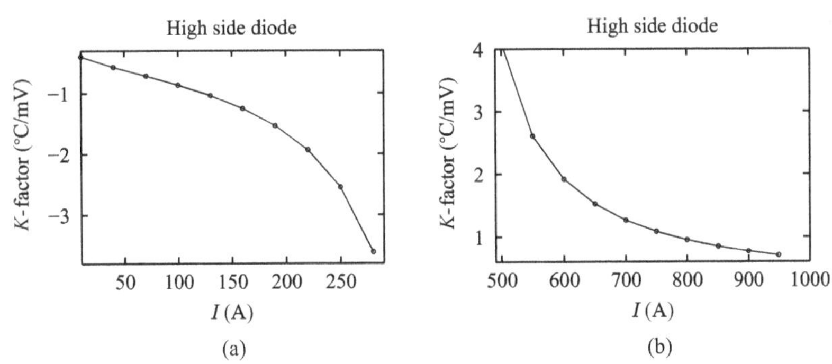

Figure 9.23 Current and temperature dependency at (a) NTC and (b) PTC for high-side diode

9.5.3.2 Real-time measurement

The real-time $v_{ce,on}$ measurement technique is presented in Section 9.3.1. Figure 9.24(a) is shown for current and forward voltage drop measured at 500 A_{Peak} and at 6 Hz frequency.

9.5.4 Estimating temperature in converter operation

Temperature dependency is more linear farther from NTC and PTC cross-over current, as depicted in Figures 9.22 and 9.23. Figure 9.24(b) shows $v_{ce,on}$ variation from calibration at 25 °C, 125 °C, and during loading at 500 A. The plotted $v_{ce,on}$ measured during loading is corrected for temperature estimation due to a change in effective resistance. A calibrated voltage at a low current is used to obtain the correction voltage, which depends on the device's location inside the pack and power dissipation [36]. The time and spatial distribution of temperature depend on thermal impedance and power dissipation during both switching and conduction. In the load current method, the temperature dependency is determined at all current ranges, which allows estimating temperature on each switching; however, because of non-linear temperature dependency and ohmic contribution, the estimated temperature has errors, especially for current close to cross-over level. The temperature gradient that appears during loading in interconnection changes the effective resistance from the calibration conducted at constant temperature. This effect should be compensated to avoid a measurement error. This is achieved by removing the voltage contribution in an ohmic part. Normally, the module manufacturer specifies equivalent resistance from a terminal to the chip, which is 0.25 mΩ at 25 °C for the tested module. In fact, this resistance remains no longer same during power dissipation and also varies with the sample. Here, close to 0.30 mΩ is determined using a direct measurement method to find Δv_{ce} at 25 °C for $I_{DUT,H}$. A linear approximation shall be considered [34] to remove the voltage drop in series

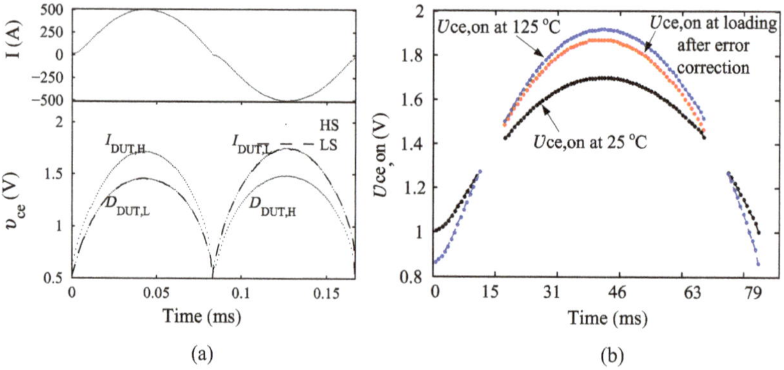

Figure 9.24 Junction temperature estimation: (a) I_L, $v_{ce,on}$ and v_{FD} measurement for one cycle at 500 Apeak; (b) calibrated and measured $v_{ce,on}$ to estimate $T_{avg,space}$

interconnection. The temperature-dependent effective series interconnection can be obtained using the Shockley model with a series resistor as described in [34]. Here, the correction is used for the reference voltage which is calibrated at 25 °C. Using Δv_{ce} in (9.6), $T_{avg,space}$ can be determined.

$$T_{avg,space} = T_{ref} + T_{cor} + K\left(v_{ce,meas} + \Delta v_{err} - v_{ref}(T_{ref})\right) \tag{9.6}$$

where T_{ref} is a calibrated temperature, and $v_{ref}(T_{ref})$ is the calibrated voltage at the reference temperature. $T_{cor}/\Delta v_{err}$ are corrections for the calibration approach. Updated (K – factor) and $v_{ref}(T_{ref})$ compensates the influence of aging on the temperature estimation. T_{cor} is the heat up during calibration, which has to be added to the reference temperature. The influence of this parameter can be minimized by shortening the calibration time [34]. Δv_{err} is the correction in forward voltage from a homogeneous temperature field to an online inhomogeneous field. The Δv_{err} is obtained from T_j, corresponding baseplate temperature (T_b), and change in interconnection resistance (R_{slope}) at the corresponding load current as given in (9.7). Here a scaling factor (SF) is required, which changes depending upon the location and packaging of power modules [34]. Hence calibration of individual modules is the most in this method. However, it is necessary to test on more samples/different packaging to find the sensitivity of this method.

$$\Delta v_{err} = \left(T_j - T_b\right) \times R_{slope} \times I_L \times SF \tag{9.7}$$

9.5.5 Temperature measurement using direct method

Infrared (IR) thermography is an established method for observing temperature fields in semiconductors. A new generation thermal camera (*X*8400*SC*) with a higher resolution is used. This method requires a special preparation of an open module and the modification of the test setup.

9.5.5.1 Setup and sample preparation

The visibility in a chip area of the PM is covered by a low inductance bus bar in a converter stack, as shown in Figure 9.25(a). Hence, the DUT is pulled out physically and connected to the same test rig with additional physical support to allow

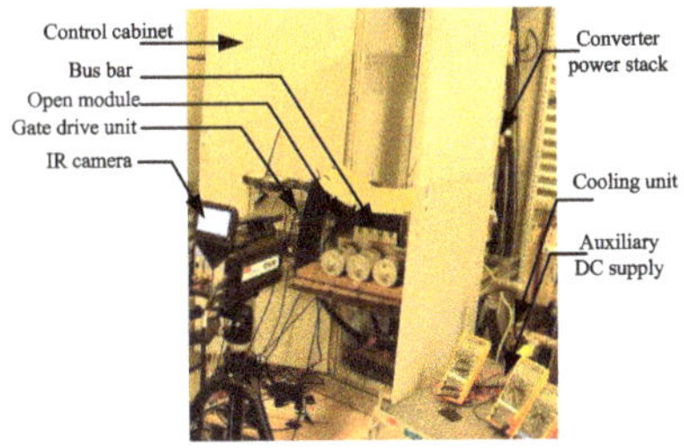

(a)

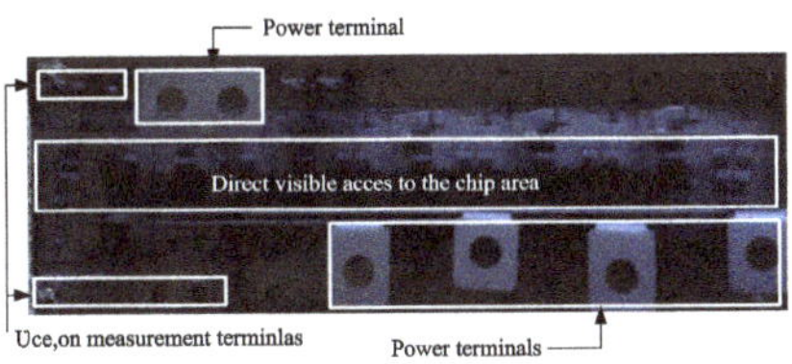

(b)

Figure 9.25 IR thermography: (a) test setup; (b) an open DUT power module

direct visibility. The open module is stressed at the same thermal and electrical parameters in the converter, as shown in Figure 9.25(a). A cooling pipe is extended and the same thermal interface is used in the DUT.

The internal bus bar in this open module is bent opposite, so that the chip and electrical interconnections are visible without interference, as exhibited in Figure 9.25(b). High-temperature matte black paint with 0.954 emissivity is used on a surface of the sample. 50 μm thickness of paint is quoted on the surface. The suitable thickness is applied after investigating different thicknesses of paint using an IR spectrometer [34]. Voltage insulation is used in between positive and negative bus bars in order to avoid spark during high voltage switching.

9.5.5.2 IR temperature field measurement

The temperature field is monitored from the IR camera on each individual section, when the converter is active. The electrical parameters are sampled at the f_{sw} (2.5 kHz), whereas the thermal image is captured at 714 Hz.

Figure 9.26 shows the temperature field in a single section, as shown in 9.18 when the $D_{DUT,H}$ is hotter. Figure 9.26(b) shows the temperature gradient contour on the corresponding surface.

Similarly, Figure 9.27(a) shows the temperature field when the $D_{DUT,L}$ is hotter. Figure 9.27(b) shows the temperature gradient contour on the corresponding surface.

9.5.6 Estimated temperature evaluation

The $v_{ce,on}$ includes a contribution from a series interconnection, as shown in Figure 9.28(a). On the other hand, the temperature on the surface of the device changes slowly because of a larger thermal time constant compared to an electrical signal. The thermal impedance can be determined by the Caur in Figure 9.28(b) and the foster models. The temperature keeps rising even after the current starts falling from the peak. For device operating at 500 A_{Peak}, the temperature reaches its peak at close to 350 A in the falling side, showing a nearly 20 ms delay in the thermal response. The thermal response of the applied paint is not known. In the IR

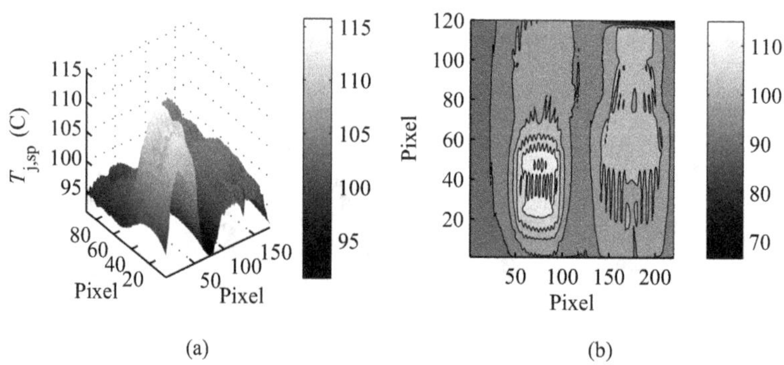

Figure 9.26 IR measurement in spacial temperature distribution at 500 A *peak for one section when high side diode $D_{DUT,H}$ is at maximum temperature: (a) temperature field and (b) a temperature contour*

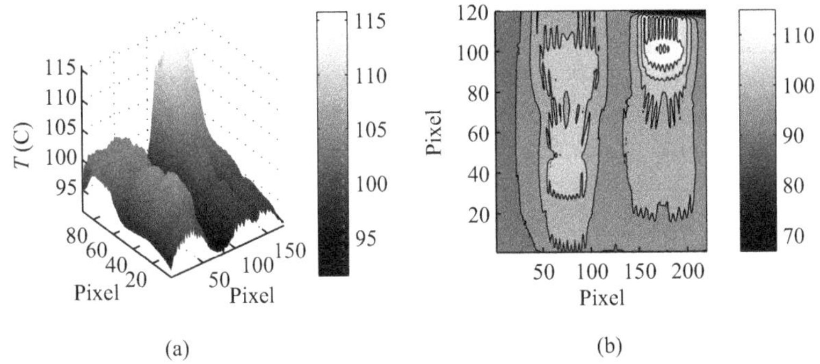

Figure 9.27 IR measurement in spacial temperature distribution at 500 A peak for one section when low side diode $D_{DUT,L}$ is at maximum temperature: (a) temperature field and (b) a contour

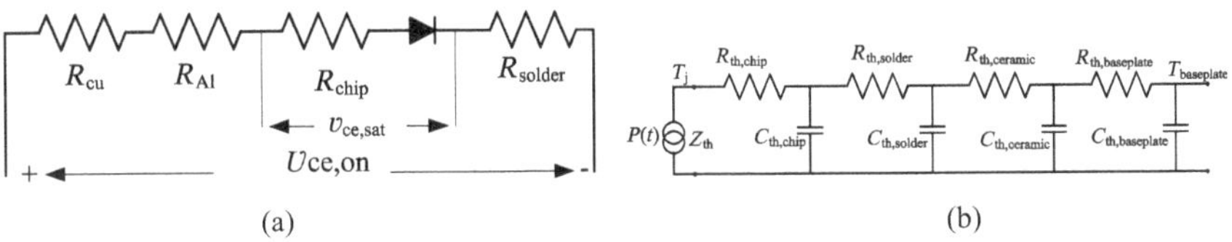

Figure 9.28 Figure showing (a) a layout of ohmic part in the on-state of IGBT and (b) a Caur model for an IGBT

Table 9.3 Temperature measurement from $v_{ce,on}$–load current and IR thermography (the peak temperature shown in table reaches after some delay of current peak)

Current	**v_{ce}–load current method**	**IR thermography**				
$I_{c,peak}$	T_{est}	$T_{baseplate}$	T_{avg}	T_{max}	T_{min}	$\Delta T_{chip}=T_{max}-T_{min}$
(A)	(°C)	(°C)	(°C)	(°C)	(°C)	(°C)
500	100	88	99.5	102.6	92.9	9.7
450	81	71	79.6	82	74.8	7.2
400	110	101.6	110.4	113.7	104.5	9.2
350	88	82	89	92	85	7

measurement, the delay in the rise in temperature may also include the paint's thermal response.

Table 9.3 gives a temperature comparison at different currents at 6 Hz and baseplate temperatures. The baseplate temperature is also monitored by IR thermography. The estimated temperature is shown when the current is at the peak for a given value with respect to the time. The IR measurement results are presented for maximum, minimum, and average temperature on the $I_{DUT,H}$ surface, which includes temperature variation in bond wires and chips, when the temperature reaches its peak. The temperature gradient across the chip (ΔT_{chip}) is also tabulated for corresponding current. Except at close to the peak current level, the estimated temperature deviates greater than 5 °C because of non-linear dependencies and ohmic contribution. With the presented results, it is difficult to conclude explicitly which temperature the estimated value represents. However, with small error depending on the current level, the estimated temperatures are close to average in spatial temperature distribution, ($T_{\text{avg,space}}$), for the current close to peak level. Table 9.3 shows the estimated temperature for $I_{\text{DUT},H}$ when the current reaches its peak for different currents at 6 Hz. Similarly, the IR thermography measurements are shown when the temperature is at peak together with the baseplate temperature for the same loading. When the current is at peak, the estimated temperature is close to the IR measurement; however, when the device reaches its peak temperature, the estimated temperature falls much faster, giving large errors greater than 5 °C from the estimation. The estimated temperature is displayed only for a current level above 400 A in Figure 9.29.

The temperature distribution in space is depicted in Figure 9.30(a) and (b) for $I_{\text{DUT,H}}$, when the chip surface is hotter.

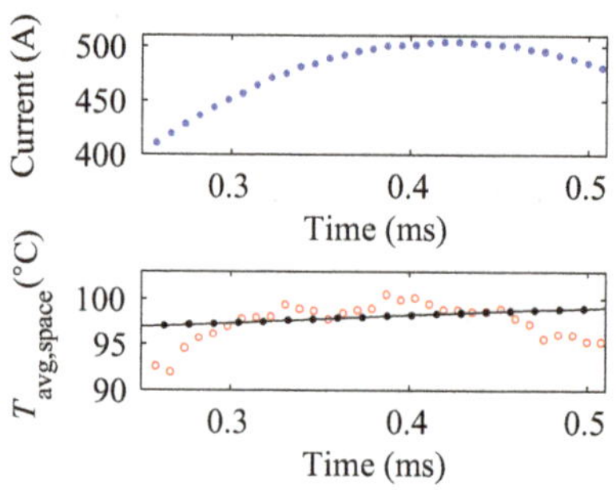

Figure 9.29 Temperature compared between estimation and direct measurement at 500 A_{Peak} *for* $I_{\text{DUT,H}}$

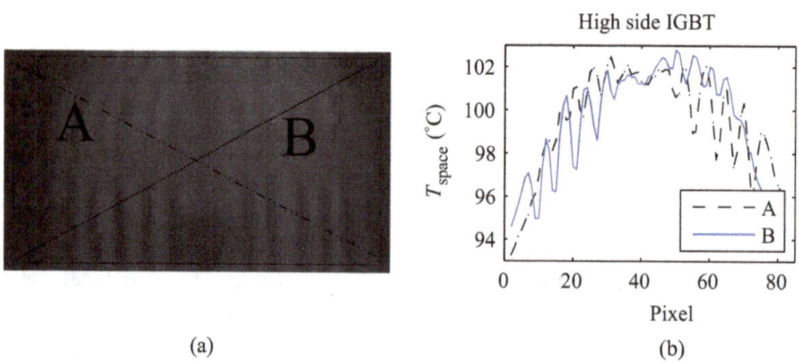

Figure 9.30 Spacial temperature T_{space} *variation is shown for high-side IGBT* $I_{\text{DUT,H}}$ *when temperature is at peak: (a) an IR image; (b) temperature distribution along lines A and B*

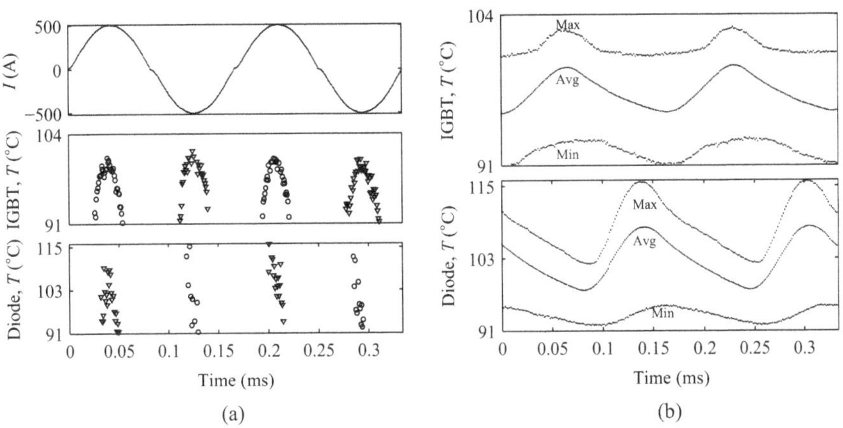

Figure 9.31 Junction temperature variation in converter operation: (a) an estimated temperature and (b) direct measured by IR camera

Estimated temperatures are shown for both IGBTs and diodes in half-bridge in Figure 9.31(a). Similarly, the direct measured temperature including maximum, average, and minimum variation in chip area are shown in Figure 9.31(b) for high-side IGBT and diode. The estimated temperature is shown for current above 400 A for IGBT, whereas for diode above 460 A. On the other hand, the IR measurement is shown for two full electrical cycles.

9.6 Processing of state monitoring data

State data are defined as instantaneous values of system-energy-related variables, such as temperatures, voltages, currents, flow, and speed, as well as system health-related parameter values such as thermal and electrical impedance.

9.6.1 Basic types of state data handling

Data handling is divided into four basic types, namely:

- Online control,
- Online monitoring,
- Event-driven data logging,
- Mission profile data logging.

Online control based on-state monitoring can be used for derating (or uprating) of power production, including complete shut-down of a power converter. In the case of more parallel subsystems, e.g., wind turbines in a wind farm, and the state data can be used to control load-sharing and wear-out in order to optimize the operation. Derating control can make a system more robust against tripping/ stopping, e.g., due to over-temperature on hot days, and perhaps even allow

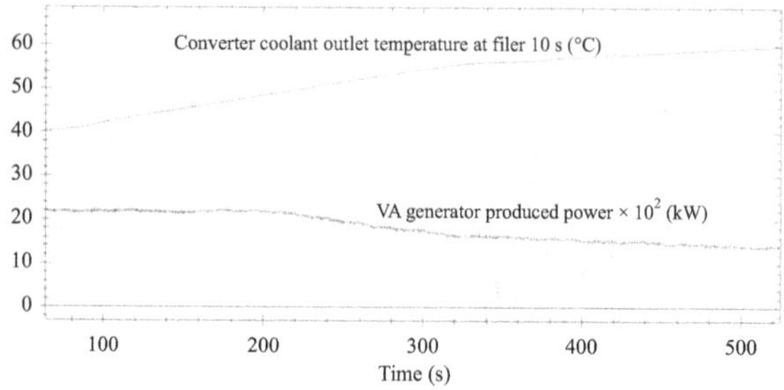

Figure 9.32 Converter power derating control response to an increasing coolant temperature

closer to the design limit, at the same time. In this way, intelligent use of state data during operation can lead to better reliability and/or better system utilization. An example of derating control in a power converter is shown in Figure 9.32, where the power output is reduced smoothly with increasing coolant temperature, as an alternative to simple tripping (operation stop) at a predefined temperature level [37].

Online monitoring is defined as online streaming of selected signals/variables from a system. The source of data can be a number of similar subsystems (e.g., wind turbines), for manual and/or automated monitoring, handled in supervisory control and data acquisition (SCADA) system. The automated monitoring could also include various statistical analysis across a population of similar subsystems. The monitoring typically includes alarm and warning functionalities.

Event-driven data logging involves typically high time resolution snapshots of important data for a predefined time window around an event, much like a single-shot oscilloscope record. For a power converter, such logging could contain hundreds of signals, variables, and states (most of which are pure software) for a few seconds before and after an event, e.g., a fault ride-through situation for a wind power converter, from which data are logged for possible post-processing and analysis. Such data logging with full information (full sampling rate) about all signals and states, and the most important variables, are crucial for an efficient post-event analysis. For example, to analyze the root cause of trip or bad performance. Simultaneous logging at different time scales and sampling times can also be made for different parts of the system, e.g. electrical, thermal, and mechanical parts.

Mission profile data logging is done in order to get a picture of the operation life history of the system as shown in Figure 9.33. This can be done in several ways, including simple counting of cycles, rain-flow counting, and multi-dimensional stressor records. An example of a simple and efficient way to log the load profile history of a power converter is to sort the main load-related-variables, such as torque, current, voltage, and temperature, in one-dimensional bins, say, 10–20, distributed over the possible operation range. Then the mission profile is simply built by continuously counting up in each actual bin, for each selected variable, for

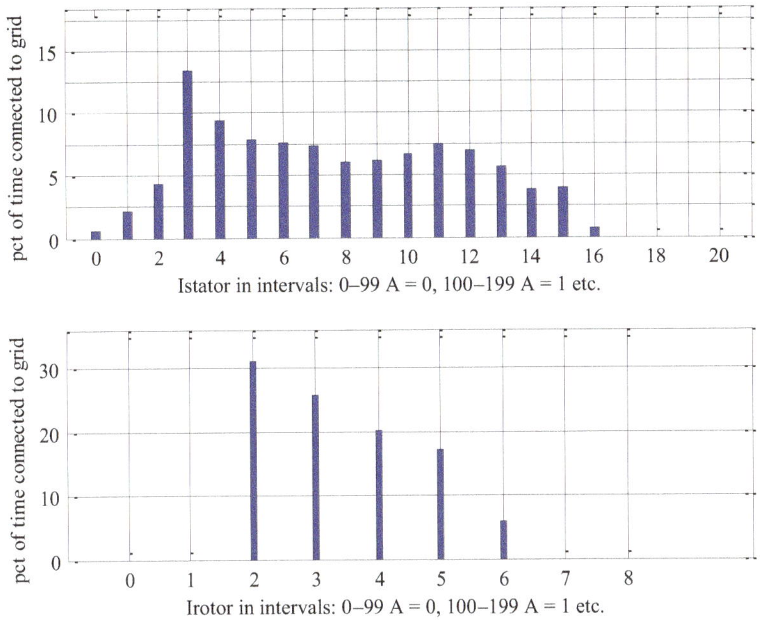

Figure 9.33 Example of mission profile data logging from a wind turbine in the field

every control period. In this way, every millisecond of operation can actually be recorded.

9.6.2 Application of state monitoring

State monitoring data can be used in all of a product's life phases, from development to production, operation, and finally to post-mortem analysis. In the development phase, the state monitoring data can be used for design verification and monitoring of selected signals, e.g., during accelerated testing. In the production test, the state monitoring data can be used to detect weak/wrong components and bad interfaces. It might also be possible to replace big and expensive burn-in test facilities with small, smart, and much cheaper test facilities based on online monitoring of state data. During the lifetime, the data can be used in the control (derating/uprating) and for protection purposes, and the smart service (predictive maintenance). Logged and saved state monitoring data are of high importance in root cause analysis in the case of a failed unit (post-mortem analysis). A couple of examples of the use of state monitoring data in a setup suited for production test setup are shown below. The T_j estimation (based on $v_{ce,on}$) is used in the product during product burn-in testing to detect a missing thermal interface material or too high switching losses, as shown in Figure 9.34(a) and (b) [38].

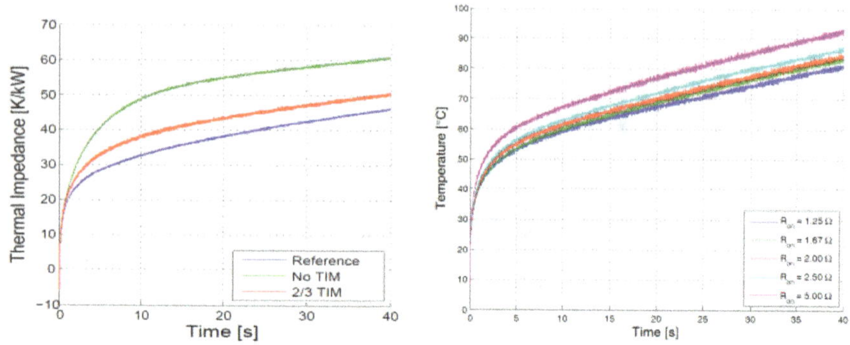

(a) (b)

Figure 9.34 Increase of junction temperature as a function of (a) the wrong amount of thermal interface material (TIM) and (b) increased switching losses (wrong turn-on resistor)

9.7 Conclusion

This chapter presents a power cycling method capable to emulate the closest field stresses in power modules using a full-scale converter. This method could enable to optimize end of life investigation of power modules for laboratory test as well as for field applications. A rise in $v_{\text{ce,on}}$ and v_{FD} can be used as a precursor parameter to understand aging in power modules in a real time. The presented online voltage monitoring technique is suitable for state of health monitoring as well as to generate alarming signals to make smart decision on power derating, operation and maintenance, removal of faulty parts to prevent catastrophic failures, etc.

Often understanding of failure mechanism is difficult to access after the catastrophic failure of the device. However, a real-time aging monitoring of the device could indicate the process of initiation of failures which can be used to understand the mechanism in post-mortem analysis.

Knowing the temperature of the chip is a key parameter to optimize electrical and thermal design as well as reliability studies of power modules. Although different temperature measurement methods are identified, it is still a challenge to apply and measure a real-time temperature variation in a converter. A $v_{\text{ce,on}}$–load current is a promising method to use in a real time without the modification in a structure and a converter control. However, several parameters are needed to consider such as aging, contribution from effective series resistance, and discrepancies in power module production to estimate the chip temperature accurately.

Smart handling of large amounts of data can be achieved by separating them into different categories, as explained in Section 9.6.1 for different stages in a product's lifetime, from design to end of life. The use of state monitoring data ensures a more robust design, screening of infant mortality failures, safer operation, and better post-mortem analysis.

Acknowledgments

The presented work is conducted under Intelligent and Efficient Power Electronic (IEPE) and Center of Reliable Power Electronics (CORPE) frameworks at the Department of Energy Technology, Aalborg University, Denmark.

References

[1] J. Lutz, H. Schlangenotto, U. Scheuermann, and R. D. Doncker, *Semiconductor Power Devices*, Springer, Berlin, 2011.

[2] P. Cova and F. Fantini, "On the effect of power cycling stress on IGBT modules," *Microelectronics Reliability*, vol. 38, no. 6–8, pp. 1347–1352, 1991.

[3] M. Held, P. Jacob, G. Nicoletti, P. Scacco, and M. -H. Poech, "Fast power cycling test of IGBT modules in traction application," *International Conference on Power Electronics and Drive Systems*, vol. 1, pp. 425–430, 1997.

[4] M. Ciappa, "Selected failure mechanisms of modern power modules," *Microelectronics Reliability*, vol. 42, pp. 653–667, 2002.

[5] K. B. Pedersen, L. H. Ostergaard, P. Ghimire, V. Popok, and K. Pedersen, "Degradation mapping in high power IGBT modules using four-point probing," *Microelectronics Reliability*, 2015, doi:10.1016/j.microrel.2015.05.011.

[6] P. Ghimire, K. B. Pedersen, S. Rannestad, B. S. Munk-Nielsen, and F. Blaabjerg, "Field oriented IGBT module wear out test and its physics of failure analysis," *Under Peer Review at TPEL* 2015.

[7] V. Smet, F. Forest, J.-J. Huselstein, *et al.*, "Ageing and failure modes of IGBT modules in high-temperature power cycling," *IEEE Transactions on Industrial Electronics*, vol. 58, no. 10, pp. 4931–4941, 2011.

[8] J. Due, S. Munk-Nielsen, and R. Nielsen, "Lifetime investigation of high power IGBT modules," in *14th European Conference on Power Electronics and Applications (EPE)*, vol. 1, 2011.

[9] R. Nielsen, J. Due, and S. Munk-Nielsen, "Innovative measuring system for wear-out indication of high power IGBT modules," in *Energy Conversion Congress and Exposition (ECCE)*, 2011.

[10] M. S. Bobi, in M. Sanz Bobi (eds.), *Operation and Maintenance of Renewable Energy Systems: Experiences and Future Approaches (Green Energy and Technology)*, Springer, Berlin, 2014.

[11] A. de Vega, P. Ghimire, K. Pedersen, *et al.*, "Test setup for accelerated test of high power IGBT modules with online monitoring of v_{ce} and v_f voltage during converter operation," in *Power Electronics Conference (IPEC-Hiroshima 2014 - ECCE-ASIA)*, 2014.

[12] G. Consentino, M. Laudani, G. Privitera, *et al.*, "Effects on power transistors of terrestrial cosmic rays: Study, experimental results and analysis," in *Applied Power Electronics Conference and Exposition (APEC)*, 2014.

[13] I. Trintis, B. Sun, J. Guerrero, S. Munk-Nielsen, F. Abrahamsen, and P. Thøgersen, "Dynamic performance of grid converters using adaptive dc voltage control," in *Power Electronics and Applications (EPE'14-ECCE Europe)*, 2014.

[14] *How to Measure Lifetime for Robustness Validation – Step by Step*, ZVEI - German Electrical and Electronics Manufacturers' Association e.V. Std., Rev. 1.9.

[15] V. Smet, F. Forest, J. Huselstein, A. Rashed, and F. Richardeau, "Evaluation of vce monitoring as a real-time method to estimate aging of bond wire-IGBT modules stressed by power cycling," *IEEE Transactions on Industrial Electronics*, vol. 60, no. 7, pp. 2760–2770, July 2013.

[16] P. Ghimire, S. Beczkowski, S. Munk-Nielsen, B. Rannestad, and P. Thogersen, "A review on real time physical measurement techniques and their attempt to predict wear-out status of IGBT," in *15th European Conference on Power Electronics and Applications (EPE)*, Sept. 2013, pp. 1–10.

[17] S. Beczkowski, P. Ghimire, A. de Vega, S. Munk-Nielsen, B. Rannestad, and P. Thøgersen, "Online vce measurement method for wear-out monitoring of high power IGBT modules," in *15th European Conference on Power Electronics and Applications (EPE)*, Sept 2013, pp. 1–7.

[18] P. Ghimire, A. de Vega, S. Beczkowski, S. S. Munk-Nielsen, B. Rannestad, and P. B. Thøgersen, "Improving power converter reliability: Online monitoring of high-power IGBT modules," *IEEE Industrial Electronics Magazine*, vol. 8, no. 3, pp. 40–50, Sept 2014.

[19] G. Coquery and R. Lallemand, "Failure criteria for long term accelerated power cycling test linked to electrical turn off {SOA} on {IGBT} module: A 4000 hours test on 1200a, 3300v module with alsic base plate," *Microelectronics Reliability*, vol. 40, no. 810, pp. 1665–1670, 2000.

[20] A. H. C. Herold, J. Lutz, and M. Thoben, "Thermal impedance monitoring during power cycling tests," in *PCIM Europe*, VDE VERLAG GMBH, Berlin, 2011.

[21] L. Feller, S. Hartmann, and D. Schneider, "Lifetime analysis of solder joints in high power {IGBT} modules for increasing the reliability for operation at 150°C," *Microelectronics Reliability*, vol. 48, no. 89, pp. 1161–1166, 2008.

[22] L. Ran, S. Konaklieva, P. McKeever, and P. Mawby, "Condition monitoring of power electronics for offshore wind," *Engineering & Technology Reference*, 2014.

[23] P. Tavner, *Offshore Wind Tturbine Reliability, Availability And Maintenance*, IET, Stevenage, 2012.

[24] K. B. Pedersen, P. K. Kristensen, V. Popok, and K. Pedersen, "Micro-sectioning approach for quality and reliability assessment of wire bonding interfaces in {IGBT} modules," *Microelectronics Reliability*, vol. 53, no. 9–11, pp. 1422–1426, 2013.

[25] K. Pedersen, D. Benning, P. Kristensen, V. Popok, and K. Pedersen, "Interface structure and strength of ultrasonically wedge bonded heavy

aluminium wires in si-based power modules," *Journal of Materials Science: Materials in Electronics*, 2014.

[26] J. Goehre, M. Schneider-Ramelow, and U. Geibler, and klaus Dieter lang, "Interface degradation of al heavy wire bonds on power semiconductors during active power cycling measured by the shear test," in *Conference on Integrated Power Elctronic Systems (CIPS)*, 2010.

[27] N. Hansen, "Hallpetch relation and boundary strengthening," *Scripta Materialia*, vol. 51, no. 8, pp. 801–806, 2004.

[28] R. Bayerer, T. Herrmann, T. Licht, J. Lutz, and M. Feller, "Model for power cycling lifetime of IGBT modules- various factors influencing lifetime," in *5th International Conference on Integrated Power Systems (CIPS)*, March 2008, pp. 1–6.

[29] R. John, O. Vermesan, and R. Bayerer, "On-road evaluation of advanced hybrid electric vehicles over a wide range of ambient temperatures," in *IMAPS High Tempertaure Electronics Network (HiTEN 2009)*, 2009.

[30] C. Busca, R. Teodorescu, F. Blaabjerg, *et al.*, "An overview of the reliability prediction related aspects of high power IGBTs in wind power applications," *Microelectronics Reliability*, vol. 51, pp. 1903–1907, 2011.

[31] M. Bartram, J. von Bloh, and R. W. D. Doncker, "Doubly-fed-machines in wind-turbine systems: Is this application limiting the lifetime of IGBT-frequency-converters?" in *IEEE 35th Annual Power Electronics Specialists Conference (PESC)*, 2004.

[32] Y. Avenas, L. Dupont, and Z. Khatir, "Temperature measurement of power semiconductor devices by thermo-sensitive electrical parameters: A review," *IEEE Transactions on Power Electronics*, vol. 27, no. 6, pp. 3081–3092, 2012.

[33] N. Baker, M. Liserre, L. Dupont, and Y. Avenas, "Improved reliability of power modules: A review of online junction temperature measurement methods," *IEEE Industrial Electronics Magazine*, vol. 8, no. 3, pp. 17–27, 2014.

[34] P. Ghimire, K. B. Pedersen, I. Trintis, B. Rannestad, and S. Munk-Nielsen, "Online chip temperature monitoring using v_{ce}–load current and IR thermography," in *IEEE Energy Conversion Congress and Exposition (ECCE)*, 2015.

[35] P. Ghimire, K. B. Pedersen, A. R. D. Vega, B. Rannestad, S. Munk-Nielsen, and P. B. Thøgersen, "A real time measurement of junction temperature variation in high power IGBT modules for wind power converter application," in *8th International Conference on Integrated Power Systems (CIPS)*, Feb 2014, pp. 1–6.

[36] X. Perpina, J. F. Serviere, J. Saiz, D. Barlini, M. Mermet-Guyennet, and J. Millan, "Temperature measurement on series resistance and devices in power packs based on on-state voltage drop monitoring at high current," *Microelectronics Reliability*, 2006.

[37] P. Thøgersen, "Converter solutions for wind power," in *EWEA*, 2012.

[38] S. D. Snerskov, K. L. Frederiksen, A. B. Jørgensen, E. Iciragiye, A. E. Maarbjerg, and N. Christensen, "Novel screening procedure for wind turbine power converters," *Master's thesis*, School of Engineering and Science, Aalborg University, Aalborg, December 2014.

Chapter 10

Stochastic hybrid systems models for performance and reliability analysis of power electronic systems

Sairaj V. Dhople[1], *Philip T. Krein*[2] and *Alejandro D. Domínguez-García*[2]

In this chapter, we propose stochastic hybrid systems (SHS) models for performance and reliability analysis of power electronic systems. The state space of an SHS is composed of (i) a discrete state that describes the possible configurations of the system and (ii) a continuous state that captures a physics-based or behavioral model of the system associated with performance metrics of interest. Transitions of the discrete state are random and occur at rates that can be functions of the continuous state. Similarly, the evolution of the continuous state is discrete-state dependent. Transitions of the discrete state are accompanied by reset maps that determine how the transitions affect the continuous state. The proposed framework expands on conventional reliability models such as Markov models by providing the capability to monitor a dynamical system that captures some notion of the performance of the power electronic system. Applications of the framework are demonstrated to model the expected accumulated revenue of a two-inverter residential-scale photovoltaic system.

10.1 Introduction

Stochastic hybrid systems (SHS) are a class of stochastic processes with a state space composed of a discrete state and a continuous state. The transitions of the discrete state are random, and the rates at which these transitions occur are, in general, a function of the value of the continuous state. For each value that the discrete state takes—referred subsequently as *modes* of the system—the evolution of the continuous state is described by a stochastic differential equation. The vector fields that govern the evolution of the continuous state in each mode depend on the

[1]Department of Electrical and Computer Engineering, University of Minnesota, USA
[2]Department of Electrical and Computer Engineering, University of Illinois at Urbana-Champaign, USA

operational characteristics of the system in that mode. Reset maps associated with mode transitions define how the discrete and continuous states map into post-transition discrete and continuous states. The SHS-based framework we outline is built on theoretical foundations developed in [1,2].

The SHS formalism described above is well suited to model system dynamics in a variety of uncertain environments. The set in which the discrete state takes values describes the possible configurations that the system can adopt. For instance, in the context of power electronic systems reliability modeling, in addition to one (or more) nominal (non-faulted) operational modes, other operational modes include those that arise due to faults (and repairs) in components that comprise the system. Modes may also model discretizations of uncertain generation and load values in several applications including renewable-based resources, energy storage, and electric vehicle systems [3–5]. The continuous state captures the evolution of variables associated with the system's performance. For instance, physics-based models could include inductor currents and capacitor voltages, network voltages and currents, and electrical frequency as the continuous states in the SHS. Similarly, behavioral models could be formulated to describe system economics, expended repair costs, availability, energy yield, and incentives to participate in demand-response programs. Reset maps in this context enable and improve modeling accuracy since they can describe instantaneous impacts of mode transitions (arising from failures and repairs in the constituent elements of the power electronic system) on the continuous states.

An SHS-based model is completely characterized by the combined distribution of the continuous and discrete states. However, the coupling between the discrete and continuous states—given the generality afforded to the transition rates and reset maps in the model formulation—renders the problem of obtaining the combined distribution analytically intractable in most practical applications. In fact, the combined distribution can only be recovered in a few special cases. For instance, if the evolution of the discrete state does not depend on the continuous state, the model boils down to a continuous-time Markov chain; the probability distribution in this case is fully characterized and specified by the Chapman–Kolmogorov equations [6]. However, given the difficulty in obtaining the combined distribution of the discrete and continuous states, we focus on a method to compute any arbitrary number of their raw moments. To this end, we leverage the approach outlined in [1,2] and demonstrate how to formulate a family of nonlinear ODEs, the solutions of which yield the moments of the continuous state. In lieu of the availability of the complete distribution, the moments are useful in many contexts pertaining to dynamic risk assessment in the system. For instance, the moments can be used to compute bounds on the probabilities of events when the continuous states do not satisfy performance specifications.

The SHS framework encompasses a variety of commonly used stochastic modeling and analysis tools including: (i) jump linear systems (linear flows with no jumps in the continuous state); (ii) discrete-space continuous-time Markov chains (no continuous state and constant/time-varying transition rates for the discrete state); (iii) Markov reward models (constant rate of growth in the continuous state);

and (iv) piecewise deterministic Markov processes (no diffusion terms in the stochastic differential equations that govern the evolution of the continuous state). Given the generality offered by the SHS modeling formalism, they have been applied to study a host of systems such as communication networks, financial systems, air-traffic management systems, bulk power systems, and biological systems (e.g., see [7] and references therein).

The goals of this chapter are to provide a concise introduction to SHS (Section 10.2.1), demonstrate how the moments of the continuous state can be recovered (Section 10.2.2) and leveraged for dynamic risk assessment (Section 10.2.4), and establish the links between SHS and Markov reliability models (Section 10.2.5). As an application example, we demonstrate how the SHS-based approach can be leveraged to model the accumulated revenue in operating a photovoltaic (PV) system under an uncertain environment characterized by failures and repairs in the constituent inverters (Section 10.3).

10.2 Fundamentals of stochastic hybrid systems

This section begins with a brief overview of stochastic hybrid systems (SHS). For a particular class of SHS, leveraging the results in [1,2], we demonstrate how to formulate a family of ODEs, the solutions of which can yield the moments of the discrete and continuous states of the SHS.

10.2.1 Evolution of continuous and discrete states

In the most general sense, an SHS is a combination of a continuous-time, discrete-state stochastic process $Q(t) \in \mathcal{Q}$, coupled with a continuous-time, continuous-state stochastic process $X(t) \in \mathbb{R}^n$. Let $\mathcal{Q}_i^+ \subseteq \mathcal{Q}$ denote the set of all modes that $Q(t)$ can transition to, given that $\Pr\{Q(t) = i\} = 1$; similarly, let $\mathcal{Q}_i^- \subseteq \mathcal{Q}$ denote the set of all modes from which $Q(t)$ can transition to mode i. The evolution of $Q(t)$ and $X(t)$ can be described with the aid of the following functions:

$$\lambda_{ij}(x,t),\ \lambda_{ij} : \mathbb{R}^n \times \mathbb{R}^+ \rightarrow \mathbb{R}^+, \tag{10.1}$$

$$\phi_{ij}(q,x),\ \phi_{ij} : \mathcal{Q} \times \mathbb{R}^n \rightarrow \mathcal{Q} \times \mathbb{R}^n. \tag{10.2}$$

The λ_{ij}'s are the *transition rates* that govern the times when the system switches from mode i to mode j, and the ϕ_{ij}'s are the *transition reset maps* that tell us how the discrete and continuous states change when there is a reset.*

We now provide an intuitive description of how the discrete and continuous state evolve in an SHS. Without loss of generality, as a particular example consider the SHS system in mode i at time t, i.e., $\Pr\{Q(t) = i\} = 1$. In a small time interval

*We point out that the notation adopted for reset maps is slightly cumbersome; in particular, for the $i \rightarrow j$ transition, it follows that $\varphi_{ij}(q,\cdot) = \varphi_{ij}(i,\cdot) = (j,\cdot)$. Nevertheless, we persist with this notation for clarity.

$[t, t+\tau)$ the probability of a transition out of mode i is given by

$$\sum_{j \in \mathcal{Q}_i^+} \lambda_{ij}(X(t), t)\tau + o(\tau), \tag{10.3}$$

and the probability of a particular $i \to j$ transition is given by

$$\lambda_{ij}(X(t), t)\tau + o(\tau). \tag{10.4}$$

If the $i \to j$ transition occurs, the new values of Q and X (i.e., the initial conditions for the post-transition evolution) are defined to be

$$\phi_{ij}(Q((t+\tau)^-),\ X((t+\tau)^-)) = \phi_{ij}(i, X((t+\tau)^-)) = (j, X(t+\tau)), \tag{10.5}$$

where $f(t^-) := \lim_{s \to t} f(s)$. The probability that no transition out of state i occurs in the time interval $[t, t+\tau)$ is given by

$$1 - \tau \sum_{k \in \mathcal{Q}_i^+} \lambda_{ik}(X(t), t). \tag{10.6}$$

Between transitions, $X(t)$ evolves according to

$$\frac{\mathrm{d}}{\mathrm{d}t} X(t) = f(Q(t),\ X(t),\ t), \tag{10.7}$$

where $f : \mathcal{Q} \times \mathbb{R}^n \times \mathbb{R}^+ \to \mathbb{R}^n$. In general, the evolution of the continuous state can be governed by a stochastic differential equation. For details of this more general setting, readers are referred to [1,2].

The SHS model described above affords flexibility and generality to model a variety of stochastic phenomena of interest in power electronic systems. With a particular emphasis on reliability modeling, the elements of the set $\mathcal{Q}$ index different *operational modes*, including the nominal (non-faulted) mode and any modes that arise due to faults (and repairs) in the components comprising the power electronic system. Similarly, $X(t)$ denotes the underlying dynamic states of the power electronic system of interest. The states of $X(t)$ could be derived from physics-based models. In this case, they could represent, for instance, inductor currents and capacitor voltages in the power converter model. Based on the desired modeling resolution, the dynamics in (10.7) could represent an averaged or switching time-scale model of the power converter. Alternately, one could investigate other behavioral models that describe a particular attribute of interest for the power electronic system under study. For example, in Section 10.3, we provide a numerical case study focused on PV system economics. In this setting, $X(t)$ represents the accumulated revenue of the PV system.

10.2.2 Test functions, extended generator, and moment evolution

The evolution of the discrete and continuous states in the general SHS model described in Section 10.2.1 is tightly coupled. In particular, the vector field f that governs $X(t)$ (see (10.7)) is discrete-state dependent. Concurrently, transitions of the discrete state depend on the value of the continuous state, since the transition rates, λ_{ij} in general are functions of the continuous state, $X(t)$ (see (10.1)). This tight interplay challenges the analysis of SHS, and indeed, it is intractable to obtain the distribution of the discrete and continuous states in closed form except in some elementary cases. Therefore, we focus instead on computing the moments of the discrete and continuous states; the general procedure for this is described next.

Consider the SHS model described in Section 10.2.1. We define a *test function*, $\psi(q,x)$, $\psi : \mathcal{Q} \times \mathbb{R}^n \to \mathbb{R}$, a linear operator given by

$$
\begin{aligned}
(L\psi)(q,x) = {} & \frac{\partial}{\partial x}\psi(q,x)\cdot f(q,x,t) \\
& + \sum_{i,j\in\mathcal{Q}} \lambda_{ij}(x,t)(\psi(\phi_{ij}(q,x)) - \psi(q,x)),
\end{aligned}
\tag{10.8}
$$

where $\partial\psi/\partial x \in \mathbb{R}^{1\times n}$ denotes the gradient of $\psi(q,x)$ with respect to x, $\lambda_{ij}(x,t)$ is the transition rate for the $i \to j$ transition, and $\phi_{ij}(q,x)$ denotes the corresponding reset map for the discrete and continuous states. The definition of the test function and the generator above follows from [1,2,8]. The evolution of the expected value of the test function $\mathbb{E}[\psi(Q(t),X(t))]$, is governed by Dynkin's formula, which can be stated in differential form as follows [1,8]:

$$
\frac{d}{dt}\mathbb{E}[\psi(Q(t),X(t))] = \mathbb{E}[(L\psi)(Q(t),X(t))]. \tag{10.9}
$$

Dynkin's formula indicates that the time rate of change of the expected value of a test function evaluated on the stochastic process is given by the expected value of the generator. Given the definition of the generator in (10.8), this makes intuitive sense. The first term in (10.8) captures the total derivative of the test function with respect to time, and the second term captures the impact of incoming and outgoing transitions on the test function [9].

By judicious choice of test functions, (10.9) can be used to obtain ODEs that describe the evolution of relevant conditional moments of interest. From this, the law of total expectation will yield the desired moments of the continuous states. For the SHS model where the discrete state $Q(t)$ takes values in the set $\mathcal{Q}$, we define the following family of test functions:

$$
\psi_i^{(m)}(q,x) := \begin{cases} x^m & \text{if} \quad q = i \\ 0 & \text{if} \quad q \neq i \end{cases}, \forall i \in \mathcal{Q}, \tag{10.10}
$$

where

$$
\begin{aligned}
m &:= (m_1, m_2, \ldots, m_n) \in \mathbb{N}^{1\times n}, \\
x^m &:= x_1^{m_1} x_2^{m_2} \ldots x_n^{m_n}.
\end{aligned}
\tag{10.11}
$$

With this definition in place, note that the mth order conditional moment of the continuous state—conditioned on the discrete state being in mode i-is given by the expected value of the test function. In particular, we have $\forall i \in \mathcal{Q}$:

$$\mu_i^{(m)}(t) := \mathbb{E}[\psi_i^{(m)}(q,x)] = \mathbb{E}[X^m(t)|Q(t) = i]\pi_i(t), \tag{10.12}$$

where $\pi_i(t)$ denotes the occupational probability of mode i, i.e.,

$$\pi_i(t) := \Pr\{Q(t) = i\}. \tag{10.13}$$

10.2.3 Evolution of the dynamic-state moments

Now that various conditional moments are defined, we explain how the law of total expectation is applied to obtain the moments of $X(t)$ from them. We then derive ODEs that govern the evolution of the conditional moments of $X(t)$ [2,8].

Suppose we want to compute $\mathbb{E}[X^m(t)]$, for some $m \in \mathbb{N}^{1\times n}$. Applying the law of total expectation, it follows that this is given by

$$\mathbb{E}[X^m(t)] = \sum_{i\in\mathcal{Q}} \mathbb{E}[X^m(t)|Q(t) = i]\pi_i(t) = \sum_{i\in\mathcal{Q}} \mu_i^{(m)}(t). \tag{10.14}$$

Therefore, at each time t, to obtain $\mathbb{E}[X^m(t)]$, we need to know the conditional moments of $X(t)$, $\mu_i^{(m)}(t)$, $\forall i \in \mathcal{Q}$. Dynkin's formula (10.9) then yields ODEs that govern the evolution of $\mu_i^{(m)}(t)$. In particular, the evolution of $\mu_i^{(m)}(t)$, $\forall i \in \mathcal{Q}$, is given by

$$\dot{\mu}_i^{(m)}(t) = \frac{d}{dt}\mathbb{E}[\psi_i^{(m)}(q,x)] = \mathbb{E}[(L\psi_i^{(m)})(q,x)]. \tag{10.15}$$

Simulating a family of relevant ODEs of the form in (10.15), and applying (10.14) yields the desired moment of interest.

10.2.4 Leveraging continuous-state moments for dynamic risk assessment

While the combined distribution of the discrete and continuous states would completely characterize the SHS, we have already noted how this is intractable to recover. Nonetheless, the moments of the continuous states convey important information about the distribution, and in fact, upper bounds on the probability that the power electronic system dynamic states satisfy certain performance requirements can be obtained with a few lower-order moments.

Suppose performance requirements establish the maximum and minimum values that the continuous state $x(t)$ can take at any time by $\mathcal{R}_x := [x^{\min}, x^{\max}]$. In spite of mode transitions (including those triggered by failures and repairs), we are interested in studying whether the continuous state satisfies the performance requirements, i.e., we are interested in answering whether $x(t) \in \mathcal{R}_x$, $\forall t$. This can be accomplished by establishing the following probabilistic notion of risk, $\rho_x(t)$, which quantifies the probability that the continuous state does not confirm to the performance requirements at time $t > 0$,

$$\rho_x(t) := \Pr\{X(t) \notin \mathcal{R}_x\} = 1 - \Pr\{x^{\min} \le X(t) \le x^{\max}\}. \tag{10.16}$$

The moments of $X(t)$, i.e., $\mathbb{E}[X^m(t)]$, $m \in \mathbb{N}^+$ that can be obtained from (10.14) to (10.15) can be utilized to establish an upper bound on $\rho_x(t)$ using moment inequalities. For example, consider the following Chebyshev-based moment inequality [10], which yields an upper bound on $\rho_x(t)$, that we denote by $\overline{\rho}_x(t)$:

$$\rho_x(t) \leq 1 - \frac{4\left(\left(\mathbb{E}[X(t)] - x^{\min}\right)\left(x^{\max} - \mathbb{E}[X(t)]\right)\right)}{\left(x^{\max} - x^{\min}\right)^2} - \frac{4\sigma_{X(t)}^2}{\left(x^{\max} - x^{\min}\right)^2} =: \overline{\rho}_x(t), \tag{10.17}$$

where $\sigma_X(t)$ is the standard deviation of $X(t)$,

$$\sigma_{X(t)} := \left(\mathbb{E}[X^2(t)] - (\mathbb{E}[X(t)])^2\right)^{1/2}. \tag{10.18}$$

Essentially, (10.17) indicates how upper bounds on the probability that the dynamic states do not meet a priori specified performance specifications can be obtained simply from a few lower-order moments, the evolution of which is recovered from the solution of a nonlinear ODE.

One approach to compute $\rho_x(t)$ would be based on repeated Monte Carlo simulations. In each simulation, the transition rates would determine when mode transitions are triggered. Repeated simulations would yield the distribution of the continuous state $X(t)$, from which (10.16) could be numerically computed. This approach is indeed easy to conceptualize and implement; however, it is computationally burdensome and accuracy is directly related to the number of simulations. On the other hand, the SHS-based alternative is analytical and repeated simulations are not required. Also, the bound in (10.17) is conservative in the sense that the actual probability of violating a performance objective is always lower. More precise estimates of $\rho_x(t)$ can be computed if higher-order moments are known.

10.2.5 Recovering Markov reliability and reward models from SHS

A major appeal of SHS is that a wide variety of stochastic modeling frameworks can be recovered as special cases of the most general SHS formalism. In this section, we demonstrate how Markov reliability models and Markov reward models [11,12] can be recovered as special cases of the most general SHS formalism described in Section 10.2.

10.2.5.1 Continuous-time Markov chains and Markov reliability models

A continuous-time discrete-state stochastic process $Q(t)$ is called a continuous-time Markov chain (CTMC) if it satisfies the Markov property:

$$\begin{aligned}\Pr\{Q(t_r) = i | Q(t_{r-1}) = j_{r-1}, \ldots, Q(t_1) = j_1\} \\ = \Pr\{Q(t_r) = i | Q(t_{r-1}) = j_{r-1}\},\end{aligned} \tag{10.19}$$

for $t_1 < \ldots < t_r$, $\forall i, j_1, \ldots, j_{r-1} \in \mathcal{Q}$, and for $r > 1$ [6]. The chain Q is said to be *homogeneous* if it satisfies

$$\Pr\{Q(t) = i | Q(s) = j\} = \Pr\{Q(t-s) = i | Q(0) = j\}, \forall i, j \in \mathcal{Q}, 0 < s < t. \tag{10.20}$$

With the states of the CTMC, i.e., the entries of the set $\mathcal{Q}$, denoting different operational modes of the system of interest, we recover a Markov reliability model. Transitions of the discrete state are triggered by failures and subsequent repair actions that aim to restore functionality [11]. Continuous-time Markov chains are commonly used for system reliability and availability modeling in many application domains. In addition to power and energy systems [13–18], these include: computer systems [19], communication networks [20], electronic circuits [21,22], and phased-mission systems [23,24].

The problem of interest in Markov reliability models is to determine the distribution of the discrete state, $Q(t)$, at any instant of time $t > 0$. Recall, from (10.13) that the occupational probability of mode i is denoted by $\pi_i(t)$. Let us denote the entries of the column vector of occupational probabilities by $\{\pi_q(t)\}_{q \in \mathcal{Q}}$. The evolution of $\pi(t)$ is governed by the Chapman–Kolmogorov equations [12]:

$$\frac{d}{dt}\pi(t) = \Lambda\pi(t), \tag{10.21}$$

where $\Lambda \in \mathbb{R}^{|\mathcal{Q}| \times |\mathcal{Q}|}$ is the Markov chain generator matrix that is composed of component failure and repair rates. In particular, let

$$\lambda_{ij}(t), \quad \lambda_{ij} : \mathbb{R}^+ \to \mathbb{R}^+, \tag{10.22}$$

denote the transition rate for the $i \to j$ transition. Then the generator matrix, Λ, is constructed as follows:

$$[\Lambda]_{ij} = \begin{cases} \lambda_{ji}(t) & \text{if} \quad i \neq j, j \in \mathcal{Q}_i^- \\ -\sum_{\ell \in \mathcal{Q}_i^+} \lambda_{i\ell}(t) & \text{if} \quad i = j. \end{cases} \tag{10.23}$$

The evolution of the ith occupational probability from (10.21), (10.23) is therefore given by

$$\dot{\pi}_i(t) = \sum_{j \in \mathcal{Q}_i^-} \lambda_{ji}(t)\pi_j(t) - \sum_{k \in \mathcal{Q}_i^+} \lambda_{ik}(t)\pi_i(t). \tag{10.24}$$

Markov reliability and availability models can be readily recovered from the general SHS formulation in Section 10.2 by ignoring the continuous states, $X(t)$, and the reset maps $\phi(\cdot)$. When the continuous states are ignored, the transition rates in (10.1) are now either constants or functions of time, thus recovering the formulation in (10.22). Therefore, using (10.15) we can recover the Chapman–Kolmogorov differential equations that govern the occupational probabilities of the CTMC that underlies the Markov reliability model. To this end, choosing $m = (0, 0, \ldots, 0)$ in

(10.15) recovers the discrete-state occupational probabilities

$$\mu_i^{(0,0,\ldots,0)}(t) = \Pr\{Q(t) = i\} = \pi_i(t). \tag{10.25}$$

Subsequently, the moment ODEs in (10.15) boil down to

$$\dot{\mu}_i^{(0,\ldots,0)}(t) = \sum_{j\in Q_i^-} \lambda_{ji}(t)\mu_j^{(0,\ldots,0)}(t) - \sum_{k\in Q_i^+} \lambda_{ik}(t)\mu_i^{(0,\ldots,0)}(t), \tag{26}$$

which are precisely the Chapman–Kolmogorov differential equations for the occupational probabilities of the CTMC (10.24).

10.2.5.2 Markov reward models

A Markov reward model comprises a Markov chain $Q(t)$ taking values in the set Q (which describes the possible system operational modes) and an *accumulated reward* $X(t)$, which captures some performance measure of interest. The most commonly studied Markov reward models are rate-reward models (see, e.g., [25,26], and the references therein). The accumulated reward in rate-reward models evolves according to

$$\frac{dX(t)}{dt} = f(Q(t)),$$

where $f : Q \to \mathbb{R}$ is the (discrete-state-dependent) reward growth rate. Impulses in the accumulated reward capture one-time effects due to failures or repairs of components in the system. The Markov reward formalism can also be recovered as a special case of the most general SHS formulation; in particular, with the choice $f(q,x,t) = f(q)$ in (10.7), we recover the Markov reward modeling framework.

10.3 Application of SHS to PV system economics

This case study demonstrates how the SHS framework can be applied to model the accumulated revenue in a residential-scale PV system with multiple inverters. The sample PV system is the Gable Home: a net-zero, solar-powered house built for the US Department of Energy's 2009 Solar Decathlon [27]. The PV electrical system consists of forty 225 W mono-crystalline modules. The dc power sourced by the PV modules is converted to utility-compatible ac power by two 5 kW grid-tied inverters.

The state-transition diagram that illustrates the reliability model for the inverters in the system is depicted in Figure 10.1. The CTMC that describes the reliability model takes values in the set $Q = \{0, 1, 2\}$. In operational mode 2, both inverters are functioning, in operational mode 1, a single inverter is functioning, and in operational mode 0, both inverters have failed. The failure rate, repair rate, and common-cause failure rate are denoted by λ, μ, and λ_c, respectively. From the state-transition diagram in Figure 10.1, it follows that the transition rates are

$$\begin{aligned} &\lambda_{21} = 2\lambda, \quad \lambda_{20} = \lambda_c, \\ &\lambda_{10} = \lambda, \quad \lambda_{01} = \lambda_{12} = \mu. \end{aligned} \tag{10.27}$$

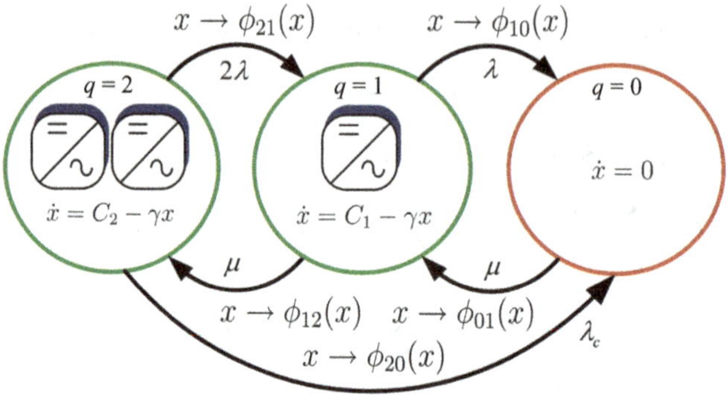

Figure 10.1 State-transition diagram for the PV-system reliability model

The reward of interest is the accumulated revenue of operating the PV system, denoted by $X(t)$. The constant rate at which the accumulated revenue grows in the operational mode is denoted by C_i \$/yr. Additionally, we factor in a degradation rate (that captures natural wear and tear and depreciation) denoted by γ in each operational mode. The dynamical systems that govern the evolution of the accumulated revenue in the three modes are specified by

$$f(q,x) = \begin{cases} C_2 - \gamma x & \text{if} \quad q = 2, \\ C_1 - \gamma x & \text{if} \quad q = 1, \\ 0 & \text{if} \quad q = 0. \end{cases} \tag{10.28}$$

Transitions due to failures are associated with impulses that model one-time expenses in replacing or repairing the inverters. In particular, the impulse change in accumulated revenue as a result of a failure transition from operational mode i to mode j is denoted by C_{ij} \$.[†] The reset maps that describe how the discrete and continuous states are affected by the transitions are given by

$$\begin{aligned} &\phi_{21}(q,x) = (1, x - C_{21}), \quad \phi_{20}(q,x) = (0, x - C_{20}), \\ &\phi_{10}(q,x) = (0, x - C_{10}), \quad \phi_{01}(q,x) = (1, x), \quad \phi_{12}(x) = (2, x). \end{aligned} \tag{10.29}$$

The problem of interest is to determine the moments of the accumulated revenue of the PV system, i.e., $\mathbb{E}[X(t)]$, $\mathbb{E}[X^2(t)]$. We address this problem with the SHS-based framework. To this end, begin by defining test functions for each state of the CTMC:

$$\psi_i^{(m)}(q,x) = \begin{cases} x^m & \text{if} \quad q = i \\ 0 & \text{if} \quad q \neq i \end{cases}, \; i \in \mathcal{Q} = \{0, 1, 2\}. \tag{10.30}$$

[†]The cost parameters could be modeled to be time-dependent to factor inflation or cash flow streams. For instance, following along the model in [13], we could model $C_i(t) = C_i e^{-\delta t}$ and $C_{ij}(t) = C_{ij} e^{-\delta t}$; where δ is the discount rate that represents future costs by a discounted value [13].

From (10.8), the extended generators are given by

$$\left(L\psi_0^{(m)}\right)(q,x) = -\mu\psi_0^{(m)}(q,x) + \lambda\left(\psi_1^{(1)}(q,x) - C_{10}\psi_1^{(0)}(q,x)\right)^m + \lambda_c\left(\psi_2^{(1)}(q,x) - C_{20}\psi_2^{(0)}(q,x)\right)^m, \tag{10.31}$$

$$\left(L\psi_1^{(m)}\right)(q,x) = mc_1(t)\psi_1^{(m-1)}(q,x) - m\gamma\psi_1^{(m)}(q,x) - (\lambda+\mu)\psi_1^{(m)}(q,x) + 2\lambda\left(\psi_2^{(1)}(q,x) - C_{21}\psi_2^{(0)}(q,x)\right)^m + \mu\psi_0^{(m)}(q,x), \tag{10.32}$$

$$\left(L\psi_2^{(m)}\right)(q,x) = mc_2(t)\psi_2^{(m-1)}(q,x) - m\gamma\psi_2^{(m)}(q,x) - (2\lambda+\lambda_c)\psi_2^{(m)}(q,x) + \mu\psi_1^{(m)}(q,x). \tag{10.33}$$

Applying Dynkin's formula in (10.15) to (10.31)–(10.33), we obtain the following set of differential equations for the conditional moments of mth order,

$$\frac{d}{dt}\mu_0^{(m)}(t) = -\mu\mu_0^{(m)}(t) + \lambda\left((-1)^m C_{10}^m\pi_1(t) + \sum_{k=0}^{m-1} m\,k\,\mu_1^{(m-k)}(t)(-1)^k C_{10}^k\right) \tag{10.34}$$

$$\frac{d}{dt}\mu_1^{(m)}(t) = mc_1(t)\mu_1^{(m-1)}(t) - m\gamma\mu_1^{(m)}(t) - (\lambda+\mu)\mu_1^{(m)}(t) \tag{10.35}$$

$$+ 2\lambda\left((-1)^m C_{21}^m\pi_2(t) + \sum_{k=0}^{m-1} m\,k\,\mu_2^{(m-k)}(t)(-1)^k C_{21}^k\right) + \mu\mu_0^m(t),$$

$$\frac{d}{dt}\mu_2^{(m)}(t) = mc_2(t)\mu_2^{(m-1)}(t) - m\gamma\mu_2^{(m)}(t) - (2\lambda+\lambda_c)\mu_2^{(m)}(t) + \mu\mu_1^{(m)}(t), \tag{10.36}$$

where $\pi_0(t)$, $\pi_1(t)$, and $\pi_2(t)$ are the occupational probabilities of the different modes. The m-order moment of the accumulated revenue is given by

$$\mathbb{E}[X^m(t)] = \mu_0^{(m)}(t) + \mu_1^{(m)}(t) + \mu_2^{(m)}(t). \tag{10.37}$$

Notice that substituting $m = 0$ in (34)–(36) recovers the Chapman–Kolmogorov equations: $\dot{\pi}(t) = \Lambda\pi(t)$, where $\pi(t) = [\pi_0(t), \pi_1(t), \pi_2(t)]^{\mathrm{T}}$, and Λ is given by:

$$\Lambda = \begin{bmatrix} -\mu & \lambda & \lambda_c \\ \mu & -(\lambda+\mu) & 2\lambda \\ 0 & \mu & -(2\lambda+\lambda_c) \end{bmatrix}.$$

For illustration, we consider the following simulation parameters. The transition rates are assumed to be $\lambda = 0.1\ \text{yr}^{-1}$, $\lambda_c = 0.001\ \text{yr}^{-1}$, and $\mu = 30\ \text{yr}^{-1}$ [16]. The impulse costs are assumed to be a fraction, ρ, of the upfront inverter installed cost, $C_{\text{inverter}} = \$2850$; with this model we have $C_{21} = C_{10} = \rho C_{\text{inverter}}$, and $C_{20} = 2\rho C_{\text{inverter}}$. We set the nominal value of ρ to be 6%. The nominal degradation rate γ, is set to be 0.7% following [28]. The rate at which revenue is accumulated for the two-inverter system, C_2, is assumed to be 1125\$/yr, and we assume that $C_1 = C_2/2$. This is computed using the National Renewable Energy Laboratory PVWatts Calculator[‡] for the location of Springfield, IL, assuming total system losses of 14%, inverter efficiency of 96%, and a dc-ac size ratio of 1.1 (given the inverter rating of 5 kW).

We demonstrate the impact of the model parameters on the expected accumulated revenue of the PV system. For comparison, we also model the evolution of the accumulated revenue with an alternate investment of $2C_{\text{inverter}}$ that accrues revenue at a discount rate d. Figure 10.2 plots the expected accumulated revenue with the PV system and also depicts the accumulated revenue with the alternate revenue for discount rates of $d = 1\%$, 3%, 5%. The intersection of the curves provides an idea of when the PV-inverter investment is expected to be competitive with the alternate investment, i.e., the expected payback time for the inverter cost. Figure 10.3 plots the expected accumulated revenue with the PV system for

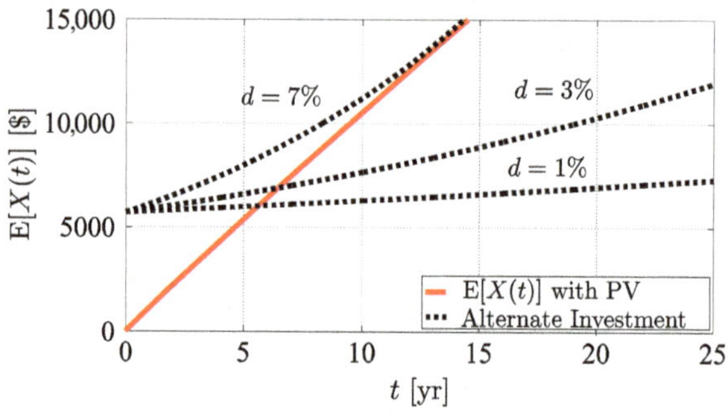

Figure 10.2 Comparing the PV investment with an alternative investment that accrues value at discount rates $d = 1\%, 3\%, 7\%$

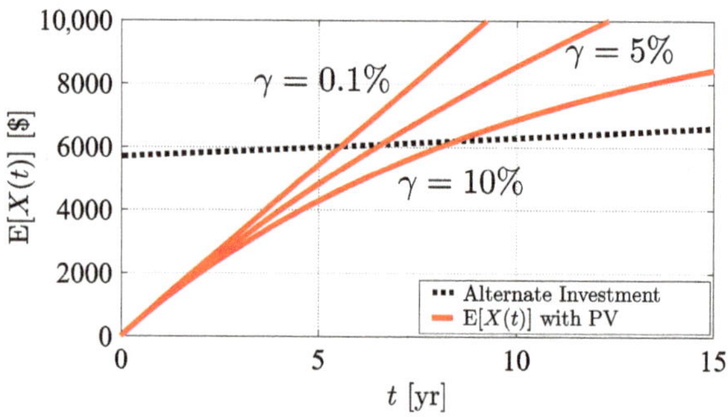

Figure 10.3 Comparing the PV investment for different degradation rates, $\gamma = 0.1\%, 5\%, 10\%$ with the alternative investment

[‡][Online] Available from: http://pvwatts.nrel.gov/

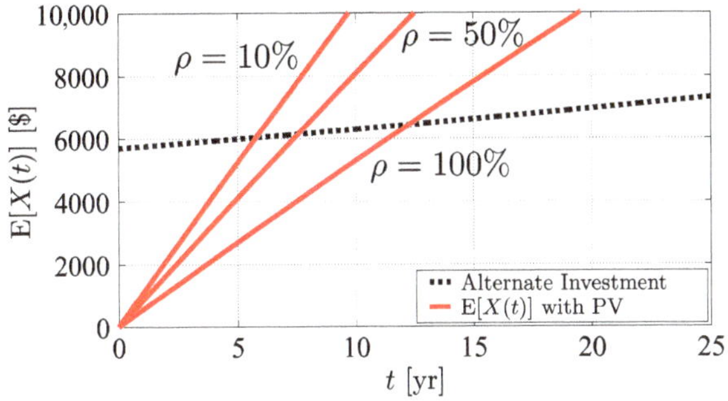

Figure 10.4 Comparing the PV investment with an alternative investment as the fraction of upfront inverter costs expended in repair is varied from $\rho = 10\%$ *to* $\rho = 100\%$

degradation rates of $\gamma = 0.1\%, 5\%, 10\%$, and also depicts the accumulated revenue with the alternate investment for a nominal discount rate of $d = 1\%$. For instance, the expected payback time only increases by around 3 years for an increase in degradation rate by two orders of magnitude. With the proposed repair model, the expected payback time more than doubles as the fraction of upfront inverter costs expended in repairs varies from 10% to 100% (Figure 10.4).

10.4 Concluding remarks

This chapter introduced an SHS-based framework to analyze the performance and reliability of power electronic systems. Expanding on conventional reliability models that only include a discrete-state space, the SHS formulation includes continuous states from physics-based or behavioral models of a power converter. Analytical methods based on Dynkin's formula were leveraged to obtain the moments of the continuous state. We demonstrated how the moments can be utilized in dynamic risk assessment. Applications of the framework to model the expected accumulated revenue in a PV system were presented.

Acknowledgments

The work of the author Sairaj V. Dhople was supported in part by the National Science Foundation (NSF) under Award CyberSEES-1442686. The work of the author Alejandro D. Domínguez-García was supported in part by NSF under CAREER Award ECCS-CAR-0954420.

References

[1] J. P. Hespanha, "A model for stochastic hybrid systems with application to communication networks," *Nonlinear Analysis, Special Issue on Hybrid Systems*, vol. 62, no. 8, pp. 1353–1383, 2005.

[2] J. P. Hespanha, "Modelling and analysis of stochastic hybrid systems," *IEE Proceedings-Control Theory and Applications*, vol. 153, no. 5, pp. 520–535, 2006.

[3] J. Endrenyi, *Reliability Modeling in Electric Power Systems*. New York, NY: Wiley, 1978.

[4] O. Ardakanian, S. Keshav, and C. Rosenberg, "Markovian models for home electricity consumption," in *Proceedings of the 2nd ACM SIGCOMM Workshop on Green Networking*, 2011, pp. 31–36.

[5] S. Koch, J. L. Mathieu, and D. S. Callaway, "Modeling and control of aggregated heterogeneous thermostatically controlled loads for ancillary services," in *Proceedings of the Power Systems Computation Conference*, August 2011.

[6] G. Grimmett and D. Stirzaker, *Probability and Random, Processes*. Oxford: Oxford University Press, 1992.

[7] A. R. Teel, A. Subbaraman, and A. Sferlazza, "Stability analysis for stochastic hybrid systems: A survey," *Automatica*, vol. 50, no. 10, pp. 2435–2456, 2014.

[8] M. H. A. Davis, *Markov Models and Optimization. Boundary Row*, London: Chapman and Hall, 1993.

[9] S. V. Dhople, Y. C. Chen, L. DeVille, and A. D. Dominguez-Garcia, "Analysis of power system dynamics subject to stochastic power injections," *IEEE Transactions on Circuits and Systems I: Regular Papers*, vol. 60, no. 12, pp. 3341–3353, 2013.

[10] K. Steliga and D. Szynal, "On Markov-type inequalities," *International Journal of Pure and Applied Mathematics*, vol. 58, no. 2, pp. 137–152, 2010.

[11] R. A. Sahner, K. S. Trivedi, and A. Puliafito, *Performance and Reliability Analysis of Computer Systems*. Norwell, MA: Kluwer Academic Publishers, 2002.

[12] M. Rausand and A. Hoyland, *System Reliability Theory*. Hoboken, NJ: Wiley Interscience, 2004.

[13] G. J. Anders and A. M. Leite da Silva, "Cost related reliability measures for power system equipment," *IEEE Transactions Cron Power Systems*, vol. 15, no. 2, pp. 654–660, 2000.

[14] P. M. Anderson and S. K. Agarwal, "An improved model for protective-system reliability," *IEEE Transactions on Reliability*, vol. 41, no. 3, pp. 422–426, 1992.

[15] A. M. Bazzi, A. D. Dominguez-Garcia, and P. T. Krein, "Markov reliability modeling for induction motor drives under field-oriented control," *IEEE Transactions on Power Electronics*, vol. 27, no. 2, pp. 534–546, 2012.

[16] S. V. Dhople and A. D. Dominguez-Garcia, "Estimation of photovoltaic system reliability and performance metrics," *IEEE Transactions on Power Systems*, vol. 27, no. 1, pp. 554–563, 2012.

[17] S. V. Dhople, A. Davoudi, A. D. Dominguez-Garciia, and P. L. Chapman, "A unified approach to reliability assessment of multiphase dc-dc converters

in photovoltaic energy conversion systems," *IEEE Transactions on Power Electronics*, vol. 27, no. 2, pp. 739–751, 2012.

[18] H. Behjati and A. Davoudi, "Reliability analysis framework for structural redundancy in power semiconductors," *IEEE Transactions on Industrial Electronics*, vol. 60, no. 10, pp. 4376–4386, 2013.

[19] M. L. Shooman and A. K. Trivedi, "A many-state Markov model for computer software performance parameters," *IEEE Transactions on Reliability*, vol. R-25, no. 2, pp. 66–68, 1976.

[20] D. F. Lazaroiu and E. Staicut, "A Markov model for availability of a packet-switching computer network," *IEEE Transactions on Reliability*, vol. R-32, no. 4, pp. 358–365, 1983.

[21] J. T. Blake, A. L. Reibman, and K. S. Trivedi, "Sensitivity analysis of reliability and performability measures for multiprocessor systems," *Proceedings of the ACM Sigmetrics*, 1988, pp. 177–186.

[22] M. Tainiter, "An application of a Markovian model to the prediction of the reliability of electronic circuits," *IEEE Transactions on Reliability*, vol. R-12, no. 4, pp. 15–25, 1963.

[23] K. Kim and K. S. Park, "Phased-mission system reliability under Markov environment," *IEEE Transactions on Reliability*, vol. 43, no. 2, pp. 301–309, 1994.

[24] D. Wang and K. S. Trivedi, "Reliability analysis of phased-mission system with independent component repairs," *IEEE mlTransactions on Reliability*, vol. 56, no. 3, pp. 540–551, 2007.

[25] G. Horvath, S. Racz, A. Tari, and M. Telek, "Evaluation of reward analysis methods with MRMSolve 2.0," *Proc. International Conference on Quantitative Evaluation of Systems*, pp. 165–174, 2004.

[26] G. Horvath, S. Racz, and M. Telek, "Analysis of second-order Markov reward models," in *Proc. International Conference on Dependable Systems and Networks*, June 2004, pp. 845–854.

[27] S. V. Dhople, J. L. Ehlmann, C. J. Murray, S. T. Cady, and P. L. Chapman, "Engineering systems in the gable home: A passive, net-zero, solar-powered house for the U.S. Department of Energy's 2009 Solar Decathlon," in *Proc. Power and Energy Conference at Illinois (FECI)*, 2010, pp. 58–62.

[28] D. C. Jordan, R. M. Smith, C. R. Osterwald, E. Gelak, and S. R. Kurtz, "Outdoor PV degradation comparison," in *Proc. IEEE Photovoltaic Specialists Conference*, June 2010, pp. 2694–2697.

Chapter 11
Wind turbine power electronic system reliability

Peter Tavner[1]

11.1 Introduction

Modern turbines for converting wind energy to electrical energy have been developed since the 1970s, in step with the development of modern power electronics. This has played a major part in wind turbine (WT) speed control of the drive train and facilitating speed control through the pitch control system. These technologies have enabled a steady increase in wind turbine power output and in its connection to the wider extra high voltage (EHV) grid.

This chapter will outline the major power electronic architectures that have been used in wind turbines and then describe the reliability of those innovations. The terminology used here is given at the end of the chapter.

This is particularly important for the growing number of offshore wind turbines, where early information has suggested that significant power electronic failures caused increased wind turbine downtimes, because of lengthy mobilisations to clear relatively minor faults. This is an important current area of wind power development.

Figure 11.1(a) shows the nacelle of a relatively early wind turbine, in this case, an indirect drive, variable speed machine with variable pitch blades, one of the most common architectures used by the industry, particularly in the size range 1–5 MW. It shows at 4 the gearbox, whilst 5 shows a variable speed Doubly Fed Induction Generator, whose rotor slip-rings are connected to the output of a partially rated power converter, housed in the base of the wind turbine tower and 2 indicates the power electronic pitch drives for individual turbine blades. Figure 11.1(b) shows a more modern, large operational offshore wind turbine and its associated substation transmitting power to shore.

[1]Department of Engineering, Durham University, UK

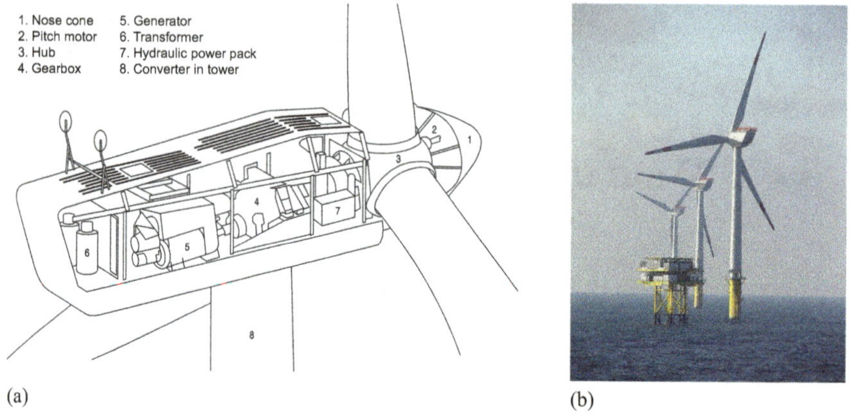

Figure 11.1 Modern wind turbines, including their power electronic converters. (a) Diagram of a 2 MW wind turbine nacelle highlighting important components. (b) View of Alpha Ventus Offshore Wind Farm with 3 off 5 MW WTs in the North Sea off the coast of Germany together with their associated electrical substation.

11.2 Review of main wind turbine power electronic architectures

11.2.1 Onshore and offshore

11.2.1.1 Main power conversion

The main wind turbine conversion architectures are shown in Figure 11.2 where the development can be seen from the earliest fixed-speed electrical generators down to the more complex variable-speed turbines now in wide-scale operation.

Power electronics has played a major role in the development of each architecture.

Type A turbine drives were available in the 1970s. Their main motivation is to use a fully rated power electronics soft starter, Figure 11.3, to limit squirrel cage induction generator starting current, installed in parallel with power correction capacitors to compensate for the large generator magnetising current.

Early wind turbines were stopped using disc brakes but as turbine ratings increased above 200 kW these brakes became unacceptably large and blades began to be pitched to slow the turbine, the disc brake then being used for final turbine stopping and parking.

Type B turbine drives, in the late 1980s, started the exploitation of power electronics in the turbine, for power ratings from 200 to 1000 kW, by switching rotor resistance into the induction generator wound rotor to control its speed, again in parallel with a thyristor soft starter and power correction capacitors.

A typical soft starter arrangement for a Type A or B turbine drive is shown in Figure 11.3.

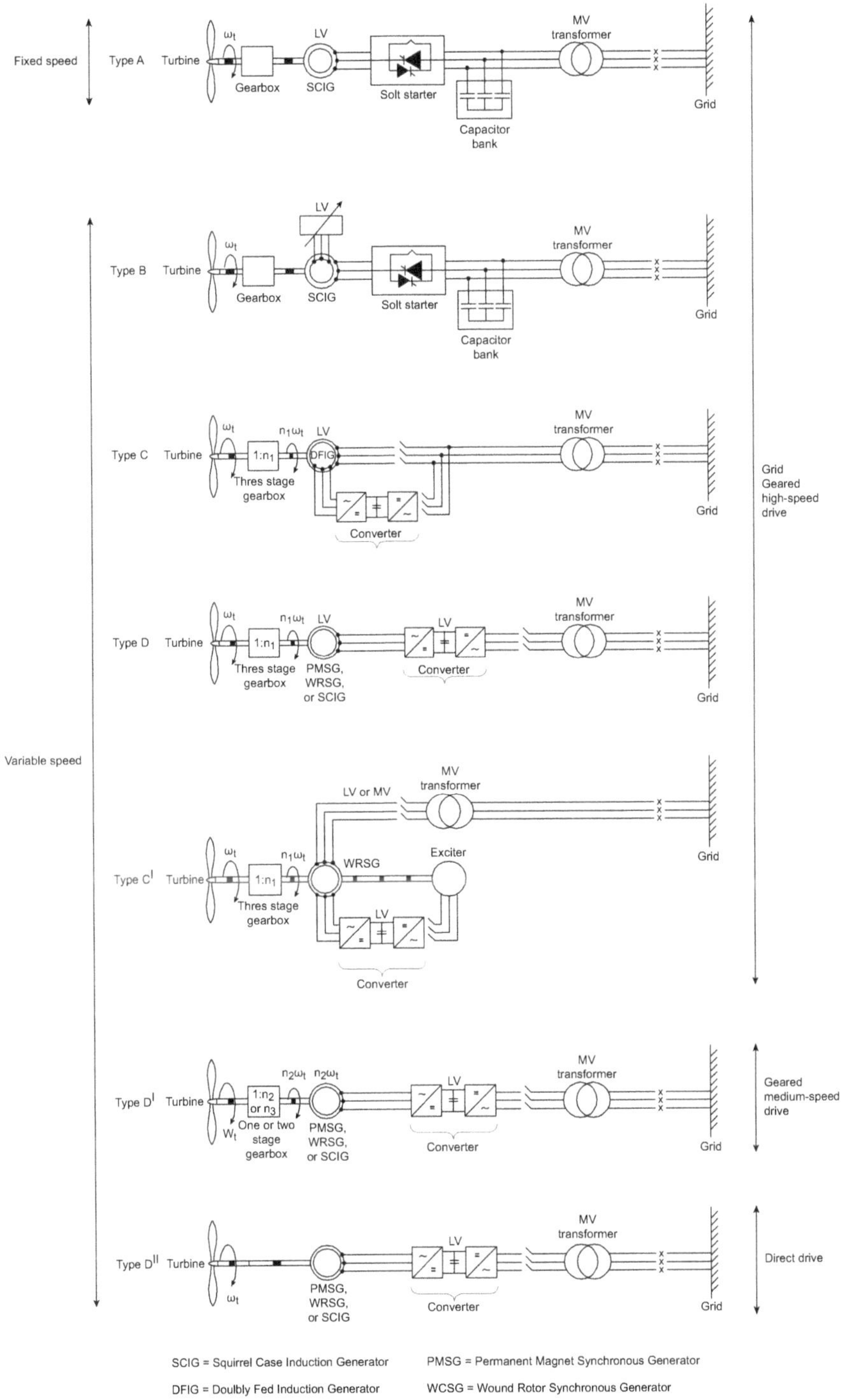

Figure 11.2 Examples of WT conversion architectures showing power electronic converters, categorised from Type A to Type D with specific special arrangements

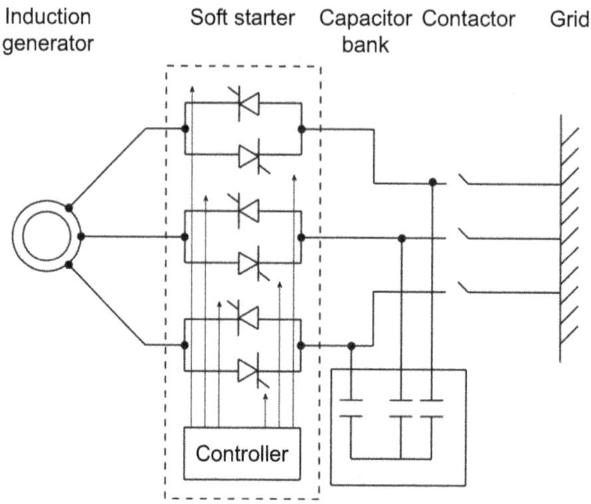

Figure 11.3 Typical soft starter architecture for a fixed speed or low-range variable speed WT with induction generator, using thyristors

Type C turbine drives were developed in the late 1980s, as turbine sizes increased, and concerns arose about wasted energy in generator rotor resistances. Early attempts were made, on turbines of around 1 MW, to recover energy from the rotor winding using partially rated current source converters with diode rectifiers and thyristors but giving only a limited speed range.

By 1990 fully reversible voltage source converters (VSC) with insulated gate bipolar transistors (IGBTs) with diodes were becoming available as partially rated converters (PRC) and full generator control, over a useful aerodynamic speed range, became possible for turbines in the range 1–2 MW. This has proved to be one of the most enduring large wind turbine architectures, although it had serious early limitations, from 1990 to 2000, associated with poor reliability and grid incompatibility. However, extensive development by converter manufacturers, wind turbine designers, and university research has resolved most of these limitations and this is now the most widely fitted wind turbine speed control technology:

- In Europe and USA offered by Alstom, Areva, Enercon, Gamesa, GE, Nordex, Siemens, Vestas, and other manufacturers;
- In China by Dong Fang Dian Ji, Goldwind, and other manufacturers;
- In Japan by Mitsubishi and other manufacturers.

A typical converter arrangement for a Type C wind turbine is shown in Figure 11.4, with the necessary series and bypass contactors. A crow-bar device is fitted on the rotor-side of the converter, to short-circuit the DFIG rotor, in the event of a grid-side disturbance, by absorbing rotor stored energy and preventing generator-side inverter damage during grid faults, at the time one of the most serious reliability problems of this technology.

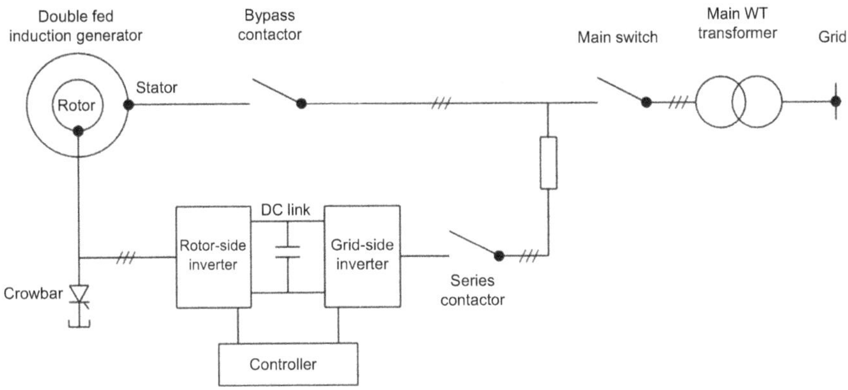

Figure 11.4 Typical Type C DFIG wind turbine architecture, showing a reversible, partially rated, voltage source converter incorporating two inverter bridges

Most DFIG converters for 1.5–3 MW Type C converters operated with a 690 V DC link with chopper DC link voltage control, incorporating two inverter H-bridges with six IGBT diodes. However, larger wind turbines ≥3 MW, use medium voltages, incorporating two 12 IGBT or integrated gate commutated thyristor (IGCT) inverter H-bridges. Such converters operate both above and below the mean generator speed, which means that the converter operates bi-directionally from 0 to 60 Hz, depending on the generator pole number. However, operation at low frequencies can cause H-bridge thermal problems. The anticipated benefit of the geared turbine concept was that it utilised more standardised, high-speed generators with a partially rated converter (PRC), thereby saving cost as shown in Polinder *et al.* [14].

Type D turbine drives were also developed in the late 1980s, initially using the simple induction generator of Types A and B but with a fully rated converter (FRC). This technology had been offered by Siemens and now other turbine manufacturers. 1.5–2.3 MW Type D converters operating using a 690 V DC link with chopper control and two inverter H-bridges, with six IGBT diodes. However, for wind turbines ≥3 MW higher DC link voltages are being used with two 12 IGBT or IGCT inverter H-bridges at up to 1200 V and above. The converter operates uni-directionally from 10 to 60 Hz, depending on the generator pole number. Crow-bar systems on the generator side are unnecessary for the FRC.

Figure 11.5 shows the early growth in the application of these technologies, taken from the authoritative Hansen *et al.* [10].

In the early stages of European Wind Power National Governments, in particular the UK, Germany, and the Netherlands, demanded disclosure by operators of operational experience, in return for financial support, Figure 11.5 was based on this data.

However, since that time wind turbine manufacturers and operators have been reluctant to disclose reliability information in the public domain. So it has not been possible to bring Figure 11.5 up to date. However, there is strong competition between wind turbine manufacturers and their operators to improve turbine

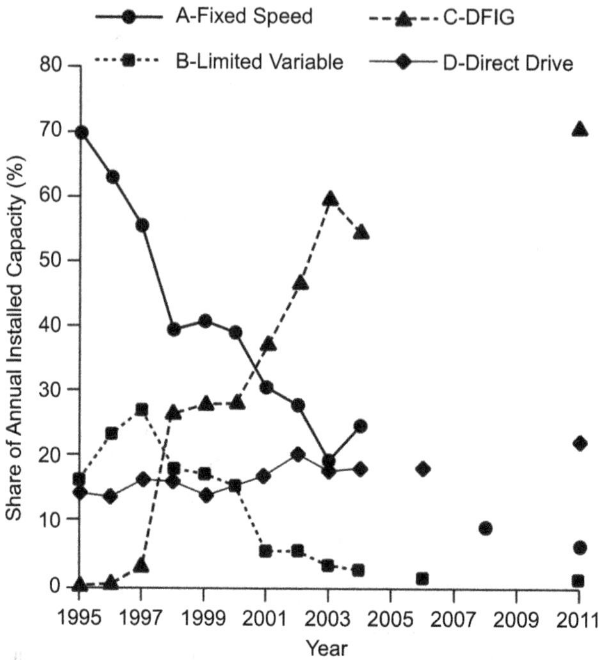

Figure 11.5 Change in the share of WT drive architectures with time. Source: Hansen and Hansen [10]

performance and potential customers are given access by manufacturers to that up-to-date data.

Other modern wind turbine power electronic converter architectures have been under development, shown in Figure 11.2, but have met with less significant applications, as follows:

Type C′, a development of the DFIG but using a synchronous machine and allowing grid connection at a high voltage, offered by Inge Team;

However, by far the most widely applied technology in the current wind industry is Type D, in two forms:

- Types D′ and D″ in Figure 11.2 are now proving by far the most common configurations for present large-scale wind turbines up to 14 MW in output with multiple 12 IGBT or IGCT inverter H-bridges and DC-link voltages are used with two.
- The medium-speed version of the high-speed Type D′ with a reduced gearbox permitting the use of a permanent magnet or wound rotor synchronous generator or an induction generator utilising a fully rated converter (FRC). This architecture is currently offered by Areva, Siemens, Gamesa, Vestas, and GE. Interestingly a high-speed version of this architecture was first offered in 1988 by Enercon as the E33, the brain-child of Alois Wobben, a pioneer of power electronics in the wind industry;

- The low-speed Type D″ version of the high-speed Type D′ allowing the use of a permanent magnet or wound rotor synchronous generator with a fully rated converter (FRC). This type of turbine has become increasingly popular, as shown in Figure 11.5, and there is considerable interest in its relative efficiency, economy, and reliability, Polinder *et al.* [14]. An anticipated benefit of the direct drive concept is that by avoiding the use of the gearbox it should prove more reliable but there are other potential benefits, for example, lower losses in low wind. Research by Spinato *et al.* [20] has shown that the aggregate failure rates of generators and converters in direct-drive wind turbines are generally greater than the aggregate failure rate of gearboxes, generators, and converters in geared wind turbines. Therefore, the price paid by direct drive wind turbines for the reduction of failure rate by the elimination of the gearbox is a substantial increase in the failure rate of electrical-related subassemblies, particularly in the converter. Carroll *et al.* [2] have used a more extensive and up-to-date survey to explore this hypothesis. The wound rotor synchronous generator Type D″ architecture is offered by Enercon, whilst the permanent magnet generator Type D″ architecture is offered by Siemens, Alstom, and Goldwind.

Wind turbine power converter technologies are in continuous development, in terms of power output, cooling, control, and reliability. Power converters are arranged in parallel stacks to provide redundancy in the event of individual converter module failures.

It is likely that as wind turbine power outputs rise and power electronic costs fall, power electronics will play an even greater part in wind energy. This will not only be in the wind turbine itself but also in the transmission systems from the larger wind farms, particularly those installed offshore, perhaps more than 100 km from the coast.

11.2.1.2 Pitch systems

Wind turbine pitch systems were developed in the 1980s, initially to allow the blades to brake the turbine and later developed to control wind turbine power output. Early systems used a single hydraulic ram to operate a yoke and rotate all the blades together in their pitch bearings. This was developed further to operate an individual ram for each blade, with a backup hydraulic accumulator, to allow a turbine to stop even with a faulty ram or loss of the hydraulic power unit. Electrical pitch systems evolved in the 1990s, based upon the advantages of improved actuation, controllability, and battery lightness, to support emergency operations in the event of a power failure.

Figure 11.6 shows a typical electrical actuator for one blade of a three-bladed wind turbine, using a series-wound DC motor-driven gearbox to rotate the blade in the pitch bearing, the DC motor being powered by a 2-quadrant IGBT chopper. The converter is supplied with DC from a full-wave rectifier, which also charges the backup battery, allowing the blades to be pitched to brake the turbine even under failed power conditions, in the event of wind turbine mains failure.

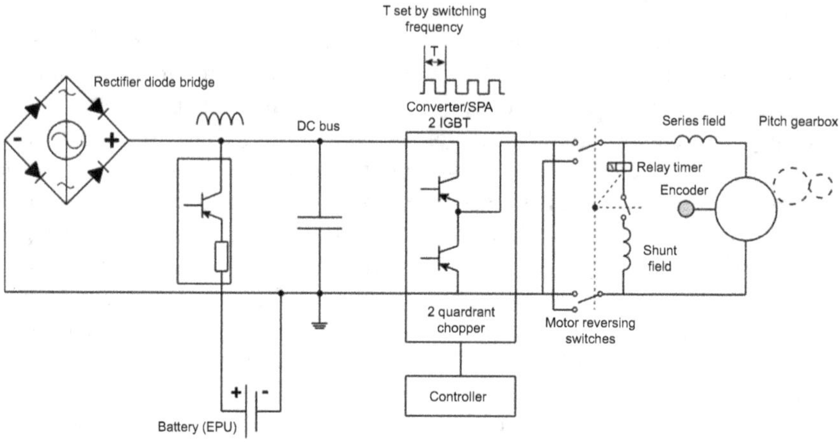

Figure 11.6 Typical electrical blade pitch actuator system, showing a 2-quadrant IGBT chopper converter

More sophisticated methods of pitch control, including AC motor-driven, have been developed as wind turbine power outputs, blade lengths, and weights have increased, which require more powerful pitch control methods.

11.3 Public domain knowledge of power electronic converter reliabilities

11.3.1 Architecture reliability

The following describes a systematic review of literature and data available on wind turbine converter reliability, prior to carrying out a brief meta-analysis of these data sources.

There has been some criticism of wind turbine reliability analysis on the basis that the quantity of data available in the public domain may not render statistically significant results; this is certainly a concern. However, in the absence of large volumes of operator reliability data, due to operator and OEM IP limitations, the author has used the data sources available and the best methods currently in use. Therefore there is a corollary that the results presented here may need to be reviewed when OEMs and operators release sufficient data to meet statistical significance criteria.

Figure 11.7 shows failure rates and downtimes of the main subassemblies of onshore wind turbines in Europe, taken from four different public domain surveys, noting their size, which is large. The power electronics are not specifically identified but are subsumed into the Electric System & Electric Control, with the highest failure rates. However, their downtime is not necessarily high, and this characterises power electronics reliability, a high failure rate but easy to repair, as borne out by Spinato *et al.* [20] and Faulstich *et al.* [7].

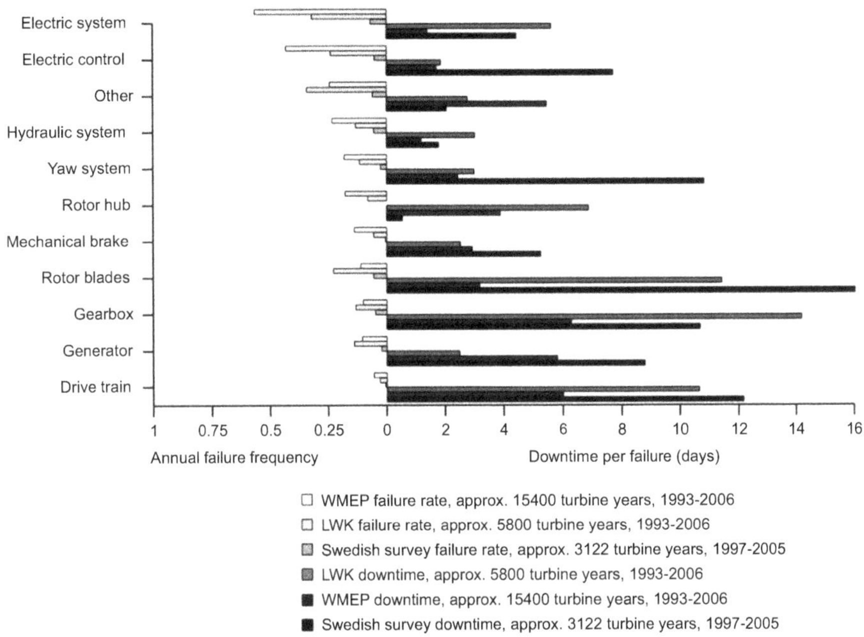

Figure 11.7 Typical failure rates and downtimes for onshore wind turbines in Europe. Sources: Windstats [24], Hahn et al. [9], Ribrant et al. [18]

Before moving to consider individual wind turbine power electronic converter failure rates, it is necessary to pay particular attention to the failure rates of different turbine architectures.

Public domain LWK wind turbine failure data [12] allows turbine models to be grouped according to size and architecture.

Figure 11.8 from [20] summarises the failure rates over 11 years for 12 wind turbine models in the LWK population, concentrating on drive train subassemblies, segregated by turbine conversion architecture and control configuration. The figure shows the relationship between failure rates of blades, pitch mechanism, and electronics as turbine architecture and control configurations change.

With fixed-speed, stall-regulated turbines a significant number of failures are concentrated in the blades and gearbox, as would be expected from such a turbine where transient torques, due to turbulence, are un-damped by variable speed control. With the introduction of variable speed, pitch-regulated machines for smaller wind turbines, Figure 11.8(a), the pitch mechanism now appears as a failure mode, as expected. However, the pitch mechanism reduces blade and generator failure rates but strangely not gearbox failure rates. This is confirmed for larger wind turbines, Figure 11.8(b), where blade, generator, and gearbox failure rates reduce.

However, the Enercon E40, a direct drive wind turbine, is an exception as there are no gearbox failures but generator failures were high. The reduction in blade failures is more noticeable with the larger Enercon E66 direct-drive wind turbine, Figure 11.8(c).

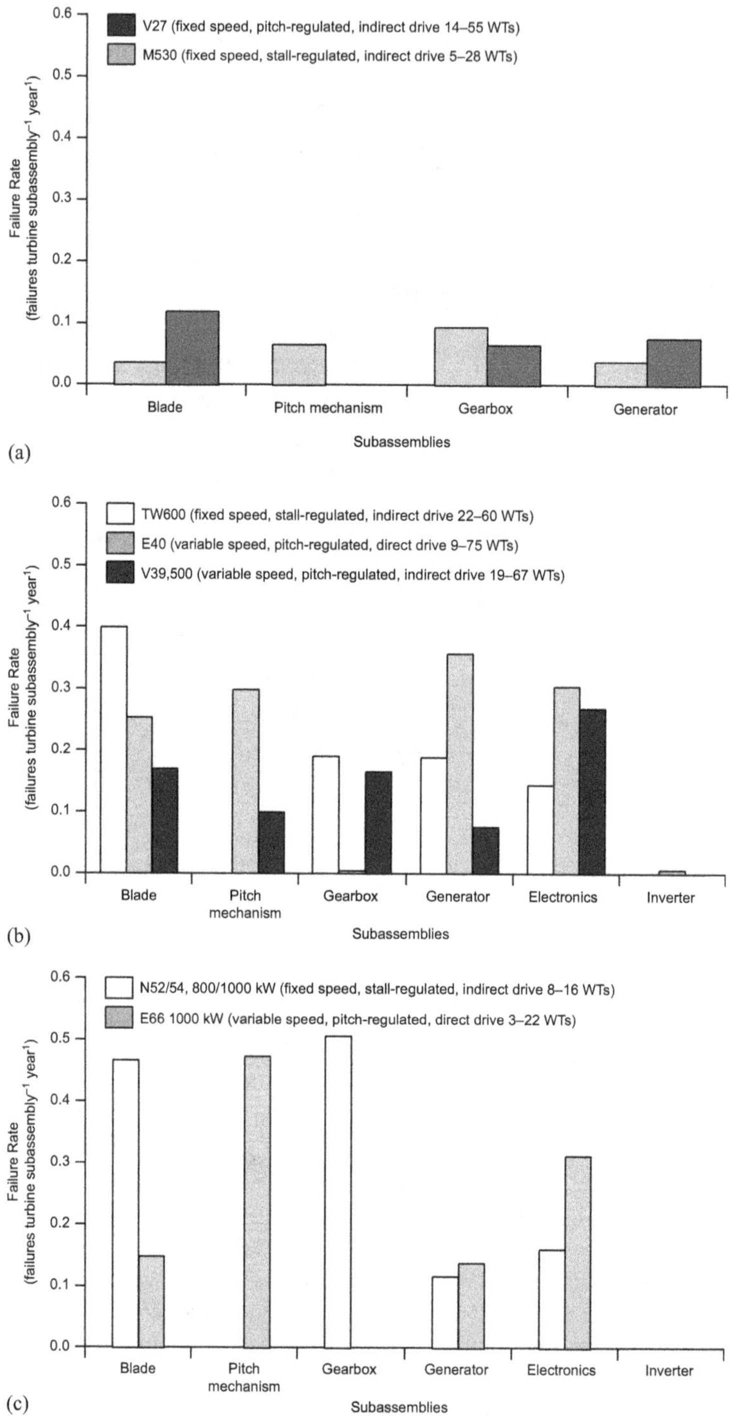

Figure 11.8 LWK project, analysis of failure intensity distributions of three different groups of early European wind turbine designs, upper (a), middle (b), and lower (c). This focuses on blades, pitch, gearbox, generator, electronics, and inverters. Stall-regulated turbines to the left, variable speed, pitch-regulated turbines to the right. Source: [20]

An even more noticeable conclusion from Figure 11.8 is that the increased adoption of power electronics converters, to achieve variable speed operation, resulted in increased failure rates in that assembly.

In other words, the technological advance of variable speed and pitch control, whilst conferring greater energy extraction and noise reduction benefits, introduces new failure modes, not least in the power electronic converters.

The industry addresses this issue by providing redundancy in the power electronics control system, permitting turbines to continue in operation when power electronic components are down, but resulting in a high level of power electronic module change-out over time.

The following section considers some of the monitoring information coming from converters.

11.3.2 SCADA data

Signal condition and data acquisition (SCADA) has a significant role to play in monitoring wind turbine converters as a significant proportion of wind turbine SCADA signals and alarms, logged every 10 min, are coming from the power electronics.

It could be said that there is more SCADA data coming from wind turbine power electronics than can be dealt with satisfactorily by OEMs or operators, and yet this data is almost certainly the key to improved power electronics reliability.

SCADA data has been used to predict wind turbine converter assembly failures and to investigate alarm showers from wind turbine controller alarm indication, Qiu *et al.* [15]. This work adopted a physics of failure approach using a normalised cumulative percentage of selected grid, turbine, and converter alarms plotted in Figure 11.9 against calendar time, during an extended period of operation for two specific variable speed wind turbines of ~2 MW, chosen at random from the same wind farm. These wind turbines were of Type C and had a converter architecture similar to Figure 11.4. The SCADA alarms monitored include those from this converter, as shown in Figure 11.9, and in particular monitored the grid-side and generator-side inverters and their IGBTs.

Figure 11.9 shows:

- The impact of two grid voltage dip incidents on the two wind turbine alarm patterns, on days 39,200 and 39,500. The same patterns were observed on all the other wind turbines in the wind farm on the same days;
- Serious grid voltage dips of >75% caused more than ten converter or inverter alarms during the period investigated;
- Converter general alarms strongly correlated with grid voltage dip alarms, indicating this as a converter failure root cause;
- The steps observed in the normalised cumulative alarm percentage indicate alarm triggers with long cumulative duration. The numerous alarms in these steps were accompanied by inverter IGBT alarms, giving advance warning of converter bridge failures.

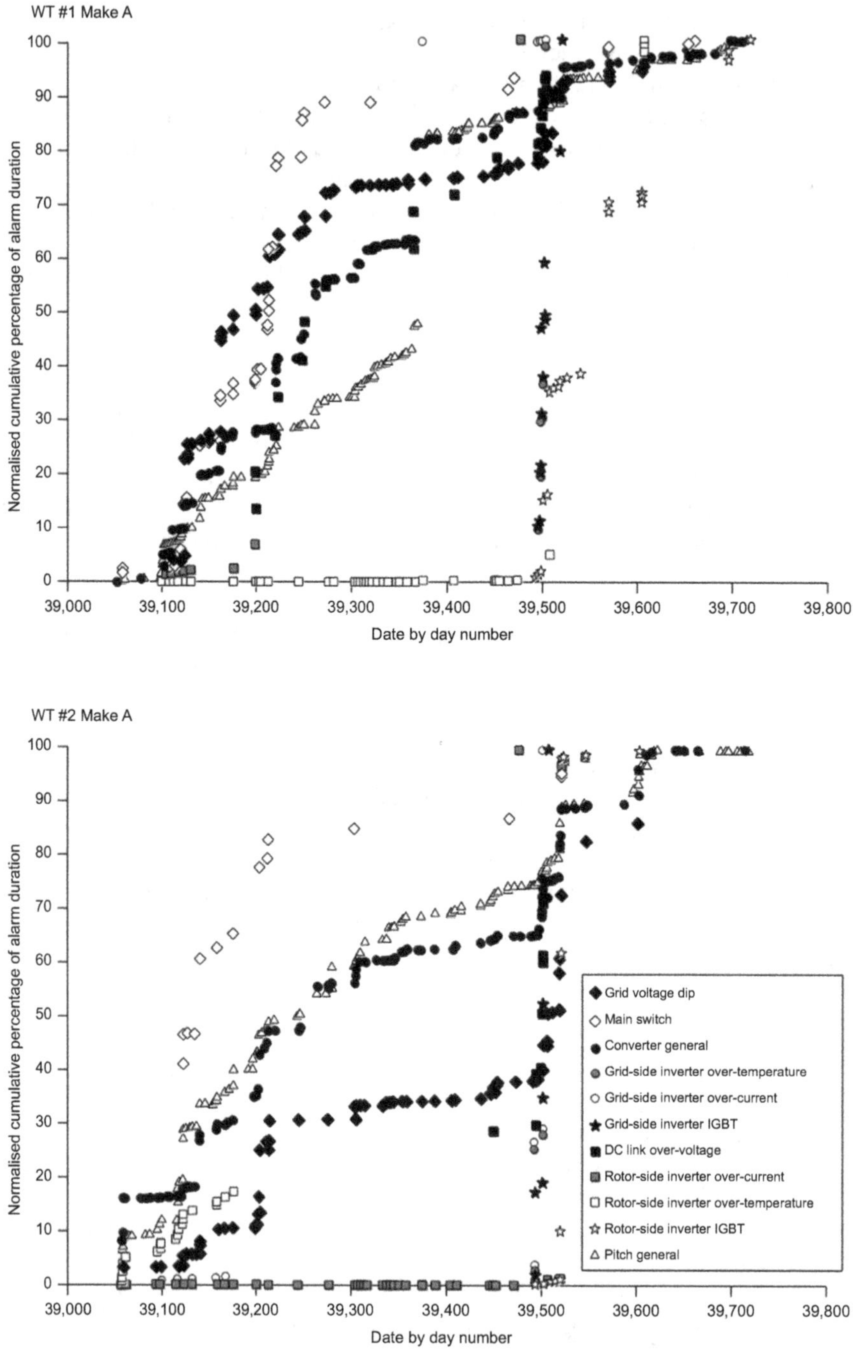

Figure 11.9 Normalised cumulative alarm percentage duration versus calendar time, for two WTs in the same wind farm. Source: Qiu et al. [15]

In total 15–20 alarm triggers were observed in each wind turbine for both these incidents. Therefore, for a wind farm with 30–35 wind turbines, 450–700 alarms could be triggered simultaneously by such incidents. With the probable repetition of some alarms, this suggests a possible alarm rate >1000 per 10 min, suggesting a need to optimise wind turbine alarms, particularly those from the power electronic converters.

In this respect, further data has been collected by Qiu *et al.* [15] on the significance of alarms on different wind turbine architectures. Table 11.1 is a summary of alarms from wind turbines in seven wind farms operating Type B and C turbines in Europe and America and its results are startling.

Firstly, the average number of wind farm SCADA alarms was very significant and could not be handled systematically by any wind farm operator.

Secondly, the average alarm rate per turbine per 10 min for the 1.67 MW Type C turbine was greater than that for the 1.0 MW Type B turbine, as would be expected for a more complex variable speed technology.

Thirdly the most startling fact is the very high peak alarm rates at all wind farms, almost certainly the result of alarm showers arising from grid voltage disturbances, as shown in Figure 11.9, which itself was taken from a wind farm operating the same Type C turbines. The volume of these alarm showers is largely driven by the main converter and pitch control alarms.

Qiu *et al.* [15] show that SCADA alarms do give warnings, as intended, and can identify root causes. Further work by Chen *et al.* [3,4] on pitch control systems has shown that alarms alone cannot give prognosis but that SCADA signal analysis arising with these alarms can give significant prognosis time.

Table 11.1 SCADA alarms over 2 years for seven wind farms deploying Type B and C WTs

Alarm key performance indicators		**Europe, geared drive, variable speed, 1.6 MW WTs, Type C**						**USA, geared drive, fixed 2-speed, 1. 0 MW WT, Type B**
Total WT numbers		**Wind Farm 1**	**Wind Farm 2**	**Wind Farm 3**	**Wind Farm 4**	**Wind Farm 5**	**Wind Farm 6**	**Wind Farm 7**
		13	**15**	**31**	**30**	**30**	**34**	**153**
Survey turbine-years					306			308
Per 10-min average alarm rates	Per Wind Farm	4	8	11	10	10	21	10
	Per WT	0.34	0.50	0.37	0.35	0.32	0.61	0.07
Per 10 min maximum alarm rates	Per Wind Farm	391	1143	636	1570	439	541	289
	Per WT	30.1	76.2	20.5	52.3	14.6	15.9	1.9

11.3.3 Converter reliability

A power electronic converter is a complex assembly with a large number of components, typical public domain industrial converter reliability experience is given in Table 11.2.

It can be difficult for operators to record converter assembly failures accurately as they may be unable to assign a turbine failure unequivocally to the converter because it is more complex compared to a generator or gearbox, where failures are usually straightforward. The electronics and converter assembly failure rates in Figure 11.8 subsume the converter. This means that caution must be exercised when considering recorded converter failures.

Converter failures from the public domain have been aggregated in Table 11.3, see columns 3 and 4 for data from the LWK survey [11], and the data plotted in time for specific architecture turbines.

Figure 11.10, Spinato *et al.* [20], shows the reliability results for three LWK converters, exhibiting the early part of the bath-tub curve. Figure 11.11 gives a full bath-tub curve showing the full range of early failures, intrinsic failures, and deterioration.

For one case from Figure 11.10, TW 1500 converter, failure intensities are falling, that is a power law process (PLP) with exponent $\beta<1$, reflecting reliability improvement. However in the case of the Enercon E66 and E40 converters, the failure intensities are higher, improve with time, and become nearly flat with β close to 1.

Industrial converter failure rate, λ, data, Table 11.2, ranges between 0.045 and 0.200 failure/unit/yr. The lower limit arises from a specific analysis of relatively small converters, Spinato [19], but such a low value of failure rate cannot be applicable to wind turbine converters of such ratings, therefore the upper limit, 0.2 failures/unit/yr, is shown as the industrial level in Figure 11.10.

More recent work has tracked the distribution of wind turbine failure rates due to the converter, as shown in Table 11.3, where failures are compared between different surveys. These show converter failure rates ranging from 0.106 to 2.630 failures/unit/yr, those shown in Figure 11.10 are within this range but more than ten times that in Table 11.2.

It is important to point out here that the figures in Table 11.3 represent wind turbine stoppages ascribed by an operator due to converter faults, arising from many converter alarms, as shown in Section 11.3.2. They do not define the converter fault location or the associated failure rates. However, these could be estimated based on a knowledge of converter assembly reliabilities and this has been done in the lower rows of Table 11.3 where inverter bridge and DC-link failures dominate.

Table 11.2 Reliability of converters from industrial experience

Assembly	**Failure rate, λ, failures/assembly/*r***	**MTBF, h**	**Source**
Converter	0.0450–0.2000	43,800–195,000	Spinato [19]

Table 11.3 Distribution of converter assembly failures from various surveys of onshore WTs

			From WMEP data [8]		From LWK data [12]		From Carroll *et al.* [2]		From ReliaWind data, [17] and Wilkinson *et al.* [23]	
			1998–2000	1989–2006	1993–2006		2005–2010		2007–2011	
Survey turbine-years			**209**	**1028**	**5719**	**679**	**9110**	**1200**	**366**	**1**
Survey WT technology			Large 0.8–2 MW WTs	Large and small 0.3–2 MW WTs	Large and small 0.2–2 MW WTs	DFIG and PRC 1.5 MW 0.6–0.8 MVA converter (Tacke 1.5 s) and WRSG and FRC 0.5 and 1.5 MW 0.6-2 MVA converter (E40 and E66)	DFIG and PRC 1.5–2.5 MW WT 0.6–0.8 MVA converter	PMG and FRC 1.5–2.5 MW WT 2–3 MVA converter	Predicted data from a DFIG and PRC2 MW stereotype WT 0.7 MVA converter	
Failure rate, λ, failures/unit/yr	Whole WT		5.23	3.60	1.92	2.60	–	–	Not disclosed for confidentiality reasons	23.37
	Converter total		1.000	0.450	0.220	0.320	0.106	0.593	Not disclosed for confidentiality reasons	2.630
%	Converter as % of WT		19.1	12.4	11.6	12.2	–	–	11.6	11.3
Estimated failure rate, λ, failures/unit/yr	Estimated location of tde faults	Converter control unit	0.070	0.031	0.016	0.022	0.007	0.042	–	0.184
		Series contactor	0.090	0.040	0.020	0.028	0.010	0.053	–	0.237
		Grid-side filter	0.030	0.013	0.007	0.009	0.003	0.018	–	0.079
		Grid-side inverter	0.189	0.085	0.042	0.060	0.020	0.113	–	0.500

(Continues)

Table 11.3 (Continued)

		From WMEP data [8]		From LWK data [12]		From Carroll *et al.* [2]		From ReliaWind data, [17] and Wilkinson *et al.* [23]	
		1998–2000	**1989–2006**	**1993–2006**		**2005–2010**		**2007–2011**	
Survey turbine-years		**209**	**1028**	**5719**	**679**	**9110**	**1200**	**366**	**1**
	Pre-charge circuit	0.060	0.027	0.013	0.019	0.006	0.036	–	0.158
	DC-link capacitor	0.110	0.049	0.024	0.035	0.012	0.065	–	0.289
	Chopper circuit	0.060	0.027	0.013	0.019	0.006	0.071	–	0.158
	Generator-side inverter	0.189	0.085	0.042	0.060	0.020	0.113	–	0.500
	Crow-bar circuit	0.060	0.027	0.013	0.019	0.006		–	0.158
	Generator-side filter	0.030	0.013	0.007	0.009	0.003	0.018	–	0.079
	Bypass contactor	0.090	0.040	0.020	0.028	0.010	0.053	–	0.237
	Auxiliaries	0.025	0.011	0.006	0.008	0.003	0.015	–	0.066
		Measured and estimated failure rates						Measured failure rates from a survey and predicted failure rates from an FMEA	

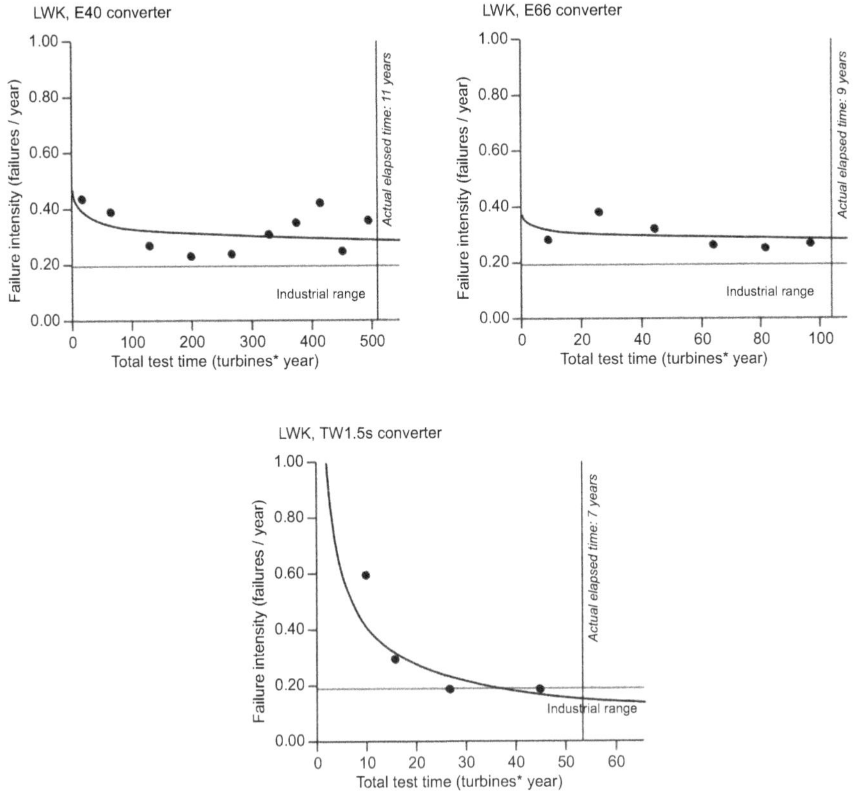

Figure 11.10 Failure intensity, λ(t), variation of converter subassemblies, in LWK population, using Power Law Process model. Source: Spinato et al. [20]

An alternative converter reliability survey from Carroll *et al.* [2], also shown in Table 11.3, studied the reliability differences of 2222 wind turbines over the first 5 years of their operation, separating PRCs in doubly fed induction generator (DFIG) drive trains (Type C) from fully rated converters (FRC) in permanent magnet generator (PMG) drive trains (Type D″).

The DFIG configuration had a sample size building to 1822 turbines over 5 years, whilst the PMG configuration had a sample size building to 400 turbines over three years, the FRC and PRC converters being constructed by the same converter manufacturer.

The results of Table 11.3 are summarised graphically in Figure 11.12.

The surveys show clear reliability differences between them, indicating failure rates increase with rating and with the way the converter operates.

Despite the limitations of the data, Figures 11.7, 11.8, 11.10, and 11.12, and Table 11.3 give a clear and consistent picture of wind turbine power electronic

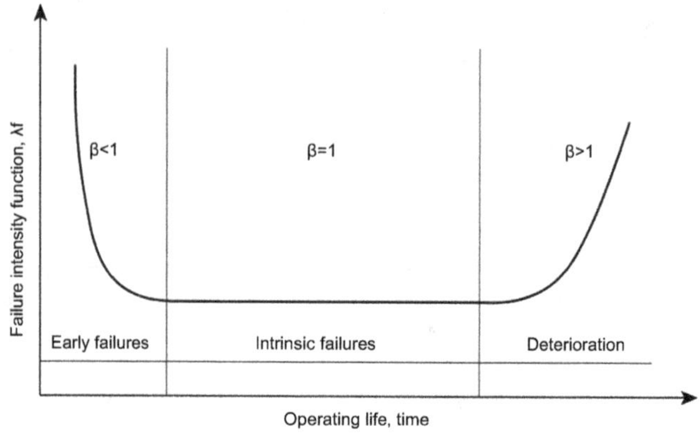

Figure 11.11 Bath-tub curve, showing the variation in failure intensity, λ(t), with fife for a repairable assembly, such as a power electronic converter, where the value of β defines the differences between early, intrinsic, and deterioration failures

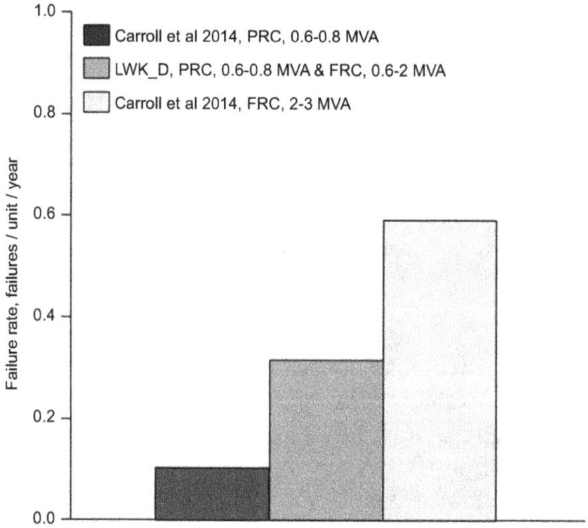

Figure 11.12 Summary of Table 11.3 showing PRC and FRC converter failure rates against industrial experience

converter failures across a variety of public domain surveys plus an indication of where faults arise.

The result from the ReliaWind survey [17] is a special case, as it was a converter failure rate prediction from an FMEA performed on a 2 MW, DFIG, PRC, Type C stereotype wind turbine, giving a rather high result as it considered failures leading to downtime of $\geq$1 h, whereas the other surveys only considered faults causing downtimes $\geq$24 h.

11.4 Reliability FMEA for each assembly and comparative prospective reliabilities

11.4.1 Introduction

This Section proposes a method for considering power electronic system reliabilities during design and development in an organised way, to guide the reader towards those areas where improvements can be made.

True measured failure rates of power electronics in the wind industry are currently considered to be intellectual property and are not available in the public domain.

However, the systematic review and meta-analysis of Table 11.3 has shown a range of valuable reliability results from public surveys and this Section will show how by drilling down into that information one can assemble a reasonable picture of how different power electronic architectures survive in service. This work could become the basis of future, more reliable power converters.

The methodology uses a failure modes and effects (FMEA) approach, based on Arabian *et al.* [1], to break down the structure of the assembly to be considered in a reliability block diagram (RBD). This was developed by Delorm [5] for analysing tidal stream devices, containing power electronic converters, and was demonstrated in Delorm *et al.* [6].

The FMEA structure will be populated with appropriate surrogate failure rate data, obtained from various sources, for example, WMEP [9], IEEE [11], LWK [12], Carroll *et al.* [2], and Windstats [24], and can be adjusted for the environmental conditions in which the assembly and components are found, in accordance with MIL-HDBK [13].

The object of the process is to demonstrate the relative prospective reliabilities of a variety of power electronic architectures, from this information one can select more reliable structures and improve those components and subassemblies which are predicted to be unreliable. The assemblies to be considered in this chapter are summarised below, and the nomenclature used in the tables is taken from VGB (2007) [22].

11.4.2 Assemblies

11.4.2.1 Soft start converters

The approach described has first been used in Table 11.4 to estimate the failure rate of a soft starter and power factor correction module.

11.4.2.2 Partially rated converters

The approach has then been used in Table 11.5 to estimate the failure rate of a PRC with one single channel.

Table 11.4 Soft starter predicted failure rates

Reliability characteristics for soft starter and power factor correction capacitors			**Failure rates**				
			Soft Starter with six thyristors and power factor correction capacitors, assembly failure rate estimate based on surrogate data				
Assemblies	Subassemblies	VGB Code	Subassembly failure rate source	Subassembly failure rate estimate λi_FREcon	Subassembly Qt	Assembly failure rate source	Assembly failure rate estimate λi_FREcon
Series contactor	Series contactor	MKC10 QA001	Carroll *et al.* [2]	0.0100	1	Carroll *et al.* [2]	0.0100
Soft starter	Thyristors	MKC10 BG QBA11	IEEE Gold Book	0.0016	6	IEEE Gold Book	0.0094
Power factor correction capacitors	Power factor correction capacitors	MKC10 BUA10	Carroll *et al.* [2]	0.0120	3	Carroll *et al.* [2]	0.0360
Control unit	Control unit	MKC10 KF001	Carroll *et al.* [2]	0.0074	1	Carroll *et al.* [2]	0.0074
Assembly total failure rate, λ_{tot} (failures/unit/yr)							0.0728
Reliability survivor function, $R_{(1yr)}$, %							93.0%

Table 11.5 Partially rated converter predicted failure rate

Reliability characteristics for converter and circuit breaker			**Failure rates**		
			LV partially rated converter (PRC) with six IGBT/diodes per inverter in one single channel, assembly failure rate estimate based on surrogate data		
Assemblies	Subassemblies	VGB code	Subassembly *Qt*	Assembly failure rate source	Assembly failure rate estimate *λi_FREcon*
Circuit breaker	Circuit breaker	AAG10	1	IEEE Gold Book	0.018
Series contactor	Series contactor	MKC10 QA001	1	Carroll *et al.* [2]	0.010
Grid-side filter	d*V*/d*t* filter	MKY10 BFA20	1	Carroll *et al.* [2]	0.003
Grid-side inverter	Inverter firing circuits Inverter IGBTs Inverter switching diodes	MKY10 BF QBA10 MKY10 BF QBA11 MKY10 BF QBA12	1 6 6	Carroll *et al.* [2]	0.020
DC link	DC-link capacitor Braking chopper resistor Chopper firing circuit DC-link switching IGBT DC-link switching diode	MKY10 BUA10 MKY10 BUA20 MKY10 BU QBA10 MKY10 BU QBA11 MKY10 BU QBA12	1 1 1 1 1	Carroll *et al.* [2] Carroll *et al.* [2]	0.012 0.006
Generator-side Inverter	Inverter firing circuits Inverter IGBTs Inverter switching diodes Crow-bar circuit	MKY10 BG QBA20 MKY10 BG QBA21 MKY10 BG QBA22 MKY10 BG QBA23	1 6 6 1	Carroll *et al.* [2] Carroll *et al.* [2]	0.020 0.006
Generator-side filter	dV/dt filter	MKY10 BGA20	1	Carroll *et al.* [2]	0.003
Series contactor	Bypass contactor	MKC10 QA002	1	Carroll *et al.* [2]	0.010
Cooling unit	Cooling unit	MKY10 EC001	1	IEEE Gold Book	0.006
Control unit	Control unit	MKY10 KF001	1	Carroll *et al.* [2]	0.007
Assembly total failure rate, λ_{tot} (failures/unit/yr)				This total failure rate is consistent with Spinato *et al.* [20] and Carroll *et al.* [2]	0.121
Reliability survivor function, $R_{(1yr)}$, %					88.6%

11.4.2.3 Fully rated converters, induction, or PM generators

The approach has then been used in Table 11.6 to estimate the failure rate of an FRC with one single channel.

Table 11.6 Fully rated converter predicted failure rate

Reliability characteristics for converter and circuit breaker			**Failure rates**		
			MV fully rated converter (FRC) with 12 IGCT/diodes per inverter in one single channel, assembly failure rate estimate based on surrogate data		
Assemblies	Subassemblies	VGB code	Based upon an MV converter with 12 IGCT/ diodes per inverter	Assembly failure rate source	Assembly failure rate estimate $\lambda ias\_FREenv = Q^{*}\lambda i\_FREcon * \pi Ei$
Circuit breaker	Circuit breaker	AAG10	1	IEEE Gold Book	0.0176
Grid-side filter	dV/dt filter	MKY10 BFA20	1	Carroll *et al.* [2]	0.0180
Grid-side inverter	Inverter firing circuits	MKY10 BG QBA10	1	Carroll *et al.* [2]	0.2172
	Inverter IGCTs	MKY10 BG QBA11	12		
	Inverter switching diodes	MKY10 BG QBA12	12		
DC link	DC-link capacitor	MKY10 BUA10	1	Carroll *et al.* [2]	0.0650
	Braking chopper resistor	MKY10 BUA20	2	Carroll *et al.* [2]	0.1375
	Chopper firing circuit	MKY10 BU QBA10	1		
	DC-link switching IGCT	MKY10 BU QBA11	2		
	DC-link switching diode	MKY10 BU QBA12	2		
Generator-side Inverter	Inverter firing circuits	MKY10 BG QBA20	1	Carroll *et al.* [2]	0.2172
	Inverter IGCTs	MKY10 BG QBA21	12		
	Inverter switching diodes	MKY10 BG QBA22	12		
Generator-side Filter	dV/dt filter	MKY10 BGA20	1	Carroll *et al.* [2]	0.0180
Cooling Unit	Cooling unit	MKY10 EC001	1	IEEE Gold Book	0.0058
Control Unit	Control unit	MKY10 KF001	1	Carroll *et al.* [2]	0.0420
Assembly total failure rate, λ_{tot} (failures/unit/yr)					0.7383
Reliability survivor function, $R_{(1yr)}$, %					412.8%

A similar analysis can be made if the FRC has two parallel equally rated channels, improving the failure rate by the introduction of redundancy.

11.4.2.4 Pitch system converters

Finally, the approach has been used in Table 11.7 to estimate the failure rate of three axes of an electric pitch system.

11.4.3 Summary

Table 11.8 summarises the failure rates and reliability survivor functions of the various wind turbine converters considered in the previous paragraphs. The

Table 11.7 Pitch system converters predicted failure rate

Reliability characteristics for 3-axis electric pitch system			**Failure rates**		
3-axis pitch converters with 2 IGBT/diode choppers per converter, assembly failure rate estimate based on surrogate data					
Assemblies	Subassemblies	VGB code	Subassembly *Qt*	Assembly failure rate source	Assembly failure rate estimate *λi_FREcon*
Series contactor	Series contactor	MDA20 QA001	1	Carroll *et al.* [2]	0.0100
Full-wave rectifier, BF	Rectifying diodes	MDA21-23 QBA11	4	Carroll *et al.* [2]	0.0062
DC-Bus, BU	DC-buscapacitor	MDA21-23 BUA10	1	Carroll *et al.* [2]	0.0181
	Charger switching transistor	MDA21-23 BU QBA11	2		
	Battery	MDA21-23 BUA11	1		
Motor chopper, BG	Chopper firing circuit	MDA21-23 BG QBA10	1	Carroll *et al.* [2]	0.0030
	Chopper IGBTs	MDA21-23 BG QBA11	2		
Series motor	Series motor	MDA21-23 MA001	1	Tavner *et al.* [21]	0.0471
Motor gearbox	Motor gearbox	MDA21-23 MDK20	1	Spinato [19]	0.1033
Control unit	Control unit	MDA21-23 KF001	1	Carroll *et al.* [2]	0.0074
Assembly total failure rate, λ_{tot} (failures/unit/yr)				Obtained on the assumption that two out of three blade axis pitch systems must operate	0.1952
Reliability survivor function, $R_{(1yr)}$, %					82.3%

Table 11.8 Summary of WT converter failure rates

	Table	Assembly total failure rate, λ_{tot} (failures/unit/yr)	Reliability survivor function, $R_{(1yr)}$, %
Soft starter with six thyristors and power factor correction capacitors Assembly failure rate estimate based on surrogate data	12.4	0.063	94%
LV partially rated converter (PRC) with six IGBT/diodes per inverter, assembly failure rate estimate based on surrogate data	12.5	0.121	89%
3-Axis pitch converters with 2 IGBT/ diode choppers per converter, Assembly failure rate estimate based on surrogate data	12.7	0.195	82%
MV fully rated converter (FRC) with 12 IGCT/diodes per inverter in two parallel PRC Assembly failure rate estimate based on surrogate data	Based on 12.6	0.402	67%
MV fully rated converter (FRC) with 12 IGCT/diodes per inverter in one single channel Assembly failure rate estimate based on surrogate data	12.6	0.738	48%

reliability survivor function shows the percentage of 100 converters of each type that would survive for longer than 1 year. The table shows the following:

- The soft starter and power factor correction module is the most reliable wind turbine converter.
- The PRC Type C LV converter is the next most reliable wind turbine converter.
- The 3-axis pitch converter system is the next most reliable wind turbine converter but its lack of reliability is dominated by electromechanical components, the pitch motors, and gearboxes, rather than power electronic components.
- The FRC Type D MV converter is the least reliable wind turbine converter.
- The FRC Type D MV converter with two parallel channels improves the above reliability by 40%.

11.5 Root causes of failure

It is difficult for wind farm operators and wind turbine OEMS to determine precisely what wind turbine converter failure root causes; however, the manufacturers of power electronics do have some knowledge in this respect.

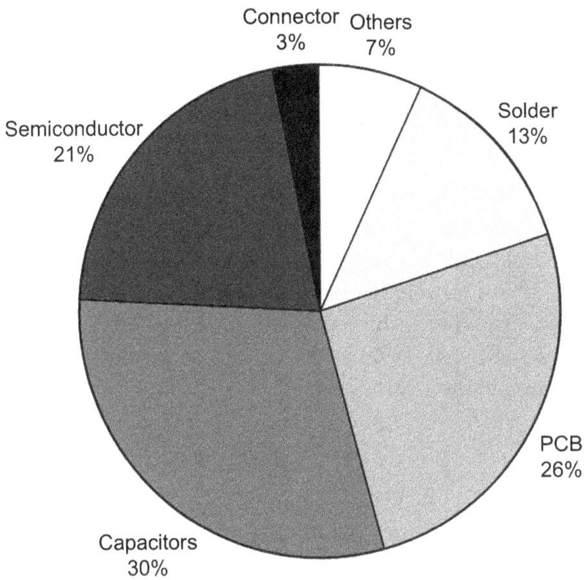

Figure 11.13 Summary of power electronic component failure root cause. Source: Wolfgang [25]

A subjective industry-based survey of reliability in power electronic converters, based on information from manufacturers and users, is presented by Yang *et al.* [27] and gives some proposals for the root cause. The results of this survey, which concentrated on components, are similar to the root cause proposals by Wolfgang [25] shown in Figure 11.13.

The estimates of Tables 11.4–11.7 in Section 11.4 give some more guidance towards converter failure root cause location.

11.6 Methods to improve WT converter reliability and availability

11.6.1 Architecture

Table 11.8 shows that the architecture of a WT converter can improve its reliability, most importantly by dividing the power flow into parallel redundant channels.

This has become an ongoing process in wind turbine nacelles as WT power outputs, particularly offshore, have risen above 8 MW per turbine with the prospect offshore of large turbines actually reaching 20–25 MW, in which parallel converter channels with in-built redundancy can and will enable much higher reliability figures. The disadvantage of this approach is that it raises converter costs. However, experience in the wind and other energy industries shows that the costs of redundancy deliver much stronger reliability and reduce operational costs.

11.6.2 Thermal management

Initial experience for cooling wind turbine power electronic converters used air as the coolant, because of its simplicity and the costs and inconvenience of providing water supplies to a wind turbine nacelle.

However, thermal stresses, particularly in the inverter bridges, have been shown to play an important role in converter reliability.

Therefore, modern wind turbine converters above 2 MW must now be fitted with water cooling, and such systems have become an important part of guaranteeing wind turbine converter reliability.

11.6.3 Control

The estimates of Tables 11.4–11.7 in Section 11.4 suggest that controllers are a significant contributor to converter failure rates, in addition to the influence of inverter and DC-link components, shown in Figure 11.13.

Improvements in controller reliability can be achieved by more rigorous quality control testing of controller manufacture, on completion before incorporation in the wind turbine.

11.6.4 Monitoring

Wind turbines are probably the most highly monitored power generation sources connected to the electrical grid.

However, the volumes of signals generated by wind farm SCADA and condition monitoring systems are large, and wind farm operators need to manage that data to provide clear and simple displays to operators and maintainers if they want to achieve the best reliable performance from their assets.

Considerable work has been done on the condition monitoring of power electronics, for example as described by Yang *et al.* [26].

Recent work has shown that the prognosis of converter faults can be achieved by improved, better SCADA alarm handling with accurate consequent analyses of SCADA signals, as referred to previously.

The author suggests that these initiatives are more likely to be immediately effective than the application of the methods proposed by Yang *et al.* [26].

11.7 Power electronic applications in modern large wind farm installations

Finally, examples of modern power electronic technologies currently being used in wind power are shown in Figure 11.14, both in the wind turbines themselves and also in the means to transfer energy by DC transmission from sea to shore.

(a)

(b)

Figure 11.14 *Modern offshore wind power technology. (a) Three 8.4 MW offshore floating wind turbines, 20 km off west coast of Viana do Castelo, Portugal, incorporating power electronic converters. (b) An offshore wind farm HVDC converter station transmitting power to shore via DC cables.*

11.8 Conclusions

The following conclusions can be drawn about wind turbine power electronic conversion reliability from the data presented in this chapter:

- Power electronic converters make a significant contribution, 10%–20%, to the unreliability of wind turbines, with converter failure rates ranging between 0.063 and 0.738 failures/unit/yr;
- The most significant power electronic assemblies in wind turbines are the main drive train converter, the 3-axis blade electric pitch converter systems, and the soft start/power factor correction systems;

- A fully rated drive train converter has been shown to be the least reliable at 0.738 failures/unit/yr, a partially rated drive train converter being more reliable at 0.121 failures/unit/yr;
- It appears that currently the larger the rating of the main drive train power electronic converter the greater the failure rate, although that may be mitigated by ongoing developments and improved manufacture;
- The 3-axis blade electric pitch converter systems had a reliability of 0.195 failures/unit/yr, their unreliability being dominated by their electromechanical components, the motors, and gearboxes;
- The soft start and power factor correction system are shown to be the most reliable power electronic systems in wind turbines, with 0.073 failures/unit/yr;
- Current monitoring is more than adequate for the power electronic converters installed but alarm rates are excessive with relatively little use being made of the alarms or signals to improve assemblies reliability.

11.9 Recommendations

To improve reliability in WT converter design, manufacture, and operation, attention needs to be given to the following recommendations:

- Answers to the questions raised above.
- Improving converter thermal management to lower power electronic device temperatures.
- Reducing high failure rates by design, through redundancy, for example, the reliability of main drive train converters can be raised by using 2, 3, or more parallel channels, increasing redundancy, although this adds to cost but will increase reliability.
- Reducing early life failures by more thorough factory testing of converter subassemblies, particularly controllers. Commitment to full converter heat runs at the end of manufacture, as is done with smaller rated converters.
- Reducing high converter alarm numbers by minimising nuisance alarms and improved alarm management to provide better fault prognosis, for example using the methods described in this chapter.

Power electronic conversion technology is in continuous development and as costs fall, reliability will rise and downtimes fall, as has characterised other power electronic applications.

From this limited extract of public domain data, it is quite clear that wind turbine converters deserve reliability attention from OEMs and operators and the availability of larger datasets of information is encouraged so that more statistically significant observations can be made.

The most valuable information in this chapter is the following tables and figures:

- Tables 11.3–11.8, which show recorded prospective failure rates from various types of power electronic converters in wind turbines;

- Figures 11.10–11.13, showing the range of failure rates in wind turbines;
- Whilst these data are historic their table formats can be used by modern wind turbine operators to record data and identify any weaknesses in their electronic systems.

The following questions arise from this chapter about wind turbine power electronic conversion reliability:

- Why do wind turbine converter failure intensities improve with time **after** installation as seen in Figure 11.10?
- Why are WT converter failure intensities generally **higher** than values given from normal industrial experience, Figures 11.10 and 11.12?
- Why does the failure rate **rise** with converter rating, Figure 11.12?
- In view of the reliability prediction of Section 11.4 how can wind turbine main converter reliability be **improved** by parallel redundancy, Table 11.8?

Terminology

Most of the terms used in this chapter, adapted from RIAC [16] and are described as follows:

Term	Definition
Availability, *A*	The probability that a converter will be available to operate for a time t and is generally quoted as a percentage. A high-availability converter has only short periods of shutdown time due to failure or maintenance.
Failure	When a device fails to perform its energy conversion function. Failure is complete and does not imply partial functionality.
Failure mode	The manner in which final failure occurred, for example: • IGBT failure; • Or DC-link capacitor failure.
Failure mechanism	The physical manner in which a failure process progresses from the root cause to the failure mode, for example for the two failure modes above: • Bond wire failure leading to the IGBT failure. • Excessive temperature over an extended period of time of a DC-link capacitor.
Failure mode and effects analysis, *FMEA*,	A subjective analysis tool, defined by standards, that uses a qualitative approach to identify potential failure modes, their root causes and the associated risks in the design, manufacture, or operation of a converter.
Failure intensity, $\lambda(t)$	The failure intensity is the rate at which failures occur in a converter varying according to the operating condition and age of a converter so is a function of time, sometimes called the hazard function. Failure Intensity can be constant for part of the converter life, when it is termed failure rate, λ, and in

(Continues)

(Continued)

Term	Definition
	this book, this is expressed as failures/converter/year. It is the objective of the maintenance engineer of the converter to keep the failure rate low, constant, and predictable.
Failure sequence duration	The time from root cause to failure mode. This may be a period of seconds, minutes, days, months, or weeks and can depend on:- • The failure mode itself • The operating conditions of the converter • The ambient conditions
Mean time to failure, *MTTF*	The expected value of that and successive *TTFs*. *MTTF* does not include the time to repair as a result of a failure. The *MTTF* is usually given in hours.
Mean time to repair, *MTTR*	Time to repair, *TTR*, time measured from instant of first failure to instant when converter is available for operation again. *MTTR* is average of that and successive *TTRs* and can be averaged over a number of converters in a population. *MTTR* is usually given in hours.
Mean time between failures, *MTBF*	Under the hypothesis of minimal repair, that is repair brings a part back to the condition before failure, time between failure, *TBF*, is the time measured from the instant of installation of part to the instant after the first failure when part is available for operation again. Average of that and successive *TBFs* is *MTBF* and is averaged over the population. *MTBF* is the sum of the *MTTF* and *MTTR*. The *MTBF* or θ usually given in hours.
Part	Entity treated in the process of design, engineering, operation, maintenance and demolition. Could be a system, subsystem, assembly, subassembly or component of a WT. A converter is an assembly of a WT.
Random failures	Failures occurring during the useful life period of a WT system, they occur unpredictably
Redundancy	The existence of one or more means, not necessarily identical, for accomplishing one or more functions of a system: • Active redundancy has all subsystems operating simultaneously • Standby redundancy has alternate means activated upon failure of a subsystem
Reliability, *R*	• The probability that a part can operate without failure for a time t and in this book is generally quoted as a probability percentage • The duration or probability of failure-free performance under stated conditions • The probability that a part can perform its intended function for a specified interval under stated conditions. For non-redundant parts, this is equivalent to the first definition

(Continues)

(*Continued*)

Term	Definition
	Reliability as a function *R(t)* is sometimes known as the survivor function because it indicates what proportion of the starting population survives at a particular time *t*. A high-reliability part has few failures, a high mean time between failures, a high percentage reliability, and a low failure rate.
Reliability model	A model for the identification of the reliability of a system that integrates part interrelations for reliability analysis and assessment
Reliability prediction	A measure for estimation of figures of merit for product reliability performance
Root cause	The manner in which a failure mode was initiated. For example for the two failure mode cases given above: • Bond wire failure • Excessive temperature
Root cause analysis, *RCA*	The analysis investigating the failure modes and underlying root causes of a failure.
System	A complete piece of machinery able of performing a required function on its own, for example, a wind turbine.
Time to failure, *TTF*	The time measured from the instant of installation of the part to the instant of failure.

Abbreviations

Abbreviation	Explanation
DC	Direct current
DFIG	Doubly fed induction generator
EHV	Extra high voltage
FMEA	Failure modes and effects analysis
FRC	Fully rated converter
HV	High voltage
IG	Induction generator
IP	Intellectual property
LV	Low voltage
LWK	Landwirtschaftskammer Schleswig-Holstein German WT database
MTBF	Mean time between failures
MTTR	Mean time to repair
MV	Medium voltage
OEM	Original equipment manufacturer
PCB	Printed circuit board
PLP	Power law process, equation representing subassembly failure intensity
PMSG	Permanent magnet synchronous generator
PRC	Partially rated converter
RBD	Reliability block diagram
SCIG	Squirrel cage induction generator

(Continues)

(*Continued*)

Abbreviation	Explanation
VGB	German Power Standards Company
VSC	Voltage source converter
WT	Wind turbine
WMEP	Wissenschaftlichen Mess- und Evaluierungsprogramm database
WRSG	Wound rotor synchronous generator
WSD	Windstats database for Germany
WSDK	Windstats database for Denmark

Variables

Symbol	Explanation
A	Availability, $A = MTBF/(MTBF + MTTR)$, %
C	Capacity or load factor, %
η	Efficiency
$\lambda(t)$	Failure intensity or hazard function, can be represented by a PLP or Weibull function, failures/year
λ	Failure rate, failure intensity when hazard function is constant with time, failures/year
$\mu(t)$	Repair rate of an assembly, $MTTR=1/\mu$, hour
N	Speed of a machine rotor, rev/min
N_{tot}	Total number of device subassemblies
P	Device power output, Megawatts
p	Integer number of pole pairs
$R(t)$	Reliability survivor function, the probability that a part will perform its intended function for a specified interval under stated conditions, hazard function λ constant with time, failures/converter/yr
$\theta(t)$	MTBF of a assembly, MTBF$\theta=1/\lambda$, hour

Acknowledgements

The author acknowledges assistance in this work from his own former PhD students Fabio Spinato, Chris Crabtree, Tatiana Delorm, Bin Di Chen, and Donatella Zappala but also from other research students and post-doctoral research workers including Hooman Arabian-Hoseynabadi.

References

[1] Arabian-Hoseynabadi, H., Oraee, H., and Tavner, P.J. (2010). Failure modes and effects analysis (FMEA) for wind turbines, *International Journal of Electrical Power & Energy Systems*, 32(7), 817–824.

[2] Carroll, J., McDonald, A., and McMillan, D. (2014). Reliability comparison of wind turbines with DFIG and PMG drive trains, *IEEE Transactions on Energy Conversion*, 30(2), 663–670, DOI 10.1109/TEC.2014.2367243.

[3] Chen, B.D., Matthews, P.C., and Tavner, P.J. (2013). Wind turbine pitch faults prognosis using a-priori knowledge-based ANFIS, *Journal Expert Systems with Applications*, 40(17): 6863–6876.

[4] Chen, B.D., Matthews, P.C., and Tavner, P.J. (2015). Automated on-line fault prognosis for wind turbine pitch systems using supervisory control and data acquisition, *IET Proceedings of Renewable Power Generation*, 9(5): 503–513.

[5] Delorm, T.M., (2013). Tidal stream devices: Reliability prediction models during their conceptual and development phases, *PhD Thesis*, Durham University.

[6] Delorm, T.M., Zappala, D., and Tavner, P.J. (2011). Tidal stream device reliability comparison models, *Proceedings of the IMechE Part O: Journal of Risk and Reliability*, 226(1): 6–17.

[7] Faulstich, S., Hahn, B., and Tavner, P.J. (2011). Wind turbine downtime and its importance for offshore deployment, *Wind Energy*, 14(3): 327–337.

[8] Faulstich S., Durstewitz M., Hahn B., Knorr K., and Rohrig K. (2008). Windenergie Report, Institut für solare Energieversorgungstechnik, Kassel, Germany.

[9] Hahn, B., Durstewitz, M., and Rohrig, K. (2007). Reliability of wind turbines. *Proceedings of the Euromech Colloquium, Oldenburg*, Springer, Berlin, pp. 329–332.

[10] Hansen A.D., and Hansen L.H., (2007). Wind turbine concept market penetration over 10 years (1995–2004). *Wind Energy*, 10(1):81–97.

[11] IEEE, Gold Book: (1990). *Recommended Practice for Design of Reliable Industrial and Commercial Power Systems*, IEEE Press, Piscataway, NJ.

[12] Landwirtschaftskammer (LWK), Schleswig-Holstein, Germany: http://www.lwksh.de/cms/index.php?id¼1743, last accessed 8th February 2010.

[13] MIL-HDBK-217F (1991). *Military Handbook. Reliability Prediction of Electronic Equipment.* US Department of Defense, Washington, DC.

[14] Polinder H., van der Pijl, F.F.A., de Vilder, G.J., and Tavner, P.J. (2006). Comparison of direct-drive and geared generator concepts for wind turbines, *IEEE Transaction on Energy Conversion*, 21(3): 725–733.

[15] Qiu, Y., Feng, Y., Tavner, P.J., Richardson, P., Erdos, G., and Chen, B.D. (2012). Wind turbine SCADA alarm analysis for improving reliability, *Wind Energy* 15(8): 951–966;

[16] RIAC (2005). *System Reliability Toolkit, A Practical Guide for Understanding & Implementing a Program for System Reliability*, US Department of Defense, Washington, DC.

[17] Relia Wind, (2011). Deliverable D.2.0.4a-Report, Whole System Reliability Model, available from ReliaWind website.

[18] Ribrant, P.J.J., and Bertling L.M. (2007). Survey of failures in wind power systems with focus on Swedish wind power plants during 1997–2005, *IEEE Transaction on Energy Conversion*, 22(1): 167–173.

[19] Spinato F. (2008). The reliability of wind turbines, *PhD Thesis*, Durham University.

[20] Spinato, F., Tavner, P.J., van Bussel, G.J. W., and Koutoulakos, E. (2009). Reliability of wind turbine sub-assemblies. *IET Proceedings of the Renewable Power Generation*, 3(4): 1–15.

[21] Tavner P.J., Ran, L., Penman J., and Sedding, H. (2006). *Condition Monitoring of Rotating Electrical Machines*, IET, Stevenage.

[22] VGB PowerTech (2007). Guideline, Reference designation system for power plants (RDS-PP); Application explanation for wind power plants guideline, Reference designation system for power plants (RDS-PP), VGB-B 116 D2, VGB PowerTech, Essen, Germany.

[23] Wilkinson, M.R., Hendriks, B., Spinato, *et al.* (2010). Methodology and results of the ReliaWind reliability field study, Proceedings of European Wind Energy Conference, EWEC2010, Warsaw.

[24] Windstats quarterly newsletter, part of Wind Power Weekly, Denmark: www.windstats.com, last accessed 8th February 2010.

[25] Wolfgang, E. (2009). Examples for failures in power electronics systems, ECPE, *Reliability of Power Electronic Systems Tutorial*, July, Prague, Czech Republic.

[26] Yang, S., Xiang, D., Bryant, A., Mawby, P., Ran, L., and Tavner, P.J. (2010). Condition monitoring for device reliability in power electronic converters—a review, *IEEE Transactions on Power Electronics*, 25(11): 2734–2752.

[27] Yang, S., Bryant, A., Mawby, P., Xiang, D., Ran, L., and Tavner, P.J. (2011). An industry-based survey of reliability in power electronic converters, *IEEE Transaction on Industry Applications*, 47(3): 1441–1451.

Chapter 12

Fault-tolerant adjustable speed drive systems

Prasad Enjeti[1], Pawan Garg[1] and Harish Sarma Krishnamoorthy[1]

12.1 Introduction

In this chapter fault tolerance in adjustable speed drive (ASD) systems is examined in detail. The chapter begins with a study of important factors affecting the reliability of ASDs. Then this chapter examines various power converter configurations employed in ASD systems and their modifications to enhance fault tolerance.

One of the most effective ways to boost motor-actuated system efficiency is the use of an adjustable speed drive (ASD) to replace mechanical equipment such as control valves and gears. Nearly 38% of all electrical energy is consumed by motor-driven equipment such as pumps, compressors, and fans [1]. As shown in Figure 12.1, an ASD is part of a modern motor drive system (MDS) along with the motor and the input supply. The use of MDS is widespread in applications such as water pumping stations, the petrochemical industry, the cement industry, the mining industry, steel plants, financial markets, and so on. In a number of these applications, MDS is an integral part of a continuous process [2,3]. Any interruption of the manufacturing process is expensive because of production loss and follow-up costs. Therefore from a system perspective, the availability of MDS in such applications has an immense financial impact, and interruptions could offset the benefits of a power electronic system. More recently, the application of motor drives in electric vehicles and other critical applications like military and hospitals have also brought up concerns regarding personnel safety upon unexpected system shutdown. This, along with the financial reasons, has triggered interest in the study of failure modes of ASDs [4].

A motor drive is a complex system that typically operates in harsh conditions. This can lead to failures due to stress or mechanical wear and tear due to aging (Figure 12.2). The relationship between the unscheduled shutdown of ASDs and power interruption phenomena like voltage sags and swells has also been studied extensively [4]. The various factors that affect the reliability of individual subsystems of an MDS are summarized in Figure 12.2. In this chapter, only the factors affecting the reliability of ASDs are investigated in detail.

[1]Department of Electrical and Computer Engineering, Texas A&M University, USA

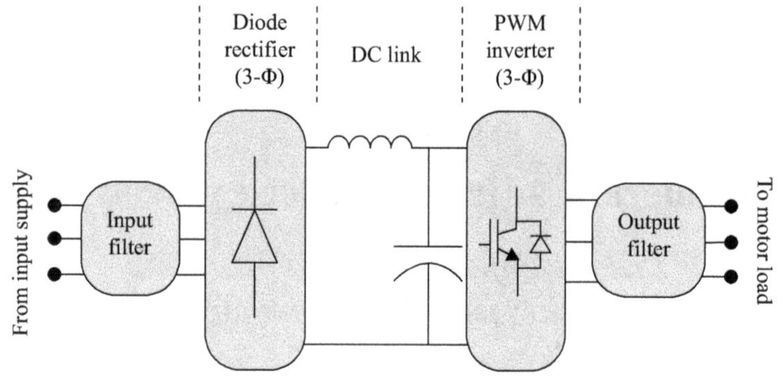

Figure 12.1 Conventional two-level 3 phase MDS topology [2,3]

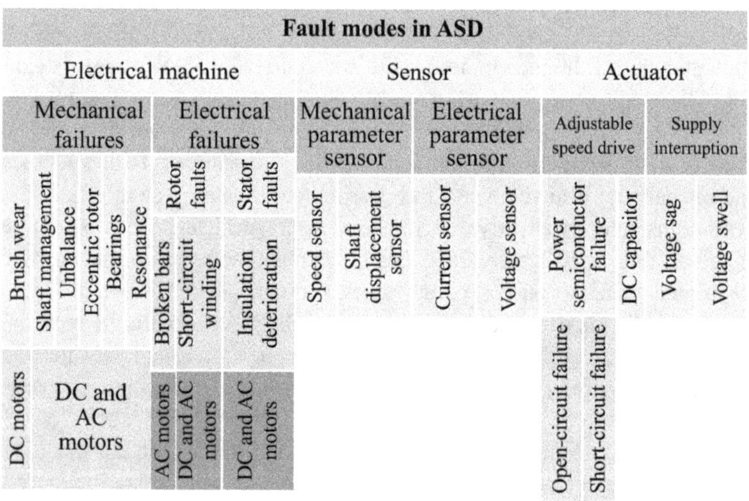

Figure 12.2 Factors affecting failure of a motor drive system [4]

12.2 Factors affecting ASD reliability

A typical ASD (shown in Figure 12.1) consists of three subsystems: (1) a front-end rectifier, (2) a DC link, and (3) an inverter. This topology uses diodes and Insulated Gate Bipolar Transistors (IGBTs) or Metal Oxide Field Effect Transistors (MOSFETs) as power semiconductor devices and electrolytic/film capacitors for the DC bus. As the conventional topology does not include redundancy, failure of any one of the three subsystems results in ASD failure. According to data obtained from field experience, ~35% of ASD failures are attributed to power electronics circuit failure [4]. The power switching components have been identified to have greater failure rates than other components in the drive system. Up to 40% of the 3-phase inverter failures in the field are a result of power semiconductor failures. These factors that affect the availability of

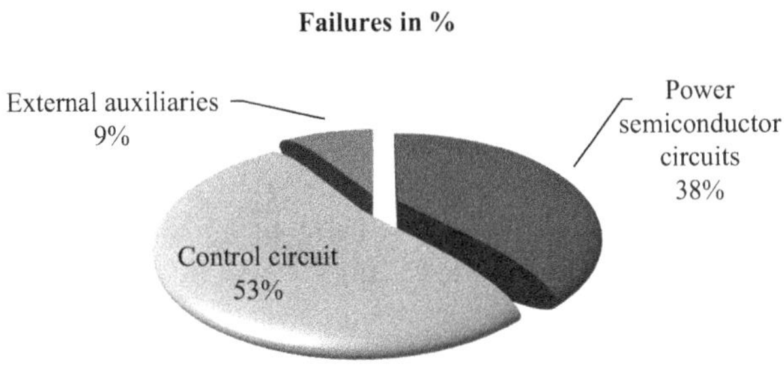

Figure 12.3 Contribution of different subsystems in ASD failures [4]

a motor drive system are briefly discussed in this section. Figure 12.3 shows the contribution of different subsystems in ASD failure [4].

12.2.1 Power semiconductor devices

Various power semiconductor device options, such as MOSFETs, IGBTs, and SCRs, are available to designers in discrete and module packages for motor drive applications. To reduce parasitics and optimize space utilization, modern drives use highly integrated power modules that utilize different materials in their construction. The differences in the thermal and mechanical properties of these materials lead to nonuniform thermomechanical stresses and consequently fatigue-induced failures. Another mode of failure for high-power IGBT modules is voltage breakdown or avalanche mode failure. Other failure mechanisms for IGBTs due to parasitic bipolar junction transistor (BJT) latch-up, breakdown due to high dV_{CE}/dt spikes during turn-off, and thermal breakdown have been studied.

12.2.2 Electrolytic capacitors (ECs)

Electrolytic capacitors are widely used in the DC link between the front-end rectifier and the inverter. ECs provide a stiff DC bus and a limited energy reserve for low-voltage ride-through (LVRT). The ECs are expensive and prone to failure in field operation. Considerable work has been done to explore improvements in capacitor technology and optimize DC link size by topology and control design [5,6]. Some of the automotive motor drives use film capacitors in place of ECs to improve system reliability.

12.2.3 Other auxiliary factors

12.2.3.1 Motor

The installation of an ASD can adversely affect the reliability of a motor. Motor failures also affect the availability of a motor drive system. According to data from commercial installation, bearing and winding failures together account for 70% of machine failures [7,8].

12.2.3.2 Winding failure

High-frequency pulse width modulated (PWM) inverter-based ASD with long lead cables can cause large voltage overshoots at the motor terminals [8,9]. The magnetic wire insulation life curves shown in [10] illustrate the effect of increasing cable length and switching frequency. Prolonged voltage stress leads to gradual deterioration of the insulation material and ultimately results in catastrophic failure [8,9].

12.2.3.3 Bearing current

The application of PWM techniques is usually accompanied by high-frequency common mode neutral to ground voltage. This voltage can induce a high-frequency motor bearing current that gradually erodes the bearing races and eventually leads to mechanical failures [8,9]. Electrical grounding plays a crucial role in determining the generation mechanism and magnitude of bearing currents.

12.2.3.4 Input supply disturbances

Based on power quality surveys and standards, an input voltage drop by more than 13% and longer than one-half cycle can lead to MDS trips. Although a trip protects the power electronic components during a sag event, it can lead to production loss. In continuous process industrial applications, this could have a significant financial impact [6,11].

12.2.3.5 Sensor failure

ASDs use sensors to accomplish closed-loop control. The sensors can experience abnormalities such as intermittent connection, complete outage, DC bias in measurement, and gain variation [12]. Incomplete or spurious information can lead to closed-loop instability.

Due to the growing demand for reliable motor drive systems, extensive research has been dedicated to the study of fault-tolerant design of ASDs. This chapter provides a comprehensive overview of state-of-the-art techniques for fault tolerance. Fault tolerance is discussed with respect to semiconductor device failures. The failures considered are (i) open circuit and (ii) short circuit failures. Intermittent gate misfiring failure will have an effect similar to open or short circuit failure. In the subsequent discussion, the terms semiconductor device, IGBT, MOSFET, and switch are used interchangeably.

12.3 Fault-tolerant ASD system

For this discussion, a fault-tolerant ASD system is a system that does not fail or shutdown when a component or subsystem fails [13]. From the perspective of the end user, the two most important aspects of fault-tolerant ASD systems are the degree of fault tolerance and post-fault performance. The degree of fault tolerance is the number of failures that cause a system to shut down. The ASD performance in post-fault operation is important to consider, as in some ASD systems a "limp-home" approach is allowable after faults.

Fault-tolerant operation has three stages: (1) fault diagnosis; (2) fault isolation; and (3) control or hardware reconfiguration. The diagnosis of failures in an ASD application is a vast area in itself and is considered beyond the scope of the current discussion. This chapter explores the third aspect in more detail. Fault isolation is discussed briefly.

12.4 Converter fault isolation stage in fault-tolerant system design

The fault isolation process is the electrical isolation of a faulted switch or a malfunctioned leg. This ensures that the failure does not propagate or result in cascading failures. Hence, fault isolation is an important step in achieving fault-tolerant operation. As the goal is the prevention of cascading failures, the following requirements are important to consider while designing an isolation scheme: (1) speed of isolation; (2) fault coverage, (3) precision of isolation scheme, (4) impact on the system in normal operation, and (5) cost and complexity.

It is noted that the leg short circuit fault implies that both S_{1a} and S_{2a} have failed shorts (Figure 12.4). Referring to Figure 12.4, the leg open circuit failure is defined as the case when point "a" is disconnected from "A." In [14], five typical isolation schemes are discussed for two-level converters. Three of them are shown in Figure 12.5. These isolation schemes can be extended to other converter topologies as well. These techniques are listed below.

- Strategy (a) in Figure 12.5(a): If the switch S_{1a} fails (short or open), it can be isolated by implementing the following sequence of steps.
  - Turn off S_{a2} to avoid damaging S_{a2} by any large current transient.
  - Turn on T_n to create a shoot-through that blows fuse F_p.

The capacitors C_p and C_n are designed to minimize the isolation time. They help in turning off the triac T_p and T_n.

- Strategy (b) (Figure 12.5(b)): If the switch S_{a2} fails (short), it is isolated by following steps.
  - Turn off S_{a1} to avoid damaging S_{a1} by any large current transient.
  - Turn on the triac or bidirectional switch, S_{rn}. This will cause a shoot-through in the top DC bus and blow fuse F_p.

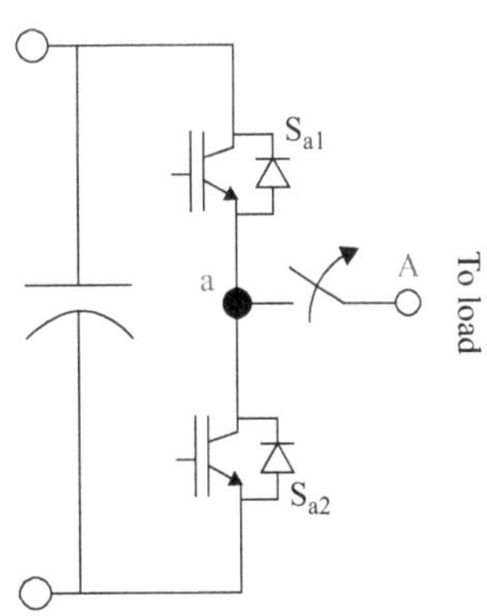

Figure 12.4 Half bridge in a conventional two-level topology

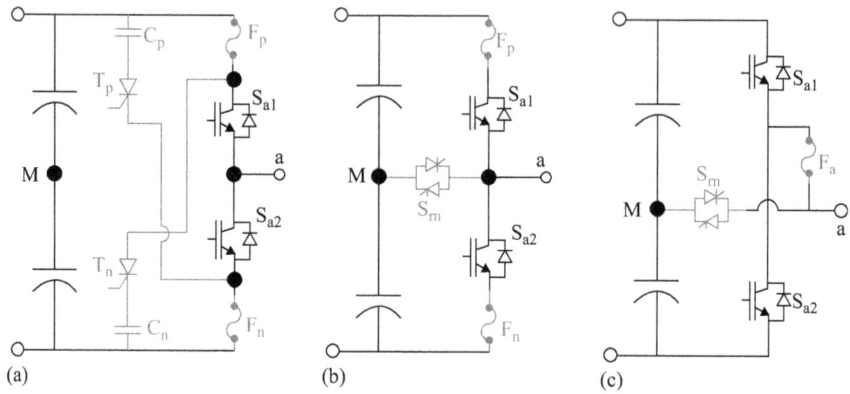

Figure 12.5 Strategies (a), (b) and (c) are three typical fault isolation techniques used in two-level converters [14]

In the case of open switch failure, the gating signals to the switch are disabled and triac S_{rn} is turned on. This clamps the output voltage to the DC bus mid-point, and the current in the faulted phase decays.

- Strategy (c) (Figure 12.5(c)): If switch S_{a2} fails (open or short), the leg for phase "a" can be isolated according to the following steps.
  - If S_{a1} fails to open, then S_{a2} and S_{rn} are turned on. This will cause a short circuit of a bottom bus capacitor and blow fuse F_a. The inverter leg-a is now isolated.
  - If S_{a1} fails short, then:
    (a) Turn off S_{a2}.
    (b) Turn on S_{rn}. This will short the top bus capacitor and will blow fuse F_a open. The inverter leg-a is isolated.

There are two other variations of these three techniques that are discussed in [14]. The operation of these techniques is similar to the three strategies discussed above.

Section summary: Three different fault isolation schemes have been discussed. They are summarized in Table 12.1.

12.5 Control or hardware reconfiguration stage in fault-tolerant system design

A number of fault-tolerant systems have been published in the literature. Although the landscape of control or hardware reconfiguration techniques is diverse, for the sake of discussion these methods are broadly divided into three categories: (a) topological, (b) software, and (c) redundant hardware techniques. These three methods are illustrated in greater detail in the subsequent sections. The section on topological techniques discusses circuits that inherently have redundant switching

Table 12.1 Summary of fault isolation schemes

Technique	Fault isolation coverage	Precision of isolation	Disadvantages
Strategy (a)	Open/short failure of S_{a1} and/or S_{a2} (any fault can be converted to open phase failure)	• Only isolates the faulty switch	• Fuses on the DC bus increase parasitic inductance in the path • Relatively large number of components required to implement • Two capacitors, C_p and C_n, need to be sized for large voltage amplitude • The i^2t value of the fuse must be lower than the withstanding capability of semi-conductor devices (T_p and T_n)
Strategy (b)	Short failure of S_{a1} and/or S_{a2}	• Only isolates the faulty switch • In case of open circuit fault, the inverter leg output is clamped to the DC bus mid-point	• Fuses on the DC bus increase parasitic inductance in the path • In case of open circuit fault, the switch stays connected electrically. However, in post-fault operation, the inverter phase output is clamped to the DC bus mid-point • Access to the DC bus mid-point is required
Strategy (c)	Open/short failure of S_{a1} or S_{a2}	• Isolates the faulted leg and clamps the inverter leg output to the DC bus mid-point	• It cannot handle the simultaneous failure of S_{a1} and S_{a2} • Access to the DC bus mid-point is required

states. In some failure modes, these topologies can be operated with a degraded output performance without any significant modification. However, in other cases, software techniques can be used to obtain more optimal operation. Redundant hardware techniques involve the addition of redundant hardware along with the original system. Although adding hardware increases system cost, it provides advantages in post-fault operation. The organization of this section is described in Figure 12.6. This chart also gives the reader a quick overview of existing fault-tolerant ASD techniques.

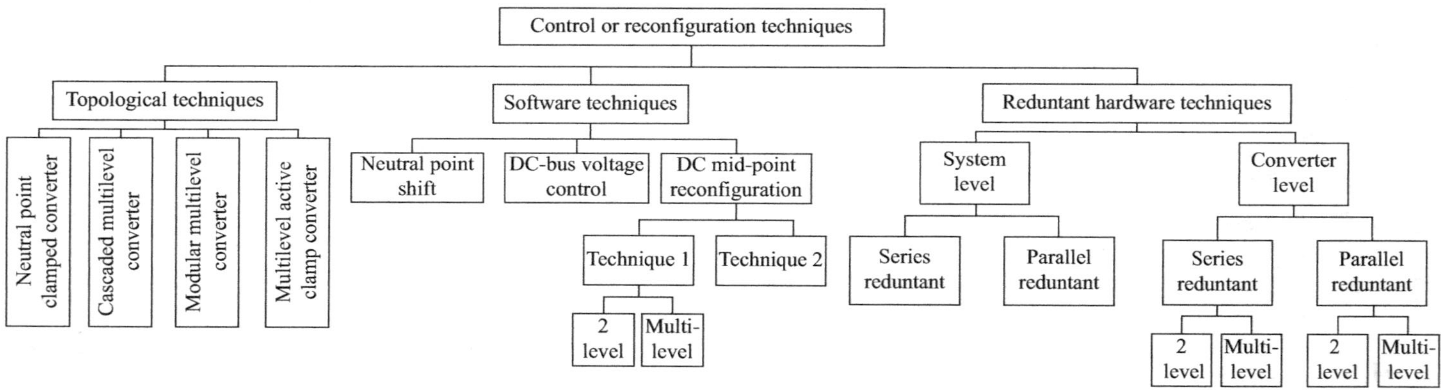

Figure 12.6 Different categories under the control or reconfiguration stage of fault-tolerant system

For the subsequent sections, the following assumptions have been made for the inverter in un-faulted operation:

- Inverter phase voltage output of 1 p.u. is the maximum phase voltage obtained at the modulation index of 1.15 (with third harmonic injection).
- Inverter phase output current of 1 p.u. is the maximum phase current.
- In post-fault operation, the voltage and current (in p.u.) are calculated for the condition in which the inverter is operated to avoid overvoltage and overcurrent stress of the devices.

The software and topological techniques are closely related. Both methods require changes in control after the fault has been detected and isolated. For this discussion, topological techniques refer to the use of topologies with redundant switching states. These topologies can continue to operate after fault with minimal change in control strategy. The software technique refers to the modification of the control strategy to achieve an optimum operating point with respect to voltage, total harmonic distortion (THD), or any other objective. An advantage of topological and software techniques is that algorithm modifications can be implemented in an existing system with little or no change. The redundant hardware techniques often require careful thought in the design stage and are difficult to implement in an existing installation. This allows redundant hardware techniques to achieve enhanced post-fault performance.

12.5.1 Topological techniques

Multilevel inverters operate by introducing additional levels in a two-level converter by adding more switches. These additional switches are involved in regular operation and create redundant output switching states/options to obtain the same output voltage. A space vector PWM approach is typically used to avoid states involving the failed device while maintaining a balanced output voltage. The presence of redundant states can be utilized in fault-tolerant operation, which is discussed next for some popular topologies.

12.5.1.1 Neutral point clamped topology

The 3-level neutral point clamped (NPC) topology and its switching state diagram have been shown in Figure 12.7. In NPC topology, three types of output switching states are available, positive (S_{a1} and S_{a2} are on), negative (S_{a3} and S_{a4} are on), and zero (S_{a2} and S_{a3} are on). It is observed in Figure 12.7(a) that once S_{a2} has failed short, the negative state (−) at the output of phase A will be unavailable as it will short circuit the bottom dc bus capacitor. The post-fault switching state diagram is shown in Figure 12.7(b) (ii)–(vii). From Figure 12.7(b) (iii), it can be concluded that the inverter can continue to function as long as the unavailable states are avoided. Since these states fall on the output perimeter of the hexagon, that is, large vectors (− + −), (− + +), and (− + +), the maximum modulation index is reduced to half of the normal operating value [15]. It can also be observed that there are a few states that lead to overvoltage stress on healthy devices. The switching state diagram for open circuit failure of each device (S_{1a}, S_{2a}, D_{5a}) is different from their respective short circuit failure switching states. It is also noted that the modification of the PWM

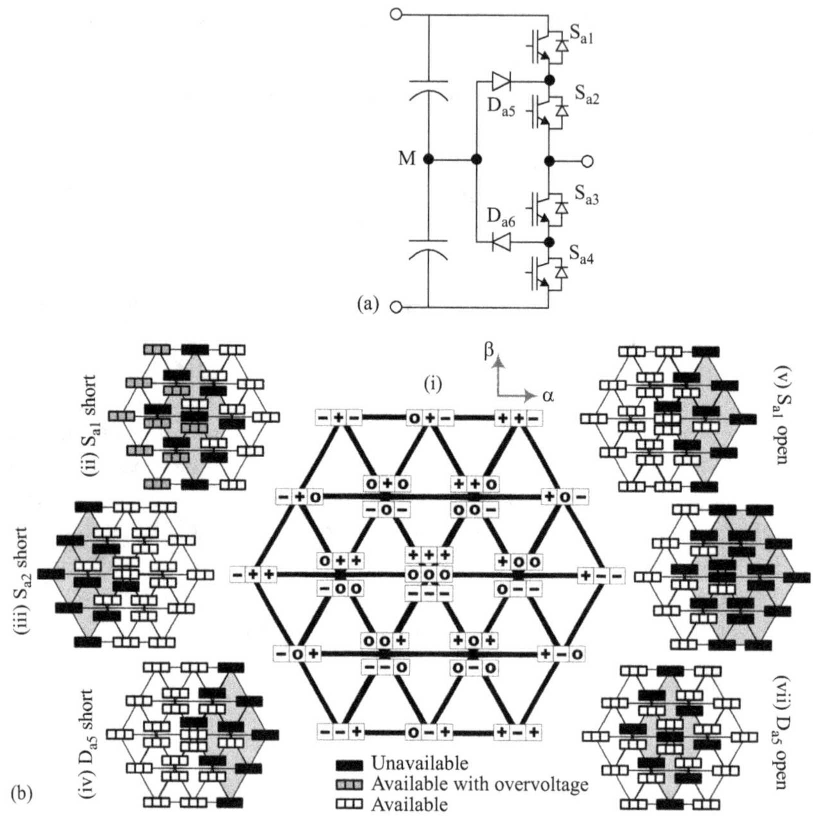

Figure 12.7 (a) Neutral point clamped topology with A1 semiconductor failed short and (b) switching state diagrams (including states that are "available," "unavailable," and "available with overvoltage"): (i) normal operation, (ii)–(vii) for different failure modes. The relative position of switching states established in (i) holds true for (ii)–(vii) [15 16]

strategy to avoid unavailable states leads to the DC bus mid-point imbalance. Some other challenges include spurious fault detection [16] and overrating of device voltage to full DC bus voltage [17]. Also, with all the drawbacks mentioned above, the conventional NPC can only compensate for switch short circuit-type failures.

A conventional NPC topology, as shown in Figure 12.7, can be modified to utilize the inherent redundant states, as shown in Figures 12.8 and 12.9. The topology in Figure 12.8 is obtained by adding an isolation circuit to a conventional NPC topology. The clamp diodes Da5 and Da6 are replaced with IGBTs to obtain Figure 12.9. In normal conditions, the new redundant switching states are utilized to balance power loss between devices [18]. The single switch short and open circuit failure cases are discussed separately.

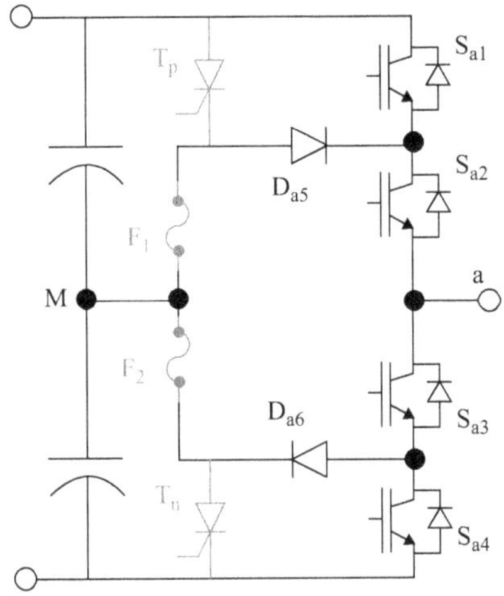

Figure 12.8 Conventional NPC topology modified for fault tolerance [18]

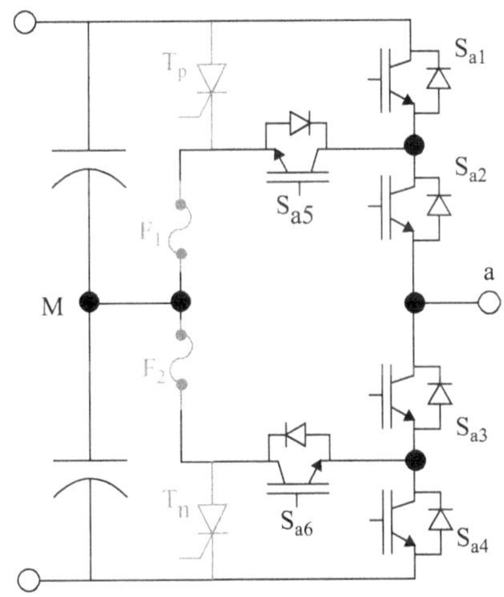

Figure 12.9 Active-NPC topology modified for fault tolerance [14]

In the case of the single switch short circuit failure in Figure 12.8, the faulty leg can be reduced to a two-level half bridge. If S_{a1} fails short, then the top DC bus capacitor will experience a short circuit through Da6 every time S_{a2} and S_{a3} are turned on. To avoid this issue and achieve controlled isolation, the thyristor, T_n, is turned on to blow fuse F_2. The faulted leg is now operated in the two-level mode, that is, the zero state (S_{a2} and S_{a3} on) is avoided. The fault coverage is limited to the switch short circuit failure of one switch among the top (S_{a1}, S_{a2}, D_{a5}) and/or bottom (S_{a3}, S_{a4}, D_{a6}) switches. The maximum available output voltage is preserved, but the output voltage THD performance degrades. However, the devices need to be rated for full DC bus voltage. In contrast, the topology in Figure 12.9 can be operated as a three-level leg after a single switch short circuit fault. For example, if S_{1a} fails short, thyristor T_n is turned on to blow fuse F_2. Once F_2 is blown, the zero state can be obtained by the combination of switches S_{a2} and S_{a5} being turned on. The other switching states remain unchanged. The healthy devices have to block full DC bus voltage. The output voltage can be maintained at 1 p.u. in post-fault operation. Another post-fault operating mode available in the case presented in Figure 12.9 is to connect the phase mid-point to the DC bus mid-point by turning on S_{a2} and S_{a5}, and then to apply the DC mid-point reconfiguration technique 1 discussed in Section 12.5.2.3. In this case, the output voltage is reduced, but healthy devices are not subjected to overvoltage stress. Both topologies (shown in Figures 12.8 and 12.9) can withstand single short circuit failure in multiple phases if healthy devices are rated to block the full DC bus voltage. The inverter can keep operating with faulty legs operating in two-level mode for Figure 12.8 and 3-level mode for Figure 12.9.

In the case of single switch open circuit failure for Figure 12.9, the DC bus mid-point is connected to the phase output. For example, if the switch Sa2 fails to open, then Sa3 and Sa6 can be turned on to clamp the phase voltage to the DC bus mid-point. Once the reconfiguration process is complete, there are two options. In post-fault operation, the DC mid-point reconfiguration technique 1 discussed in Section 12.5.2.3 is used. In this case, devices are not required to be overrated. The maximum modulation index is reduced to half the original value.

In a T-type 3-level inverter, a fault can be handled by reducing the faulty phase to a two-level inverter leg when one of the middle two switches fails to open. The operation of the inverter is similar to a two-level inverter in the post-fault condition. A different method is adopted in case of the top and bottom switch failures [19].

12.5.1.2 Cascade multilevel converter (CMC) topology

The cascaded multilevel converter structure was initially proposed for static var compensation [20]. Since then, it has been widely used in high-power motor drive applications [21,22]. A typical CMC topology is shown in Figure 12.10. The topology has inherent module-level redundancy. If a module experiences failure (such as A1), the failed module is bypassed (with S1 for A1) and a fault-tolerant control scheme is implemented for post-fault operation [23]. There are two options at this point. In the first approach, the corresponding healthy modules of two other phases (B1 and C1) are bypassed. This will allow the inverter to maintain a balanced output voltage. However, this approach results in suboptimal utilization of the inverter. The CMCs can effectively use the principle of redundant switching states. Another approach is to implement fault-tolerant operation on the basis of space vector modulation redundancy. As in the case of the NPC converter, when the CMC experiences a failure, some of the space vectors become unavailable. In the post-fault operation, the same output voltage can be obtained by using different switching states. A selection of the appropriate switching states needs to be made. In [24,25], space vectors have been defined in a 60° g-h coordinate system to reduce computational

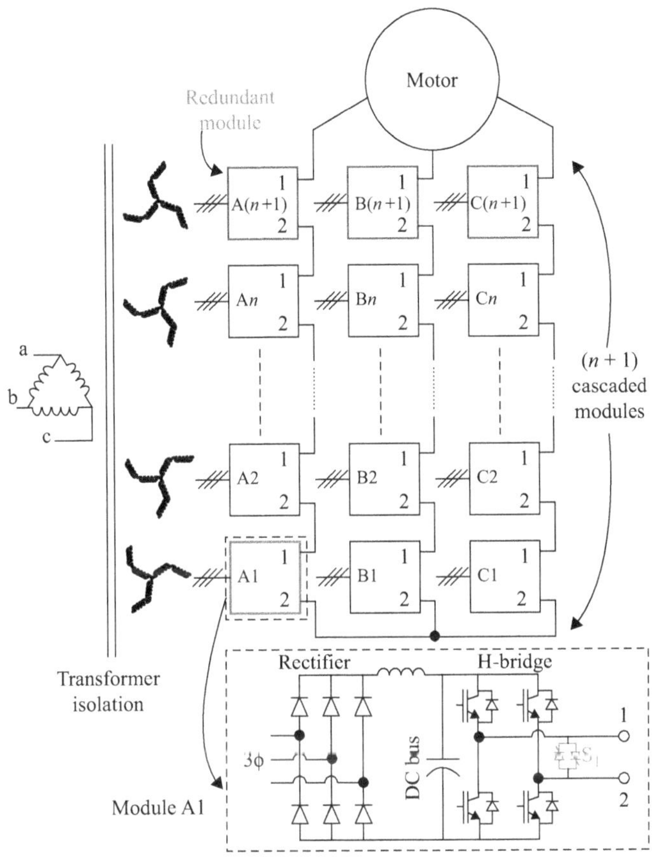

Figure 12.10 Cascade multilevel converter (CMC) topology with (n+1) modules in motor drive application [23]

complexity. Based on this coordinate system, a large small alternation (LSA) modulation scheme has been proposed to achieve the lowest THD in post-fault operation. To maximize the balanced inverter output voltage, the neutral point shift method can also be adopted. This approach is discussed in the section 12.5.2.1.

12.5.1.3 Modular multilevel converter (MMC) topology

The basic operation of a modular multilevel converter (Figure 12.11) is introduced in [27]. The voltage balancing control and application of MMC in motor drive applications are discussed in [28,29]. Other variations of the topology and their applications are discussed in [30]. Modular multilevel converters (MMCs), like CMCs, utilize the module-level redundancy of the topology for post-fault operation. When a device (S_{11}, S_{12}) fails (open/short), the failed submodule can be bypassed and a redundant submodule can be introduced. For example, if S_{11} fails short, the switch S_{12} is turned off. The bypass triac S_1 is turned on to blow the fuse F_1. In subsequent post-fault

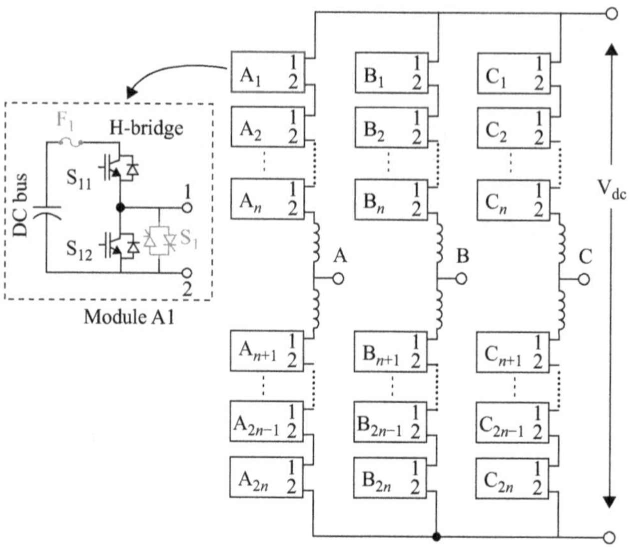

Figure 12.11 Modular multilevel converter (MMC) topology with 2n modules per phase [26]

operations, the triac S_1 remains turned on to bypass the failed module A1. If redundant submodules are not present, it will result in an unbalanced output voltage. To resolve this issue, a fault-tolerant control strategy is introduced in [26], which utilizes carrier rotation–based modulation to achieve voltage balance. Once a fault is detected, the reference signals to the submodules of all three phases are adjusted. The reconstructed modulating signals result in balanced line–line output voltage.

12.5.1.4 Multilevel active clamped (MAC) topology

As discussed in the preceding two sections for CMC and MMC, the redundancy in three-phase switching combinations enables these inverters to produce the same output voltage even in post-fault operation. The MAC topology incorporates this redundancy in certain switching states into a single phase. In case of failure, this redundancy enables the controller to choose an alternate conduction path to retain the same output voltage. This redundancy, however, only exists for a certain output level or for semiconductor devices. For example, if S_{n11} fails, there is no alternate path and level 1 is lost. Also, in some other failure scenarios, healthy devices may be stressed to a higher blocking voltage. Figure 12.12 shows the switching states of different semiconductor devices to obtain level 4 at the output, O. However, in the modified switching states there are two devices that experience overvoltage (2 V) stress. In [31], a modified switching strategy is proposed to eliminate switching states that cause overvoltage stress.

Section summary: Several multilevel converter topologies provide structural redundancy, which can be utilized to continue operation after semiconductor device failure. The topologies discussed in this subsection are summarized in Table 12.2.

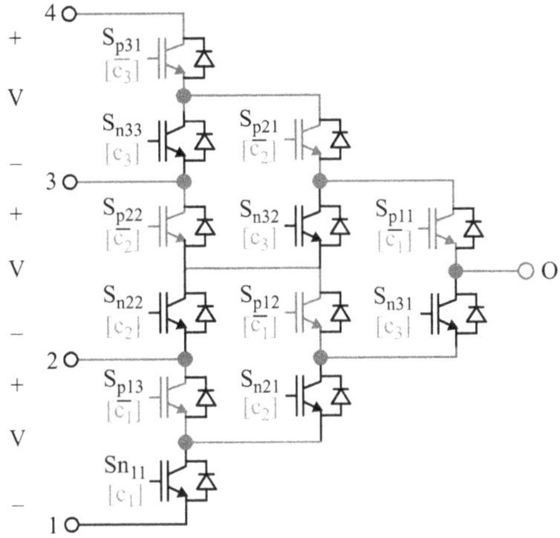

Figure 12.12 Normal switching state to obtain level 4 at the output of MAC (light gray color represents the current path)

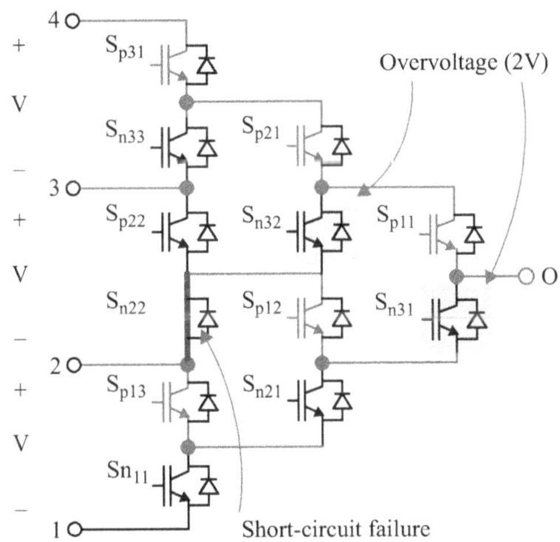

Figure 12.13 Modified switching state to obtain level 4 results in overvoltage at Sn32 and Sn31 (light gray color represents the current path) [31]

12.5.2 Software techniques

12.5.2.1 Neutral point shift

The module-level redundancies of CMC and MMC topologies can be extremely useful in applications where satisfactory process throughput can be achieved under degraded post-fault operation. This approach provides an economically optimal solution [2]. This subsection discusses a neutral point shift method for post-fault

Table 12.2 Summary of topology-based techniques

Topology	Advantages	Disadvantages
Neutral point clamped (NPC)	• Lower voltage devices can be used • Reduction to a two-level topology is possible with appropriate fault isolation circuit	• DC bus voltage balance in post-fault operation is a challenge • Spurious fault detection has been reported • Device may experience over-voltage in certain fault conditions • Post-fault operation without any modification and overvoltage stress is possible in certain fault conditions with reduced output voltage (ex. A1 open circuit failure)
Cascaded multilevel converter (CMC)	• Modular construction • Easy fault isolation (only bypass relay/antiparallel thyristor) • If large numbers of cells are cascaded, bypassing (due to failure) one cell has a minimal effect on system performance.	• Greater control complexity • In post-fault operation, the output voltage is lower than normal operation
Modular multilevel converter (MMC)	• Modular construction • Easier fault isolation (only bypass relay/antiparallel thyristor) • If large numbers of cells are cascaded, bypassing (due to failure) one cell has minimal effect on system performance • Multiple DC sources are not required	• Bulky electrolytic capacitors make the modular cells less reliable • Greater control complexity • In post-fault operation, the output voltage is lower than in normal operation
Multilevel active clamp (MAC)	• Provides multiple conduction paths in each phase • Intraphase redundancy can be achieved for certain switching states	• Device experiences overvoltage in certain fault conditions • A few levels can only be achieved through one switching combination and hence become unattainable when the corresponding device fails • Post-fault control schemes can be complex

operation. It has also been referred to as fundamental phase shift compensation (FPSC).

The neutral point shift technique is typically applied to multilevel converters. The redundancy built in the structure of CMC topology can be utilized by employing the neutral point shift (NPS) technique. Once a fault is detected, the faulty module in the converter is bypassed as discussed in Section 12.5.1.2. As an unequal number of modules is operating in the three phases, it leads to unbalanced output voltage at the three phases of the inverter. As mentioned earlier, there are mainly two choices for post-fault operation. The first choice is to bypass an equal number of modules in all three phases. This approach helps to maintain a balanced output. However, the output voltage is reduced and the converter is under-utilized. The second and more optimal approach is the NPS technique. In NPS, to achieve balanced line-to-line voltage at the output under unbalanced phase voltage conditions, the angle between the phases is adjusted [23]. The neutral point of this adjusted system is different from the neutral point of the balanced condition. Hence, this is called the neutral point shift method. This technique is only applicable if the neutral of the inverter is not connected to the motor neutral.

The traditional technique can be illustrated with the help of the example shown in Figures 12.14 and 12.15. Let the angle between phasors of $V_a - V_b$, $V_b - V_c$, and $V_c - V_a$ after faults be α, β, and γ, respectively. Since the line-to-line voltages are balanced after the NPS, (12.1)–(12.5) are as follows.

$$V_{ab} = V_a^2 + V_b^2 - 2 \cdot V_a \cdot V_b \cdot \cos\alpha \tag{12.1}$$

$$V_{bc} = V_b^2 + V_c^2 - 2 \cdot V_b \cdot V_c \cdot \cos\beta \tag{12.2}$$

$$V_{ca} = V_c^2 + V_a^2 - 2 \cdot V_c \cdot V_a \cdot \cos\gamma \tag{12.3}$$

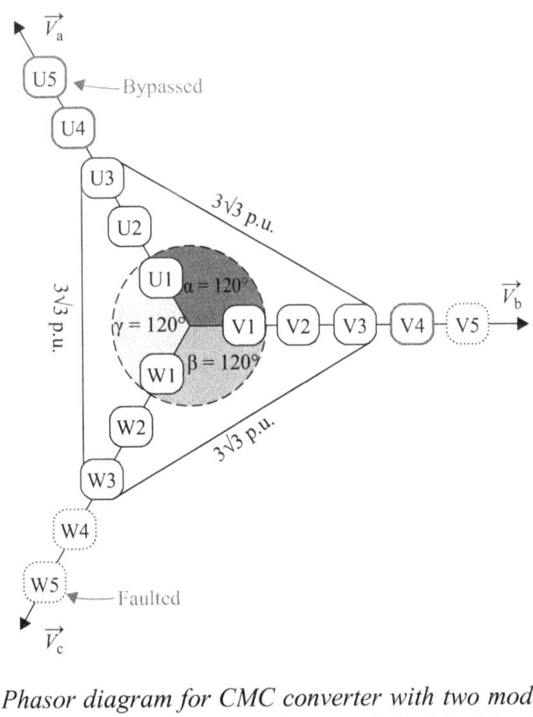

Figure 12.14 Phasor diagram for CMC converter with two modules failed in phase C (W4 and W5) and one module failed in phase b (V5)

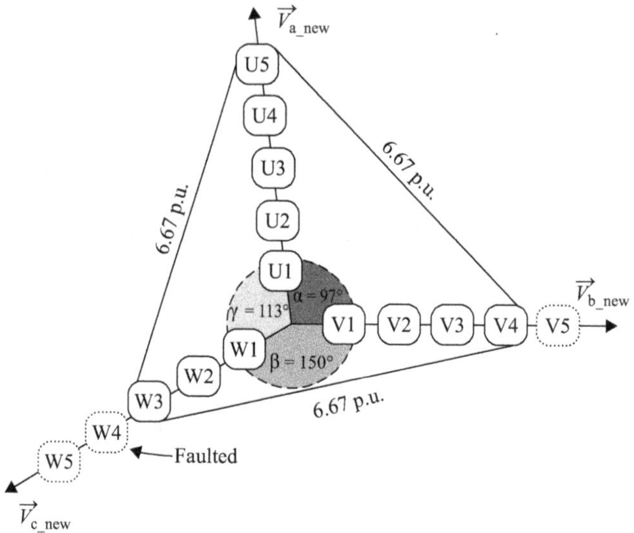

Figure 12.15 Three-phase voltage phasors for CMC after applying the NPS method with all healthy modules functioning [32 33]

$$V_{ab} = V_{bc} = V_{ca} \tag{12.4}$$

$$\alpha + \beta + \gamma = 360^\circ \tag{12.5}$$

The values of α, β, and γ can be obtained by solving nonlinear trigonometric equations [32]. It should be noted that this method may not yield solutions in some cases. Also, in certain fault scenarios, the shifted neutral point may lie outside the triangle formed (Figure 12.16) by the line-to-line voltage phasors, which leads to suboptimal solutions. In such cases, a different approach is adopted (Figure 12.17). The two smaller phasors (i.e., the phases with a greater number of failed modules) are positioned 180° apart. So let us assume that phase b and phase c each have more failed modules than phase a. Then phasor b and phasor c are placed 180° apart, that is, $\beta = 180^\circ$. The next step is to solve for the third phasor magnitude (V_{a_new}) and then the remaining phase-to-phase angles α and γ. The solutions can be obtained from (12.6) and (12.7).

$$V_{a_new} = \sqrt{{V_b}^2 + {V_c}^2 + V_b \cdot V_c} \tag{12.6}$$

$$\alpha = \sin^{-1}\left[\frac{\sqrt{3}}{2 \cdot V_{a_new}} \cdot (V_b + V_c)\right],\ \gamma = 180 - \alpha, \beta = 180 \tag{12.7}$$

This technique always yields a solution. However, this technique should only be used when a solution is not obtainable through the traditional approach or when suboptimal solutions are obtained from the traditional approach.

Another technique has been proposed to extend the modulation range. This is obtained by a balanced third harmonic injection [14]. Balance in the third harmonic

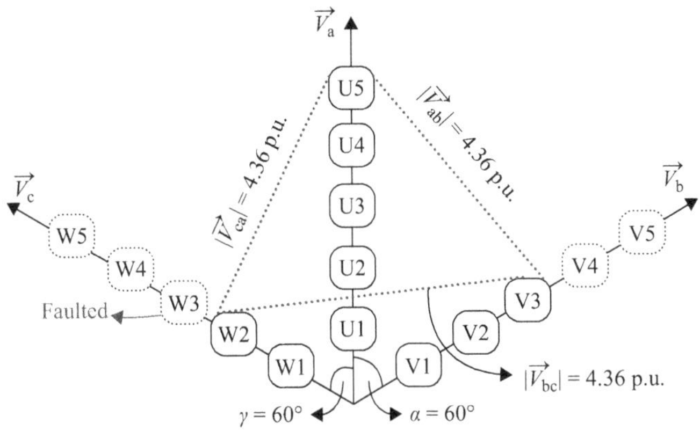

Figure 12.16 Solution from the NPS method where the neutral point lies outside the triangle that leads to suboptimal solutions [32]

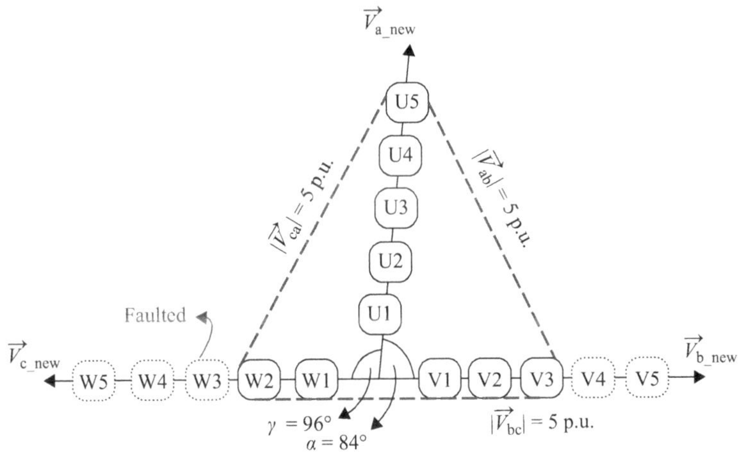

Figure 12.17 Modified voltage phasor diagram with β=180 and adjusted values of α and γ leads to higher line-to-line voltage [32]

is essential to achieve cancellation in line–line voltages. The magnitude of this third harmonic injection is governed by the phase with the minimum available voltage after the fault. For example, in Figure 12.18, upon failure of module W3, modules U1, U2, V1, V2, W1, and W2 can stay balanced with third harmonic injection. The line-to-line rms voltage then becomes $2\times1.15\sqrt{3}$ (as shown in Figure 12.19). The phase shift between the remaining cells, in this case, U3 and V3, can be calculated using the NPS technique. The resultant line-to-line voltage is ~96% of the pre-fault line-to-line voltage (Figure 12.19).

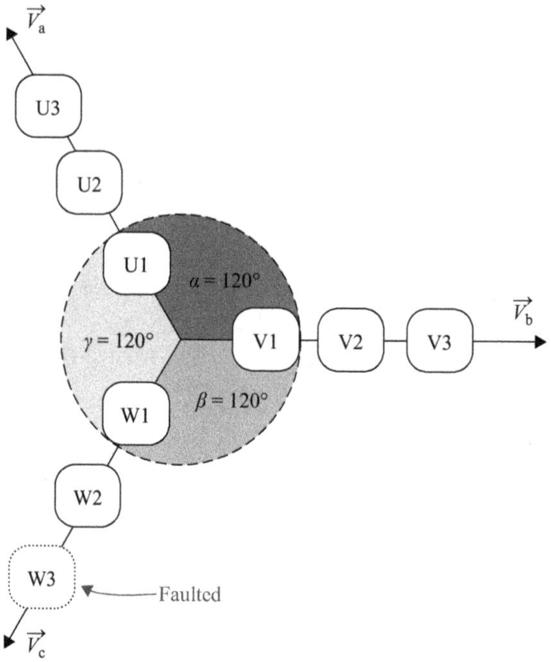

Figure 12.18 Balanced CMC system with one faulty module (W3) in phase c

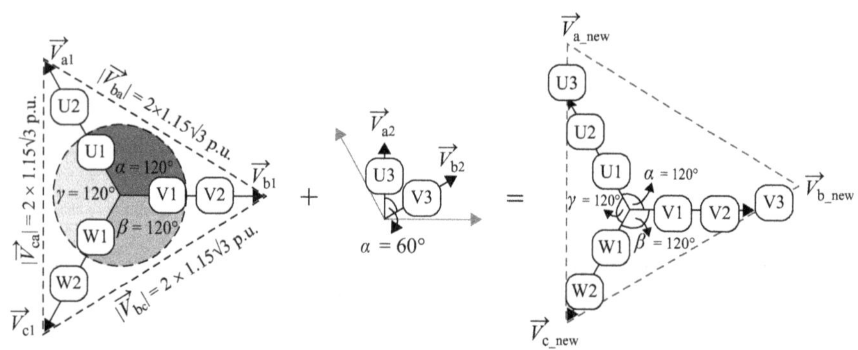

Figure 12.19 Technique for third harmonic injection to enhance the modulation range of CMC system [14]

One important precaution mentioned in [33] is regarding the change in load power factor due to the neutral point shift. The zero-sequence voltage of the fundamental frequency can create a negative real power in a phase. This has been mathematically shown in [32]. The new load power factor depends on the load current (or load power factor) and the angle between the inverter and load voltage.

A negative real power can cause the destruction of a nonregenerative module. This limits the operation to a certain range of load power factor values. Other modulation techniques have also been explored for closed-loop control, such as peak-reduction methods and space vector modulation. In [34], a generalized carrier modulation method based on a geometric approach is presented. Nonfundamental zero-sequence components are added to achieve modified modulation signals. This method obtains solutions in all cases within the linear modulation region. Although not discussed here, the fault-tolerant operation of asymmetric CMC has also been explored.

An equivalent of the neutral point shift technique, referred to as DC bus mid-point reconfiguration, has been discussed in Section 12.5.2.3.

12.5.2.2 DC bus voltage control

As discussed in the previous section, the inherent structural redundancy of CMC has been widely explored for fault-tolerant operation. One approach to obtain optimal output voltage is the NPS technique. Another parameter available for control is the DC bus voltage [18]. This technique has been typically employed in STATCOM applications. The H-bridge is operated as a rectifier to absorb real power for charging the DC bus to the pre-fault value. A regenerative module-based CMC converter is proposed in [35] for motor drive applications. In the regenerative module, the three-phase diode rectifier front end is replaced by a 3-phase PWM rectifier. This gives an additional degree of freedom to reach the pre-fault output voltage level. The NPS method is first utilized to obtain the optimal phase shift for the maximum possible output line–line voltage. If this line–line voltage is lower than the pre-fault line–line voltage, then the DC bus voltage is increased to obtain the desired output line–line voltage. The voltage stress on the devices can be reduced by sharing this increase in DC bus voltage between all the healthy modules in a phase. However, the devices and DC bus need to be overrated to handle the increased voltage stress (and, if applicable, current stress).

12.5.2.3 DC mid-point reconfiguration

The conventional 3-phase two-level topology can be operated in a degraded output mode under fault conditions. One key difference between this technique and the previously discussed methods is that this approach can only be used when a common DC bus is present. In CMC, a common DC bus does not exist. So this technique cannot be used. The DC mid-point reconfiguration can be achieved in two different ways.

Technique 1—DC bus mid-point to inverter phase reconfiguration

Two level: In the first approach shown in Figure 12.20, the DC bus mid-point is connected to the faulted phase using a bidirectional switch represented by a triac. If S_{ap} develops a failure (open or short), the isolation process of S_{ap} is triggered and the gating signals to the other device in the leg, S_{an}, are disabled immediately after fault detection. The controller then turns on the reconfiguration triac, S_{ra}, to link the DC bus mid-point to the faulted phase. The reduced system thus obtained is similar to the

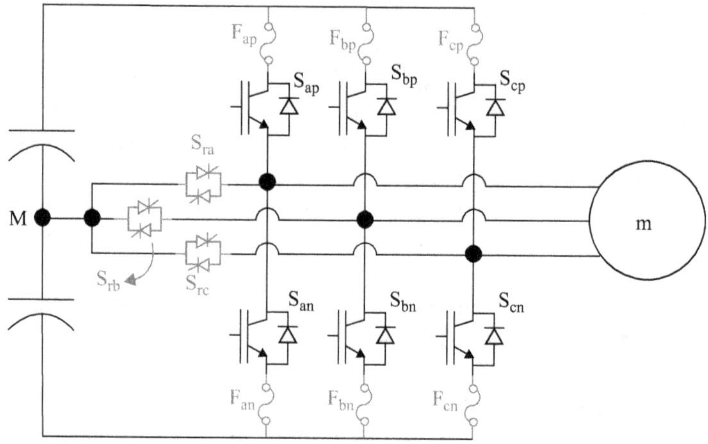

Figure 12.20 DC mid-point reconfiguration technique used in a three-phase motor drive application [36]

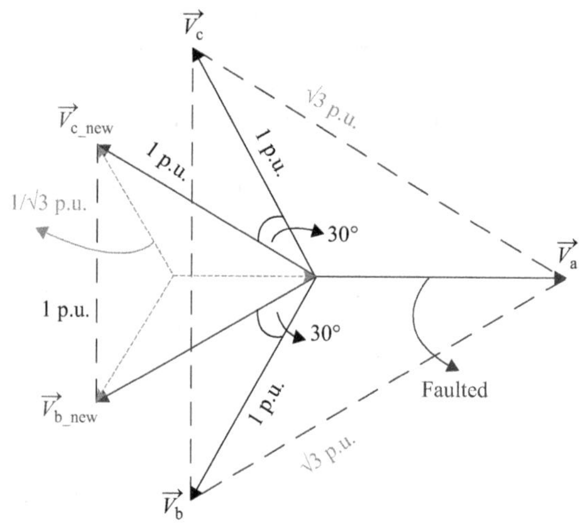

Figure 12.21 Modified inverter phasor diagrams for phase B, phase C after failure of phase A, and DC point reconfiguration for a carrier-based modulation scheme

B4 inverter or 4-switch inverter discussed in [36]. As a result of this reconfiguration, the faulted phase current (phase a) is dependent on the two healthy phase currents (phase b and phase c). The two inverter phase currents can be controlled to obtain balanced positive sequence stator currents. In the carrier-based modulation strategy, the phase angle of the healthy phase voltages (b and c) is adjusted as shown in Figure 12.21 to create balanced line-to-line output voltages. However, as shown in

Figure 12.21, in the post-fault operation the inverter phase voltage is reduced to 0.5 p.u., which is half the maximum value (of 1 p.u. at a modulation index of 1.15 obtained with third harmonic injection). This may be understood better considering the case of the space vector PWM strategy shown in Figure 12.22. The line-to-line voltages in both the carrier- and space-vector-based PWM strategies are reduced to $1/\sqrt{3}$ of the normal operating conditions. The possible instantaneous phase voltages in normal and post-fault (phase C faulted and connected to mid-point "o") are shown in Figure 12.23. During normal operation, only one unique combination of phase voltages of phases A, B, and C is possible (Vdc/2, Vdc/2, −Vdc/2) or (p, p, n). All other combinations can be derived by symmetry. While in post-fault operation, two unique combinations (p, n, o) and (p, p, o) are possible (with other variations like (n, n, o)).

Because of the reduction in motor phase voltages, the system will operate in field weakening mode beyond approximately half the rated speed. The current capability of the inverter remains the same as the pre-fault condition. Hence, the full-rated motor torque can be produced even in post-fault operation. The system can output half the rated power in this reconfigured state.

Multilevel: This DC bus mid-point reconfiguration technique has also been used in NPC inverters. Two major topologies that enable this technique are shown in Figures 12.24 and 12.25. In the first topology (Figure 12.24), a bidirectional switch is used to connect the output to the DC bus midpoint [37]. A variation of the first topology is active-NPC topology, which was briefly discussed earlier (Figure 12.25). The clamp diodes in conventional NPC topology are replaced with IGBTs [38]. During any device failure, the output is connected to the DC bus mid-point using S_{a5}/S_{a6}. As in a two-level inverter, the output voltage in these two

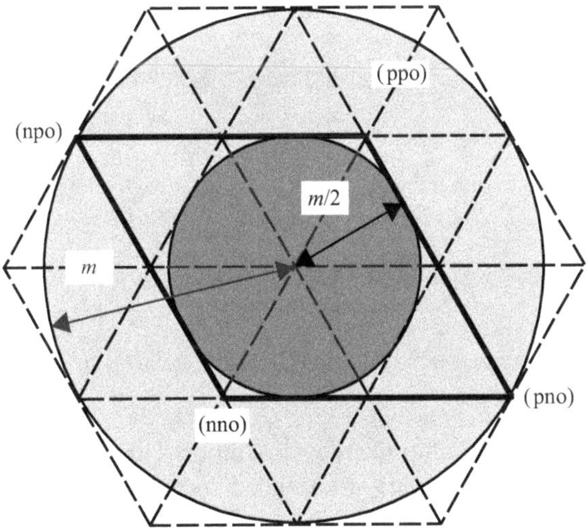

Figure 12.22 Space-vector-based modulation in post-fault condition after one phase is connected to DC bus mid-point. Maximum modulation index reduced to half (from $2/\sqrt{3}$ to $1/\sqrt{3}$)

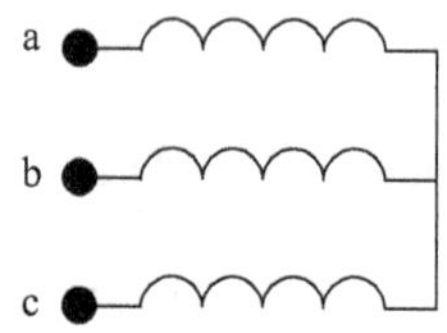

	V_{ao}	V_{bo}	V_{co}	V_{an}	V_{bn}	V_{cn}
Faulted (pno)	$V_{dc}/2$	$-V_{dc}/2$	0	$V_{dc}/2$	$-V_{dc}/2$	0
Faulted (ppo)	$V_{dc}/2$	$V_{dc}/2$	0	$V_{dc}/6$	$V_{dc}/6$	$-V_{dc}/3$
Normal (ppn)	$V_{dc}/2$	$V_{dc}/2$	$-V_{dc}/2$	$V_{dc}/3$	$V_{dc}/3$	$-2V_{dc}/3$

Figure 12.23 Possible instantaneous inverter output phase voltages w.r.t. "o" (DC bus mid-point) and motor phase voltages w.r.t. "n" (motor neutral)

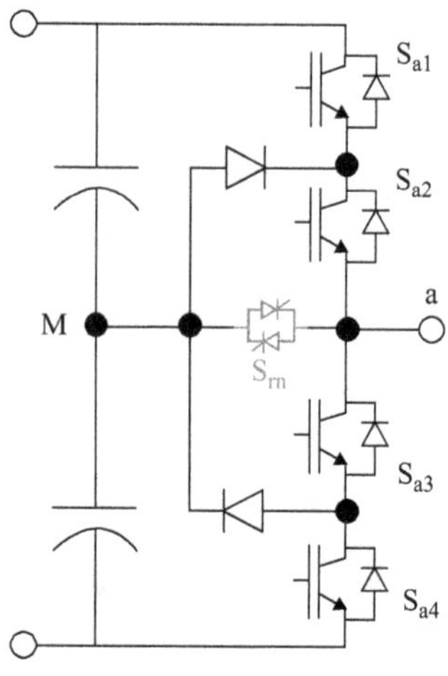

Figure 12.24 NPC converter leg modified for DC mid-point reconfiguration [37]

approaches will be reduced after the reconfiguration. However, the devices are not oversized, and the fault coverage is extended to switch open and short circuit failure.

Technique 2—DC bus mid-point to motor neutral reconfiguration

DC bus mid-point reconfiguration can also be performed with a motor neutral that forms technique 2. Once a fault is detected (see Figure 12.26), the faulted switch

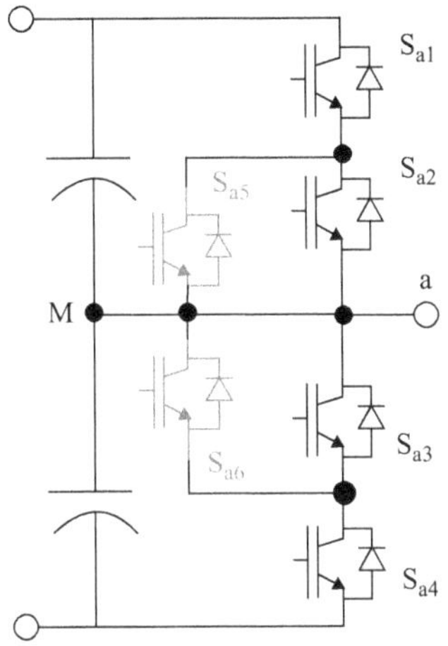

Figure 12.25 NPC topology modified (also referred as active NPC) to enable DC mid-point reconfiguration [38]

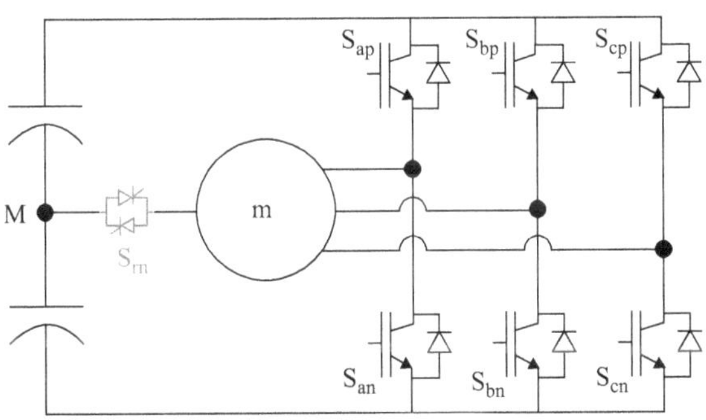

Figure 12.26 Technique 2 reconfigures the DC mid-point to connect to the motor neutral in post-fault operation [39]

(S_{ap}) is isolated, the gating signal to the other switch (S_{an}) in the leg is disabled, and the reconfiguration triac or bidirectional switch (S_{rn}) is turned on to connect the DC mid-point to the motor neutral. The control of the remaining two healthy phases (phase b and phase c) is modified [39]. In the post-fault operation, the torque can be preserved by adjusting the magnitude and phase angle of the current in healthy phases b and c, as shown in Figure 12.27. If the post-fault current is assumed to be 1 p.u., then the torque-producing current is reduced to $1/\sqrt{3}$ p.u. It should be noted that, in this case, the neutral current is $\sqrt{3}$ times the original phase current value. Also, as the system is now unbalanced, the third harmonic voltage injection is not possible. This causes a reduction in the voltage capacity of the system from 1 p.u. to 0.866 p.u.

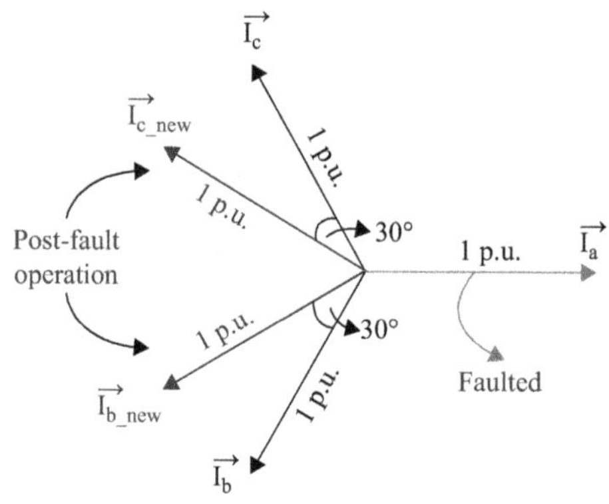

Figure 12.27 Current phasor diagram in the DC mid-point reconfiguration technique 2 to preserve torque output

Section summary: Software-based techniques allow post-fault operation for a small cost. The three such techniques discussed in this subsection are summarized in Table 12.3.

12.5.3 Redundant hardware techniques

The redundant hardware techniques can be divided further into subcategories: (a) system level and (b) converter level. The definitions and some of the techniques proposed in the literature are discussed below.

12.5.3.1 System level

As the name implies, system-level redundant hardware techniques utilize series or parallel converters that are added to the existing system to introduce redundancy. The terms *series* and *parallel* signify the type of connection of the redundant hardware with respect to the load. Both these approaches are being widely used in the industry.

In the series inverter/cascaded inverter approach, a series-connected inverter is added at the system level [40–43]. This topology is also known as dual converter topology, and it has been proposed for open-winding induction machines [44] as well as a general-purpose multiphase machine with an independent phase drive [41]. This effectively translates to a single-phase inverter driving each phase of the motor. From Figure 12.28, it can be observed that each phase voltage can have three states: +Vdc, −Vdc, and 0. These switching states are similar to a three-level inverter. The presence of redundant states allows fault-tolerant operation. The fault-tolerant capability of the series inverter approach shown in Figure 12.28 is limited to a single open/short switch fault and a phase leg open circuit fault. In post-fault operation, the power handling capacity of the topology is reduced. This approach is typically implemented in the case of multiphase motors ($\geq$3 phase) [41]. In [45], an analysis of different fault-tolerant strategies for dual converter topologies with a common DC link has been presented. A variation of this topology has been proposed with two cascaded

Table 12.3 Summary of software-based techniques

Technique	Topologies	Advantages	Disadvantages
Neutral point shift	Cascaded multilevel converter (CMC)	• Optimal output voltage can be achieved in post-fault operation • Achieves balanced line-to-line voltages • All healthy modules participate in post-fault operation • Can be implemented in the form of look-up table to minimize real-time computation	• Only used where motor neutral is not connected to the inverter star-point • Input current THD and output voltage THD increases in post-fault operation • Nonregenerative drives can experience DC bus voltage surge for some load conditions • Unequal loading of healthy modules in post-fault operation if rated power is required • Full output voltage is not be attainable in post-fault operation (if not overrated) • Load power factor after the neutral point shift depends on the load and the shift in neutral point
DC bus voltage control	Cascaded multilevel converter (CMC)	• Increases the available post-fault voltage • Achieves balanced line–line voltage	• Can only be used with active front-end topologies • Increase in DC bus voltage would require the semiconductor device to withstand overvoltage stress
DC bus mid-point to inverter phase reconfiguration (technique 1)	Two-level and multilevel convertersNot applicable to CMC and MMC	• Balanced output voltage • Useful for both two-level and multilevel converters (typically NPC)	• Access to DC bus mid-point is required • Line frequency (60 Hz/ 50Hz) current flows through the DC bus capacitor • Output line-to-line voltages are reduced to $1/\sqrt{3}$ p.u. • Output power reduced to half

(Continues)

Table 12.3 (Continued)

Technique	Topologies	Advantages	Disadvantages
			• Lower phase voltage results in field weakening at half the rated speed
DC bus mid-point to motor neutral reconfiguration (technique 2)	Two-level and multilevel convertersNot applicable to CMC and MMC	• Balanced output voltage • Minimum reconfiguration hardware required • Can handle motor winding open failure	• Access required to both DC bus mid-point and motor neutral • Fourth wire between motor neutral – DC bus mid-point increases system cost • Third harmonic injection is not possible • Large neutral current ($\sqrt{3}$ times the phase current, if phase current is limited to 1 p.u. in post-fault operation) • In post-fault operation, output torque is reduced (without current overrating)

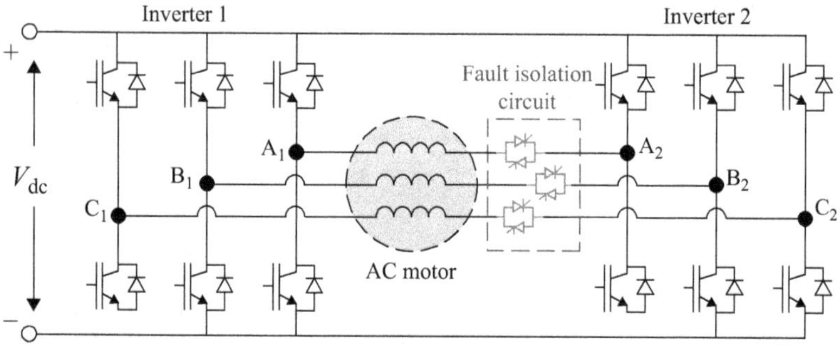

Figure 12.28 System-level redundancy—series converter [40]

inverters with unequal DC bus [46]. The concept of parallel connected redundancy/interleaved converters is popular in not only motor drive applications but also in DC/DC converters, AC/DC converters, active power filters, DFIGs, and so on. The paralleling of power converter modules is also adopted in distributed power system/

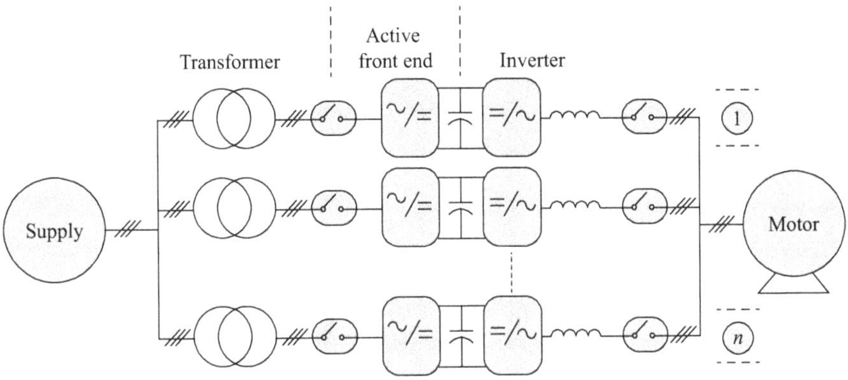

Figure 12.29 System-level redundancy—parallel converter [48]

microgrid scenarios. Interleaved converters are typically used to support higher loads [47], achieve better input power factor, lower current THD, smaller passive (capacitor/magnetic) components, and higher light load efficiency. In addition to all these advantages, they can also provide inherent redundancy.

A system with parallel redundancy can be operated in two ways (see Figure 12.29). In one case, only subsystem 1 is operated in normal operation. When a fault occurs, the faulty subsystem 1 is isolated and subsystem inverter 2 is used to replace subsystem 1. Another approach is to operate both subsystems simultaneously under normal conditions. This approach enables high power levels and excellent grid and load harmonics along with redundancy. However, in this approach, load sharing and circulating currents become a challenge. Depending on the application, different control approaches have been proposed to achieve dynamic load sharing [49–51] and minimize circulating currents [52–55]. In motor drives, the interleaved approach has been reported in multi-megawatt rating drives for the oil and gas industry [48,56,57]. Another strategy uses interleaved inverters that share a common DC link and drive the same load [47,58]. This strategy has the same challenges as the interleaved subsystem approach.

12.5.3.2 Converter level

The converter-level redundancy approaches introduce redundant hardware (single or multiple) to the existing converter topology to enable fault-tolerant operation. Additional power semiconductor devices or legs can be employed in series or parallel redundancy with the existing converter topology for achieving higher reliability.

Redundant series switches

Two-level: The series stacking of power semiconductor devices as shown in Figure 12.30 is used in high-voltage applications such as HVDC transmission systems and industrial drives [59,60,62]. The series connection provides a stack of semiconductor devices with higher blocking voltage. When one of the devices in the stack fails, it can be bypassed (shorted) and the remaining devices

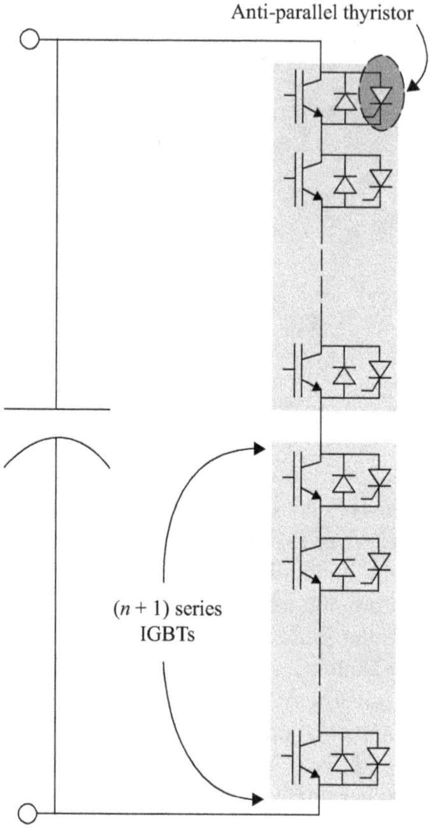

Figure 12.30 Series redundant switches with antiparallel thyristors in medium voltage 2-level converters applications [59–61]

in the stack can continue operation. It is noted that in post-fault operation, the remaining healthy devices in the stack block higher voltage. Some challenges with series connection of power semiconductor devices include static and dynamic voltage sharing, gate drive delays, device parameter spread, thermal cycling, risk of rupture, and undefined failure mode after short circuits and high dV/dt during commutation [59–61]. Press pack IGBTs have been developed to overcome some of these challenges [63]. The press pack fails short after the failure of the IGBT/diode [64]. This enables interruption-free operation in post-fault conditions. However, it has been reported that a failed press pack could subsequently fail in an open circuit condition over a period of time due to aging and the formation of intermetallics, leading to poor conductivity [65]. Another method proposed to address the open circuit failure is a parallel connected thyristor [66].

Multilevel: Flying capacitor multilevel converter (FCMC) topology has inherent redundant states. In [67], a modified FCMC topology is proposed that achieves series redundancy similar to redundant series switches (Figure 12.31).

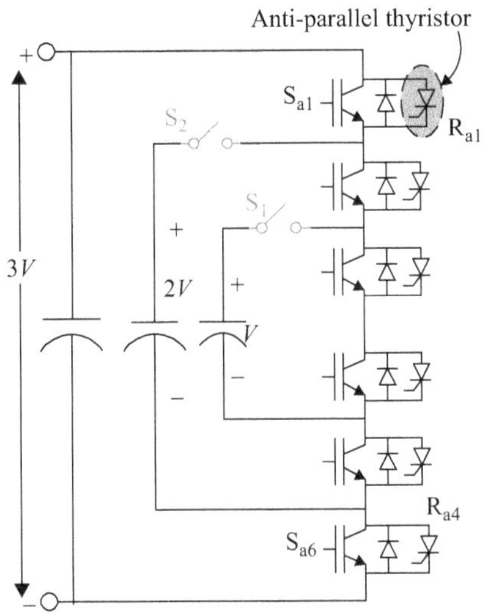

Figure 12.31 Modified FCMC with series switches (S1 and S2) and antiparallel thyristors for fault tolerance [67]

Under a single switch fault, the proposed topology with n-cells is reconfigured and reduced to $(n - 1)$ cells. The faulty switch is bypassed using the antiparallel thyristor. The inherent redundant states are further exploited by the modulation scheme for maintaining capacitor voltage balance in post-fault operation.

Redundant parallel switches

Two-level: Parallel power semiconductors are typically used to increase power handling capability [68]. The enhancement of reliability using parallel semiconductors can be achieved in two ways: stand-by/offline redundant switch and online redundant switch [69]. For example, stand-by parallel switch-based redundancy for a matrix converter has been described in [70]. As shown in Figure 12.32, a bidirectional switch (S_R) is added in parallel (along with an auxiliary reconfiguration circuit) to the matrix converter to replace a faulty switch in post-fault operation.

In [71], offline redundant parallel switches have been added to a two-level three-phase inverter to improve reliability (Figure 12.33). In case of open or short switch failures, the redundant switch S_{Rp} (or S_{Rn}) replaces the high side (or low side) switch. The single-pole double-throw (SPDT) relay works as the isolation and reconfiguration circuit. The antiparallel diodes enable smooth commutation of current and prevent large di/dt transients. In a variation of this topology, a shunt-connected redundant half-bridge is added to the converter to enable fault tolerance at the leg level [72,73]. This is also referred to as phase-redundant topology (Figure 12.34). After a semiconductor switch failure, the entire half-bridge containing the faulty switch is isolated and replaced with the redundant leg. For example, if S_{ap} falls short, then S_{an} is turned on to blow fuses F_{ap} and F_{an}

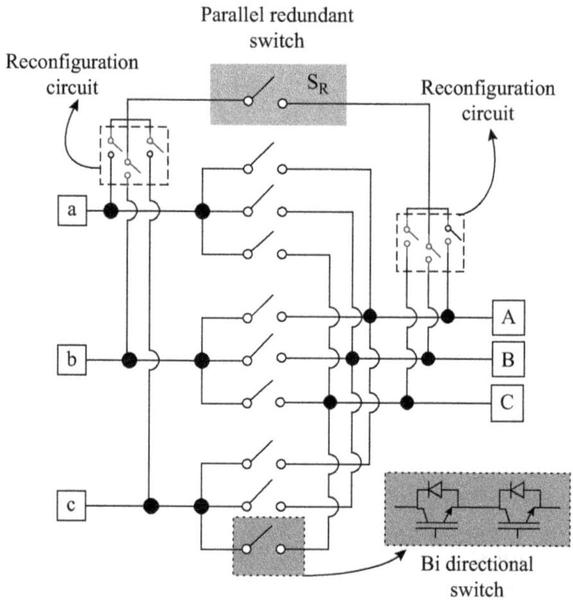

Figure 12.32 Parallel redundant bidirectional switch added to a matrix converter [70]

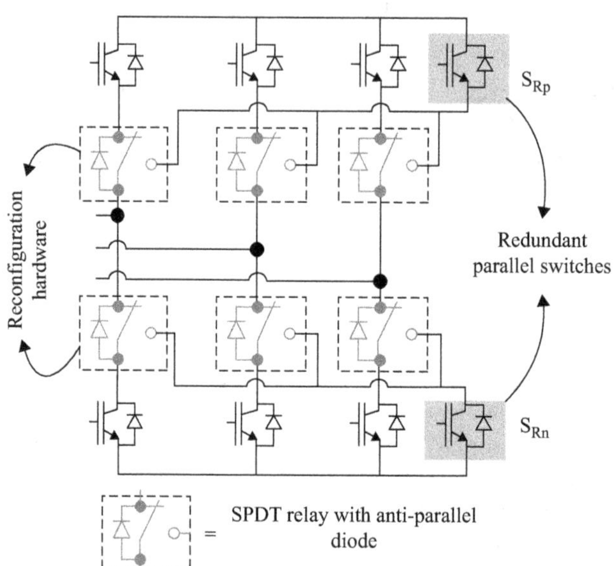

Figure 12.33 Offline redundant parallel switches in a two-level converter [71]

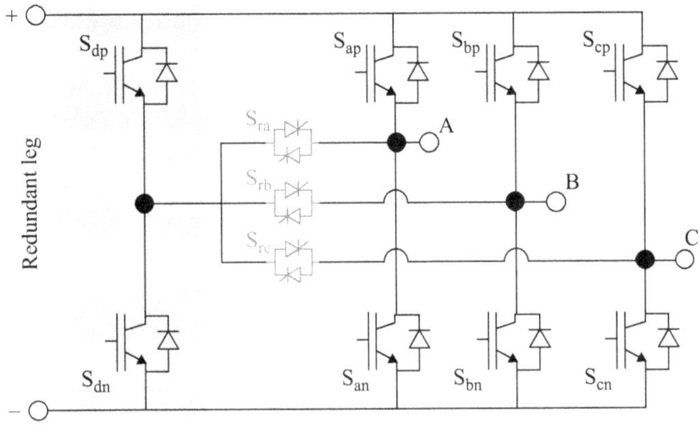

Figure 12.34 Phase-redundant approach for two-level converters [72,73]

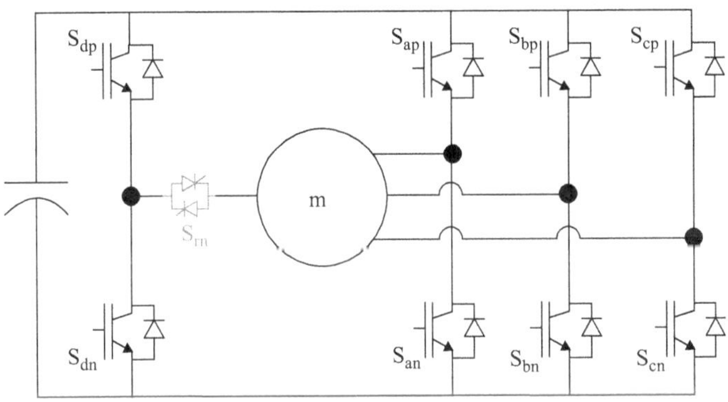

Figure 12.35 Redundant leg connected to the motor neutral using the isolation scheme shown in Figure 12.5(a) [40]

to isolate the faulty leg. Then triac S_{ra} is turned on, and gating signals to S_{ap} and S_{an} are, respectively, applied to S_{dp} and S_{dn}. For open circuit failure of switch S_{ap}, the gating signals to S_{ap} and S_{an} are disabled and respectively applied to switches S_{dp} and S_{dn}. The stand-by redundant phase is shared by all three phases to reduce cost.

Instead of a single stand-by half-bridge for all three phases, a stand-by redundant leg can be added individually to each phase. Another popular variation shown in Figure 12.35 involves an offline redundant leg that is connected to the motor neutral in post-fault operation [40]. This is referred to as a double-switch redundant topology. The control of this topology has been discussed in [74–76]. The modulation strategy discussed in [77] for single-phase induction motors can be extended to three-phase machines to achieve a post-fault voltage of 1 p.u. As in the case of the DC bus mid-point reconfiguration technique 2 discussed earlier, the torque-producing current is reduced to $1/\sqrt{3}$ p.u., and hence the post-fault output power is reduced to $1/\sqrt{3}$ p.u. The neutral current is 3 p.u. The advantages of this topology include a lack of DC bus balancing challenges and an increase in line–line voltage to 1 p.u. compared to the topology in Figure 12.26. In a different control strategy presented in [75], failures can be tolerated in all three top or bottom switches but not a combination of both. The post-fault phase currents are unipolar. The post-fault voltage space is the same as the pre-fault voltage space, but as with the previous modulation scheme, the torque-producing current is reduced to $1/\sqrt{3}$ p.u. The neutral current is also $\sqrt{3}$ p.u.

The redundant leg added to the three-phase inverter can be utilized in normal operation to reduce common mode voltage. From [4], it can be seen that it is also possible to compensate for ASDs' utility input supply disturbances (discussed in section 12.2.3.4) with some modifications to the topology in Figure 12.34. The complete ASD system topology is shown in Figure 12.36. The fault coverage is similar to the topology shown in Figure 12.34.

Multilevel: Online redundant parallel switch approach has been analyzed for multilevel active clamp topology (Figure 12.37). In [31], parallel switches (S_{Rp3}, S_{Rp2}, S_{Rp1}, S_{Rn3}, S_{Rn2}, S_{Rn1}) are added to the open circuit critical diagonal of the MAC topology to achieve fault tolerance to single switch open circuit failure. In case of a single switch short circuit or multiple open switch failures, the converter operation is described with possible loss of output levels. The healthy switches may have to block a larger voltage in post-fault operation.

The redundant parallel leg approaches have also been proposed for NPC converters. The simplest approach is the one similar to the topology in Figure 12.34. Another method is to add an online redundant parallel leg. Two different topologies, which are shown in Figures 12.38 and 12.39, use a redundant leg that participates in pre-fault operation as well. In the first approach, a fourth leg with a flying capacitor (FC) structure is added to a standard 3-phase NPC inverter. If the fourth leg is used in normal operation, it provides a stiff neutral point voltage. The voltage on the flying capacitor can be controlled by choosing appropriate switching patterns for the FC leg. In pre-fault conditions, this frees up the modulation strategy from the task of neutral point balancing and allows the healthy 3-phase NPC inverter control to improve output voltage THD or loss performance. In case of a semiconductor device failure, the hybrid four-leg converter can be reconfigured as a standard NPC converter [78]. If phase A has a failed switch, the fuses F_a and F_d need to be blown using switches on the respective converter legs. Then triacs S_{ra} and S_{dn} are turned ON. Finally, the control strategy is updated to configure the redundant leg to replace the faulted phase. A similar principle is used for the topology shown in Figure 12.39 as well [79]. The redundant leg approach has also been proposed for a T-type 3-level converter [80].

Section summary: Redundant hardware schemes are extensively used in applications where retaining full output power in post-fault operation is critical. All

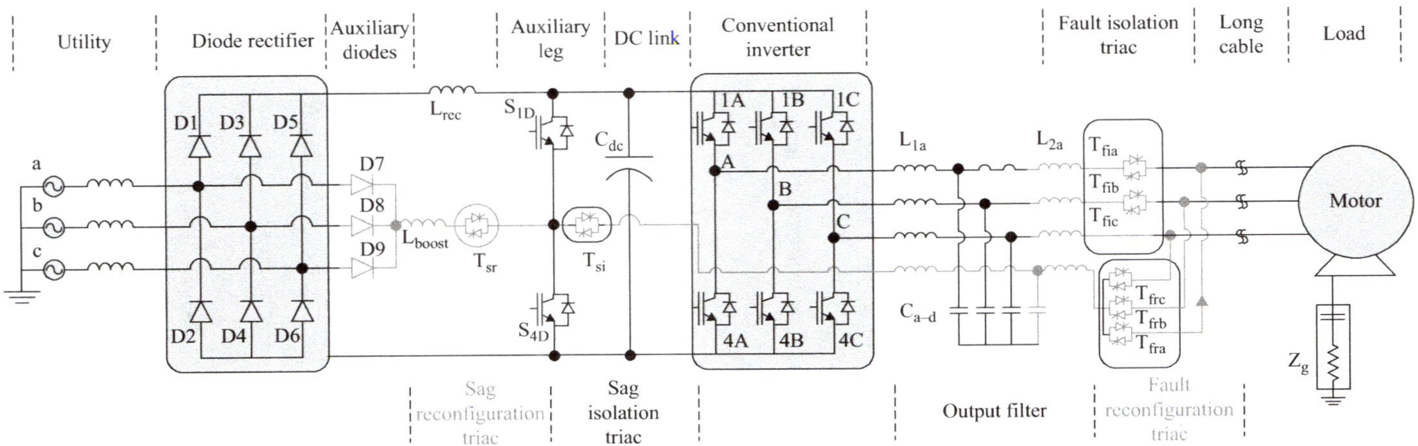

Figure 12.36 Fault-tolerant ASD topology with common mode suppression and input voltage sag compensation [4]

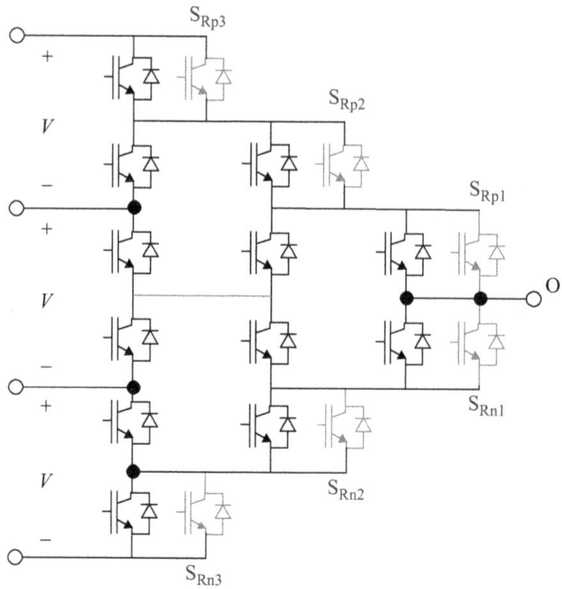

Figure 12.37 Redundant parallel switches added to multilevel active clamp converter [31]

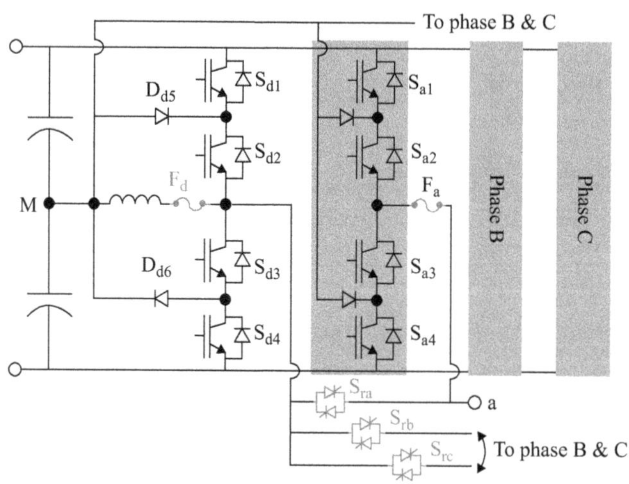

Figure 12.38 Modified NPC converter with fourth flying capacitor leg [78]

the redundant hardware schemes fall under one of two categories: (a) system level and (b) converter level. These schemes can further be classified into series or parallel configurations. The key points from this section have been summarized in Table 12.4.

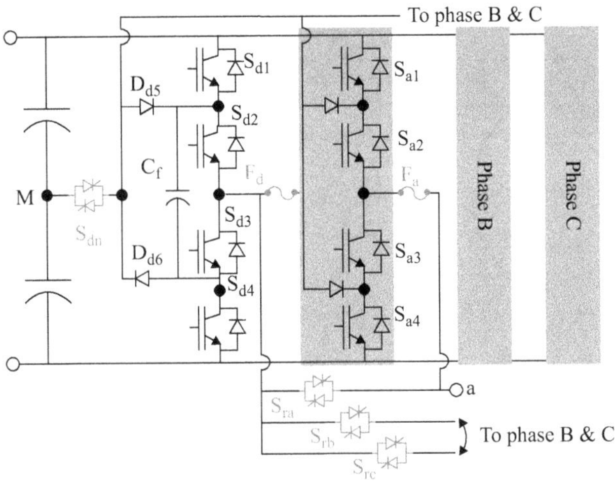

Figure 12.39 Modified NPC converter with a parallel redundant leg [79]

Table 12.4 Summary of redundant hardware-based techniques

Technique	Topologies	Advantages	Disadvantages
System-level series	Two-level and multilevel topologies	• High-quality output waveform (due to more than two levels) • Lower common mode and higher output voltage can be achieved	• May have higher conduction losses • At least twice the number of devices
System-level parallel	Two-level and multilevel	• Redundant branches can be online or on stand-by • Input and output power quality improves when the redundant branch participates in normal operation • Fault isolation process is simplified	• Circulating current can be an issue • n-times the number of devices (n being the number of converters/branches in parallel) • Static and dynamic load sharing is challenging • Expensive, as entire branch is isolated after a single failure • Post-fault power output reduces (if redundant branch is operated online)
Converter-level series	Two-level and multilevel	• Higher blocking voltage can be achieved with lower-rating devices	• Series switches increase the conduction loss

(Continues)

Table 12.4 (Continued)

Technique	Topologies	Advantages	Disadvantages
	converters	• Less expensive than system-level redundancy	• Healthy devices block greater voltage in post-fault operation than normal operation • Static and dynamic voltage sharing are challenging • High di/dt can be an issue; balancing gate drive delays is critical
Converter-level parallel	Two-level and multilevel converters	• Redundant switches can be implemented as stand-by or online • Higher current rating can be achieved by paralleling switches • Online redundant hardware can be utilized for objectives such as neutral point balancing or common mode reduction, freeing up the modulation strategy to pursue other objectives (THD or power loss) • Less expensive than system-level redundancy	• Fault isolation is complex • Complex reconfiguration process requires precise timing/control • Current sharing is challenging when many devices are connected in parallel

12.6 Conclusion

Adjustable speed drives are used in a variety of applications, from electric vehicles to offshore oil rigs, and they operate in moderate to harsh environments. These harsh operating conditions, coupled with aging, can lead to failure in motor drive systems, including semiconductor components. In continuous processes and some other critical applications, an unplanned shutdown is very expensive and a potential safety hazard. This has led to further research in the area of fault-tolerant motor drive systems.

In this chapter, fault-tolerant drives have been broadly classified into topological-, software-, and redundant-hardware-based techniques. It has been shown that each of these approaches is different from the other in terms of post-fault performance, fault coverage, and cost. Also, the advantages and disadvantages of fault-tolerant converter topologies have been discussed along with their applications. A brief summary is presented in Table 12.5.

Table 12.5 *Fault-tolerant topologies: a summary*†

Topology (Figure #)	Fault coverage					Added hardware		Split DC bus	Overrating of additional hardware	Remarks
	One SC	Leg SC	One OC	Leg OC	Motor winding fault	Switches	Fuses			
Two-level conventional	✖	✖	✖	✖	✖	–	–	☺	–	• No fault tolerance capability
Two-level (Figure 12.20)	✔	✔	✔	✔	✖	3	6	☹	☺	• Leg open circuit failure (see Figure 12.4) can only be covered if the failure is between the leg terminal and the triac terminal • DC bus mid-point voltage regulation is required for successful post-fault operation • In post-fault operation, output phase voltage is reduced to half while current capability remains 1 p.u.
Two-level (Figure 12.26 using isolation scheme in Figure 12.5(a))	✔	✔	✔	✔	✔(open)	1	0	☹	☹ (Triac S_m rated to $\sqrt{3}$ p. u. current)	• Current overrating is required for post-fault operation at rated torque (else torque reduced to $1/\sqrt{3}$) • Access to motor neutral is required • A large current flows through the motor neutral to DC bus mid-point (triac rated for $\sqrt{3}$ p.u. current) • Motor electrical frequency (low frequency) current flows through DC bus capacitors

(Continues)

Table 12.5 (*Continued*)

Topology (Figure #)	Fault coverage					Added hardware		Split DC bus	Overrating of additional hardware	Remarks
	One SC	Leg SC	One OC	Leg OC	Motor winding fault	Switches	Fuses			
Two-level (Figure 12.33)	✔	✔	✔	✖*	✖	5	6 SPDT relay	☺	☺	• Output phase voltage reduced to 0.866 p.u. • Mechanical relays are slow • Losses due to the additional hardware can be minimized • No stress on the redundant devices while in normal operation • No degradation of output voltage and current in post-fault operation • This topology can cover two open switch failures in a phase leg • No isolation scheme is required • *Cannot work for Leg OC, the way leg OC is defined
Two-level (Figure 12.34)	✔	✔	✔	✔	✖	5	6	☺	☺	• No capacitor voltage balancing issues • Full output voltage and current can be obtained without overrating the additional hardware
Two-level (Figure 12.35 using isolation scheme Figure 12.5(a))	✔	✔	✔	✔	✔(open)	6	6	☺	☹ (S_{dp}, S_{dn}, and S_{rn} are rated to $\sqrt{3}$ p.u. current)	• Large neutral current ($\sqrt{3}$ p.u. current) • Output voltage can be increased to 1 p.u. or pre-fault value with third harmonic injection

(Continues)

										• Complex control is required for post-fault operation • Access to DC bus mid-point is not required • No DC bus voltage balancing is necessary • Can tolerate single open switch failure in multiple phases of either multiple top or bottom switches (not a combination of top and bottom)
NPC conventional (Figure 12.7)	✖	✖	✖	✖	✖			☹		• Post-fault operation is possible without overvoltage stress in certain fault conditions (ex. open circuit fault A1) with reduced output voltage • Since the effects of failure of open/short switch fault of inverter devices (say A1, A2, and Da5 in phase A) are dissimilar, coverage of open/short circuit failure for one device (say A1) does not qualify as fault coverage for the overall converter (unlike two-level inverter). Leg short circuit is less probable
NPC (Figure 12.24)	✔	✖	✔	✔	✖	3	0	☹	☺	• DC mid-point reconfiguration technique 1 is used in carrier-based PWM or the modulation index is reduced in space vector techniques. In both cases, output voltage is lower than pre-fault voltage • Leg short circuit is less probable

(Continues)

Table 12.5 (Continued)

Topology (Figure #)	Fault coverage					Added hardware		Split DC bus	Overrating of additional hardware	Remarks
	One SC	Leg SC	One OC	Leg OC	Motor winding fault	Switches	Fuses			
NPC (Figure 12.8)	✔	✖	✖	✖	✖	3	6	☹	☹ (T_p and T_n according to the fuse rating)	• Full output power can be delivered by reducing the topology to the equivalent of a two-level converter • Output voltage THD increases and common mode voltage may increase • Healthy devices will experience overvoltage stress • Leg short circuit is less probable
Active NPC (Figure 12.9)	✔	✖	✔	✖	✖	3	6	☹	☺	• Full output power can be delivered with single switch failure (open/short) in multiple phases, but devices experience overvoltage stress • Operating the faulted phase as two-level leg • One OC failure coverage is possible, control needs to be modified to DC mid-point reconfiguration technique 1 • Leg short circuit is less probable
Active NPC (Figure 12.25)	✔	✖	✔	✖	✖	3	6	☹	☺	• Modulation index is reduced to half or NPS method is utilized • Devices do not experience overvoltage • Leg short circuit is less probable

(Continues)

NPC (Figure 12.38 and Figure 12.39)	✔	✖	✔	✔	✖	9 and 8	4	☹	☺	• Full output power after fault • Semiconductor do not need to be oversized • Redundant leg can provide a stiff neutral point voltage in normal operation • Leg short circuit is less probable
Dual bridge (Figure 12.28)	✔	✖	✔	✔	✔(open)	3	0	☺	☺	• The system losses vary based on whether a redundant converter is operated online or offline. • In post-fault condition, independent control of currents in healthy phases is possible, which allows application of 2-phase control methods
MAC (Figure 12.12)	✖	✖	✖	✖	✖					• Intraphase redundancy can be achieved • Complex post-fault control scheme • Certain output voltage levels become unattainable even with a single switch failure
MAC modified (Figure 12.37)	✖	✖	✔	✖	✖	2*(m-1) (no. of levels)	5		☺	• Redundant switching states and added switches help in handling single switch open circuit faults • Can handle certain short circuit switch failures • Complex post-fault control scheme
CMC (Figure 12.10)	✔	✖	✔	✔	✖	3*m (no. of cascaded modules)	0		☺	• SC and OC here are for the cells or modules

(Continues)

Table 12.5 (*Continued*)

Topology (Figure #)	Fault coverage					Added hardware		Split DC bus	Overrating of additional hardware	Remarks
	One SC	Leg SC	One OC	Leg OC	Motor winding fault	Switches	Fuses			
										• Four options available to obtain balanced output voltage • Redundant modules or bypass health cells to balance voltage • Use redundant switching states in space vector PWM • Use NPS or FPSC techniques • DC bus voltage control • Output voltage reduces in first three options. The reduction depends on the number of failed cells
MMC (Figure 12.11)	✔	✔	✔	✔	✖	3*m (no. of cascaded modules)	3*m		☺	• Bulky electrolytic capacitors lower the cell reliability • Due to the bulky DC link capacitors, DC bus-side fuse is necessary to avoid catastrophic outcomes from a single switch short circuit failure • Post-fault control is complex • Output voltage in post-fault operation is lower than in normal operation

References

[1] P. Waide and C. U. Brunner. (2011, May 1). *Energy-efficiency policy opportunities for electric motor-driven systems*. Available: https://www.iea.org/publications/freepublications/publication/EE_for_ElectricSystems.pdf
[2] D. Eaton, J. Rama, and P. Hammond, "Neutral shift [five years of continuous operation with adjustable frequency drives]," *Industry Applications Magazine, IEEE*, vol. 9, pp. 40–49, 2003.
[3] R. A. Epperly, F. L. Hoadley, and R. W. Piefer, "Considerations when applying ASDs in continuous processes," *IEEE Transactions on Industry Applications*, vol. 33, pp. 389–396, 1997.
[4] P. Garg, S. Essakiappan, H. Krishnamoorthy, and P. Enjeti, "A fault tolerant 3-phase adjustable speed drive topology with active common mode voltage suppression," *Power Electronics, IEEE Transactions on*, vol. 30, pp. 2828–2839, 2014.
[5] S. J. Castillo, R. S. Balog, and P. Enjeti, "Predicting capacitor reliability in a module-integrated photovoltaic inverter using stress factors from an environmental usage model," in *North American Power Symposium (NAPS)*, 2010, 2010, pp. 1–6.
[6] J. L. Duran-Gomez, "New approaches to improve the performance of adjustable speed drive (ASD) systems under power quality disturbances," *Ph. D dissertation*, Texas A&M University, College Station, TX, 2000.
[7] A. von Jauanne and Z. Haoran, "A dual-bridge inverter approach to eliminating common-mode voltages and bearing and leakage currents," *IEEE Transactions on Power Electronics*, vol. 14, pp. 43–48, 1999.
[8] A. Von Jouanne and P. N. Enjeti, "Design considerations for an inverter output filter to mitigate the effects of long motor leads in ASD applications," *IEEE Transactions on Industry Applications*, vol. 33, pp. 1138–1145, 1997.
[9] A. Von Jouanne, D. A. Rendusara, P. N. Enjeti, and J. W. Gray, "Filtering techniques to minimize the effect of long motor leads on PWM inverter-fed AC motor drive systems," *IEEE Transactions on Industry Applications*, vol. 32, pp. 919–926, 1996.
[10] (Apr. 1998, May 1). *Effects of AC Drives on Motor Insulation*. Available: http://www.aic-controls.com/fullpanel/uploads/files/abb-technical-guide-us102-motor-cable-lengths.pdf
[11] J. L. Duran-Gomez, P. N. Enjeti, and B. O. Woo, "Effect of voltage sags on adjustable-speed drives: A critical evaluation and an approach to improve performance," *IEEE Transactions on Industry Applications*, vol. 35, pp. 1440–1449, 1999.
[12] D. U. Campos-Delgado, D. R. Espinoza-Trejo, and E. Palacios, "Fault-tolerant control in variable speed drives: A survey," *IET Electric Power Applications*, vol. 2, pp. 121–134, 2008.
[13] R. V. White and F. M. Miles, "Principles of fault tolerance," in *Conference Proceedings of Eleventh Annual Applied Power Electronics Conference and Exposition*, 1996. 1996, vol.1, pp. 18–25.

[14] W. Zhang, D. Xu, P. Enjeti, H. Li, J. Hawke, and H. Krishnamoorthy, "Survey on fault-tolerant techniques for power electronic converters," *IEEE Transactions on Power Electronics*, vol. 29, pp. 6319–6331, 2014.

[15] L. Shengming and L. Xu, "Strategies of fault tolerant operation for three-level PWM inverters," *Power Electronics, IEEE Transactions on*, vol. 21, pp. 933–940, 2006.

[16] A. K. Jain and V. T. Ranganathan, "Vce sensing for IGBT protection in NPC three level converters – Causes for spurious trippings and their elimination," *Power Electronics, IEEE Transactions on*, vol. 26, pp. 298–307, 2011.

[17] P. Jong-Je, K. Tae-Jin, and H. Dong-Seok, "Study of neutral point potential variation for three-level NPC inverter under fault condition," in *Industrial Electronics, 2008. IECON 2008. 34th Annual Conference of IEEE*, 2008, pp. 983–988.

[18] P. Lezana, J. Pou, T. A. Meynard, J. Rodriguez, S. Ceballos, and F. Richardeau, "Survey on fault operation on multilevel inverters," *Industrial Electronics, IEEE Transactions on*, vol. 57, pp. 2207–2218, 2010.

[19] C. Ui-Min, L. Kyo-Beum, and F. Blaabjerg, "Diagnosis and tolerant strategy of an open-switch fault for T-type three-level inverter systems," *Industry Applications, IEEE Transactions on*, vol. 50, pp. 495–508, 2014.

[20] P. Fang Zheng, L. Jih-Sheng, J. W. McKeever, and J. VanCoevering, "A multilevel voltage-source inverter with separate DC sources for static VAr generation," *Industry Applications, IEEE Transactions on*, vol. 32, pp. 1130–1138, 1996.

[21] J. Rodriguez, L. Jih-Sheng, and P. Fang Zheng, "Multilevel inverters: A survey of topologies, controls, and applications," *Industrial Electronics, IEEE Transactions on*, vol. 49, pp. 724–738, 2002.

[22] E. Morris and D. Armitage. (Oct. 18, 2014). *Choosing a motor control platform and drive system.* Available: http://www05.abb.com/global/scot/scot216.nsf/veritydisplay/4fb66e46af347939c1256ed800338956/$file/fact%20packs%20part2.pdf

[23] P. W. Hammond and M. F. Aiello, "Multiphase power supply with plural series connected cells and failed cell bypass," *United States Patent*, 1999.

[24] W. Sanmin, W. Bin, S. Rizzo, and N. Zargari, "Comparison of control schemes for multilevel inverter with faulty cells," in *Industrial Electronics Society, 2004. IECON 2004. 30th Annual Conference of IEEE*, 2004, vol. 2, pp. 1817–1822.

[25] W. Sanmin, W. Bin, L. Fahai, and S. Xudong, "Control method for cascaded H-bridge multilevel inverter with faulty power cells," in *Applied Power Electronics Conference and Exposition, 2003. APEC '03. Eighteenth Annual IEEE*, 2003, vol. 1, pp. 261–267.

[26] S. Ke, X. Bailu, M. Jun, L. M. Tolbert, W. Jianze, C. Xingguo, and J. Yanchao, "A modulation reconfiguration based fault-tolerant control scheme for modular multilevel converters," in *Applied Power Electronics Conference and Exposition (APEC), 2013 Twenty-Eighth Annual IEEE*, 2013, pp. 3251–3255.

[27] R. Marquardt and A. Lesnicar, "A new modular voltage source inverter topology," in *Conf. Rec. EPE*, 2003, pp. 1–50.

[28] M. Hagiwara and H. Akagi, "Control and experiment of pulsewidth-modulated modular multilevel converters," *Power Electronics, IEEE Transactions on*, vol. 24, pp. 1737–1746, 2009.

[29] M. Hagiwara, K. Nishimura, and H. Akagi, "A medium-voltage motor drive with a modular multilevel PWM inverter," *Power Electronics, IEEE Transactions on*, vol. 25, pp. 1786–1799, 2010.

[30] H. Akagi, "New trends in medium-voltage power converters and motor drives," in *Industrial Electronics (ISIE), 2011 IEEE International Symposium on*, 2011, pp. 5–14.

[31] J. Nicolas-Apruzzese, S. Busquets-Monge, J. Bordonau, S. Alepuz, and A. Calle-Prado, "Analysis of the fault-tolerance capacity of the multilevel active-clamped converter," *Industrial Electronics, IEEE Transactions on*, vol. 60, pp. 4773–4783, 2013.

[32] P. Lezana and G. Ortiz, "Extended operation of cascade multicell converters under fault condition," *Industrial Electronics, IEEE Transactions on*, vol. 56, pp. 2697–2703, 2009.

[33] P. W. Hammond, "Enhancing the reliability of modular medium-voltage drives," *Industrial Electronics, IEEE Transactions on*, vol. 49, pp. 948–954, 2002.

[34] F. Carnielutti, H. Pinheiro, and C. Rech, "Generalized carrier-based modulation strategy for cascaded multilevel converters operating under fault conditions," *Industrial Electronics, IEEE Transactions on*, vol. 59, pp. 679–689, 2012.

[35] P. Lezana, G. Ortiz, and J. Rodriguez, "Operation of regenerative cascade multicell converter under fault condition," in *Control and Modeling for Power Electronics, 2008. COMPEL 2008. 11th Workshop on*, 2008, pp. 1–6.

[36] H. W. van der Broeck and J. D. Van Wyk, "A comparative investigation of a three-phase induction machine drive with a component minimized voltage-fed inverter under different control options," *Industry Applications, IEEE Transactions on*, vol. IA-20, pp. 309–320, 1984.

[37] S. Farnesi, P. Fazio, and M. Marchesoni, "A new fault tolerant NPC converter system for high power induction motor drives," in *Diagnostics for Electric Machines, Power Electronics & Drives (SDEMPED), 2011 IEEE International Symposium on*, 2011, pp. 337–343.

[38] L. Jun, A. Q. Huang, L. Zhigang, and S. Bhattacharya, "Analysis and design of active NPC (ANPC) inverters for fault-tolerant operation of high-power electrical drives," *Power Electronics, IEEE Transactions on*, vol. 27, pp. 519–533, 2012.

[39] L. Tian-Hua, F. Jen-Ren, and T. A. Lipo, "A strategy for improving reliability of field-oriented controlled induction motor drives," *Industry Applications, IEEE Transactions on*, vol. 29, pp. 910–918, 1993.

[40] B. A. Welchko, T. A. Lipo, T. M. Jahns, and S. E. Schulz, "Fault tolerant three-phase AC motor drive topologies: A comparison of features, cost, and

limitations," *IEEE Transactions on Power Electronics*, vol. 19, pp. 1108–1116, 2004.
[41] T. M. Jahns, "Improved reliability in solid-state AC drives by means of multiple independent phase drive units," *Industry Applications, IEEE Transactions on*, vol. IA-16, pp. 321–331, 1980.
[42] B. C. McCrow, A. G. Jack, D. J. Atkinson, and J. A. Haylock, "Fault tolerant drives for safety critical applications," in *New Topologies for Permanent Magnet Machines (Digest No: 1997/090), IEE Colloquium on*, 1997, pp. 5/1–5/7.
[43] L. de Lillo, L. Empringham, P. W. Wheeler, *et al.*, "Multiphase power converter drive for fault-tolerant machine development in aerospace applications," *Industrial Electronics, IEEE Transactions on*, vol. 57, pp. 575–583, 2010.
[44] H. Stemmler and P. Guggenbach, "Configurations of high-power voltage source inverter drives," in *Power Electronics and Applications, 1993, Fifth European Conference on*, 1993, pp. 7–14.
[45] J. A. Restrepo, A. Berzoy, A. E. Ginart, J. M. Aller, R. G. Harley, and T. G. Habetler, "Switching strategies for fault tolerant operation of single DC-link dual converters," *IEEE Transactions on Power Electronics*, vol. 27, pp. 509–518, 2012.
[46] B. V. Reddy, V. T. Somasekhar, and Y. Kalyan, "Decoupled space-vector PWM strategies for a four-level asymmetrical open-end winding induction motor drive with waveform symmetries," *Industrial Electronics, IEEE Transactions on*, vol. 58, pp. 5130–5141, 2011.
[47] M. Hashii, K. Kousaka, and M. Kaimoto, "New approach to a high-power GTO PWM inverter for AC motor drives," *Industry Applications, IEEE Transactions on*, vol. IA-23, pp. 263–269, 1987.
[48] S. Schroder, P. Tenca, T. Geyer, *et al.*, "Modular high-power shunt-interleaved drive system: A realization up to 35 MW for oil and gas applications," in *Industry Applications Society Annual Meeting, 2008. IAS'08*. Piscataway, NJ: IEEE, 2008, pp. 1–8.
[49] U. Borup, F. Blaabjerg, and P. N. Enjeti, "Sharing of nonlinear load in parallel-connected three-phase converters," *Industry Applications, IEEE Transactions on*, vol. 37, pp. 1817–1823, 2001.
[50] L. Shiguo, Y. Zhihong, L. Ray-Lee, and F. C. Lee, "A classification and evaluation of paralleling methods for power supply modules," in *Power Electronics Specialists Conference, 1999. PESC 99. 30th Annual IEEE*, pp. 901–908 vol. 2, 1999.
[51] J. M. Guerrero, J. C. Vasquez, J. Matas, M. Castilla, and L. G. de Vicuna, "Control strategy for flexible microgrid based on parallel line-interactive UPS systems," *Industrial Electronics, IEEE Transactions on*, vol. 56, pp. 726–736, 2009.
[52] L. Asiminoaei, E. Aeloiza, P. N. Enjeti, and F. Blaabjerg, "Shunt active-power-filter topology based on parallel interleaved inverters," *Industrial Electronics, IEEE Transactions on*, vol. 55, pp. 1175–1189, 2008.

[53] T. Yoshikawa, H. Inaba, and T. Mine, "Analysis of parallel operation methods of PWM inverter sets for an ultra-high speed elevator," in *Applied Power Electronics Conference and Exposition, 2000. APEC 2000. Fifteenth Annual IEEE*, vol. 2, pp. 944–950, 2000.

[54] J. G. Ciezki and R. W. Ashton, "The control of parallel-connected inverters for U.S. Navy shipboard applications," 2001.

[55] H. Ming, H. Haibing, X. Yan, and H. Zhongyi, "Distributed control for AC motor drive inverters in parallel operation," *Industrial Electronics, IEEE Transactions on*, vol. 58, pp. 5361–5370, 2011.

[56] T. Geyer and S. Schroder, "Reliability considerations and fault-handling strategies for multi-MW modular drive systems," *Industry Applications, IEEE Transactions on*, vol. 46, pp. 2442–2451, 2010.

[57] R. Baccani, R. Zhang, T. Toma, A. Iuretig, and M. Perna, "Electric systems for high power compressor trains in oil and gas applications—system design, validation approach and performance," in *Annual Turbomach. Symp*, 2007.

[58] X. Lei and S. Jian, "Motor drive system EMI reduction by asymmetric interleaving," in *Control and Modeling for Power Electronics (COMPEL), 2010 IEEE 12th Workshop on*, 2010, pp. 1–7.

[59] Y. Shakweh and P. Aufleger, "Multi-megawatt, medium voltage, PWM, voltage source, sine-wave-output converter for industrial drive applications," in *Power Electronics and Variable Speed Drives, 1998. Seventh International Conference on (Conf. Publ. No. 456)*, 1998, pp. 632–637.

[60] N. Shammas, R. Withanage, and D. Chamund, "Review of series and parallel connection of IGBTs," *IEE Proceedings-Circuits, Devices and Systems*, vol. 153, pp. 34–39, 2006.

[61] Y. Shakweh and E. A. Lewis, "Assessment of medium voltage PWM VSI topologies for multi-megawatt variable speed drive applications," in *Power Electronics Specialists Conference, 1999. PESC 99. 30th Annual IEEE*, 1999, vol. 2, pp. 965–971.

[62] R. Chokhawala, B. Danielsson, and L. Angquist, "Power semiconductors in transmission and distribution applications," in *Power Semiconductor Devices and ICs, 2001. ISPSD'01. Proceedings of the 13th International Symposium on*, 2001, pp. 3–10.

[63] S. Bernet, "Recent developments of high power converters for industry and traction applications," *Power Electronics, IEEE Transactions on*, vol. 15, pp. 1102–1117, 2000.

[64] Y. Uchida, Y. Seki, Y. Takahashi, and M. Ichijoh, "Development of high power press-pack IGBT and its applications," in *Microelectronics, 2000. Proceedings. 2000 22nd International Conference on*, 2000, vol. 1, pp. 125–129.

[65] W. Rui, F. Blaabjerg, W. Huai, M. Liserre, and F. Iannuzzo, "Catastrophic failure and fault-tolerant design of IGBT power electronic converters – An overview," in *39th Annual Conference of the IEEE Industrial Electronics Society, IECON*, 2013, pp. 507–513.

[66] A. L. Julian and G. Oriti, "A comparison of redundant inverter topologies to improve voltage source inverter reliability," *Industry Applications, IEEE Transactions on*, vol. 43, pp. 1371–1378, 2007.

[67] K. Xiaomin, K. A. Corzine, and Y. L. Familiant, "A unique fault-tolerant design for flying capacitor multilevel inverter," *Power Electronics, IEEE Transactions on*, vol. 19, pp. 979–987, 2004.

[68] C. Keller and Y. Tadros, "Are paralleled IGBT modules or paralleled IGBT inverters the better choice?," in *Power Electronics and Applications, 1993, Fifth European Conference on*, 1993, vol. 5, pp. 1–6.

[69] H. Behjati and A. Davoudi, "Reliability analysis framework for structural redundancy in power semiconductors," *Industrial Electronics, IEEE Transactions on*, vol. 60, pp. 4376–4386, 2013.

[70] J. Andreu, I. Kortabarria, E. Ibarra, I. M. de Alegria, and E. Robles, "A new hardware solution for a fault tolerant matrix converter," in *Industrial Electronics, 2009. IECON'09. 35th Annual Conference of IEEE*, 2009, pp. 4469–4474.

[71] A. Cordeiro, J. Palma, J. Maia, and M. Resende, "Combining mechanical commutators and semiconductors in fast changing redundant inverter topologies," in *EUROCON – International Conference on Computer as a Tool (EUROCON), 2011 IEEE*, 2011, pp. 1–4.

[72] S. Bolognani, M. Zordan, and M. Zigliotto, "Experimental fault-tolerant control of a PMSM drive," *Industrial Electronics, IEEE Transactions on*, vol. 47, pp. 1134–1141, 2000.

[73] S. Yantao and W. Bingsen, "Analysis and experimental verification of a fault-tolerant HEV powertrain," *IEEE Transactions on Power Electronics*, vol. 28, pp. 5854–5864, 2013.

[74] S. Bolognani, M. Zordan, and M. Zigliotto, "Experimental fault-tolerant control of a PMSM drive," *IEEE Transactions on Industrial Electronics*, vol. 47, pp. 1134–1141, 2000.

[75] R. L. A. Ribeiro, C. B. Jacobina, A. M. N. Lima, and E. R. C. da Silva, "A strategy for improving reliability of motor drive systems using a four-leg three-phase converter," in *Sixteenth Annual IEEE Applied Power Electronics Conference and Exposition (APEC)*, 2001, vol. 1, pp. 385–391.

[76] M. Beltrao de Rossiter Correa, C. B. Jacobina, E. R. Cabral da Silva, and A. M. N. Lima, "An induction motor drive system with improved fault tolerance," *Industry Applications, IEEE Transactions on*, vol. 37, pp. 873–879, 2001.

[77] D. G. Holmes and A. Kotsopoulos, "Variable speed control of single and two phase induction motors using a three phase voltage source inverter," in *Industry Applications Society Annual Meeting, 1993, Conference Record of the 1993 IEEE*, 1993, vol. 1, pp. 613–620.

[78] S. Ceballos, J. Pou, E. Robles, *et al.*, "Three-level converter topologies with switch breakdown fault-tolerance capability," *Industrial Electronics, IEEE Transactions on*, vol. 55, pp. 982–995, 2008.

[79] S. Ceballos, J. Pou, J. Zaragoza, E. Robles, J. L. Villate, and J. L. Martin, "Fault-tolerant neutral-point-clamped converter solutions based on including a fourth resonant leg," *Industrial Electronics, IEEE Transactions on*, vol. 58, pp. 2293–2303, 2011.

[80] Z. Wenping, L. Guangyuan, X. Dehong, J. Hawke, P. Garg, and P. Enjeti, "A fault-tolerant T-type three-level inverter system," in *Applied Power Electronics Conference and Exposition (APEC), 2014 Twenty-Ninth Annual IEEE*, 2014, pp. 274–280.

Chapter 13

Mission profile-oriented reliability analysis and design in wind turbine and photovoltaic systems

Yongheng Yang[1], *Frede Blaabjerg*[2], *Ke Ma*[3] *and Dao Zhou*[2]

Power electronics has enabled a wide-scale adoption of renewable energy systems like wind turbines (WTs) and photovoltaics (PVs) in modern power systems. Requirements for higher efficiency and higher reliability are of concern along with this thriving increase to further reduce the cost of energy. Currently, power electronics converters for renewable energy sources (RESs) are typically targeted at a servicing period of more than 30 years. To achieve so, it is essential to design the WT and PV systems considering mission profiles, which are an inherent reflection of the RES intermittency. As such, this chapter starts with a description of the mission profiles for WT and PV systems. It is centered on how to translate the mission profiles to the loading on power electronics in those systems and thus how to design for higher reliability. The mission profile-oriented reliability design approach is demonstrated through various case studies.

13.1 Mission profile for renewable energy systems

Mission profiles are closely related to the cost and reliability performance of power electronics, as they affect the loading distribution. In RES applications, the mission profiles are relatively tough, demanding the interfacing power electronics must withstand a large amount of power (e.g., up to 18 MWs in WT systems), perform a series of complicated functions required by the grid operator, and meanwhile operate under harsh environmental conditions [1]. For instance, there might be day-night temperature swings, passing-cloud-induced irradiance changes, dust, vibration, humidity, salinity, and so forth. In this section, typical mission profiles and demands for WT and PV systems are presented.

[1]College of Electrical Engineering, Zhejiang University, China
[2]AAU Energy, Aalborg University, Denmark
[3]School of Electronics Information and Electrical Engineering, Shanghai Jiao Tong University, China

13.1.1 Operational conditions

As the input of the entire WT system, wind energy quantified by wind speed is an important factor determining the design, control, maintenance, energy yield, and also the cost of wind energy. During the construction and design phase, assumptions must be made about the wind conditions in which the WT will operate. Typically, wind classes are used in the design phase of WT systems, and the wind classes are characterized in terms of the average wind speed, extreme 50-year gust, and turbulence [2]. This is also being updated continuously [3]. Since the loading of power converters for WT systems is mainly caused by wind speed and its variations, the knowledge of wind class is of importance for reliability evaluation. Table 13.1 shows the most commonly adopted classifications of wind conditions, and an annual wind speed distribution is depicted in Figure 13.1 using a Weibull function.

A real-field 1-year wind speed profile is illustrated in Figure 13.2 [4], which is a 3-h average of the real wind speeds at 80-m hub height, and it was designated for the wind farm located near Thyborøn, Denmark, with a latitude of 56.71° and a longitude of 8.20°.

Table 13.1 Wind classes according to IEC 61400 [2]

Wind class	**I-a**	**I-b**	**II-a**	**II-b**	**III-a**	**III-b**	**IV-b**
Turbulence[a]	**18%**	**16%**	**18%**	**16%**	**18%**	**16%**	**16%**
Annual average wind speed (m/s)[b]	10		8.5		7.5		6
Extreme 50-year gust (m/s)[c]	70		59.5		52.5		42

[a]Turbulence is measured at wind speed of 15 m/s, quantifying how much the wind varies within 10 min.
[b]At hub height.
[c]Based on a 3-s average of wind speed.

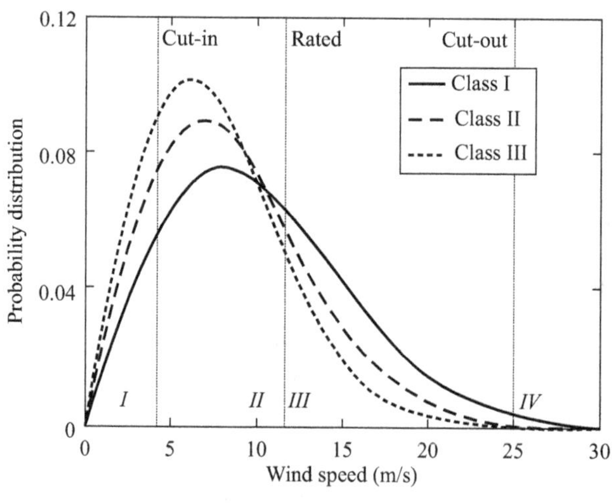

Figure 13.1 Annual wind distribution with different wind classes (Class I–Class III) defined in the IEC 61400 Standard

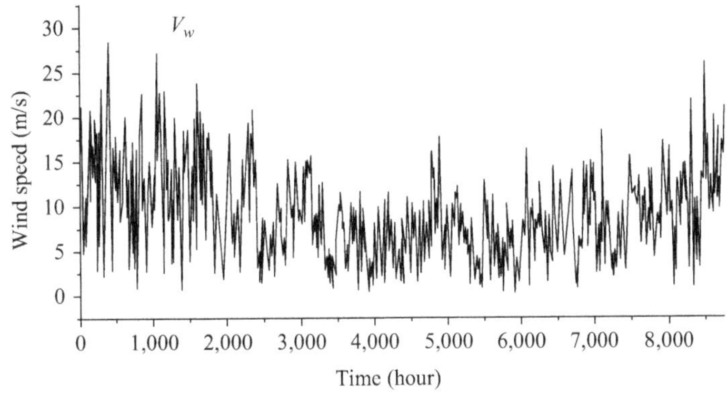

Figure 13.2 One-year mission profile of wind speed from a wind farm in Denmark (3-h averaged) with Class I roughness according to the IEC 61400 Standard

The chosen hub wind speed belongs to the IEC 61400 I-a wind class (see Table 13.1) with an average wind speed of 8.5–10 m/s. It can be seen in Figure 13.2 that the wind speed fluctuates greatly from 0 m/s up to 28 m/s. Without proper controls of the mechanical and electrical parts, this fluctuation may be transferred to the grid, eventually, inducing poor power quality and even grid instabilities. Moreover, the components in the system may suffer from large cyclical loading, which could induce failures [5–10]. Consequently, the roughness and fluctuation of the wind speeds should be considered in the control and design of the power converters for WT applications [11–13].

Similarly to the WT case, the mission profiles, including solar irradiance and ambient temperature, are also very important for the design, control, and operation of PV systems. With the accumulation of field experience and the advancement of monitoring technologies, better mission profile data are available in power electronic systems, including PV inverters [14,15]. This offers higher possibilities to predict the lifetime and also the energy production of a PV system [15–17] with high confidence.

Figure 13.3 shows yearly mission profiles for PV systems [15]. It can be observed in Figure 13.3 that both the solar irradiance and the ambient temperature vary significantly over the year, indicating the intermittency of solar PV energy. The intermittency will result in power fluctuations, requiring a dedicated maximum power point tracking (MPPT) control to optimize the energy yield [18]. It is also known that the reliability of a power electronics system strongly couples with the temperature loading [4,10,11,14–17,19–22], and it is affected by the mission profiles. As such, mission profiles at various timescales may affect the consumed lifetime differently [4]. In addition, the resultant fluctuating power may introduce grid stability issues, as reported in recent years. Therefore, those aspects related to mission profiles must be considered in the control and design of the PV systems. Furthermore, the advancement of power electronics technologies, monitoring techniques, and also intelligent control strategies, can contribute to the improvement of the lifetime and a reduction of the cost of PV energy.

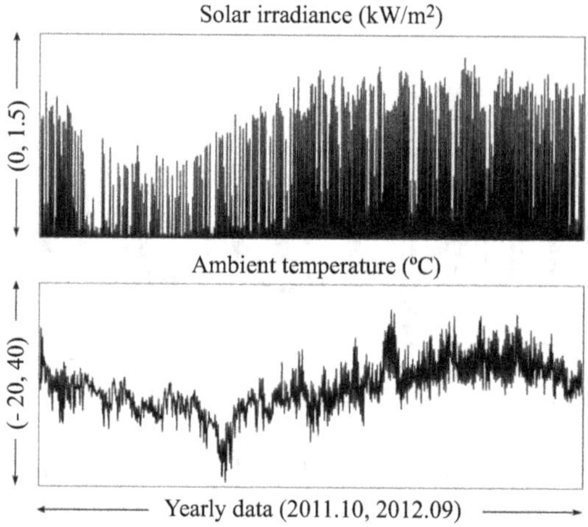

Figure 13.3 Yearly real-field mission profiles (5 mins per sampling data) for PV systems: (a) solar irradiance level and (b) ambient temperature

13.1.2 Grid-side demands

Many countries/regions, including international entities, have issued grid codes to guide the integration of RESs [23–29]. Updating the requirements is also an indirect reflection of the penetration level of RESs into the grid, and thus the requirements normally cover a wide range of voltage levels from medium voltage (MV) to ultra-high voltage (UHV). Seen from the power system point of view, grid codes are always trying to make the RESs act as a conventional power plant. Taking the WT system as an example, the power controllability, power quality, fault ride-through capability, and grid support during grid disturbances have been focused on in the grid codes. The following exemplifies certain grid codes for WT systems in selected countries, and most of the grid codes are effective either for an individual WT or for a wind farm. Notably, such grid regulations are dynamic and may change from year to year.

13.1.2.1 Active power control

According to grid codes in most countries, for example, Denmark, Ireland, and Germany, an individual WT must be able to control the active power at the point of common coupling (PCC) within a given power range. Typically, the active power is regulated in accordance with the system frequency to maintain frequency stability. Figure 13.4 demonstrates the characteristics of the frequency control through active power regulation in the Danish grid codes.

For wind farms, which are normally connected at the transmission line, the WTs as a whole should act as a conventional power plant providing a wide range of controlled active power according to the demands of the transmission system operator (TSO). Moreover, such wind farms must participate in primary and

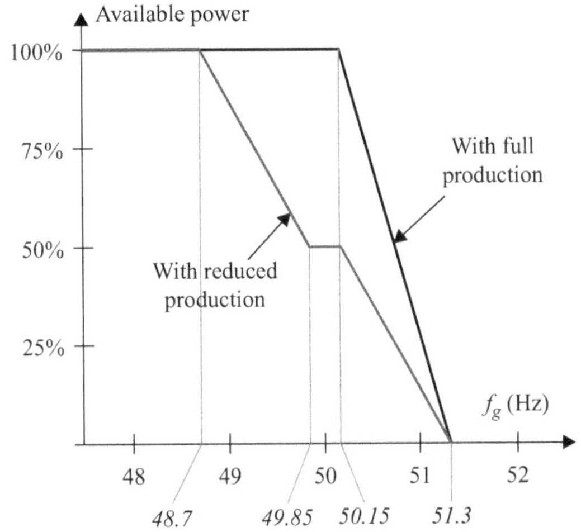

Figure 13.4 Frequency control profiles for wind turbines in Denmark [30]

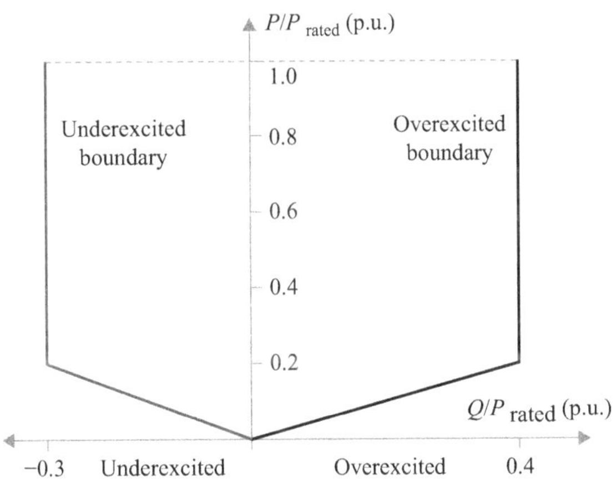

Figure 13.5 Reactive power range under different generating powers in Germany [30]

secondary control of the grid. Generally, to achieve so, these active power controls require a certain amount of reserved power, which will provide sufficient support in the case that extra active power is on request, and thereby maybe reduce the required capacity of energy storage systems.

13.1.2.2 Reactive power control

In normal operation, the reactive power delivered by a WT or a wind farm also has to be regulated within a certain range according to the grid codes. However, the specified reactive power control behaviors vary in countries. For example, as shown in Figure 13.5, the German grid codes specify the range of controlling the reactive power of the WT system against the active power output. Additionally, the TSOs will normally define the reactive power range delivered by wind farms w.r.t. the grid voltage levels. Notably, reactive power control is usually achieved with slow dynamics (at minute level) [30].

13.1.2.3 Fault ride-through capability

In the case of abnormal grid conditions, for example, voltage sags (referred to as low voltage ride-through, LVRT), additional requirements have been imposed on individual WTs and also wind farms. Specifically, the WTs must remain connected to the grid within a specified period under voltage sags. In Figure 13.6, the boundaries with various grid voltage dipping amplitudes as well as the allowable tripping time are defined for a wind farm [30]. It should be noted that the control of WT systems during asymmetrical grid faults is more challenging [31]. Moreover, under grid faults, the wind power generation system should provide reactive power (up to 100% current capacity) to support voltage recovery. Figure 13.7 shows the required amount of reactive current for wind farms against the grid voltage amplitude in Denmark and Germany [30].

Such grid codes posed great challenges to the WT systems in the past two decades and also drove the development of power electronics in wind power applications. On one hand, these requirements have increased the cost per produced kWh; on the other hand, they made wind power technology much more suitable to be utilized and integrated into the grid. It can be predicted that more stringent grid codes in the future will keep challenging the WT systems and pushing forward the technology of power electronics. For instance, the recently updated grid codes in the UK have imposed new requirements on power electronics-based systems for more active participation in the grid regulation [24,27,32].

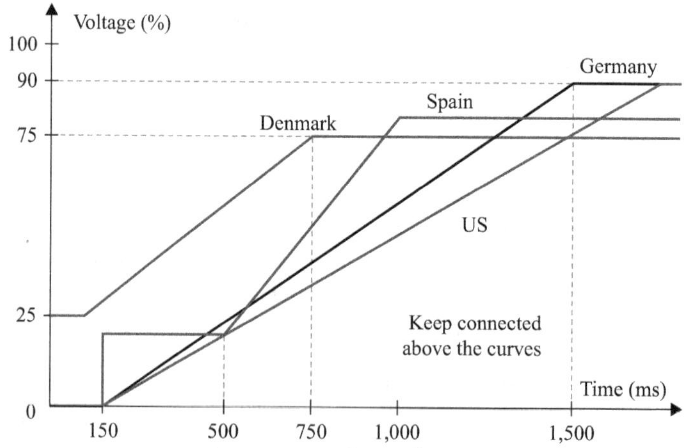

Figure 13.6 Voltage profile for low voltage fault ride-through capability of wind turbines [30]

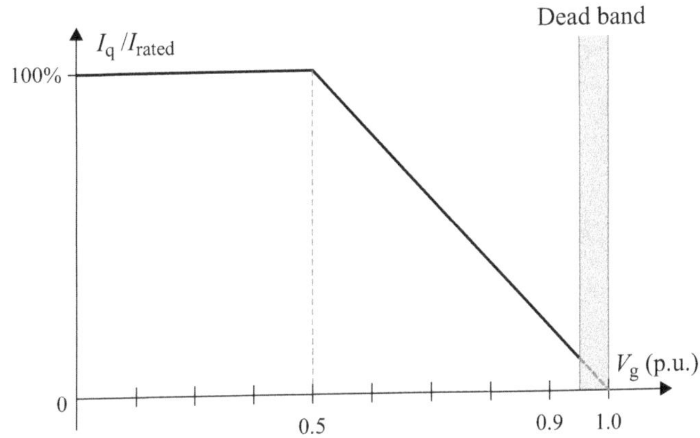

Figure 13.7 Reactive current requirements for a wind farm during grid sags by the German and Danish grid codes [30]

13.1.2.4 Grid codes for PV systems

Grid-connected PV systems are a major part of electricity generation in certain areas, while the demand for PV systems is becoming much tougher. At present, although the power capacity of a PV system is still not comparable to that of an individual WT, similar demands for WT systems are being transitioned to PV systems, since the number of large-scale PV systems is continuously increasing thanks to the matured PV technology and still declining cost of PV cells.

Nevertheless, the demands for PV systems are specified at different levels. In the case of large-scale PV systems with higher power ratings, the PV systems should not violate the grid voltage and the grid frequency by means of providing ancillary services (e.g., frequency regulation). In that case, the grid codes for PV systems are like those for WT systems, especially when the power ratings of the PV systems can be up to a few MWs for utility-scale applications. Similar requirements for WT and PV systems are already in effectiveness in certain countries [32–34]. Consequently, it calls for much flexible controllability of grid-connected PV systems [35].

13.2 Mission-profile-oriented reliability assessment

The reliability of power electronics in renewable applications is important, considering the cost of energy. As shown in Figure 13.8, the failure rate for wind power converters and PV inverters account for 13% and 37%, respectively, of all failures in WT and PV systems [22,36–38]. It further emphasizes the importance of power electronics reliability in RES systems. Thus, the reliability improvements of power electronics may effectively contribute to an extension of the energy production of RESs, being a helpful approach to further reduce the cost of energy.

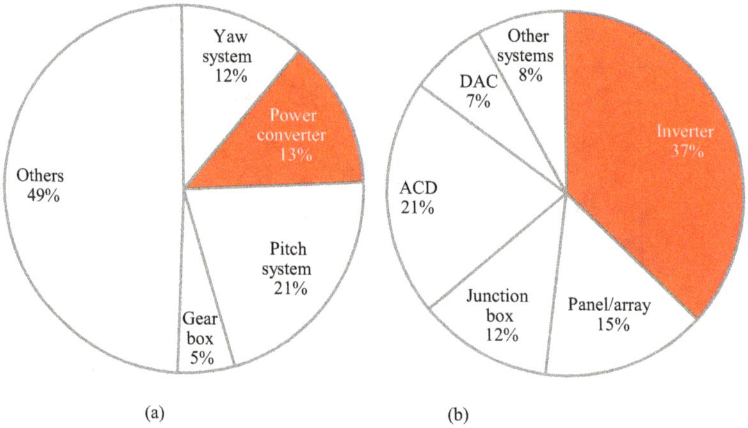

Figure 13.8 Contribution of subsystems to the failure rate of renewable energy systems: (a) WT system and (b) PV system (DAC, data acquisition system; ACD, AC disconnects) [22,36–38]

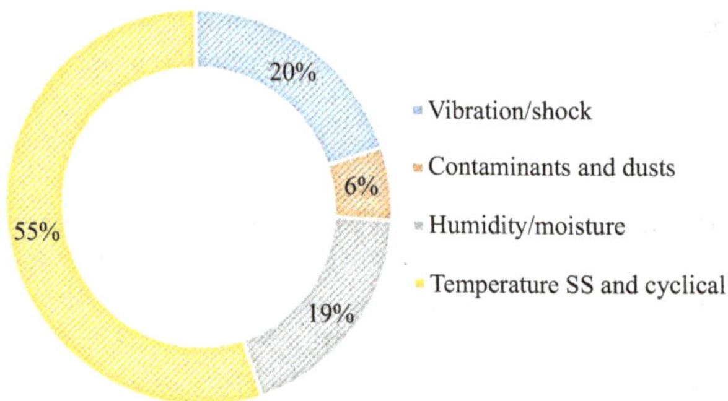

Figure 13.9 Stressor distribution in a power electronics system (SS, steady state) [4,40–43]

13.2.1 Importance of thermal stresses

Reliable, safe, and robust operation of power electronics converters is closely related to its mission profile. A mission profile is considered as the representation of all relevant conditions that a system can be exposed to throughout its entire life cycle [39]. Then, failures may happen during operation and be analyzed through the strength and stress, in which the stress factors may be the environmental (e.g., thermal, mechanical, humidity, etc.), or the functional loading (e.g., usage profiles, electrical operation). As shown in Figure 13.9, various stressors are distributed in a power electronics system, and it has been acknowledged that thermal cycling accounts for more than half of the entire failure probability [4,40–43].

13.2.2 Lifetime model of power semiconductor

To obtain thermal cycling, power cycling is first performed by power semiconductor manufacturers to model their strength. Depending on the current supplied from time to time, the power components are heated up and cooled down, repetitively. Because of various thermal expansion coefficients between semiconductor layers, the thermo-mechanical stress leads to the failure of power components. From the power cycling tests, it is known that connections between the chip and direct bonded copper (DBC), links of the bond wire, and the chip area are the most seen failure mechanisms [44–47]. To accelerate the test, the current applied for power cycling is almost equal to the rated current, and the cycling period is normally between hundreds of milliseconds to dozens of seconds. Due to the parameter variation during the manufacturing, the power cycling of an individual power component may be varied. Reliability is thus defined by B_{10} lifetime that is described as the point at which 10% of the total population will fail.

The number of power cycles can be obtained at higher junction temperature fluctuations, and then the values at lower temperature variation can be extrapolated according to the Coffin–Manson rule [48]:

$$N = A \cdot \mathrm{d}T_{\mathrm{j}}^{\alpha} \cdot \exp\left(\frac{E_{\mathrm{a}}}{k_{\mathrm{b}} \cdot T_{\mathrm{jm}}}\right) \tag{13.1}$$

in which E_{a} and k_{b} denote the activation energy and the Boltzmann constant, respectively, α and A are obtained from the LESIT project [49]. Based on the assumption of a uniform failure mechanism, the power cycles are exemplified in Figure 13.10 with the mean junction temperature of T_{jm} and the junction temperature fluctuation of $\mathrm{d}T_{\mathrm{j}}$, in which the power cycling is tested under the fixed cycling period (i.e., 1.4 s).

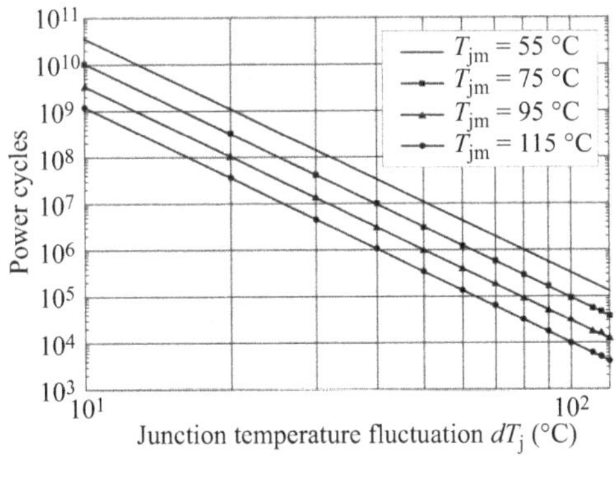

Figure 13.10 Example of fitting power cycles curve versus the mean junction temperature and the junction temperature fluctuation according to the Coffin–Manson model in (13.1)

Due to the thermal time constant difference between layers [48,50], the cycling period is also closely related to the power cycle capability, and this factor should also be considered:

$$\frac{N(t_{\mathrm{on}})}{N(0.7\mathrm{s})} = \left(\frac{t_{\mathrm{on}}}{0.7\mathrm{s}}\right)^{\beta} \tag{13.2}$$

in which t_{on} normally denotes the cycling period, $N(0.7\ \mathrm{s})$ indicates the power cycles at $t_{\mathrm{on}} = 0.7$ s, and β implies the effect of the cycling period [48].

13.2.3 Loading translation at various timescales

To assess the reliability, a stress analysis should be carried out. However, the translation from the loading profile to the thermal profile is challenging in RES systems. As shown in Figure 13.11, the thermal cycles of the power semiconductors in a wind power converter may vary from a millisecond to even a year and are caused by the electrical characteristics (alternative current at the fundamental frequency) and the environmental characteristics (changing ambient temperature and wind speed), respectively. To fully represent the thermal stress of the power component, short-term thermal cycling and long-term thermal cycling must be handled, respectively.

The first group of thermal behaviors is mainly caused by the fast and periodical alternating current in the converter, which is synchronized with the fundamental frequency of the grid or electric machines. The temperature of power devices under this disturbance swings periodically with a smaller amplitude but at a much faster speed. Steady-state simulation results are demonstrated in Figure 13.12, where the converter is operating under the rated condition [4], and T_{j} and T_{c} within 20 ms are illustrated. The junction temperature T_{j} oscillates at 50 Hz with a constant amplitude, while the case temperature T_{c} remains almost unchanged, as shown in Figure 13.12.

The second group of thermal dynamics is mainly caused by the long-term variation of the input power to the converter system. The output power of power electronics in RES applications is closely related to the wind speed and solar irradiance, which fluctuate intensively. An example is given in Figure 13.13 [4], which translates a one-year wind speed into the junction and case temperature of the power devices. It can be seen that many large and irregular thermal cycles ranging from 15°C to 90°C are identified, leading to unsatisfied lifetime consumption [4,37].

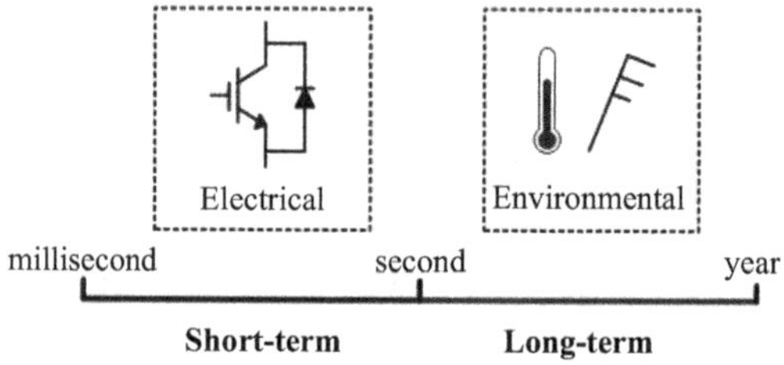

Figure 13.11 Short-term and long-term thermal cycles of the power semiconductors in a wind power converter

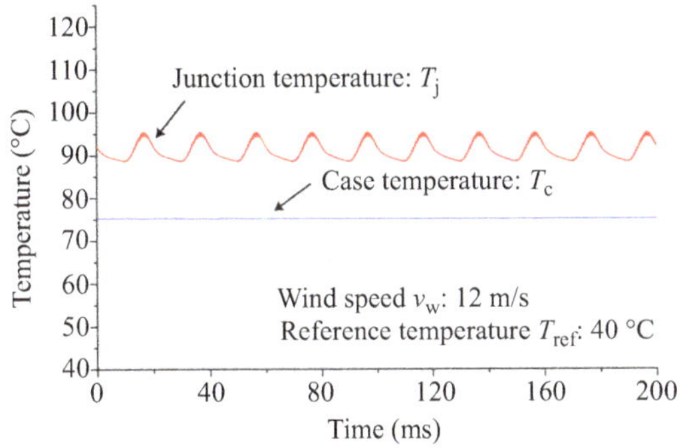

Figure 13.12 Short-term thermal behaviors inside power device when wind speed v_w = 12 m/s (0.2 s with 0.5 ms time step, T_{ref} = 40 °C)

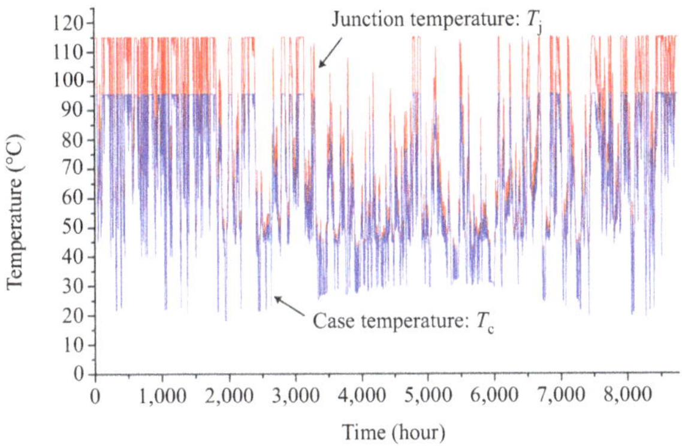

Figure 13.13 One-year thermal profile under the given mission profile in Figure 13.2 (junction temperature T_j and case temperature T_c of the IGBT, time step of 3 h)

13.2.4 Lifetime estimation approach

When the above thermal profiles are available, both the short-term and the long-term thermal cycles can be extracted in terms of the mean junction temperature, junction temperature fluctuation, cycling period, and the number of cycles. The short-term thermal cycling is counted by using the annual loading distribution (probability distribution of the annual wind speed or the solar irradiance), while the temperature information of the long-term thermal cycling can be acquired using the rain-flow counting algorithm [50–52].

To analyze the damage to the power semiconductor, the concept of the annual consumed lifetime (CL) is then introduced:

$$\mathrm{CL}_i = \frac{n_i}{N_i} \tag{13.3}$$

where n_i is the number of cycles at stress dT_{ji} under a yearly mission profile, and N_i is the corresponding number of cycles to fail according to (13.1) and (13.2).

According to Miner's rule [53], the total consumed life (TCL) per year is linearly dependent on the contributions from different temperature cycles, which can be expressed as

$$\text{TCL} = \sum_i \text{CL}_i \tag{13.4}$$

and the reciprocal of the TCL defines the B_{10}-lifetime in years.

13.3 Reliability assessment of wind turbine systems

There are several WT concepts [30,33,54–56], in which the power electronics plays quite different but key roles and have various power rating coverage of the system. In the past till now, the doubly fed induction generator (DFIG) with partial-scale power converters has been dominating the WT market, as shown in Figure 13.14(a), but the configuration of synchronous generators (SGs) with full-scale power converters is taking over the market [30,54–56], as demonstrated in Figure 13.14(b). The back-to-back power converters are denoted as the grid-side converter and the rotor-side converter in the DFIG system, but the grid-side and the generator-side converters are in the permanent-magnetic SG (PMSG) system, as shown in Figure 13.14. Compared to the DFIG concept, the PMSG system offers much more flexibility and controllability, but on the other hand, is more costly.

In the following, reliability assessment is exemplified on a 2-MW DFIG-based partial-scale WT system and a 2-MW PMSG-based full-scale WT system, respectively. Considering the short-term thermal cycling, the lifetime between the back-to-back power converters is estimated and compared for both configurations. Then, the influence of the grid codes and the wind profile on reliability is also evaluated.

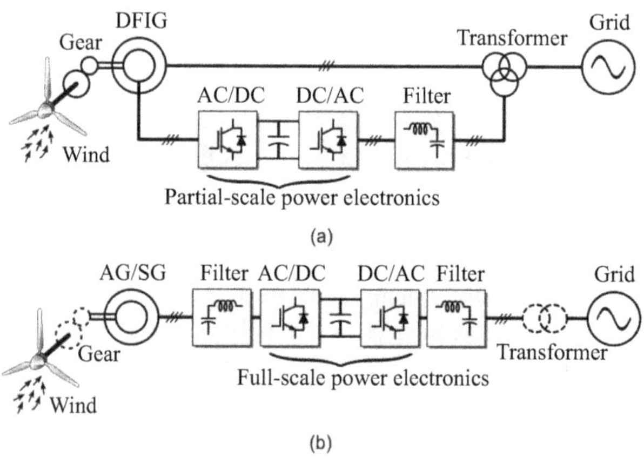

Figure 13.14 Mainstream wind turbine systems in the wind power market: (a) DFIG system; (b) PMSG system (AG, asynchronous generator; SG, synchronous generator) [33]

13.3.1 Lifetime estimation for wind power converter

To estimate the lifetime of the power converter, its "weakest" points should be identified, as the most stressed power semiconductor (the IGBT or the diode) of a power module dominates the lifetime. As mentioned in [47], it has been concluded that regardless of topologies in the WT system, the IGBT is the most loaded in the grid-side converter, while the diode has the worst thermal performance in the rotor-side converter or the generator-side converter. Therefore, the two categories of power devices will be focused on in the next sections.

As described in [47], the thermal profile of the most stressed power semiconductor can be obtained. Accordingly, the power cycles of B_{10}-lifetime are calculated according to (13.1) and (13.2). The results are shown in Figure 13.15 for the

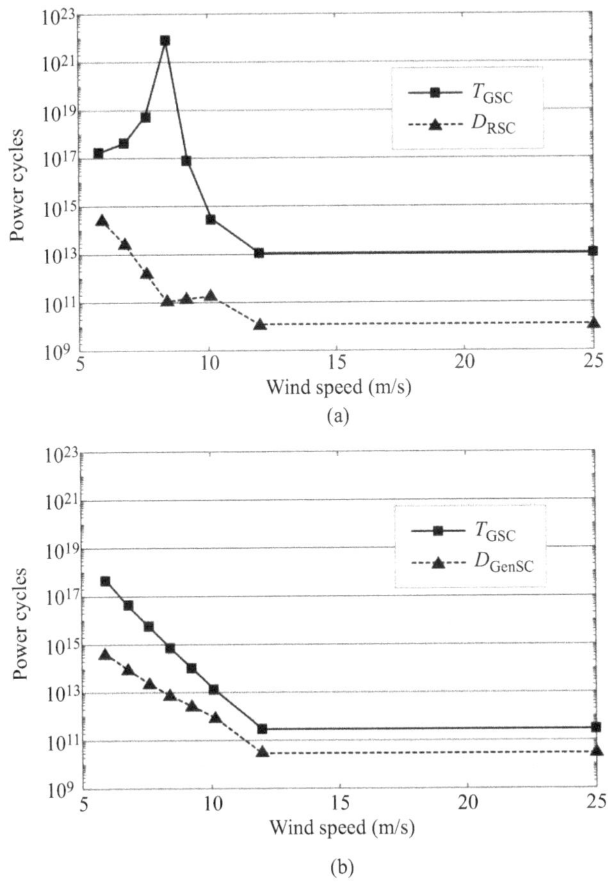

Figure 13.15 Power cycles of the most stressed power semiconductor at various wind speeds: (a) the DFIG system (T_{GSC}, IGBT of the grid-side converter; D_{RSC}, diode of the rotor-side converter) and (b) the PMSG system (T_{GSC}, IGBT of the grid-side converter; D_{GenSC}, diode of the generator-side converter)

back-to-back power converters in the DFIG system and the PMSG system, respectively. For the DFIG system, the grid-side converter and the rotor-side converter show quite different performances. The power cycles become very high at the synchronous operation in the grid-side converter. However, the power cycles at the synchronous operation are very low in the rotor-side converter, not only due to its higher junction temperature fluctuation but also because of the very low operational frequency. By contrast, for the PMSG system, if the wind speed is below the rated one (i.e., 12 m/s), the number of power cycles is inversely proportional to the wind speed in both the grid-side converter and the generator-side converter, as shown in Figure 13.15(b).

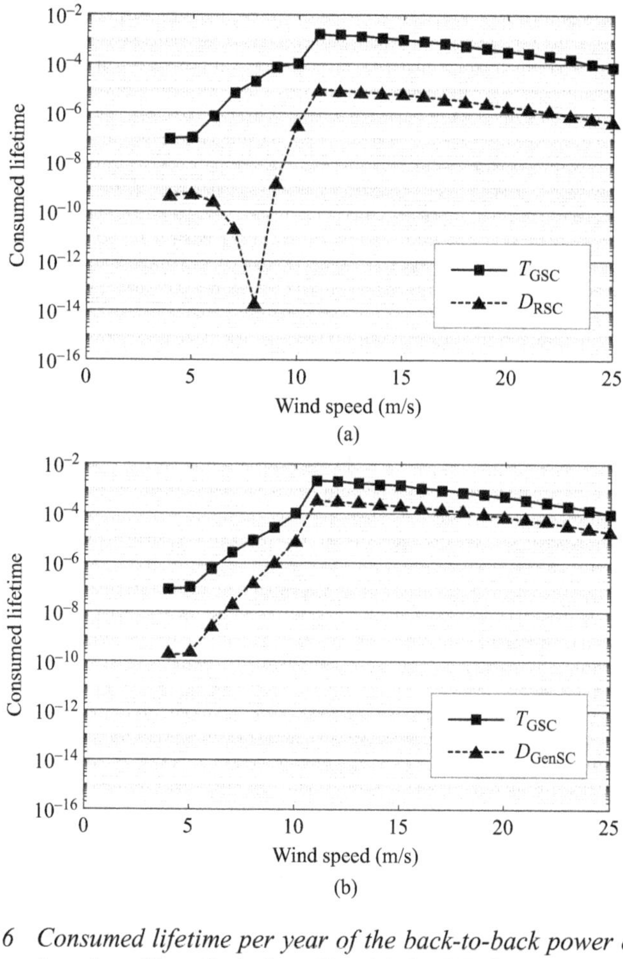

Figure 13.16 Consumed lifetime per year of the back-to-back power converter based on Class I wind profile: (a) the DFIG system (T_{GSC}, IGBT of the grid-side converter; D_{RSC}, diode of the rotor-side converter) and (b) the PMSG system (T_{GSC}, IGBT of the grid-side converter; D_{GenSC}, diode of the generator-side converter)

To evaluate the consumed lifetime per year and the total consumed lifetime, the annual wind speed distribution should be considered, where the total number of thermal cycles per year can be anticipated. The annual wind speed distribution of various wind classes is shown in Figure 13.1, and Wind Class I is considered in this case. The consumed lifetime per year is then obtained according to (13.4), and the DFIG system and the PMSG system are individually studied with the results shown in Figure 13.16(a) and (b), respectively. It is noted that the wind speed from the cut-in (4 m/s) to the cut-out (25 m/s) contributes to the most lifetime consumption. Regarding the DFIG system, the back-to-back power converters exhibit different behaviors. Specifically, the consumed lifetime of the grid-side converter changes more dramatically compared to that of the rotor-side converter at around the synchronous speed operation. About the PMSG system, the consumed lifetime turning points of both the generator-side converter and the grid-side converter are at the rated wind speed. Above the rated wind speed, the consumed lifetime continuously reduces along with the increase of the wind speed, because of the same value of the B_{10}-power cycle but the lower wind speed distribution.

Summing up the individual consumed lifetime results in the total consumed lifetime for both wind turbine configurations, as shown in Figure 13.17. It is demonstrated in Figure 13.17 that the consumed lifetime difference between the back-to-back power converters in the PMSG system is relatively lower compared to that in the DFIG system.

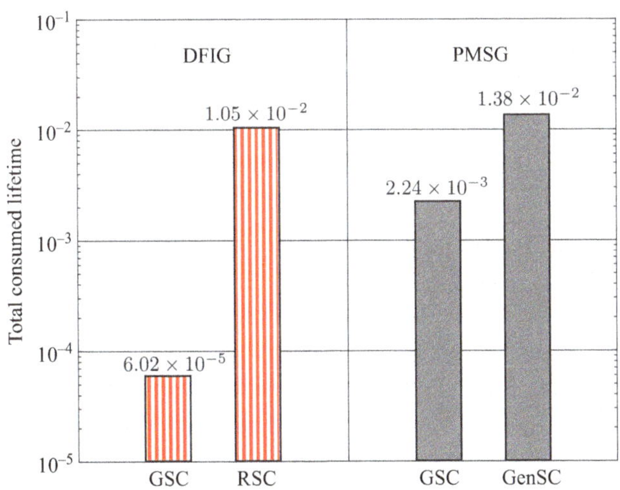

Figure 13.17 Comparison of the total consumed lifetime per year between grid-side converter (GSC) and rotor-side converter (RSC)/generator-side converter (GenSC) of the DFIG and PMSG system

13.3.2 Mission profile effects on lifetime

Since the power electronics converter of the wind turbine system may be operated at different conditions (e.g., grid codes, and wind profiles), the effect of mission profiles on reliability is demonstrated in the following. The rotor-side converter of the DFIG system is generally responsible for meeting the grid codes, while the grid-side converter of the PMSG system is obliged to fulfill the grid codes. Thus, the two power converters in the DFIG and PMSG systems are explored.

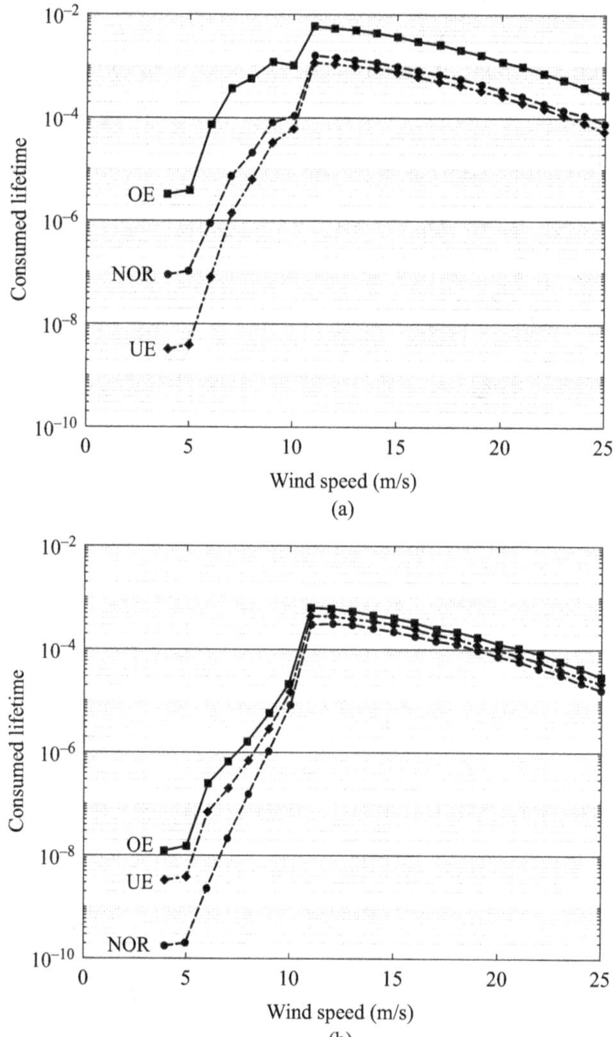

Figure 13.18 Consumed lifetime per year with various reactive power injections: (a) rotor-side converter of the DFIG system and (b) grid-side converter of the PMSG system (OE, overexcited; NOR, normal; UE, under-excited)

13.3.2.1 Grid codes

Reactive power is often preferred during LVRT operation for voltage recovery. Additionally, many countries with a strong penetration of wind power, like Germany, Denmark, and China, require WT systems also to deliver reactive power in normal operation. As shown in Figure 13.5(b), one of the grid requirements is established by the German TSO, in which up to 40% over-excited (OE) and 30% under-excited (UE) reactive power (in percentage of the rated active power) should be delivered if the active power is above 20% of the rated power. For comparison, NOR denotes the normal operation mode, which is the case if there is no reactive power exchange between the grid and the WT system.

Figure 13.18 depicts the consumed lifetime of the power semiconductors in the three cases of reactive power injection (UE, NOR, and OE). For the DFIG system, the OE operation significantly reduces the reliability of power semiconductors compared to the NOR operation, while the UE operation slightly enhances the reliability. By contrast, for the PMSG system, both the OE and the UE operation result in a higher consumption of lifetime at all wind speeds.

The effect of grid codes on the reliability cost is also measured in this case study, and the results are shown in Figure 13.19. It is seen in Figure 13.19 that, in the DFIG system, if the OE reactive power is provided all year around, the lifetime is substantially shortened to almost a quarter of that in the NOR operation. With respect to the PMSG system, the OE operation almost reduces the lifetime by 1/2 compared to the NOR operation. As for the UE operation, the DFIG system has consumed less lifetime, while the PMSG system consumes slightly more lifetime compared to that in the NOR operation of the respective system, as observed in Figure 13.19.

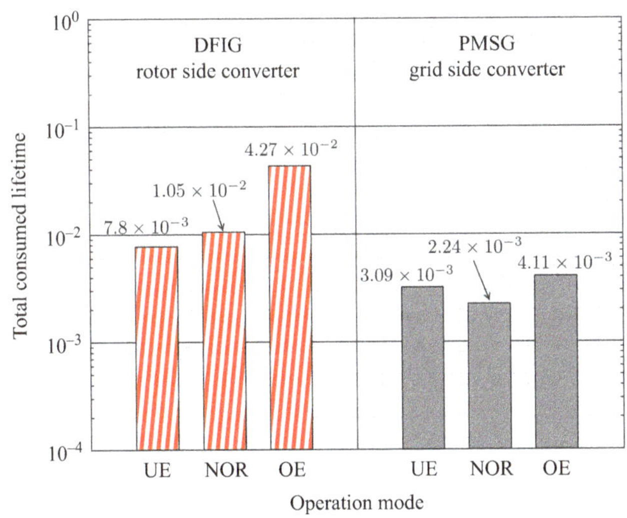

Figure 13.19 Grid codes influence on the consumed lifetime per year: A comparison of the rotor-side converter of the DFIG system and the grid-side converter of the PMSG system (OE, overexcited; NOR, normal; UE, under-excited)

13.3.2.2 Wind profile

According to the IEC 61400 Standard [2], three wind classes in Figure 13.1 will be used to compare the impact of wind speed profiles. Figure 13.20 shows the comparison results of the CL. It can be observed that a slight difference in the CL occurs from the cut-in wind speed (4 m/s) to the rated wind speed (12 m/s) in both the DFIG and the PMSG systems. Besides, since the Class I wind profile has the

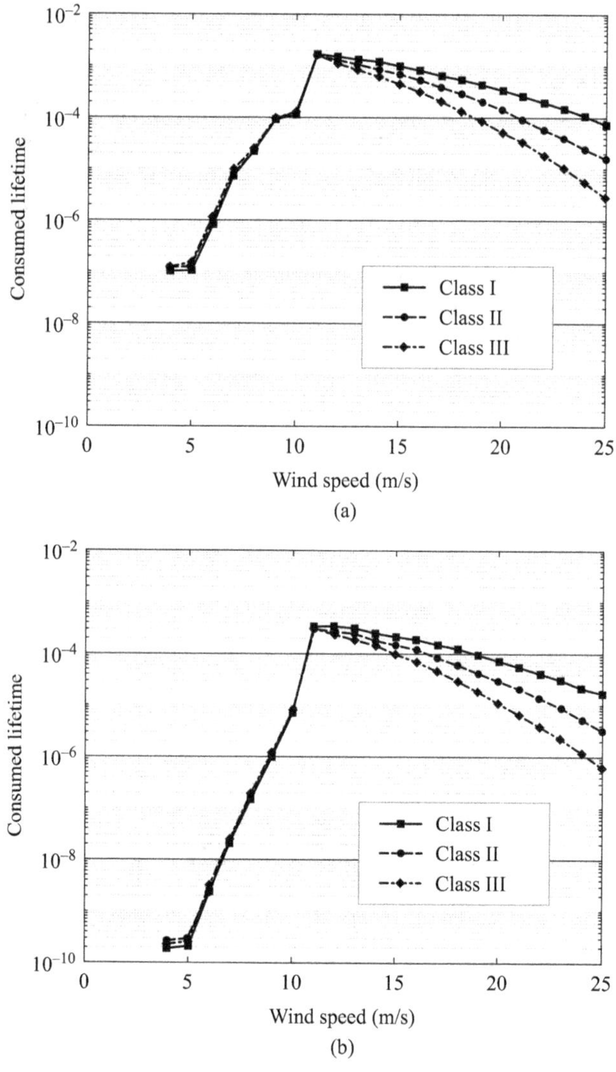

Figure 13.20 Consumed lifetime per year at various wind classes (see Figure 13.1): (a) rotor-side converter of the DFIG system and (b) grid-side converter of the PMSG system

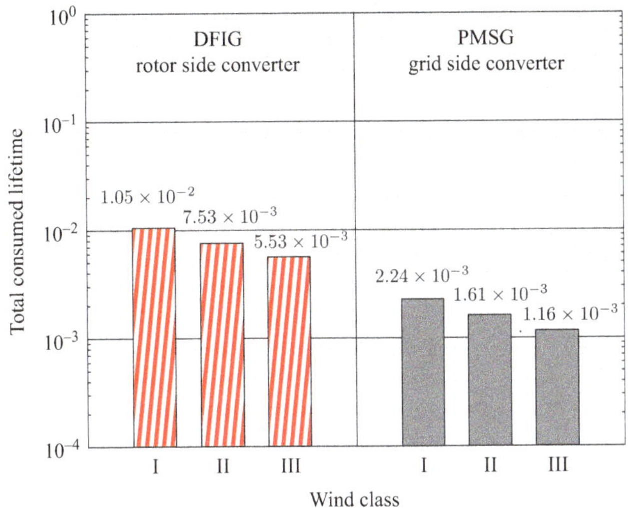

Figure 13.21 Total consumed lifetime per year at various wind classes (see Figure 13.1): a comparison of the rotor-side converter of the DFIG system and the grid-side converter of the PMSG system

highest probability from the rated wind speed to the cut-out wind speed (25 m/s), its CL becomes the highest, followed by Class II and Class III profiles. Furthermore, the CL in this case study is also presented in Figure 13.21, which shows that upper wind classes will result in a lower lifetime of the power converters in both DFIG and PMSG configurations.

13.4 Reliability assessment of photovoltaic system

Considering the harsh operating conditions, for example, voltage sags, and also the increasing customer demands, the reliability of power electronics–enabled PV systems is significantly challenged. In the following, the popular topologies used in PV systems are briefly described. Then, the thermal loading performance of a grid-connected PV system with selected transformerless inverter topologies is analyzed (mission profile translation and CL comparison), where a yearly mission profile has been adopted. A system-level reliability calculation is demonstrated on a single-phase H-bridge inverter system.

13.4.1 PV inverter candidates

Unlike wind technology, the solar PV system produces much lower power per generating unit (e.g., individual PV panel or PV string). Therefore, a common configuration of the PV system as the PV "generator" includes many panels connected in parallel and/or series to increase the output power and voltage. To maintain high efficiency and small volume, transformerless topologies are commonly adopted in PV systems. Those transformerless PV systems can be categorized according to the power levels [57], as shown in Figure 13.22. The string inverters and multistring inverters are the most used solution for the single-phase configuration. However, due to the lack of galvanic isolation, leakage currents may

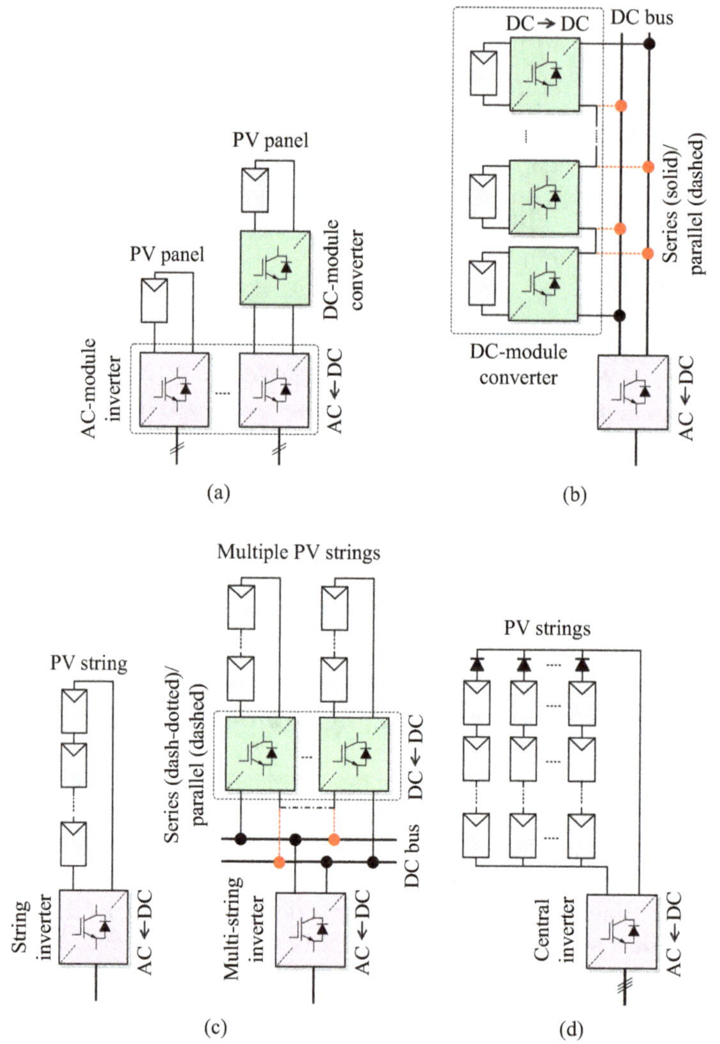

Figure 13.22 Different grid-connected PV inverter structures and the typical applications along with the power ratings. (a) Module inverter for small systems; (b) string inverter for residential applications; (c) multistring inverter for residential or commercial applications; and (d) central inverter for commercial or utility-scale PV plants.

appear between the PV panels and the ground, which will cause safety issues if not coped with properly. Thus, transformerless inverters are required to minimize the leakage current emission, resulting in a vast number of transformerless PV inverter candidates. Among those, several commercialized and popular topologies are shown in Figure 13.23, where it can be seen that those are based on the H-bridge inverter topology [58].

The H-bridge inverter is a standard solution for single-phase PV systems. However, in the case of transformerless applications, the pulse-width modulation (PWM) strategies should be specifically designed to avoid a large leakage current to the ground. Considering this, a bipolar modulation scheme can be adopted, but with that, a relatively low efficiency (<96.5%) is also observed. To improve the common-mode behavior, but also to further increase the efficiency, and to keep all the other merits given by the H-bridge inverter, a number of modified topologies have been developed by adding bypass switches either on the DC side or on the AC side [58–67], as exemplified in Figure 13.23.

As shown in Figure 13.23, the highly efficient and reliable inverter concept (HERIC) [64] uses two extra switches (S_5D_5–S_6D_6) on the AC output of the H-bridge inverter to minimize leakage currents. However, the common-mode

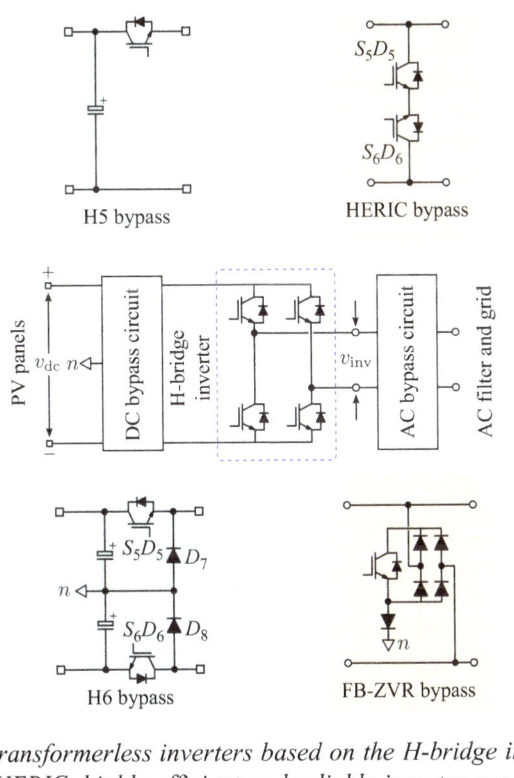

Figure 13.23 Transformerless inverters based on the H-bridge inverter topology (HERIC, highly efficient and reliable inverter concept; FB-ZVR, full bridge with zero voltage rectifier)

behavior of the HERIC is like that of the H-bridge converter with a bipolar PWM, where the voltage to ground of the PV array terminals will only have a sinusoidal shape, while the same high efficiency can be achieved as the H-bridge converter with a unipolar PWM. Thus, the HERIC topology is more suitable for single-phase transformerless PV systems in terms of efficiency (up to 98% as reported) [65].

Another single-phase transformerless inverter [61,62] also adds two extra switches (S_5D_5–S_6D_6) and two extra diodes (D_7–D_8) to the H-bridge topology, as shown in Figure 13.23. Thus, it is named an H6 inverter. The common-mode behavior of the H6 topology is similar to the HERIC topologies since the voltage to ground of the PV array has only a sinusoidal shape and the frequency is the grid frequency. It is another preferable alternative for transformerless PV systems since it can also achieve a very lower leakage current. Several commercial products are available on the market, and the reported efficiency of this H6 inverter can be up to 97%.

13.4.2 Reliability assessment of single-phase PV systems

Due to the intermittency of solar PV energy, unequal thermal loading distributions among the power switching devices in transformerless PV inverters are more likely to happen under a long-term mission [20,39,57,58,68–70]. The varying thermal loading (which appears as fluctuating junction temperature) is one of the main contributors to power electronics devices, as discussed in previous sections. With more advanced transformerless inverters coming on the market, it calls for a reliability-oriented investigation considering their mission profiles [15,71–75], which allows new thoughts for a robust design and reliable operation of PV systems. In the following, a focus has been put on translating the mission profile shown in Figure 13.3 to the corresponding device thermal loading of the H-bridge with a bipolar modulation (FB-bipolar), H6, and HERIC transformerless inverters discussed in Section 13.4.1.

Notably, a mission-profile-based analysis should be able to analyze the performance at different timescales, as shown in Figure 13.24. For short-term mission profiles (milliseconds to several seconds), thermal loading can be directly and instantaneously obtained; while for long-term mission profiles (a few minutes to several months), obtaining the instantaneous thermal loading is time consuming. Thus, look-up tables are usually adopted, as shown in Figure 13.24(b).

According to Figure 13.24, the mission profile shown in Figure 13.3 can be translated into the thermal profiles, which are closely related to the lifetime prediction and should be appropriately extracted according to a specific lifetime model. Figure 13.25 shows the translated thermal loading (i.e., the maximum junction temperature, T_{jmax}) of the semiconductors of the three transformerless PV inverters under the yearly mission profile. It can be seen from Figure 13.25 that, although the transformerless PV inverters (both the H6 and the HERIC) can maintain a higher efficiency, the thermal loading on those switching devices is unequal. Specifically, the maximum junction temperature of the extra devices in an H6 inverter is the highest, indicating high power losses. However, this is not the case for the HERIC inverter, as shown in Figure 13.25, and the above implies that more efforts should be devoted to the design of the extra switching devices in the transformerless PV inverters.

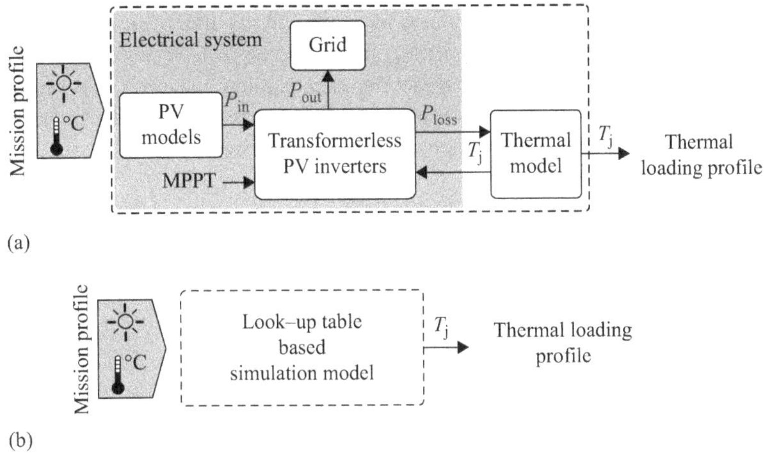

Figure 13.24 Approach of the mission profile translation to thermal loading at different timescales (MPPT, maximum power point tracking; T_j, junction temperature of the power devices). (a) Short-term mission profiles and (b) long-term mission profiles.

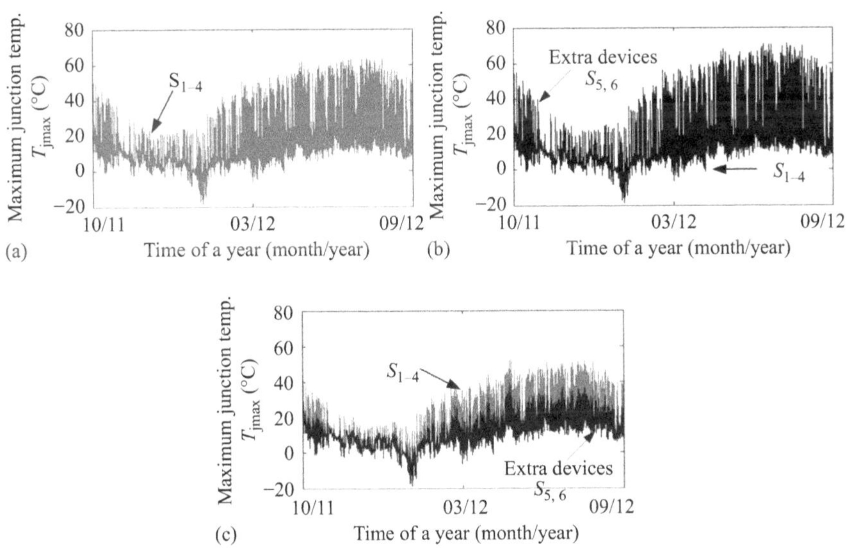

Figure 13.25 Translated thermal loading (maximum junction temperature, T_{jmax}) of the power switching devices of the three different transformerless 3-kW grid-connected PV inverters under a yearly mission profile shown in Figure 13.3. (a) Full bridge with bipolar modulation; (b) H6 inverter; and (c) HERIC inverter.

Based on the above thermal loading analysis, the lifetime of the power devices in the three transformerless PV inverters can be assessed. A rain-flow counting algorithm has been adopted to analyze the thermal loading, and the results are presented in Figure 13.26. It can be observed in Figure 13.26 that the additional devices of the H6 inverter have a larger number of cycles, which indicates that the extra devices for minimizing the leakage currents become the most critical components. Consequently, more reliable power devices are preferable when designing an H6-based transformerless PV system. For the HERIC systems, the extra devices have a smaller number of cycles compared to that of the H6 inverter, and the

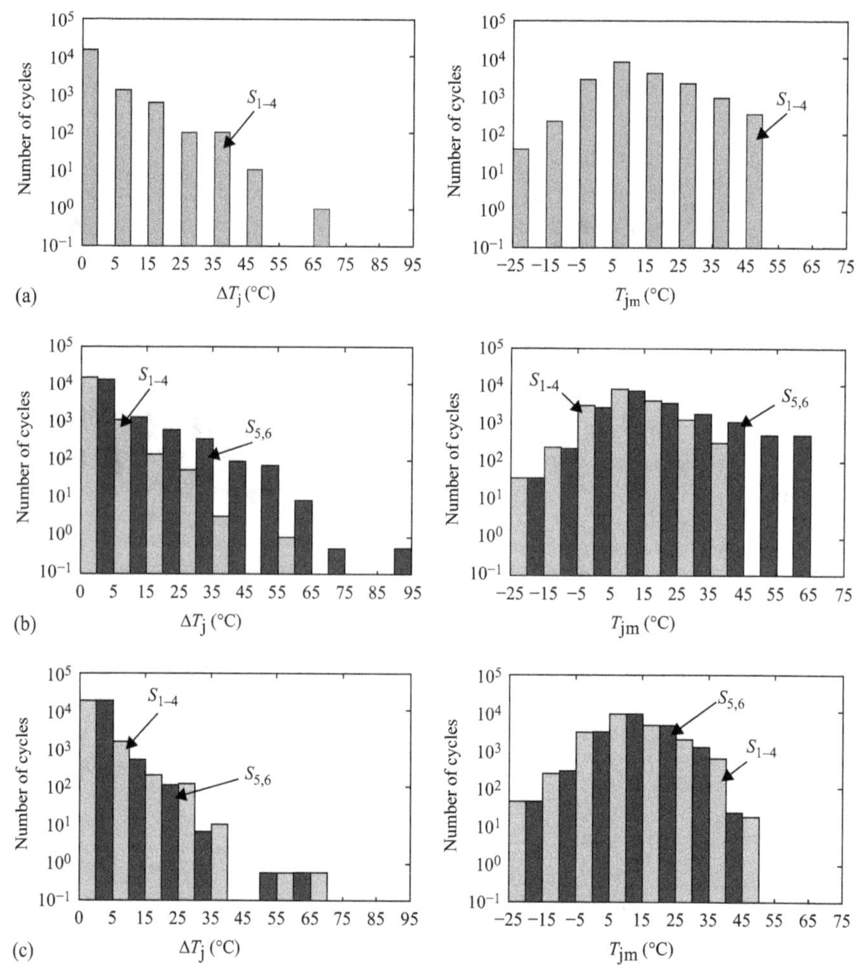

Figure 13.26 Rain-flowing counting results (number of cycles distributions at the cycle amplitude ΔT_j and the mean junction temperature T_{jm}) of the thermal loading shown in Figure 13.25. (a) FB-bipolar inverter; (b) H6 inverter; and (c) HERIC inverter.

number of cycles of the other devices is even smaller than that of the FB-bipolar inverter under the same mission profile. This means that the HERIC inverter can achieve the highest reliability under the same mission profile, while also maintaining a higher efficiency, as reported in [65].

It should be noted that the *CL* and thus the lifetime of the power switching devices according to (13.3) and (13.4) can quantitatively be obtained using the lifetime model of (13.1) and (13.2). However, those parameters are extracted under specific conditions (e.g., $0.07\ \text{s} \leq t_{\text{on}} \leq 63\ \text{s}$) for the semiconductors in [76]. This may lead to quantitative prediction errors. To reduce the parameter dependency and also the reliability model dependency, the *TCL* in (13.4) is normalized accordingly as

$$\bar{\text{TCL}} = \frac{\text{TCL}}{\text{TCL}_\text{b}} = \frac{\sum_i \frac{n_i}{(\Delta T_{\text{j},i})^{\alpha}(ar)^{\beta_1 \Delta T_{\text{j},i}}\left[C+(t_{\text{on},i})^{\gamma}\right]e^{E_\text{a}/\left(k_\text{B} T_{\text{jm},i}\right)}}}{\sum_l \frac{n_l}{(\Delta T_{\text{j},l})^{\alpha}(ar)^{\beta_1 \Delta T_{\text{j},l}}\left[C+(t_{\text{on},l})^{\gamma}\right]e^{E_\text{a}/\left(k_\text{B} T_{\text{jm},l}\right)}}} \quad (13.5)$$

in which $\bar{\text{TCL}}$ is the normalized consumed life, and TCL_b is the base TCL for normalization.

According to (13.5) and the counting results shown in Figure 13.26, Figure 13.27 shows the normalized consumed life of the three transformerless PV

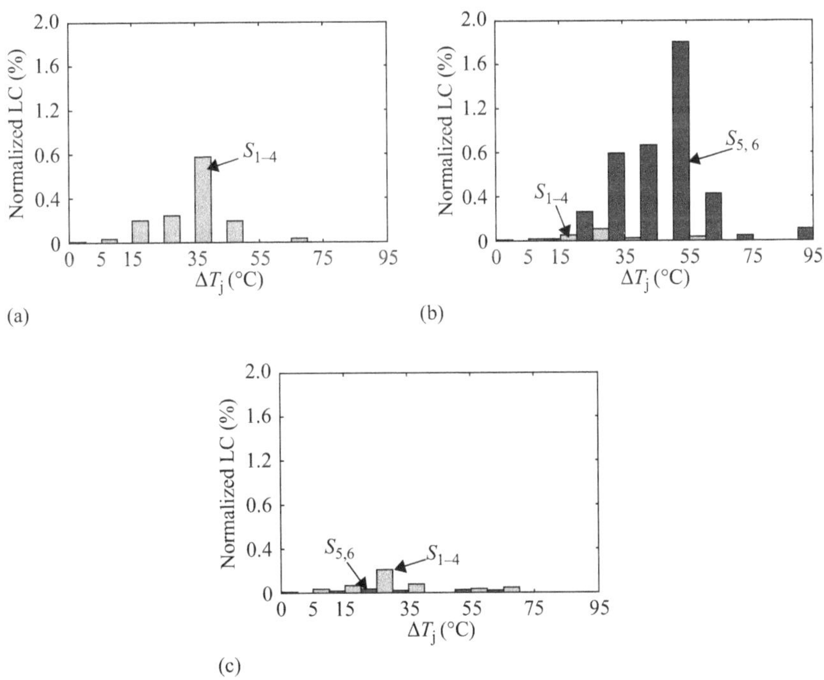

Figure 13.27 Normalized life consumption (lifetime comparison) of the three transformerless PV inverters under the same mission profile. (a) FB-Bipolar inverter; (b) H6 inverter; and (c) HERIC inverter.

inverters. The TCL of the power switching devices of the FB-bipolar (i.e., $S_{1\sim4}$) is selected as the base *TCL* for normalization. It can be seen in Figure 13.27 that the extra power switching devices of the H6 inverter consume much more life compared to the other devices and those of the HERIC inverter. This implies that the degradation of the additional devices of the H6 inverter happens much faster, and then the entire H6 system may fail to operate earlier than the other two transformerless PV inverters. This comparison further confirms that the HERIC inverter is the most promising solution in terms of reliability.

13.4.3 System-level reliability calculation of PV systems

Following the thermal loading analysis, a system-level reliability calculation can be performed. Here, a single-phase H-bridge PV system under the mission profile in Figure 13.3 is analyzed to demonstrate the system-level reliability. The flowchart of the lifetime prediction of power electronics systems considering mission profiles [57] is shown in Figure 13.28. Here, B_x lifetime implies the time when x% of the population, for example, 10,000, fail. As seen in Figure 13.28, it involves multiple steps to calculate the system-level reliability of power electronics systems:

Step 1—Mission profile translation to thermal loading
Step 2—Thermal cycling interpretation (counting)
Step 3—Lifetime model of power devices (component-level reliability)
Step 4—Monte Carlo reliability assessment
Step 5—System-level reliability analysis

The first three steps have been demonstrated in previous sections. In the following, we consider the last two steps for system-level reliability analysis.

As mentioned in the previous sections, the lifetime models of power electronics devices are obtained under specific test conditions. The predicted lifetime of power devices (component level) is then considered ideal, and all operate with the same failure rate. In practical applications, however, the power devices in a PV inverter may fail at different times. That is, uncertainties in the time-to-failure of the power electronics devices must be considered, which are mainly induced by the model parameters (i.e., tests under different conditions) and variations in the stresses. With this, in the lifetime predictions of power electronics systems, statistical values by means of the probability of failures are used. This requires performing a Monte Carlo analysis, as shown in Figure 13.28, where the Monte Carlo method introduces variations equivalently to the system parameters and then simulates the result with many samples (to ensure convergence). This can be used for lifetime predictions of power electronics devices, as demonstrated in [71,77]. The flowchart of the Monte Carlo–based system-level reliability evaluation of PV inverter systems is shown in Figure 13.29, with which a reliability block diagram method can be adopted to analyze the system-level reliability (see Figure 13.28).

Clearly, in Figure 13.29, all the parameters in the lifetime model must be modeled by a distribution function with variations. Following that, the Monte Carlo simulations can be performed with enough samples being taken randomly to calculate the power device lifetime. In this case study, a population of 10,000 samples

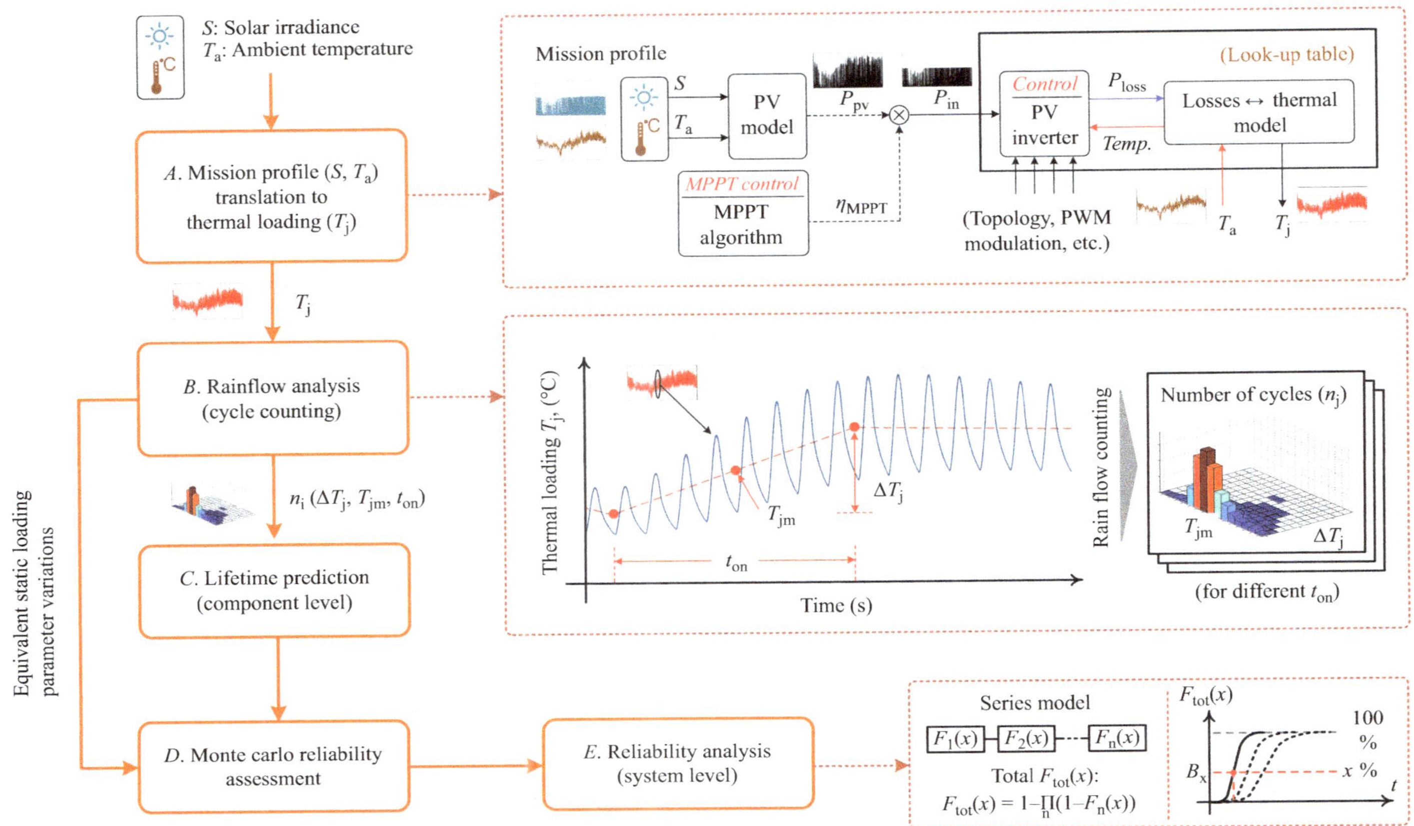

Figure 13.28 Flowchart of the mission-profile-based reliability analysis for power electronics systems (an example for PV systems)

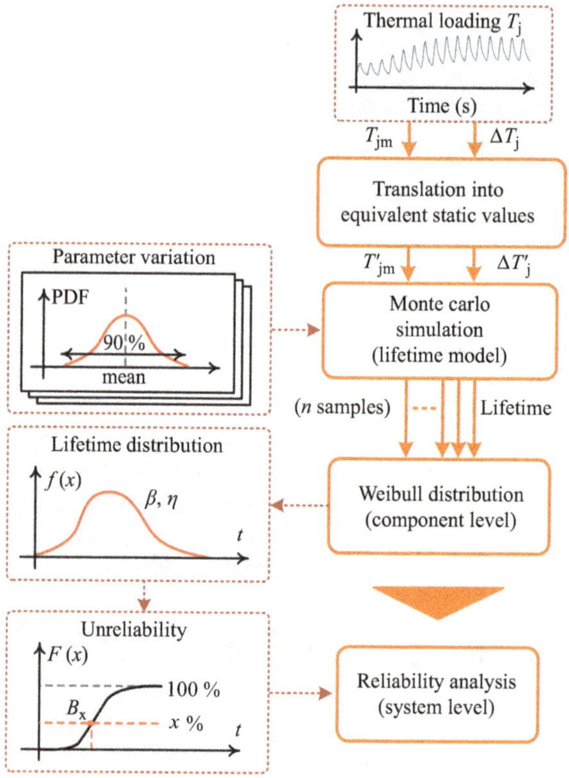

Figure 13.29 Flowchart of the Monte Carlo–based system-level reliability evaluation of PV inverter systems

is considered in the Monte Carlo simulations and the lifetime model is given in [57]. Subsequently, it gives the lifetime predictions of the power electronics devices with a certain distribution function. For power electronics, it usually follows a Weibull distribution. Then, the probability density function (PDF) of the lifetime predictions can be obtained to represent the lifetime distribution (failure distribution) $f(x)$ and its cumulative distribution function (CDF) is termed as an unreliability function $F(x)$, as shown in Figure 13.29. Notably, it can be observed in Figures 13.28 and 13.29 that the B_x lifetime can be obtained conveniently from the CDF curves. Following the above discussions and thermal loading translation, the Monte Carlo simulation results are then given in Figure 13.30. Accordingly, it is easy to obtain the B_1 and B_{10} lifetime of the power device (component level) of the H-bridge PV inverter is 42 and 74 years, respectively, under the specific mission profile. As mentioned, this is ideal, considering that the power devices have the same failure rate.

In practice, a power electronics converter has many power devices, e.g., four power devices in an H-bridge inverter. Then, using the resultant component-level

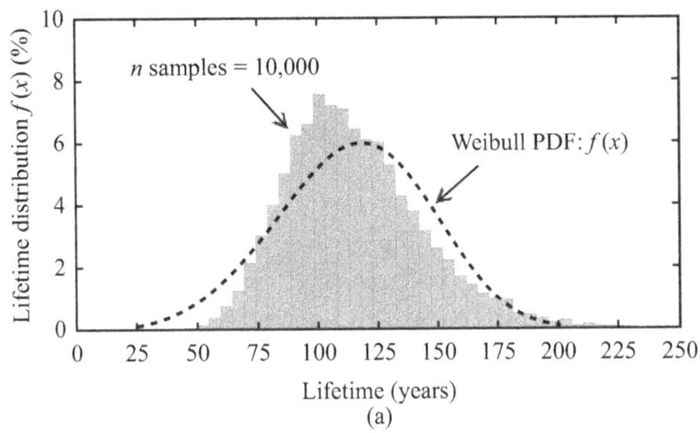

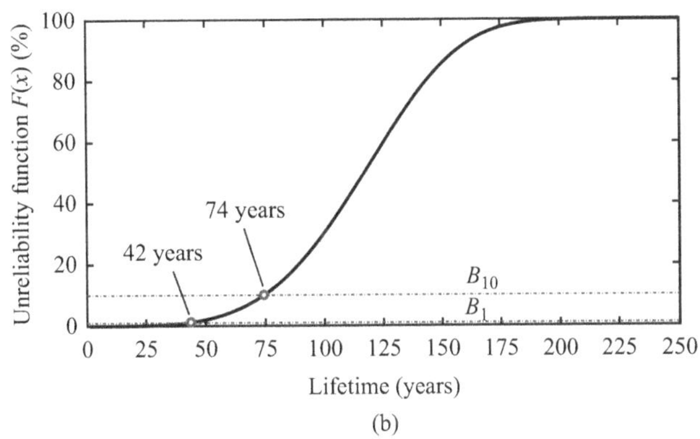

Figure 13.30 Monte Carlo simulation results (10,000 samples) of the H-bridge PV inverter system considering the mission profile in Figure 13.3: (a) lifetime distribution of one power device (i.e., the Weibull PDF) and (b) unreliability function (i.e., the Weibull CDF) at component level

function from the Monte Carlo simulations, the system-level reliability can be calculated through the reliability block diagram (see Figure 13.28). Here, in this case study, as the H-bridge PV inverter cannot function if any of its four devices fails, the total unreliability $F_{tot}(x)$ can be calculated as

$$F_{tot}(x) = 1 - \prod_{n=1}^{4} (1 - F_n(x)) \tag{13.6}$$

where $F_n(x)$ indicates the unreliability function of the nth (n = 1, 2, 3, 4) power device of the H-bridge inverter. When the bipolar modulation strategy is adopted, the loading on each power device is equal. This indicates that the unreliability functions are identical as $F(x)$ for the four power devices. Consequently, the system-level unreliability function is given as

$$F_{tot}(x) = 1 - (1 - F_n(x))^4 \tag{13.7}$$

with $F(x)$ being the unreliability function of a single power device of the H-bridge inverter. Accordingly, the total unreliability of the entire H-bridge inverter can be calculated with the component-level unreliability function shown in Figure 13.30 (b). The resultant unreliability function is shown in Figure 13.31.

For comparison, the component-level unreliability function is reproduced in Figure 13.31. As it can be seen in Figure 13.31, the B_x lifetime (system-level) can be easily obtained. Observations in Figure 13.31 indicate that the B_1 and B_{10} lifetime of the entire H-bridge PV inverter (i.e., system-level reliability) under the specific mission profile shown in Figure 13.3 is 30 and 53 years, respectively. This corresponds to 12 years and 21 years shorter than the component-level lifetime. Hence, the predicted results are more realistic with the consideration of parameter uncertainties and variations in the lifetime model (by means of the Monte Carlo analysis) and the mission profiles. The mission profile-oriented reliability analysis of power electronics systems can be of certain value to the design, control, and operation of power electronics-based RES systems.

Notably, in the above discussions, the lifetime of individual power devices and the power converters has been evaluated, while in practice the capacitors in power electronics systems are also considered very life-limiting [8,9,19]. As such, in order to design for reliability of power electronics systems, the lifetime of capacitors should also be considered. This can be done through a similar procedure presented in this chapter, while details and case studies of wind turbine and PV systems are referred to [78–82]. Moreover, the reliability of the "generator" itself (e.g., wind turbine mechanical fatigue and PV panel performance degradation) is of concern for the entire system lifetime evaluation. For example. for PV systems, the panel degradation is complicated [83,84], and it can significantly affect the loading of the power devices, thus, the reliability of the entire system [85,86]. This could also be taken into account as one parameter of the mission profiles (being more complicated mission profiles) when evaluating the entire system reliability.

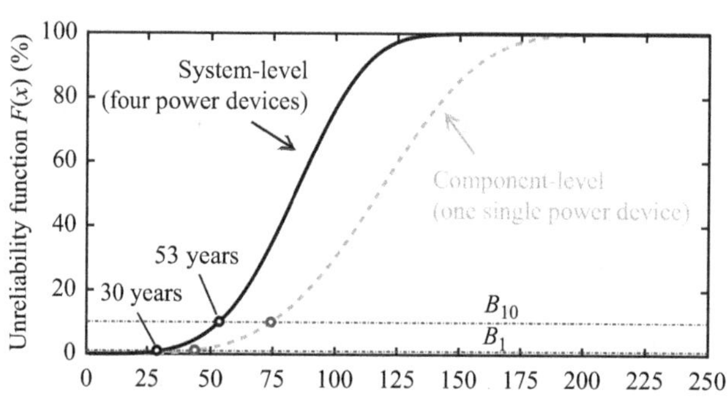

Figure 13.31 Monte Carlo simulation results (10,000 samples) of the H-bridge PV inverter system considering the mission profile in Figure 13.3: unreliability function (i.e., the Weibull CDF) at system-level

13.5 Summary

This chapter has discussed the developments and requirements for power electronics technology in RES systems, where reliability performance is especially in focus. State-of-the-art converter solutions for WT and PV systems have been presented. Principles, concepts, paradigm shifts, and practical issues about how to achieve more reliable power electronics have been briefed considering mission profiles. Several case studies on WT and PV systems are presented to illustrate how to translate mission profiles into the lifetime of power devices. It is concluded that with the thriving development of RES technology, the reliability of the interfacing power electronics becomes much more critical. As such, many emerging challenges appear and should be addressed to achieve more reliable power electronics in various applications. It is also worth mentioning that at present the reliability engineering, including reliability calculation and analysis, in power electronics is still ongoing. Thus, many other issues (e.g., testing of reliability) beyond advancing power electronics should be considered in the final product, as well as during the operation. Many joint forces are needed to further reduce the cost of energy.

References

[1] International Energy Agency (IEA), Renewables 2023 – Analysis and Forecast to 2028, Jan. 2024. (Available and assessed on July 19, 2024: https://iea.blob. core.windows.net/assets/96d66a8b-d502-476b-ba94-54ffda84cf72/Renewables_2023.pdf)

[2] Wikipedia "IEC 61400," June 2013.

[3] S. Ali, S.-M. Lee, and C.-M. Jang, "Techno-economic assessment of wind energy potential at three locations in South Korea using long-term measured wind data," *Energies*, vol. 10, pp. 1–24, 2017

[4] K. Ma, M. Liserre, F. Blaabjerg, and T. Kerekes, "Thermal loading and lifetime estimation for power device considering mission profiles in wind power converter," *IEEE Transactions on Power Electronics*, vol. 30, no. 2, pp. 590–602, 2015.

[5] E. Wolfgang, L. Amigues, N. Seliger, and G. Lugert, "Building-in reliability into power electronics systems," *The World of Electronic Packaging and System Integration*, pp. 246–252, 2005.

[6] D. Hirschmann, D. Tissen, S. Schroder, and R.W. De Doncker, "Inverter design for hybrid electrical vehicles considering mission profiles," *in Proceedings of IEEE Conference on Vehicle Power and Propulsion*, vol. 7–9, pp. 1–6, 2005.

[7] C. Busca, R. Teodorescu, F. Blaabjerg, *et al.*, "An overview of the reliability prediction related aspects of high power IGBTs in wind power applications," *Microelectronics Reliability*, vol. 51, no. 9–11, pp. 1903–1907, 2011.

[8] J. Falck, C. Felgemacher, A. Rojko, M. Liserre, and P. Zacharias, "Reliability of power electronic systems: An industry perspective," *IEEE Industrial Electronics Magazine*, vol. 12, no. 2, pp. 24–35, 2018.

[9] S. Yang, A. T. Bryant, P. A. Mawby, D. Xiang, L. Ran, and P. Tavner, "An industry-based survey of reliability in power electronic converters," *IEEE Transactions on Industry Applications*, vol. 47, no. 3, pp. 1441–1451, 2011.

[10] S. Peyghami, F. Blaabjerg, and P. Palensky, "Incorporating power electronic converters reliability into modern power system reliability analysis," *IEEE Journal of Emerging and Selected Topics in Power Electronics*, vol. 9, no. 2, pp. 1668–1681, 2021.

[11] A. Isidori, F. M. Rossi, F. Blaabjerg, and K. Ma, "Thermal loading and reliability of 10-MW multilevel wind power converter at different wind roughness classes," *IEEE Transactions on Industry Applications*, vol. 50, no. 1, pp. 484–494, 2014.

[12] K. Xie, Z. Jiang, and W. Li, "Effect of wind speed on wind turbine power converter reliability," *IEEE Transactions on Energy Conversion*, vol. 27, no. 1, pp. 96–104, 2012.

[13] X. Yu, B. Hu, K. Xie, *et al.*, "Reliability evaluation of wind power converter under fault-tolerant control scheme with quantification of temperature and humidity effects," *Journal of Modern Power Systems and Clean Energy*, vol. 11, no. 5, pp. 1700–1710, 2023.

[14] N. -C. Sintamarean, F. Blaabjerg, H. Wang, and Y. Yang, "Real field mission profile oriented design of a SiC-based PV-inverter application," *IEEE Transactions on Industry Applications*, vol. 50, no. 6, pp. 4082–4089, 2014.

[15] A. Sangwongwanich, Y. Yang, D. Sera, and F. Blaabjerg, "Mission profile-oriented control for reliability and lifetime of photovoltaic inverters," *IEEE Transactions on Industry Applications*, vol. 56, no. 1, pp. 601–610, 2020.

[16] S. E. De León-Aldaco, H. Calleja, and J. Aguayo Alquicira, "Reliability and mission profiles of photovoltaic systems: A FIDES approach," *IEEE Transactions on Power Electronics*, vol. 30, no. 5, pp. 2578–2586, 2015.

[17] Y. Yang, E. Koutroulis, A. Sangwongwanich, and F. Blaabjerg, "Pursuing photovoltaic cost-effectiveness: Absolute active power control offers hope in single-phase PV systems," *IEEE Industry Applications Magazine*, vol. 23, no. 5, pp. 40–49, 2017.

[18] B. Subudhi and R. Pradhan, "A comparative study on maximum power point tracking techniques for photovoltaic power systems," *IEEE Transactions on Sustainable Energy*, vol. 4, no. 1, pp. 89–98, 2013.

[19] H. W. M. Liserre, and F. Blaabjerg "Toward reliable power electronics - Challenges, design tools and opportunities," *IEEE Industrial Electronics Magazine*, vol. 7, no. 2, pp. 17–26, 2013.

[20] H. Huang and P. A. Mawby, "A lifetime estimation technique for voltage source inverters," *IEEE Transactions on Power Electronics*, vol. 28, no. 8, pp. 4113–4119, 2013.

[21] Y. Yang, H. Wang, F. Blaabjerg, and K. Ma, "Mission profile based multi-disciplinary analysis of power modules in single-phase transformerless photovoltaic inverters", in *Proceedings of EPE ECCE Europe*'13, 2013, pp. P.1–P.10.

[22] S. Faulstich, P. Lyding, B. Hahn, and P. Tavner "Reliability of offshore turbines–identifying the risk by onshore experience," in *Proc. of European Offshore Wind*, Stockholm, 2009.

[23] K.O. Kovanen, "Photovoltaics and power distribution," *Renewable Energy Focus*, vol. 14, no. 3, pp. 20–21, 2013.

[24] H. Luo, Y. Xiao, Y. Zhu, Y. Yang, N. P. Papanikolaou, and M. Molinas, "A review of recent requirements for inverter-based resources and grid-forming technologies," in *Proceedings of 2023 25th European Conference on Power Electronics and Applications (EPE'23 ECCE Europe)*, 2023, pp. 1–9.

[25] J. Dvoracek, J. Drapela, J. Moravek, M. Vojtek, and P. Mastny, "In-field compliance testing of an existing photovoltaic power generating module," in *Proceedings of 2024 24th International Scientific Conference on Electric Power Engineering (EPE)*, Kouty nad Desnou, Czech Republic, 2024, pp. 1–6.

[26] A. A. Aboelnaga and M. A. Azzouz, "Dual current control of renewable energy sources for recent grid code compliance and reliability enhancement," *IEEE Transactions on Sustainable Energy*, vol. 15, no. 2, pp. 1380–1395, 2024.

[27] D. Wu, G.-S. Seo, L. Xu, *et al.*, "Grid integration of offshore wind power: Standards, control, power quality and transmission," *IEEE Open Journal of Power Electronics*, vol. 5, pp. 583–604, 2024.

[28] IEEE standard for interconnection and interoperability of inverter-based resources (IBRs) interconnecting with associated transmission electric power systems," *IEEE Std 2800-2022*, pp. 1–180, 22 April 2022.

[29] "IEEE Application Guide for IEEE Std 1547™-2018, IEEE Standard for Interconnection and Interoperability of Distributed Energy Resources with Associated Electric Power Systems Interfaces," *IEEE Std 1547.2-2023 (Revision of IEEE Std 1547.2-2008)*, pp. 1–291, 20 May 2024.

[30] F. Blaabjerg and K. Ma, "Wind energy systems," *Proceedings of the IEEE*, vol. 105, no. 11, pp. 2116–2131, 2017.

[31] M. R. K. Rachi, M. A. Awal, and I. Husain, "Asymmetrical fault ride-through and power oscillation characterization for grid-tied voltage source converters," *IEEE Transactions on Industry Applications*, vol. 59, no. 4, pp. 4550–4561, ms Integration Group, Grid-Forming Technology in Energy Systems Integration, Mar. 2022. (available online, assessed on July 19, 2024: https://www.esig.energy/wp-content/uploads/2022/03/ESIG-GFM-report-2022.pdf).

[32] Y. Yang, P. Enjeti, F. Blaabjerg, and H. Wang, "Wide-scale adoption of photovoltaic energy: Grid code modifications are explored in the distribution grid," *IEEE Industry Applications Magazine*, vol. 21, no. 5, pp. 21–31, 2015.

[33] F. Blaabjerg, Y. Yang, D. Yang, and X. Wang, "Distributed power-generation systems and protection," *Proceedings of the IEEE*, vol. 105, no. 7, pp. 1311–1331, 2017.

[34] H. Kobayashi, "Fault ride through requirements and measures of distributed PV systems in Japan," in *Proc. of IEEE-PES General Meeting*, pp. 1–6, 22–26 July 2012.

[35] Z. Tang, Y. Yang, and F. Blaabjerg, "Power electronics: The enabling technology for renewable energy integration," *CSEE Journal of Power and Energy Systems*, vol. 8, no. 1, pp. 39–52, 2022.

[36] B. Hahn, M. Durstewitz, K. Rohrig "Reliability of wind turbines – Experience of 15 years with 1500 WTs", *Wind Energy*, Springer, Berlin, 2007.

[37] K. Fischer, T. Stalin, H. Ramberg, *et al.*, "Field-experience based root-cause analysis of power converter failure in wind turbines," *IEEE Transactions on Power Electronics*, vol. 30, no. 5, pp. 2481–2492, 2015.

[38] L. M. Moore and H. N. Post, "Five years of operating experience at a large, utility-scale photovoltaic generating plant," *J. Progress in Photovoltaics, Research and Applications*, vol. 16, no. 3, pp. 249–259, 2008.

[39] Y. Yang, H. Wang, F. Blaabjerg, and K. Ma, "Mission profile based multi-disciplinary analysis of power modules in single-phase transformerless photovoltaic inverters," in *Proceedings of EPE 2013*, pp. 1–10, 2013.

[40] "ZVEI - Handbook for robustness validation of automotive electrical/electronic modules," Jun. 2013.

[41] Y. Song and B. Wang, "Survey on reliability of power electronic systems," *IEEE Transactions on Power Electronics*, vol. 28, no. 1, pp. 591–604, 2013.

[42] A. M. Alcaide, G. Buticchi, A. Chub, and L. Dalessandro, "Design and control for high-reliability power electronics: State-of-the-art and future trends," *IEEE Journal of Emerging and Selected Topics in Industrial Electronics*, vol. 5, no. 1, pp. 50–61, 2024

[43] Y. Lu, E. Xiang, A. Zhu, H. Luo, H. Yang, and R. Zhao, "Mission-profile-based reliability evaluation of IGBT modules for wide-speed range electric vehicle drive using fast multistep mapping simulation strategy," *IEEE Journal of Emerging and Selected Topics in Power Electronics*, vol. 11, no. 5, pp. 5376–5390, 2023.

[44] E. Ozkol, F. Brem, C. Liu, S. Hartmann, and A. Kopta, "Enhanced power cycling performance of IGBT modules with a reinforced emitter contact," *Microelectronics Reliability*, vol. 55, no. 6, pp. 912–918, 2015.

[45] A. Wintrich, U. Nicolai, W. Tursky, and T. Reimann, *Application Manual Power Semiconductors*, Semikron ISLE Verlag, Nuremberg, 2015.

[46] W. Liu, D. Zhou, F. Iannuzzo, M. Hartmann, and F. Blaabjerg, "Separation and validation of bond-wire and solder layer failure modes in IGBT modules," *IEEE Transactions on Industry Applications*, vol. 58, no. 2, pp. 2324–2331, 2022.

[47] D. Zhou, F. Blaabjerg, M. Lau, and M. Tonnes, "Thermal cycling overview of multi-megawatt two-level wind power converter at full grid code operation," *IEEJ Journal of Industry Applications*, vol. 2, no. 4, pp. 173–182, 2013.

[48] U. Scheuermann and R. Schmidt, "A new lifetime model for advanced power modules with sintered chips and optimized Al wire bonds", in *Proceedings of PCIM 2013*, 2013, pp. 810–813.

[49] G. Zeng, L. Borucki, O. Wenzel, O. Schilling, and J. Lutz, "First results of development of a lifetime model for transfer molded discrete power devices," in *Proceedings of PCIM Europe*, Nuremberg, Germany, 2018, pp. 706–713.

[50] D. Zhou, F. Blaabjerg, M. Lau, and M. Tonnes, "Thermal profile analysis of doubly fed induction generator-based wind power converter with air and liquid cooling methods," in *Proceedings of EPE 2013*, 2013, pp. 1–10.
[51] L. R. GopiReddy, L. M. Tolbert, B. Ozpineci, and J. O. P. Pinto, "Rainflow algorithm-based lifetime estimation of power semiconductors in utility applications," *IEEE Transactions on Industry Applications*, vol. 51, no. 4, pp. 3368–3375, 2015.
[52] M Obermayr, C. Riess, and J. Wilde, "A novel online 4-point rainflow counting algorithm for power electronics," *Microelectronics Reliability*, vol. 120, pp. 1–9, 2021.
[53] B. Bertsche, *Reliability in Automotive and Mechanical Engineering*, Springer, Berlin, 2008.
[54] F. Blaabjerg and K. Ma, "Future on power electronics for wind turbine systems," *IEEE Journal of Emerging and Selected Topics in Power Electronics*, vol. 1, no. 3, pp. 139–152, 2013.
[55] F. Blaabjerg, Y. Yang, K. A. Kim, and J. Rodriguez, "Power electronics technology for large-scale renewable energy generation," *Proceedings of the IEEE*, vol. 111, no. 4, pp. 335–355, 2023.
[56] P. Catalán, Y. Wang, J. Arza, and Z. Chen, "A comprehensive overview of power converter applied in high-power wind turbine: Key challenges and potential solutions," *IEEE Transactions on Power Electronics*, vol. 38, no. 5, pp. 6169–6195, 2023.
[57] Y. Yang, A. Sangwongwanich, and F. Blaabjerg, "Design for reliability of power electronics for grid-connected photovoltaic systems," *CPSS Transactions on Power Electronics and Applications*, vol. 1, no. 1, pp. 92–103, 2016.
[58] Y. Yang, F. Blaabjerg, and H. Wang, "Low voltage ride-through of single-phase transformerless photovoltaic inverters", *IEEE Transactions on Industry Applications*, vol. 50, no. 3, 1942–1952, 2014.
[59] D. Meneses, F. Blaabjerg, O. García, and J. A. Cobos, "Review and comparison of step-up transformerless topologies for photovoltaic ac-module application," *IEEE Transactions on Power Electronics*, vol. 28, no. 6, pp. 2649–2663, 2013.
[60] S. V. Araujo, P. Zacharias, and R. Mallwitz, "Highly efficient single-phase transformerless inverters for grid-connected PV systems," *IEEE Transactions on Industrial Electronics*, vol. 57, no. 9, pp. 3118–3128, 2010.
[61] R. Gonzalez, J. Lopez, P. Sanchis, and L. Marroyo, "Transformerless inverter for single-phase photovoltaic systems," *IEEE Transactions on Power Electronics*, vol. 22, no. 2, pp. 693–697, 2007.
[62] S. R. Gonzalez, C. J. Coloma, P. L. Marroyo, T. J. Lopez, and G. P. Sanchis "Single-phase inverter circuit for conditioning and converting dc electrical energy into ac electrical," *International Patent Application*, Pub. No. WO/2008/015298, 7 Feb 2008.
[63] T. Kerekes, R. Teodorescu, P. Rodriguez, G. Vazquez, and E. Aldabas, "A new high-efficiency single-phase transformerless PV inverter topology,"

IEEE Transactions on Industrial Electronics, vol. 58, no. 1, pp. 184–191, 2011.

[64] H. Schmidt, S. Christoph, and J. Ketterer, "Current inverter for direct/alternating currents, has direct and alternating connections with an intermediate power store, a bridge circuit, rectifier diodes and a inductive choke," *German Patent DE10 221 592 A1*, 4 Dec 2003.

[65] Sunways, Yield-oriented solar inverters with up to 98% peak efficiency. Product category (available: http://www.sunways.eu/en/).

[66] SMA, SUNNY CENTRAL – High tech solution for solar power stations. Products Category Brochure (available: http://www.sma-america.com/).

[67] M. Meinhardt and G. Cramer, "Multi-string-converter: the next step in evolution of string-converter technology," in *Proceedings of EPE'01*, 2001, pp. P.1–P.9.

[68] Y. Yang, H. Wang, and F. Blaabjerg, "Reliability assessment of transformerless PV inverters considering mission profiles," *International Journal Photoenergy*, vol. 2015, pp. 1–10, 2015.

[69] T. Ryu and U. -M. Choi, "Reliability-oriented optimal DPWM strategy for single-phase five-level T-type inverter in PV systems," *IEEE Journal of Emerging and Selected Topics in Power Electronics*, vol. 11, no. 2, pp. 2227–2235, 2023.

[70] S. E. De León-Aldaco, H. Calleja, F. Chan, and H. R. Jiménez-Grajales, "Effect of the mission profile on the reliability of a power converter aimed at photovoltaic applications—a case study," *IEEE Transactions on Power Electronics*, vol. 28, no. 6, pp. 2998–3007, 2013.

[71] P. D. Reigosa, H. Wang, Y. Yang, and F. Blaabjerg, "Prediction of bond wire fatigue of IGBTs in a PV inverter under a long-term operation," *IEEE Transactions on Power Electronics*, vol. 31, no. 10, pp. 7171–7182, 2016.

[72] M. Dbeiss, Y. Avenas, H. Zara, L. Dupont, and F. A. Shakarchi, "A method for accelerated aging tests of power modules for photovoltaic inverters considering the inverter mission profiles," *IEEE Transactions on Power Electronics*, vol. 34, no. 12, pp. 12226–12234, 2019.

[73] T. Ryu, U.-M. Choi, and F. Blaabjerg, "Reliability-oriented asymmetric power device size selection of neutral-point clamped inverter considering mission profiles," *IEEE Journal of Emerging and Selected Topics in Power Electronics*, vol. 12, no. 6, pp. 5966–5976, 2024.

[74] L. Ceccarelli, R. M. Kotecha, A. S. Bahman, F. Iannuzzo, and H. A. Mantooth, "Mission-profile-based lifetime prediction for a SiC mosfet power module using a multi-step condition-mapping simulation strategy," *IEEE Transactions on Power Electronics*, vol. 34, no. 10, pp. 9698–9708, 2019.

[75] A. Anurag, Y. Yang, and F. Blaabjerg, "Thermal performance and reliability analysis of single-phase PV inverters with reactive power injection outside feed-in operating hours," *IEEE Journal of Emerging and Selected Topics in Power Electronics*, vol. 3, no. 4, pp. 870–880, 2015.

[76] U. Scheuermann, "Reliability of advanced power modules for extended maximum junction temperatures," *Bodo's Power Systems*, pp. 26–30, September, 2014.

[77] D. Zhou, H. Wang, F. Blaabjerg, S. K. Kaer, and D. Blom-Hansen, "System-level reliability assessment of power stage in fuel cell application," in Proceedings on ECCE, 2016, pp. 1–8.

[78] D. Zhou, Y. Song, Y. Liu, and F. Blaabjerg, "Mission profile based reliability evaluation of capacitor banks in wind power converters," *IEEE Transactions on Power Electronics*, vol. 34, no. 5, pp. 4665–4677, 2019.

[79] D. Zhou, H. Wang, and F. Blaabjerg, "Reactive power impacts on LCL filter capacitor lifetime in grid-connected inverter," *IEEE Open Journal of Power Electronics*, vol. 1, pp. 139–148, 2020.

[80] A. Sangwongwanich, Y. Yang, D. Sera, F. Blaabjerg, and D. Zhou, "On the impacts of PV array sizing on the inverter reliability and lifetime," *IEEE Transactions on Industry Applications*, vol. 54, no. 4, pp. 3656–3667, 2018.

[81] A. Sangwongwanich. Y. Shen, A. Chub, *et al.*, "Design for accelerated testing of DC-link capacitors in photovoltaic inverters based on mission profiles," *IEEE Transactions on Industry Applications*, vol. 57, no. 1, pp. 741–753, 2021.

[82] C. E. Silva, A. M. Alzamora, M. T. A. Êvo, and H. de Paula, "Design for reliability of DC-link capacitors for multilevel inverter topologies operating in hostile environments," *IEEE Transactions on Industry Applications*, vol. 60, no. 4, pp. 5882–5892, 2024.

[83] D. C. Jordan, B. Sekulic, B. Marion, and S. R. Kurtz, "Performance and aging of a 20-year-old silicon PV system," *IEEE Journal of Photovoltaics*, vol. 5, no. 3, pp. 744–751, 2015.

[84] N. Bosco, T. J. Silverman, and S. Kurtz, "The influence of PV module materials and design on solder joint thermal fatigue durability," *IEEE Journal of Photovoltaics*, vol. 6, no. 6, pp. 1407–1412, 2016.

[85] A. Sangwongwanich, Y. Yang, D. Sera, and F. Blaabjerg, "Lifetime evaluation of grid-connected PV inverters considering panel degradation rates and installation sites," *IEEE Transactions on Power Electronics*, vol. 33, no. 2, pp. 1225–1236, 2018.

[86] S. Bouguerra, M. R. Yaiche, O. Gassab, A. Sangwongwanich, and F. Blaabjerg, "The impact of PV panel positioning and degradation on the PV inverter lifetime and reliability," *IEEE Journal of Emerging and Selected Topics in Power Electronics*, vol. 9, no. 3, pp. 3114–3126, 2021.

Chapter 14

Reliability of power conversion systems in photovoltaic applications

Jack Flicker[1] *and Robert Kaplar*[1]

14.1 Introduction to photovoltaic power systems

A photovoltaic (PV) inverter is a balance-of-systems (BOS, i.e., every component except for the module component that controls and converts power flow through the PV system). Specifically, the inverter transforms the nominal DC power produced by the PV module to AC power, which can be transported through the electrical power grid or used on-site by various power-consuming units (Figure 14.1). As the interface between the DC and AC sides of the system, the inverter must meet rather stringent requirements for both.

Although there is a wide assortment of different inverter technologies and topologies, the basic PV inverter consists of a DC/DC converter stage in series with a DC/AC inverter stage. Inverters can be categorized in many different ways depending on the purpose. The main classifications of inverters are based on the topology of the inverter (e.g., single vs. multi-stage, isolated vs. non-isolated, single- vs. three-phase), the PV system as a whole (e.g., unipolar vs. bipolar, grounded vs. ungrounded), or power handling capabilities (utility-scale ~500 kW, residential scale ~5 kW, microinverter ~ a few hundred W).

14.1.1 DC/DC conversion

A solar cell or module can be thought of as a type of photoactive diode. Therefore, the characteristic curve for both a diode and a solar cell is a current–voltage (*IV*) curve. In the dark, a solar cell acts exactly like a regular diode. However, under illumination, the solar cell produces a current flow. This causes a negative current offset, I_L, in the *IV* curve so that the diode curve exists in the fourth quadrant. The *IV* curve in the fourth quadrant (positive voltage and negative current) has a negative product, and this negative power indicates a power output (Figure 14.2).

Several figures of merit are used to describe the *IV* curve of a solar cell under illumination. These are the short-circuit current (I_{sc}), open-circuit voltage (V_{oc}), and maximum power point (P_{mp}). The short-circuit current is defined as the output

[1]Sandia National Laboratories, Albuquerque, NM, USA

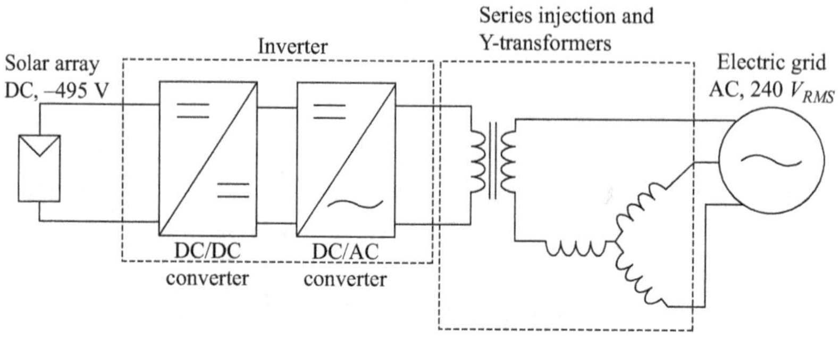

Figure 14.1 Power flow through a PV system. A solar array produces DC power at a voltage of around 500 V. The inverter transforms this DC power into AC power through some combination of DC/DC and DC/AC conversion that is then filtered and transmitted through the grid.

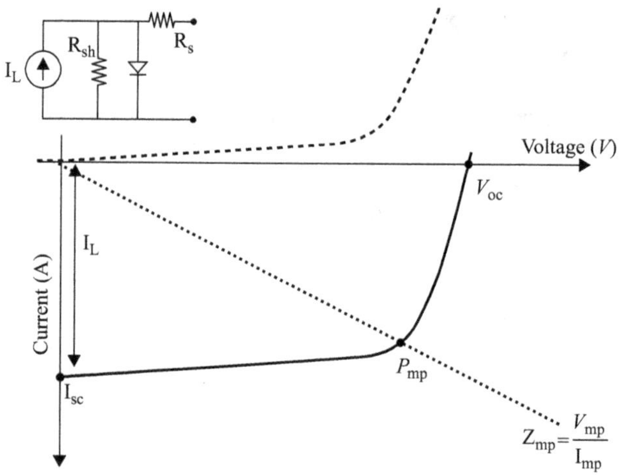

Figure 14.2 In the dark, a solar cell acts like a diode (dashed line). In the light, however, it produces a current, which shows up as an offset, I_L, in the fourth quadrant of the IV curve (solid black line). A PV cell or module is described by three points: the short circuit current (I_{sc}), the open circuit voltage (V_{oc}), and the maximum power point (P_{mp}). The circuit diagram for a cell is simply a current source in parallel with a diode, with a shunt resistance (R_{sh}) and series resistance (R_s) included to model non-idealities. The DC side of the inverter is responsible for maximizing power transfer by altering the input impedance of the inverter so that it matches the load line (Z_{mp}) through P_{mp}.

current from the cell under zero applied bias. This is the maximum possible output current from a cell for a given illumination. The open-circuit voltage is defined as the applied voltage corresponding to zero output current or the point where the *IV* curve crosses the positive voltage axis. The maximum power point is defined as the maximum output power (power is the product of the applied voltage and output current) of the cell. For well-behaved cells, this point occurs at the "knee" of the *IV* curve, or the point where the output current begins to rapidly decrease. In a circuit diagram, an ideal solar cell can be described as a simple diode in parallel with a current source; series and shunt resistances can be added to model non-idealities.

On the DC side, the inverter is responsible for ensuring maximum power transfer from the PV module through a process known as maximum power point tracking (MPPT). MPPT is accomplished by varying the input impedance of the inverter to match the maximum power load-line of the module, typically using a DC/DC converter (although there are some single-stage topologies that use a more complicated pulse-width-modulation technique to alter the input impedance while simultaneously converting from DC to AC power). There are multiple algorithms, ranging from relatively simple to complex, that can be used by the inverter for MPPT. Some of the most popular of these algorithms include constant voltage tracking (CVT), perturb and observe (P&O or mountain-climb) [1], incremental conductance (INC) [2], and periodic scanning [3]. Most of the more complex algorithms can easily reach efficiencies (known as tracking factors) of >95% [4]. For an ideal MPPT algorithm, the applied voltage to the module always results in P_{mp} from the module. Real MPPT algorithms typically cycle around P_{mp} at 0.5–5 Hz. The specific frequency and amplitude of this oscillation depend on the speed and accuracy of the control system and algorithm.

In order to decrease parasitic DC transmission losses in the PV array, the operational voltage is desired to be as high as possible (and the current as low as possible). For safety reasons, the maximum V_{oc} of the array is limited to 600 V above ground in the United States (this is dictated by the National Electric Code article 690.7 [5]; it is worth noting that some "behind the fence" utility-scale operations in the US as well as installations in Europe are not bound by the NEC and frequently operate up to 1000 V_{oc}). Due to safety de-rating of the array's V_{oc} due to temperature variation (lower ambient temperature causes bandgap widening, which increases the V_{oc} of the array [3]) and the fact that V_{mp} is ~80% of V_{oc} in silicon modules, the typical operational V_{mp} of an array is approximately 475–525 V above ground. Therefore, in addition to MPPT, the DC input stage of the inverter bucks the voltage from V_{mp} to the voltage necessary for AC injection.

Other than the MPPT controls/algorithms, one of the most important design considerations for the inverter on the DC side consists of DC bus voltage ripple (V_{ripple}). The presence of V_{ripple} in the inverter/module circuit alters the voltage point applied to the module from V_{mp} to $V_{mp} \pm V_{ripple}$ (Figure 14.3). This causes a corresponding ripple in the power output of the module so that the average power output (P_{avg}) is less than P_{mp}. Depending on the shape of the module *IV* curve, a small deviation in V_{mp} can have a large deleterious effect on P_{avg}, decreasing the utilization ratio (P_{avg}/P_{mp}) of the module.

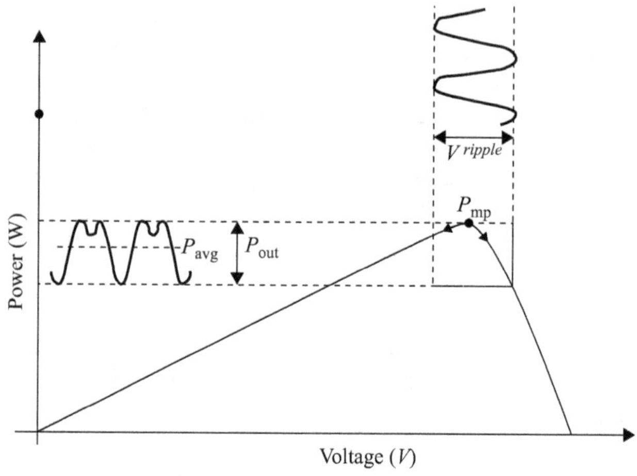

Figure 14.3 Ripple voltage (V_{ripple}) causes the operational voltage to oscillate around the maximum power point (P_{mp}) of the solar module, decreasing the average power (P_{avg}) below P_{mp} [8]

For a generic DC/DC converter, the V_{ripple} on the DC input can be calculated by solving the current flow across the bus capacitor to be [6]

$$V_{ripple} = \frac{V_{mp}}{32 \cdot L \cdot C \cdot f^2} \tag{14.1}$$

where C is the bus capacitance, L is the circuit inductance, and f is the switching frequency.

As can be seen, the ripple voltage is inversely proportional to the bus capacitance. A large capacitance is necessary to reduce V_{ripple} and maximize the utilization ratio of the array. In order to limit this voltage ripple, each inverter requires an energy storage element (either a capacitor or inductor) [7]. However, for economic and/or engineering reasons, the number or size (either physical or electrical) of these storage elements must be limited, and some magnitude of V_{ripple} can be found in all single-phase inverters at the DC side of the circuit [3]. Three-phase inverters typically have much less ripple on the DC bus due to interaction between the phase legs.

14.1.2 DC/AC conversion

On the AC side of the system, the purpose of an inverter is to inject power into the grid at the same frequency and with a small phase angle ($\varphi \approx 0$). Although the DC/AC conversion stage of the PV inverter is a relatively simple power conversion unit in a strictly topological sense, the system as a whole is a complicated switching/monitoring system with a number of responsibilities.

The main purpose of an inverter is to output power that meets power quality standards (e.g., IEEE 1547 [9] in North America or IEC 61727 [10] in Europe, as shown in Table 14.1. However, depending on the local/national ordinances and/or

Table 14.1 Electrical power standards for North America (IEEE 1547) and Europe (IEC 61727)

Category	IEEE 1547	IEC 61727
Nominal power (kW)	30	10
THD (%)	<5	<5
Power factor	>0.94	0.9 (at 50% power rating)
DC current injection (%)	<0.5	<1
Voltage range (V)	97–121 (88%–110% $V_{nominal}$)	196–253 (85%–110% $V_{nominal}$)
Frequency range (Hz)	59.3–60.5	50±1

complexity of the device, the inverter may also be required to manage the power output of the PV module, connect/disconnect from the grid, manage VAR (i.e., imaginary power), read and report system status, and monitor islanding [11].

Added inverter functionalities exponentially increase the difficulty in creating both reliable and affordable components since the necessary control schemes become more complex, and inter-component interactions become more difficult to predict. The current software on inverter units is so complex that it can be thought of as a "virtual component" of the system, which can be difficult, if not impossible, to debug in all realistic usage conditions. Software mistakes causing failures in real components are typically attributed to component failures.* This leads to a higher failure rate counting for the real components when software is the root cause of failure. As inverters become more complex and are tasked with more responsibilities (as they inevitably will be), software will become a larger and larger issue for inverter reliability.

Also, increasingly important to inverter operation is the aspect of safety monitoring and hazard mitigation. Historically, the only safety issues that inverters have been responsible for were to prevent islanded operation;† however, numerous fires due to fault events [12–16] on rooftop systems have brought to light the importance of the various fault detection technologies utilized‡ in inverters [13].

Other than the catastrophic reliability issue of an undetected fault, the greatest reliability issue for inverters is the threshold used to determine the presence of a fault. If the threshold is too aggressive, there will be "nuisance trips" that falsely detect the presence of a fault, take the inverter off-line and require "repair."

*For example, a bug in the software code may cause a shoot-through current event on one leg of the H-bridge. This causes the IGBTs to catastrophically fail and would typically be recorded as an IGBT failure by the technician, when the control was the root cause of failure.

†Anti-islanding protection requires inverters to detect the non-presence of the electrical grid and shut down power export operations, thus protecting electrical workers who may be performing maintenance down-line.

‡Partially included. Series arc-faults are required to be detected by the NEC, while parallel arc-faults are not yet required. Due to the "nuisance tripping" problems currently associated with series arc-fault detectors, there is reticence on the part of many in the industry to move forward with a parallel arc-fault detection requirement, regardless of the increased safety.

However, if the threshold is too passive, certain ground faults go undetected. The industry is currently debating proper detection thresholds to implement in order to reduce or eliminate "nuisance tripping" in all conditions (meteorological, configurational, and electrical) while also detecting the worst-case (most difficult to detect) faults possible.

14.2 Power conversion reliability in PV applications

Photovoltaic (PV) power systems have, historically, not been able to compete with conventional fossil fuels without favorable regulatory policies and financial incentives. The US Department of Energy (DOE), however, has estimated that PV systems can reach economic competitiveness with other energy sources at an installed system cost of $1/W ($0.05–0.06/kWh) [17]. With a $1/W price-point, the DOE estimates PV penetration into the US power market to be upwards of 18% by 2050.

As of 2013, the median installed cost of PV systems was $3.00/W for large utility-scale systems [18]. While most research, both historically and currently, focuses on the production costs of PV module technology, the cost of the necessary grid-connected DC-to-AC inverters has been largely ignored. As the price of PV modules drops, the price of inverters becomes more important. Inverters now constitute 8%–12% of the total lifetime PV cost [11]. As of 2010, the inverter and associated power conditioning components accounted for $0.25/W [19], well above the DOE benchmark of $0.10/W by 2017 [17].

One of the key price drivers of the inverter components is inverter reliability [20,21]. PV modules have long lifetimes, with warranties offered up to 20 years. The mean time between failure (MTBF) of these systems has been shown, in the field, to be up to 520 years for residential and 6600 years for utility systems [22]. In contrast, the inverter component has shown a field MTBF of between 1 and 16 years [22,23]. Typical warranties on these devices last only 3–5 years [20], with a few (e.g., certain units produced by PV powered) with warranties of up to 10 years. Even with the most optimistic view of inverter lifetime, it will be necessary to replace or repair an inverter multiple times over the lifetime of a PV module. Repairing the inverter is costly not only due to the necessity for replacement parts and work crews, but also due to incidental costs such as the loss of power not generated during downtime, purchasing replacement power unsupplied by the off-line system, and any performance degradation before failure identification [20].

Inverter issues are especially problematic since, depending on the PV topology, they may affect the power production of a large number of modules and/or strings. According to SunEdison, a North-American-based PV plant operator with over 750 PV plants across the world, inverter performance was responsible for 36% of total energy losses between January 2010 and March 2012 (Figure 14.4) [24].

In addition to having a number of functionalities that add complexity to the system, the inverter must also operate in relatively harsh, changing environments. Inverters may experience large temperature swings from –30 to 70°C, humidity conditions from 0 to 100%, and/or salty, corrosive environments.

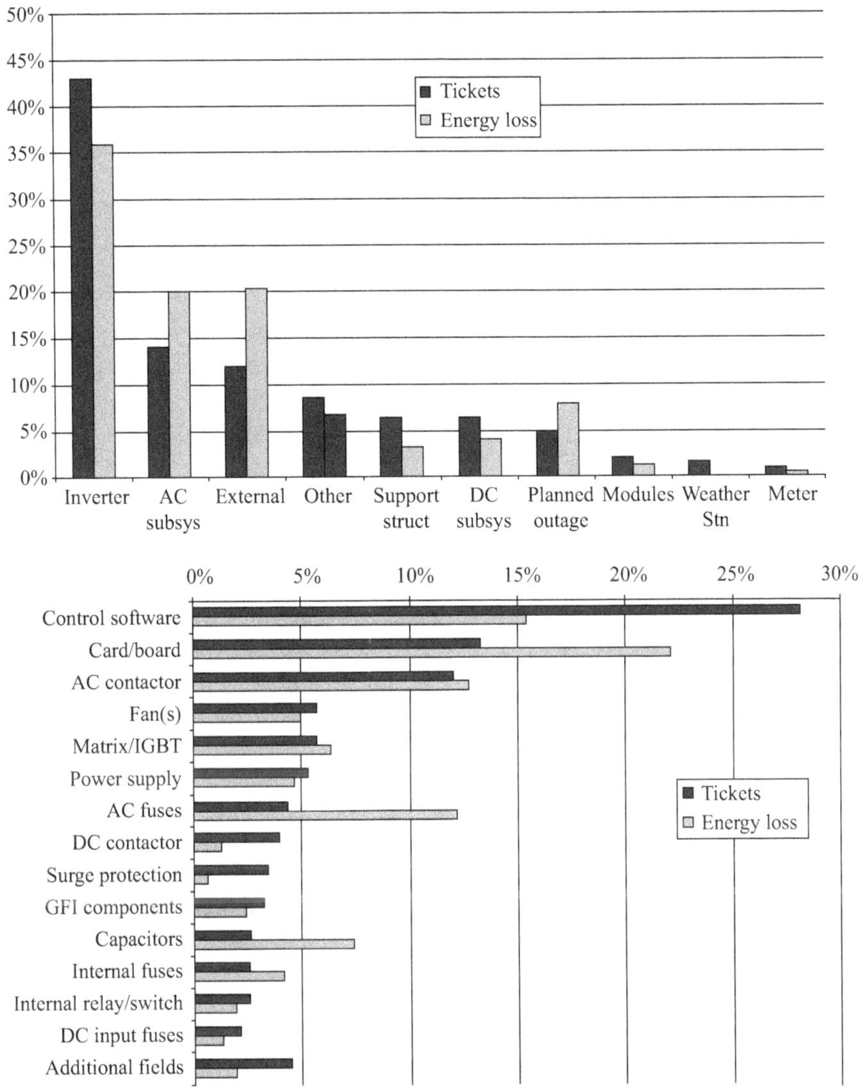

Figure 14.4. Between January 2010 and March 2012, inverter issues were responsible for 43% of maintenance tickets and 36% of energy losses at 350 PV installations operated by SunEdison (top). Inverter failure has been attributed to various internal components (bottom) [24] (Courtesy Anastasios Golnas, SunEdison).

14.2.1 Capacitors

Many consider DC bus capacitors to be the weak link in inverter reliability [11,25], decreasing inverter lifetimes by as much as 50% [26] due to constant temperature and power cycling [27] and high internal capacitor temperature [28]. According to SunEdison, over a 27-month period, capacitor issues were responsible for 7% of total energy losses [24]. There is an active debate in the industry regarding the reliability of electrolytic (elcaps) vs. thin-film capacitors in inverters, with most manufacturers moving toward metallized thin-film capacitors to increase lifetimes.

Capacitor end-of-life can be described by two basic categories of failure, catastrophic and degradation [29]. Catastrophic failure is the complete loss of functionality of the capacitor in the circuit. These failures can result in an open- or short-circuit, explosion of the capacitor can, destruction of the dielectric, harm to other electrical components, or leakage of liquids or gases from the capacitor interior. Degradation failures are failures of definition where a capacitor parameter has fallen outside of "acceptable" performance limits. These limits and parameters vary widely by manufacturer and may include a leakage current increase, equivalent series resistance (ESR) increase (typically 300% for electrolytic and 150% for metallized thin-film capacitors), dielectric value decrease (typically 40% for electrolytics), or capacitance decrease (typically 20% for electrolytics and 5%–10% for metallized thin-film [30,31]. Electrolytic capacitors tend to fail catastrophically, while metallized thin-film capacitors tend towards degradation failures.

Electrolytic capacitors have two primary modes of failure: physical and chemical. The primary physical failure mechanism for elcaps is vaporization of the electrolyte through polymer seals. A decrease in the volume of the electrolyte leads to an increase in the capacitor ESR [32]. Vaporization of the electrolyte may be driven by increased internal temperatures due to ripple current or simple evaporation through poor quality sealing. Many researchers cite electrolyte vaporization as the primary wear-out mechanism for elcaps [32,33].

Chemical failure of elcaps has been mentioned previously. Leakage current through the electrolyte drives the reformation of Al_2O_3 through the consumption of H_2O. This reaction leads to three methods of failure. The first is that reformed oxide is of a poorer quality than the oxide formed via anodization. The replacement of high-quality oxide with poor-quality oxide decreases the capacitance of the elcap over time. Second, the consumption of H_2O leads to higher ESR values due to the decreased electrolyte volume. Finally, in addition to the oxide, one of the products of the reforming reaction is H_2 gas. Since the capacitor can is airtight to prevent evaporation of the electrolyte, H_2 gas accumulates inside the canister over time, increasing the interior pressure. Once the interior pressure builds to $\sim$7 atm, the pressure relief valves release, bursting the capacitor can. After the can is exposed to air, the electrolyte readily evaporates and the capacitor fails.

With the advent of segmented electrode films, metallized thin-film capacitors seldom fail catastrophically. Over time, metallized thin-film capacitors suffer gradual degradation until a parameter falls below specification. This is known as soft failure [34].

There exist two categories of degradation in metallized thin-film capacitors [35]. The first, and most common [31], is due to clearing behavior. For each clearing event, there is a small decrease in the overall capacitance of the capacitor. With enough clearing events, the capacitor eventually fails in an open circuit. This failure is much more desirable from a safety point of view compared to the catastrophic failure modes of elcaps, which tend to short-circuit and can extensively damage both the capacitor itself as well as other electrical components [36,37].

During a clearing event, joule heating from the localized discharge vaporizes a certain volume of metallized electrode (the vaporized volume is proportional to applied voltage [38]). This discharge creates a localized plasma that interacts with the polymer surface, creating low-molecular-weight species from the degradation of polymer. Polymers that are rich in oxygen tend to form CO upon vaporization. Polymers that have a relatively low oxygen composition tend to form conductive graphitic filaments that are more likely to bridge opposing electrodes and form shorts [35]. Low-oxygen-component polymers are said to be poor clearing polymers and are more likely to clear incorrectly, shorting the electrodes and catastrophically failing.

Even for polymers with good clearing behavior, the formation of gas inside the capacitor can lead to problems. As more clearing events occur, the production of gas increases pressure inside the casing. This increase in pressure tends to increase the rate of degradation [31] via surface erosion, tracking, and polymer decomposition. This means that localized failure in one area leaves neighboring areas ripe for further electrical breakdown [35]. Excess gas can also lead to a bulging and pressure release [39] (though this usually occurs only after degradation failure has already occurred). Before pressure release, the gas may be dense enough to conduct charge between electrodes in a process known as flashover (also a catastrophic failure). To prevent this, manufacturers may choose to fill metallized thin-film capacitor windings with a liquid (liquid-impregnated) instead of air (dry-wound) [40].

The second degradation mechanism is due to direct electron impacts on polymer segments at the electrode/dielectric interface. The impact creates free radicals in the polymer that, in turn, react with absorbed H_2O or O_2. This interaction causes a chain reaction that can cause scission or cross-linking of polymer chains, deteriorating the initially good physical properties of the film [35].

14.2.2 IGBTs/MOSFETs

As the main component of the power engine for the inverter, semiconductor reliability is of primary importance for inverter performance and lifetime. Typical failure modes for an IGBT or MOSFET module are either due to thermo-mechanical degradation of the packaging or due to electrical degradation of the components. The most common thermo-mechanical failure modes result from bond-wire lift-off or deterioration of the die attach layer (Figure 14.5). Typical electrical deterioration is brought about by breakdown of the gate oxide, turn-on of the parasitic thyristor, or degradation of the package encapsulant material. Most of these degradation modes result in a positive feedback mechanism, where degradation results in an increase in junction temperature, which further accelerates degradation mechanisms in a thermal runaway mechanism [41].

An IGBT/MOSFET module is subject to cyclic thermal stresses due to both long-term temperature changes resulting from ambient temperature- and load-

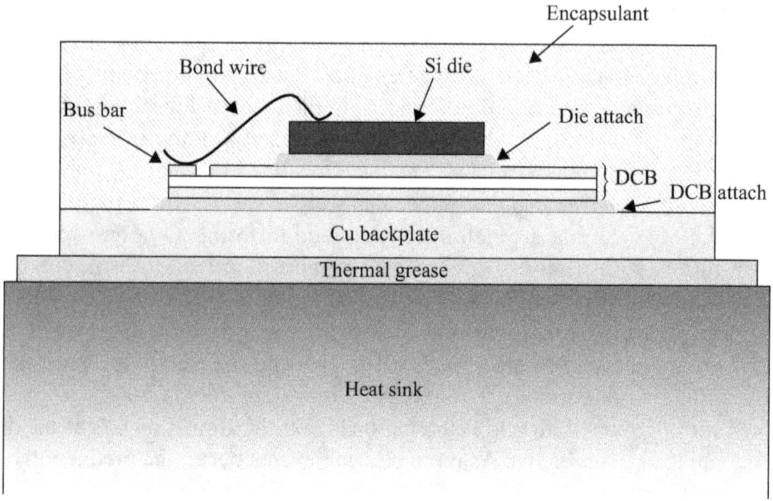

Figure 14.5. Packaging of a generic IGBT module. The module consists of a die and the elements used to control its electronic and thermal properties [45].

induced heat-sink temperature swings (thermal cycling) and short-term (~1 s) sudden variations in load conditions (power cycling). For example, in a typical IGBT-based H-bridge system, 4% of the controlled power is dissipated as heat [42] due to parasitic resistances, carrier impact ionization, and current crowing in the semiconductor junction [43]. The number of power cycles that a module may undergo is typically limited by the packaging configuration and the coefficient of thermal expansion (CTE) mismatch between adjacent layers [44]. Different mounting schemes differ slightly, but a generic package is shown in Figure 14.5.

The transistor chip (die) is attached to a double-copper-bonded (DCB) board using a die attachment material (solder, epoxy, etc.) that provides both electrical and thermal conductivity. Bus bars on the DCB board are connected to pads on the die through bond wires (typically Al). The DCB board itself is connected to a Cu backplate by another attachment layer. The Cu backplate is attached to either an air- or water-cooled heat sink via thermal grease in order to dissipate heat from the Si chip. The entire system is encapsulated in an insulator and protected with a plastic housing.

There may be as many as 800 wedge bonds connecting the bus bars to the die [46]. The disturbance of these wire bonds is the primary thermo-mechanical degradation mechanism [47–49]. Power cycling of the module quickly raises the temperature of the semiconductor junction. This temperature change is transferred to the wire bonds through conduction and ohmic heating of the wire. The CTE mismatch between the Al bond wire and the Si die (along with expansion/contraction of the bond wire) leads to mechanical stress on the Si/Al joint and bond wire failure may result [43]. These failures can result in the loss of functionality/

control of the chip or additional stress on other wire/die bonds, which must handle an increased current flow.

This increase in current density in the remaining bond wires results in an increase in $V_{CE(on)}$. Over time, as the module undergoes more power cycling, the $V_{CE(on)}$ value will tend to increase [49]. This increase in $V_{CE(on)}$ leads to an increase in power dissipation, junction temperature, and stress to the packaging. Wire bond lift-off can be mitigated through the use of Cu wires. Cu possesses both a much higher thermal conductivity and lower CTE mismatch with Si than does Al, so the temperature-cycling-induced stress is much lower [46].

The die attachment to the DCB can also suffer from deterioration due to CTE mismatch and eventual void formation and fatigue-induced cracking from shear strain due to repeated power cycling [50–53]. The die attach layer is typically made from a metal eutectic material (SnPb, SnAgCu, or AuSi). These metal solders possess excellent electrical characteristics, outstanding thermal transport properties, and high ductility (to accommodate CTE mismatch) [54]. SnPb solder is the workhorse of the electronics industry. However, due to health and environmental concerns, most companies have been moving away from solders containing Pb. Other eutectic solders are typically expensive, so many manufacturers are moving toward the use of metal-loaded epoxies. These epoxies ensure conductivity through metal concentrations just above the percolation threshold. Although this is a much less expensive attachment method, the thermal transfer and ductility properties are inferior to metal eutectic solders, especially if there is heterogeneous or incomplete metal loading or non-ideal metal/epoxy interface properties [54].

If power cycling results in cracking of the die attach layer, heat transport away from the die can be affected. This will result in a rise in junction temperature and acceleration of temperature-related stress on the module [51]. Cracking of the die attach bond also increases the "transit time" of minority carriers through the n^- body of the device (since the collector can no longer efficiently sweep out the minority carriers). This increases turn-off time, leading to a higher power dissipation rate in the junction [55]. Finally, the cracking of the die attach layer can affect electrical control of the die, either increasing parasitic resistance during the ON-state (and thus the power dissipated) or eliminating electrical control altogether if the bond is fully sheared.

While short-term power cycling is responsible for the degradation of the bond wire or die attach, the DBC attach layer is also affected by thermal cycling. This long-term temperature increase can raise the case temperature, leading to failures after around 1×10^3 cycles [56]. This compares to power cycling, which typically leads to failures at $<1 \times 10^6$ cycles [57].

After thermo-mechanical failure, the most prominent failure mechanism in IGBTs is the electrical degradation of the semiconductor device itself, primarily through gate oxide degradation. The gate oxide properties can be degraded immediately by a strong electric field, over a long period of time by a much lower applied field, or due to heating of the oxide. The primary mechanism of oxide breakdown is through the formation of defects in the oxide layer. These defects can form thermally through an electron tunnel current or thermo-mechanically. As the defect concentration increases past the percolation threshold, the defects begin to overlap and eventually form a conductive pathway through the insulating oxide layer.

Thermal formation of defects occurs because no oxide is a perfect insulator, and some conduction through the gate oxide takes place even in pristine devices, especially at high electric fields. This conduction heats the oxide and induces the formation of traps. As the trap concentration increases past the percolation threshold, the traps begin to overlap and form a conduction pathway. This formation of a conduction pathway is known as soft breakdown. Once soft breakdown occurs, a thermal runaway effect may take place, where the increase in conduction causes an increase in trap formation that, in turn, leads to further conduction [58]. If the oxide is heated enough through this thermal runaway, the oxide can physically melt and release oxygen, forming conductive Si filaments. This failure mode is known as hard breakdown [59].

Defects can also form due to the application of a large electric field, leading to very quick oxide degradation. At high field values, electrons will tunnel through the oxide and impact the gate electrode, transferring kinetic energy to electron–hole pairs [60]. Some of the holes are injected back into the oxide and can either shear an Si–O bond to form a trap (anode hole injection or "$1/\mathcal{E}$" model) [58,61,62] or break an Si–H bond at the gate/oxide interface to create a free proton (hydrogen release model) [63–65], which creates a trap by combining with an oxygen vacancy.

Even without significant electron tunneling, the oxide can be degraded by long-term exposure to low-magnitude electric fields [66]. This is due to the thermo-mechanical interaction of the SiO_2 tetrahedral structure with the applied electric field (the so-called thermochemical or "$\mathcal{E}$" model) [67–69]. In this mechanism, the nominal O–Si–O bond angle of 109° is distorted to larger angles. Above 150°, the bond is severely weakened and can break, forming an oxygen vacancy and a Si–Si bond (the so-called E′ defect) [70]. These Si–Si bonds are particularly weak and are further strained by the neighboring Si–O bond (a polar bond), which may (depending on its orientation relative to the electric field) become distorted due to the applied electric field. The distortion of the Si–O bond strains and breaks the weak Si–Si bond, which acts as a hole trap.

However, neither the anode hole injection model, nor the hydrogen release model, nor the thermochemical model can fully explain the breakdown characteristics of a transistor due to the potential difference between the drain and source [71]. In the presence of high lateral fields, electrons and holes gain sufficient energy such that they are no longer in equilibrium with the lattice [72]. These so-called "hot carriers" may create electron-hole pairs due to impact ionization in the transistor channel. The electron-hole pair separates and the hole enters the substrate while the electron enters the gate oxide and causes trap formation. While this does not typically lead to catastrophic failure, the charged traps result in changes in transistor characteristics (such as threshold voltage, drive current, and transconductance) that may impair the operation of the transistor in its circuit environment. Additionally, the presence of hot carriers can significantly decrease the time to oxide breakdown [73].

Since oxide breakdown forms a conducting pathway through the insulating oxide, once breakdown occurs, there is a large increase in the gate current. For hard breakdown, the gate current may increase by a factor of up to 10^2 in the ON-state and 10^6 in the OFF-state [74]. For soft breakdown of the oxide layer, the leakage current of the device in the OFF-state increases [75].

Oxide degradation also affects V_T due to trapped charge in the oxide. Although the V_T will decrease as temperature increases (since the bandgap of Si decreases) [46], over time the presence of a negative charge in the oxide layer causes V_T to increase [76]. This V_T shift can be used as an indicator of ongoing oxide degradation and as a predictor of the formation of a conducting percolation path [77].

For high-voltage applications, degradation of the module encapsulant can be a contributor to cumulative damage [78]. This encapsulant is usually composed of a soft (usually silicone gel) dielectric that is used to insulate the Si die, provide shock and vibration resistance, and exclude moisture and other corrosive agents. The degradation of this encapsulant layer is a product of two mechanisms, "electrical treeing" and "water treeing" [78].

Electrical treeing is a process where, under high-voltage stress, impurities, mechanical defects, gas voids, or other inhomogeneities in the polymer give rise to a large local electric field. If the field is large enough, gases within these inhomogeneities ionize in a coronal discharge. These partial discharges may emit ozone gas, high-energy ions, or UV radiation that can degrade neighboring areas of polymers and form more voids and cracks. These neighboring areas are ripe for more partial discharge instances in a positive feedback loop. Eventually, these instances of partial discharge may form partially conducting, branching, three-dimensional, tree-like figures known as electrical trees. If these trees grow unchecked, they may form a conducting pathway through the encapsulant and lead to failure. However, this failure mechanism may only be prevalent for vented trees or for those with access to free air (such as in the case of a non-hermitic seal of the IGBT module casing) [79].

A similar failure mechanism is present during water treeing [80], though the necessary voltage stress is much less than for electrical treeing [81,82]. If the encapsulant is in a humid environment [83], water may diffuse into the encapsulant and propagate due to the voltage gradient [80,81]. The end result is a tunnel filled with micro-droplets of water. These tunnels can form a conducting pathway that leads to failure, increase the local electric field and the number of partial discharges, or accelerate the formation of electrical trees [84].

14.3 Future reliability concerns

Although the concept of using semiconductor devices to harness the PV effect has been around since the 1940s [85], over the past 50 years, high generation costs compared to traditional energy sources have relegated PVs to boutique markets (i.e., highly isolated installations) where cost is less of an issue, such as in spacecraft. In reality, the terrestrial PV industry is still remarkably immature. However, as the cost of solar energy has fallen in recent years, the utilization of PV has increased exponentially. New technologies, financing schemes, and modes of operation make the solar energy landscape highly dynamic. In the following section, we outline some of these new operation/utilization modes in the solar industry and their consequences for power system reliability in the future.

14.3.1 Advanced inverter functionalities

Historically, the growth in PV generation in the utility energy generation portfolio has been driven largely by legislative mandates (especially legislation at the state level) and public relations considerations. However, the falling cost of solar energy (which decreased from 21.4 to 11.2 ¢/kWh from 2010 to 2013 [18,86]) has recently spurred a large increase in the number of operational PV plants connected to the electrical grid.

Solar (and wind) energy is, by nature, both highly variable (changing power output from moment to moment) and uncertain (deviating from forecasted power generation) [87]. For a given balancing area (there are over a hundred in the United States), the energy supply and consumer load must be closely balanced at all times [88].

Large, unexpected changes in either supply or demand can lead to excursions outside of the UL1547 standard in either voltage or frequency. For an electric energy grid that has been designed around highly consistent base and peak generation plants, the full-scale integration of these stochastic energy sources while maintaining the high reliability and low cost that consumers expect is a significant challenge. Further, since solar energy sources are not traditional rotating machines, they do not add to grid frequency response,[§] making frequency variations somewhat more problematic and more likely to occur.

Historically, the penetration ratio of renewables has been so low (<1%) that deviations in power production have been inconsequential to the operation of the grid at large. As the penetration ratio increases, however, the effect of variable and uncertain energy generation becomes more severe and the risk of grid events becomes greater. Various studies have been carried out and have found risks to the grid at penetration ratios ranging from as low as 5% (due to cloud transients) [89] to greater than 50% [90], depending on the stiffness of the grid and the geographical extent of PV arrays [91], with penetration ratios of 100% possible (under certain conditions) when paired with combinations of renewables, storage, and sheddable load [92].

Historically, in the event of a grid excursion of either voltage or frequency, utilities have asked PV inverters to disconnect from the grid and shut down while the grid situation is corrected using traditional means. However, as inverters are power conversion units that can change their voltage and frequency characteristics almost instantaneously, they can be used as a tool to stabilize the grid voltage or frequency. The next generation of PV inverters can "ride through" voltage or frequency disturbances in the grid without disconnecting as well as change power output frequency

[§]Contrary to statements that the electric grid does not possess any potential energy, traditional rotating machines resist frequency changes through small, instantaneous changes in rotation speed, so the energy supply and demand does not have to balance at each instantaneous moment. Inverters can simulate this frequency-change effect with "virtual inertia", although the speed and control schemes necessary for implementation are a challenge.

and active/reactive power flows in either autonomous (automatic response to grid events) or commanded mode in a wide variety of situations to help stabilize grid voltage or and/frequency (Figure 14.6). The combinations of these new operational modes are known as "advanced inverter functionalities," and it is thought that their widespread implementation will stabilize the grid during variable and uncertain generation conditions, allowing for greater penetration of renewable energy sources.

One method of grid voltage stabilization for a PV inverter is to change the amount of reactive power (Volts-Amps Reactive, or VAR) generated compared to active power (W), which is accomplished by changing the phase angle between the voltage and current waveforms [93]. The cosine of the phase angle between voltage and current is known as the power factor (PF). The grid is typically kept at a PF of unity in order to maximize power efficiency. However, most motors and non-linear power supplies (as well as transmission lines) are inductive in nature and tend to change the PF away from unity, which, if not corrected, may cause grid issues.

For example, on *unexpected* hot days, a balancing area may experience a larger use of air conditioners than expected. These large inductive loads coming online cause a current inrush, which lowers the line voltage and moves the PF away from unity. The lower line voltage causes even higher current in-rush as additional inductive loads come online in a positive feedback mechanism. The high current

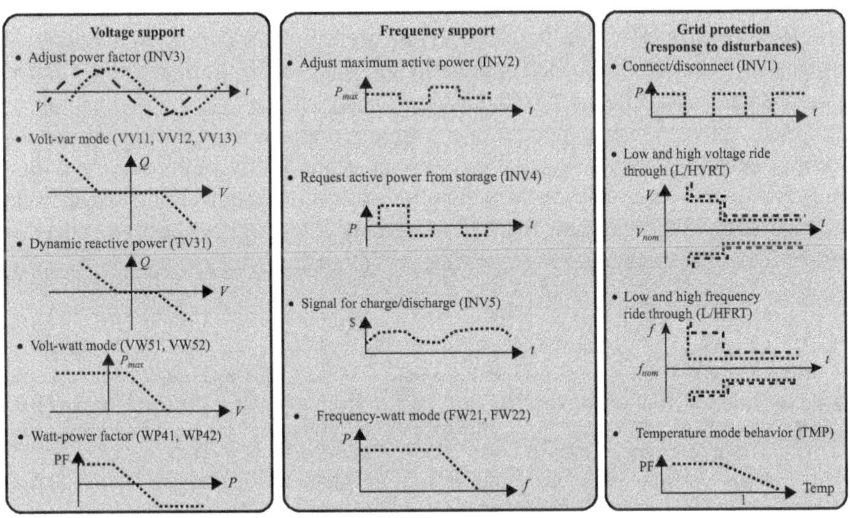

Figure 14.6. Autonomous advanced inverter functionalities as defined by IEC TC 61850-90-7 (with the exception of frequency ride-through) [94]. Advanced functionalities are split between voltage support functions (green) that alter active (P) and reactive (Q) power according to line voltage, frequency support functions (blue) that alter P, and grid disturbance functions (red) that maintain inverter connection to the grid through transient disturbance events (Courtesy Jay Johnson, Sandia National Laboratories).

flows cause the overhead lines to heat up, and thermal expansion causes them to sag. Eventually, a line will contact a tree and short to the ground. This causes more current to flow among the remaining overhead lines, which heat and sag further and eventually short to ground in a cascading failure that can lead to a blackout.[**]

Utilities have typically dealt with voltage deviations by increasing the grid capacitance to cancel out inductive loads using methods such as the insertion of capacitor banks.[††] However, capacitor banks are typically expensive and have reliability issues. Inverters, conversely, can easily and nearly instantaneously change to become inductive or capacitive through small changes in their switching schemes.

However, the usage of PV inverters in this manner is not without its penalties. Circuit designers must make educated guesses about the normal operation of the units in order to optimize efficiencies. In most cases, this involves normal operation of the inverter at a PF of unity. Running an inverter at non-unity PF tends to decrease unit efficiency (Figure 14.7). Decreased efficiency directly leads to increased operational temperature, high aging rates of components, and shorter lifetime.

For example, an inverter that operates at 96% efficiency at unity PF will drop to 91.3% efficiency at 0.55 PF. Although this may sound like a small drop in efficiency, this indicates a doubling of energy loss, which will have a significant effect on component operating temperature and will substantially reduce component lifetime.

Of course, for grid excursion events, inverters will operate in this manner for very short amounts of time, limiting the accelerated aging due to advanced inverter operational modes. However, in the future, inverters may be harnessed to operate in this way as a more regular occurrence. When the electric grid was deregulated, financial markets opened for frequency response services (grid stabilization via changes in active power generation) at various time scales. Although no market currently exists for voltage regulation services at the utility scale[‡‡] in the United States,[§§] it seems likely that voltage support will become (more) monetized in the near future. If this is the case, then the extremely low marginal cost and speed of support make inverters ideal for operating in this mode.

[**]This type of situation, in combination with a control software error, caused the Northeast Blackout of 2003. Normally, the loss of a single overhead line (known as an $N-1$ event) can be safely corrected through the management of power flows and other safety features available to the Balancing Authority. Less straightforward events ($N-2$, $N-3$, etc.) are typically much more difficult to control without loss of service for at least some of the affected area.

[††]Other methods utilized to cancel out inductive loads include Load Tap Changers (LTCs), Voltage Regulators, Static Synchronous Compensators (STATCOMs), Static VAR Compensators (SVCs,TCRs, TCSs,TRSs), Static Synchronous Series Compensators (SSSCs), Thyristor Controllers (TCSCs,TSSCs, TCSRs, TSSRs, TCPSTs, TCPRs, TCVLs, TCVRs), Unified Power Flow Controllers (UPFC), Interline Power Flow Controllers (IPFC), and others. See [95] for a detailed treatment.

[‡‡]Some private entities that have large inductive loads (e.g., hospitals, factories) may choose to operate an inverter in this way as an alternative to a capacitor bank in order to decrease the inductive nature of their loads, change their PF, and avoid additional utility charges (which can be quite costly).

[§§]Voltage regulation capabilities are compulsory in the U.S. with a fixed usage cost ($/month) and opportunity cost provided by the utility. Other countries, such as France, operate a bid price market with 2–3 year market clearing. See [96,97] for a detailed discussion of these financial markets and their operation.

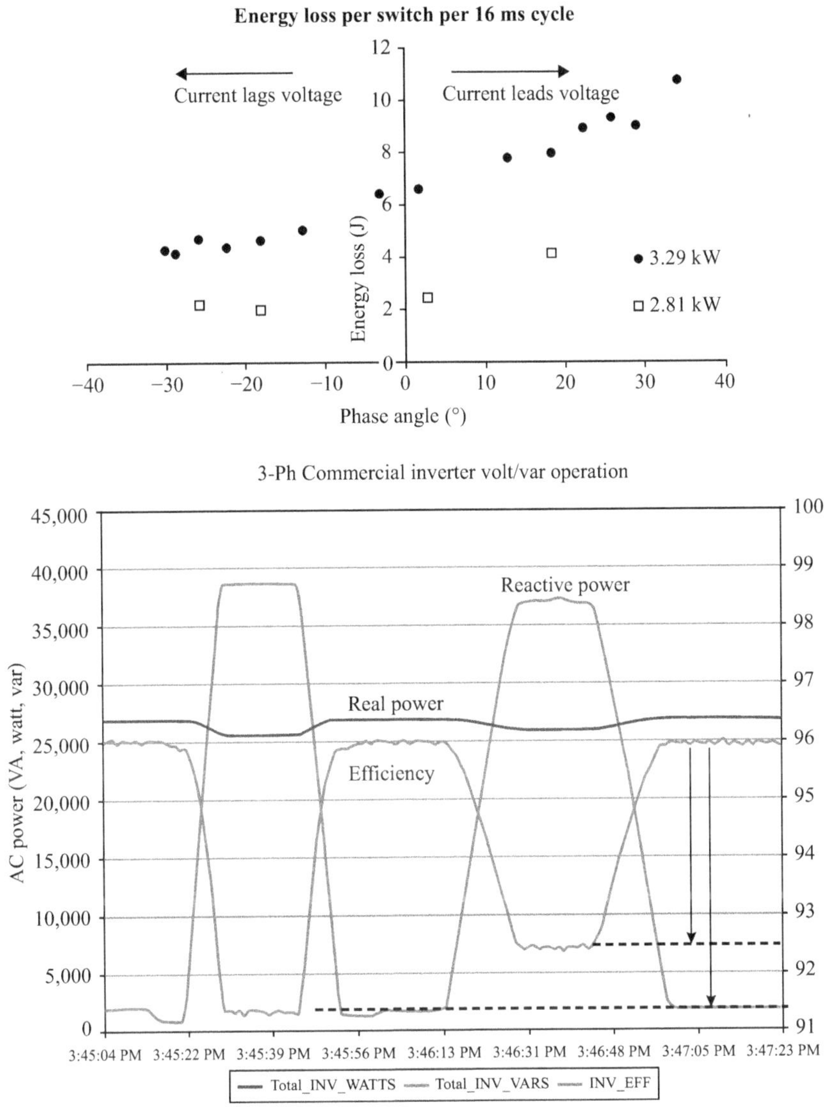

Figure 14.7. Non-unity PF operation can have significant effects on component and system losses. (Left) The average energy loss per MOSFET over a 16 ms power cycle is a function of PF. In this inverter, large positive PFs have over 60% more losses than operation at unity PF. (Right) The system as a whole shows similar effects, with efficiency decreasing at non-unity PF. These losses are directly translated into heat, increasing the operating temperature of the component and decreasing lifetime. (Courtesy Sigifredo Gonzalez, Sandia National Laboratories.)

These markets will incentivize operators to utilize inverters up to their full VA rating for all daytime hours at various PFs, subjecting the inverters to much higher losses over their operational lifetimes and, inevitably, resulting in shorter lifetimes. Since inverters do not need to export any active power, it is also conceivable for an inverter to source/sink VARs 24 hours a day. The newest generation of inverters on the market has taken the first steps required to make continuous VAR support a possibility [98].

In general, from emergence over the past five to its eventual large-scale implementation in the electric grid in the coming years, advanced inverter functionalities look to enable even higher penetration of renewable energy generation through the stabilization of grid voltage and frequency and the mitigation of renewable energy variability and uncertainty. However, it must be noted that these advanced operational modes increase power system losses and will lead to shorter unit lifetimes. It is necessary to consider the balance of added grid stability with the shorter lifetimes of the units that will result, especially as these functionalities become more monetized in the future as grid support markets continue to evolve and inverters are asked to exercise these functions on a much more regular basis.

14.3.2 Large DC/AC ratios

Although the price of newly installed PV arrays has been steadily decreasing, this has primarily been driven by falling module prices.[***] Modules are typically the most expensive capital cost for a solar installation; however, recent pricing trends dictate that as a percentage they are becoming less important to the overall cost of the plant. This indicates that there is a much smaller incremental cost to increasing the array size and, as a corollary, wasted DC power has become less important. In order to fully utilize cheap modules and offset system losses (temperature, soiling, panel degradation) over the lifetime of the system, many system designers have been adding additional modules to the array so that the ratio of power production on the DC side to that exported on the AC side (W_{DC}/W_{AC})[†††] is greater than unity [99]. Historically, this ratio has been kept at 1–1.2. However, as the price of modules continues to fall, the DC/AC ratio has risen to the 1.25–1.5 range, and even as high as 2.0 in some newer utility-scale installations.

The advantage of a high DC/AC ratio is that it makes PV power generation much more consistent. The consistency and predictability of power generation are valued by utilities and financiers nearly as much as pure power generation capability. With a very high DC/AC ratio, instead of the parabolic power generation profile that is typically seen in PV power production, the power production looks much more like the rectangle typical of normal base-load generation (Figure 14.8).

[***]This is due to a number of factors, including falling material prices, greater competition (especially from Chinese manufacturers), economies of scale, and experience curve effects.

[†††]For example, adding 6 kW_p of modules on the DC side of a 5 kW_p inverter. The modules are capable of outputting 6 kW (rated at standard test conditions, but it could be more or less depending on irradiance, temperature, and other environmental conditions), but the inverter is only capable of outputting 5 kW to the AC side, so any additional energy is not utilized.

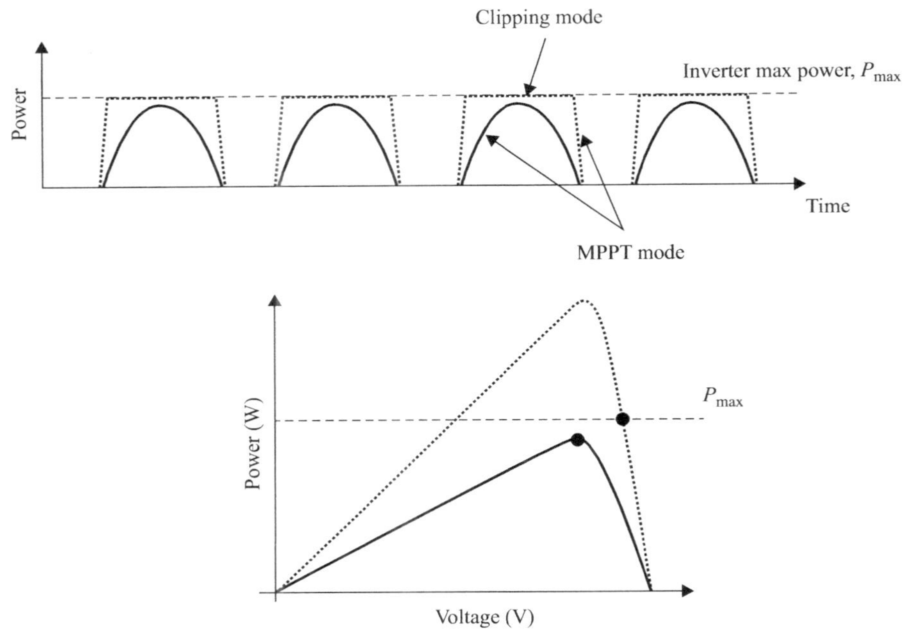

Figure 14.8 *(Top) For WDC/WAC = 1, the power output of a PV array is expected to be parabolic, since the inverter max power point tracks the array IV curve and the IV curve changes as the solar irradiance changes throughout the day (solid curve). As the WDC/WAC ratio increases (dotted curve), the power output profile looks much more rectangular, similar to a traditional base-generation power profile. In the morning and evening, the inverter is in MPPT mode. However, as the power on the DC side increases to the inverter maximum power rating, the inverter stops MPPT mode and goes into clipping mode. In this mode, the power output is equal to* P_{max}*, and, since the inverter is not closely tracking* P_{mp}*, the power output of the array is less sensitive to small changes in the array IV curve due to changes in irradiance, e.g., due to cloud cover. (Bottom) The IV curve of an array with WDC/WAC* ≈ 1 *(blue curve) with an operating point (black dot) corresponding to array* P_{mp}*. For WDC/WAC* > 1 *(red curve), the operating point is not at* P_{mp}*, but at* P_{max} *with an operating voltage higher than* V_{mp}*. This high power, high voltage (and typically high temperature) operating state is challenging for the inverter and may have a negative effect on the system lifetime.*

In the morning, for a high DC/AC ratio system, the increase in power production vs. time is much steeper than a traditional array with unity DC/AC ratio. At some point, the power output by the panels on the DC side reaches the maximum power rating of the inverter, P_{max}. The inverter is unable to convert power above its maximum power rating, so it switches from an MPPT mode, where the voltage operating point is determined by P_{mp}, to a clipping mode, where the voltage operating point ($V > V_{mp}$) is determined by the inverter P_{max}. In clipping mode, the inverter is no longer at V_{mp} of the array IV curve and is instead at some higher operating voltage. This means that changes in the array IV curve due to irradiance changes (from, for example, cloud cover) are much less noticeable in the AC power output of the inverter.

The higher the DC/AC ratio, the longer during the day the inverter is in clipping mode operation. In many utility-scale PV arrays, the inverter may operate in clipping mode at its maximum power output for 6 to 8 h per day. This clipping mode operation is an extremely challenging environment for the inverter to operate in, since it requires the inverter to operate at a high voltage, typically a high temperature, maximum-power state for many hours during the day. Compared to normal MPPT operation, the power handling components experience much greater voltage and temperature stresses for much longer time periods, leading to higher aging rates and lower lifetimes after years of operation.

14.3.3 Module-level power electronics

Traditionally, PV inverters have been centralized units connected to one or multiple strings of PV modules that range in power from 3 kW for residential-scale systems to 500 kW for utility-scale. Over the past 10 years, new technologies have emerged that deviate from this traditional topology. In order to make PV cost-competitive with traditional energy sources, economies of scale have been guiding inverter design increasingly in two directions (Figure 14.9): (1) large, centralized, utility-scale inverters (~500 kW) and (2) small, modular, module-level power electronics (MLPE, ~250 W).

MLPE (also known colloquially as AC modules), such as microinverters and DC power optimizers, are power electronic devices that are integrated[‡‡‡] with the module so that there is one power-conditioning unit per module. This sort of power handling topology offers numerous advantages on the system level. Since the power conversion and MPPT of each module are handled independently, the effect of module IV curve mismatches and partial shading of each row are eliminated. Each module can be operated at its P_{mp}, independent of every other module, increasing the overall yield of the array.

The independent and distributed architecture of MLPE is additionally useful in the case of device failures, which, as noted previously, are a significant percentage

[‡‡‡]The manner of integration can vary widely from unit to unit. Some units may be integrated directly into the module frame or junction box, while others are bolted on the module backsheet by a 3rd party. There is no consensus on what a true "AC module" is, although many companies have used such terms to advertise their newest generations of products.

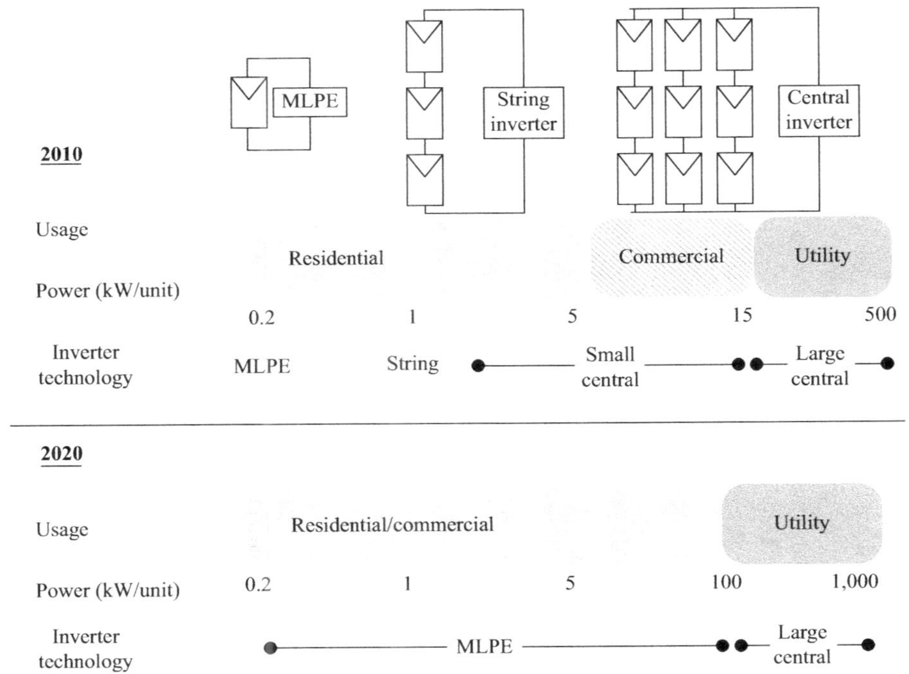

Figure 14.9. Traditionally, inverter technologies have been composed of large centralized inverters at the utility-scale and smaller string inverters at the commercial/residential scale. Over the last 10 years, economies of scale have been pushing inverter technologies towards larger power handling (~1 MW/inverter) on the utility side and smaller power handling (~250 W) on the residential/commercial side. These small power handling units are installed one per module and are known as module level power electronics (MLPE).

of component failures in an array. Unlike string or centralized system topologies, when an MLPE unit fails, it only affects the module it is attached to instead of multiple kilowatts of modules. Since each MLPE failure only interrupts the production of ~250 W, the array operator has additional freedom to optimize operations and maintenance (O&M) issues (such as technician truck rolls or replacement parts) to minimize costs. However, this is balanced by the fact that there is a large increase in the number of units, so the statistical reliability of units (especially with regard to infant mortality failures[§§§]) is of primary importance.

Second, since the voltage[***] (~70 V) and power handling requirements are relatively moderate (~250 W), the power handling topologies are similar to microelectronics and consumer-grade switch mode power supplies. Lower-rated components with lower power handling capabilities can be used to maximize lifetime compared to topologies needed for more stringent power handling requirements.

However, MLPEs suffer from a number of distinct disadvantages compared to more traditional inverter configurations. First, depending on installation configuration (especially the proximity to the PV module), they can be subjected to more extreme environments (i.e., temperature cycling) during the day, resulting in a negative impact on reliability.

As with most material failure mechanisms, the time to failure (TTF) of MLPE units is shortened by temperature effects (either maximum temperature, T_{max}, or diurnal temperature steps, ΔT, depending on the specific failure mechanism). Operational temperature of a power-handling unit is primarily a function of ambient temperature, solar irradiance, and airflow. Ambient temperature and solar irradiance are uncontrolled variables over the lifetime of the unit. However, cooling along the back-side mounted MLPE can be optimized through the chimney effect, especially on roof-mounted units where air is directed to flow between the module and the room. The presence (or lack thereof) of the free convective chimney effect, which is a function of roof slope, gap between the module and the roof, and gap between the MLPE and the module, can have a significant effect (e.g., decrease in junction temperature by 10%–15% [100]) on operational unit temperature and, therefore, TTF [101,102].

Operational temperatures of modules in rooftop settings have been of interest to module manufacturers for a number of years, and preliminary studies have been carried out to quantify the reduction in TTF (Figure 14.10). However, similar quantifiable data on module backsheet temperatures with an added power handling unit have been unavailable. A recent study [103] has shown a 35°C increase in temperature above ambient for modules with MLPE units at irradiances of 1000 W/m^2. It is

[§§§]Manufacturing variation with regards to extrinsic failure mechanisms (due to manufacturing issues) has always been a concern for PV inverters. While larger inverter units are put together in a workshop-type manufacturing process and are thus more highly susceptible to installation variation (from, for example, incorrectly torqued screws), MLPE units are installed in assembly fashion and have decreased the number of infant mortality failures.

[***]Such low operational voltage offers a safety advantage over traditional centralized inverters. The low operation reduces the danger from arc faults on the array. Since the MLPE unit is connected to or near the module, the vast majority of conductors in an MLPE array carry AC power and have no arc danger.

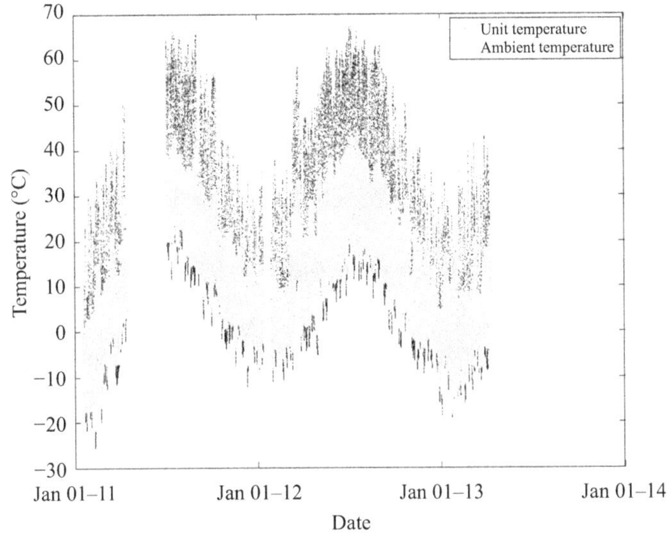

Figure 14.10 Multi-year temperature data for a microinverter unit compared to ambient temperature. Due to the power handling of the unit, the inverter unit runs hotter with larger diurnal cycles.

important to note that this study monitored module temperature, not internal MLPE temperature, which may be significantly higher.

Since MLPE units are, in many cases, sold paired with PV panels (and future direction points towards total incorporation into the module frame), customers are demanding unit lifetimes and warranties similar to that of PV modules (~25 years).

Offering a 25-year warranty for a power-handling device in a cost-competitive environment is challenging. Therefore, the statistical reliability of each device and the extension of unit lifetime are of critical importance to the continued implementation of this type of PV solution. Unlike more mature technologies (e.g., c-Si modules), the MLPE market segment is relatively nascent (large-scale implementation of MLPE units has been occurring for less than ten years) and therefore does not have long-term usage data that exists in many other industries. The 25-year lifetimes demanded by customers is an ambitious goal for a power handling unit that has been met with skepticism by many in the PV industry. At this point, without standardized reliability testing and long-term lifetime and usage data, it is unknown if MLPEs will last their claimed 25-year field lifetimes.

References

[1] O. Wasynezuk, "Dynamic Behavior of a Class of Photovoltaic Power Systems," *IEEE Transactions on Power Apparatus and Systems*, vol. 102 (9), pp. 3031–3037, 1983.

[2] K. H. Hussein, I. Muta, T. Hoshino, and M. Osakada, "Maximum Photovoltaic Power Tracking: An Algorithm for Rapidly Changing Atmospheric Conditions," *IEEE Proceedings on Generation, Transmission and Distribution*, pp. 59–64, 1995.

[3] A. Luque and S. Hegedus, *Handbook of Photovoltaic Science and Engineering*. New York: Wiley, 2011.

[4] M. A. G. de Brito, L. P. Sampaio, L. G. Junior, and C. A. Canesin, "Evaluation of MPPT Techniques for Photovoltaic Applications," *2011 IEEE International Symposium on Industrial Electronics (ISIE)*, pp. 1039–1044, 2011.

[5] National Fire Protection Association, *"National Electrical Code*, 2014 Edition, NFPA70." Quincy, MA.

[6] M. Salcone and J. Bond, "Selecting Film Bus Link Capacitors for High Performance Inverter Applications," *Electric Machines and Drives Conference, 2009. IEMDC'09*. Piscataway, NJ: IEEE International, 2009.

[7] C. Siedle and V. D. Ingenieure, *Comparative Investigations of Charge Equalizers to Improve the Long-Term Performance of Multi-Cell Battery Banks* vol. 245. Düsseldorf: VDI Verlag, 1998.

[8] J. Flicker, R. Kaplar, M. Marinella, and J. Granata, "PV Inverter Performance and Reliability: What Is the Role of the Bus Capacitor?," *IEEE Photovoltaic Specialists Conference*, Austin, TX, 2012.

[9] IEEE, *1547-IEEE Standard for Interconnecting Distributed Resources With Electric Power Systems*. Piscataway, NJ: IEEE, 2003.

[10] I. E. C. 61727, *IEC61727-Photovoltaic (PV) Systems: Characteristics of the Utility Interface*. Geneva: International Electrotechnical Commission, 1996.

[11] Y. Xue, K. C. Divya, G. Griepentrog, M. Liviu, S. Suresh, and M. Manjrekar, "Towards Next Generation Photovoltaic Inverters," *2011 IEEE Energy Conversion Congress and Exposition (ECCE)*, 2011.

[12] J. Flicker and J. Johnson, *Photovoltaic Ground Fault and Blind Spot Electrical Simulations*. Albuquerque, NM: Sandia National Laboratories, 2013. Available: http://energy.sandia.gov/wp/wp-content/gallery/uploads/SAND2013-3459-Photovoltaic-Ground-Fault-and-Blind-Spot-Electrical-Simulations.pdf.

[13] G. Ball, B. Brooks, J. Flicker, *et al.*, "Inverter Ground-Fault Detection 'Blind Spot' and Mitigation Methods," *Solar American Board for Codes and Standards*, June 2013.

[14] J. Flicker and J. Johnson, "Analysis of Fuses for 'Blind Spot' Ground Fault Detection in Photovoltaic Systems," *Solar American Board for Codes and Standards*, July 2013. Available: http://solarabcs.net/about/publications/reports/blindspot/pdfs/analysis_of_fuses-June-2013.pdf.

[15] B. Brooks, "The Ground-Fault Protection Blind Spot: Safety Concern for Larger PV Systems in the U.S.," *Solar American Board for Codes and Standards*, January 2012.

[16] H. Laukamp, "Statistical Failure Analysis of German PV Installations," *PV Brandsicherheit Workshop, Köln*, Germany, 26 January 2012 (in German).

[17] "$1/Watt Photovoltaic Systems," US Department of Energy, White Paper2010. Available: http://books.google.com/books?id=W7vIvoBzDooC&pg=PA93&dq=intitle:Semiconductor+Material+and+Device+Characterization&cd=12&source=gbs_api.

[18] D. J. Feldman, G. Barbose, R. Margolis, *et al.*, "Photovoltaic System Pricing Trends," *National Renewable Energy Laboratory*, PR-6A20-62558, 2014. Available: http://www.nrel.gov/docs/fy14osti/62558.pdf.

[19] L. Bony, S. Doig, C. Hart, and E. Maurer, *Achieving Low-Cost Solar PV: Industry Workshop Recommendations for Near-term Balance of System Cost Reductions*. Boulder, CO: Rocky Mountain Institute (RMI), 2010.

[20] A. Ristow, M. Begovic, A. Pregelj, and A. Rohatgi, "Development of a Methodology for Improving Photovoltaic Inverter Reliability," *IEEE Transactions on Industrial Electronics*, vol. 55 (7), pp. 2581–2592, 2008.

[21] B. Jablonska, H. Kaan, M. Leeuwen, and G. Boer, "PV-Prive Project at ECN: Five Years of Experience With Small-scale AC Module PV Systems," *Presented at the 20th European Photovoltaic Solar Energy Conference and Exhibition*, 2005.

[22] A. Maish, "Defining Requirements for Improved Photovoltaic System Reliability," *Progress in Photovoltaics*, pp. 165–173, 1999.

[23] A. B. Maish, C. Atcitty, S. Hester, *et al.*, "Photovoltaic System Reliability," *Conference Record of the Twenty-Sixth IEEE Photovoltaic Specialists Conference, 1997*, 1997.

[24] A. Golnas, "PV System Reliability: An Operator's Perspective," *38th Photovoltaic Specialists Conference*, Austin, TX, 2012.

[25] W. Bower and D. Ton, *Summary Report on the DOE High-Tech Inverter Workshop*. Washington, DC: DOE Office of Energy Efficiency and Renewable Energy, 2005.

[26] C. Rodriguez and G. Amaratunga, "Long-Lifetime Power Inverter for Photovoltaic AC Modules," *IEEE Transactions on Industrial Electronics*, vol. 55 (7), pp. 2593–2601, 2008.

[27] T. Von Kampen and E. Sawyer, "Reliability Considerations of Inverter/DC Link Capacitor Using PP Film and 105C Engine Coolant," *Proceedings of the International Microelectronics and Pakaging Society (IMAPS)*, 2008.

[28] S. G. J. Parler, "Thermal Modeling of Aluminum Electrolytic Capacitors," *Conference Record of the 1999 IEEE Industry Applications Conference*, 1999.

[29] C. Kulkarni, G. Biswas, X. Koutsoukos, J. Celaya, and K. Goebel, "Experimental Studies of Ageing in Electrolytic Capacitors," *Annual Conference of the Prognostics and Health Management Society*, 2010.

[30] D. Guo, "Wear Analysis and Degradation Mechanism of Film Metallized Capacitor," *Power Capacitor (China)*, vol. 2 pp. 12–15, 1995.

[31] W. Sarjeant, J. Zirnheld, and F. MacDougall, "Capacitors," *IEEE Transactions on Plasma Science*, vol. 26 (5), pp. 1368–1392, 1998.

[32] M. L. Gasperi, "Life Prediction Model for Aluminum Electrolytic Capacitors," *Conference Record of the 1996 IEEE Industry Applications Conference (IAS 1996)*, 1996.

[33] H. Ma and L. Wang, "Fault Diagnosis and Failure Prediction of Aluminum Electrolytic Capacitors in Power Electronic Converters," *31st Annual Conference of IEEE Industrial Electronics Society, 2005 (IECON 2005)*, 2005.

[34] J. Zhao and F. Liu, "Reliability Assessment of the Metallized Film Capacitors from Degradation Data," *Microelectronics and Reliability*, vol. 47 (2–3), pp. 434–436, 2007.

[35] C. W. Reed and S. W. Cichanowskil, "The Fundamentals of Aging in HV Polymer-film Capacitors," *IEEE Transactions on Dielectrics and Electrical Insulation*, vol. 1 (5), pp. 904–922, 1994.

[36] G. Buiatti, S. Cruz, and A. Cardoso, "Lifetime of Film Capacitors in Single-Phase Regenerative Induction Motor Drives," *IEEE International Symposium on Diagnostics for Electric Machines, Power Electronics and Drives, 2007 (SDEMPED 2007)*, pp. 356–362, 2007.

[37] H. Matsui, T. Fujiwara, and K. Fujiwara, "Metalized Film Capacitors with High Energy Density for Rail Vehicles," *Conference Record of the 1997 IEEE Industry Applications Conference Thirty-Second IAS Annual Meeting (IAS'97)*, 1997.

[38] B. Walgenwitz, J. H. Tortai, N. Bonifaci, and A. Denat, "Self-Healing of Metallized Polymer Films of Different Nature," *Proceedings of the 2004 IEEE International Conference on Solid Dielectrics, 2004 (ICSD 2004)*, 2004.

[39] W. Sarjeant, F. MacDougall, D. Larson, and I. Kohlberg, "Energy Storage Capacitors: Aging, and Diagnostic Approaches for Life Validation," *IEEE Transactions on Magnetics*, 1997.

[40] A. Schneuwly, P. Groning, L. Schlapbach, C. Irrgang, and J. Vogt, "Breakdown Behavior of Oil-Impregnated Polypropylene as Dielectric in Film Capacitors," *IEEE Transactions on Dielectrics and Electrical Insulation*, vol. 5 (6), pp. 862–868, 1998.

[41] S. Lefebvre, Z. Khatir, and F. Saint-Eve, "Experimental Behavior of Single-Chip IGBT and COOLMOS Devices under Repetitive Short-Circuit Conditions," *IEEE Transactions on Electron Devices*, vol. 52 (2), pp. 276–283, 2005.

[42] J. He, M. C. Shaw, J. C. Mather, and R. C. J. Addison, "Direct Measurement and Analysis of the Time-Dependent Evolution of Stress in Silicon Devices and Solder Interconnections in Power Assemblies," *IEEE Industry Applications Conference*, 1998.

[43] V. Smet, F. Forest, J.-J. Huselstesin, F. Richardeau, Z. Khatir, S. Lefebvre, and M. Berkani, "Ageing and Failure Modes of IGBT Modules in High-Temperature Power Cycling," *IEEE Transactions on Industrial Electronics*, vol. 58 (10), pp. 4931–4941, 2011.

[44] H. Ye, M. Lin, and C. Basaran, "Failure Modes and FEM Analysis of Power Electronic Packaging," *Finite Elements in Analysis and Design*, vol. 38 (7), pp. 601–612, 2002.

[45] J. Flicker, R. Kaplar, B. Yang, M. Marinella, and J. Granata, "Insulated Gate Bipolar Transistor Reliability Testing Protocol for PV Inverter

Applications," *Progress in Photovoltaics: Research and Applications*, vol. 22 (9), pp. 970–983, Sep 01 2014.

[46] M. Ciappa, "Selected Failure Mechanisms of Modern Power Modules," *Microelectronics Reliability*, vol. 42 pp. 653–667, 2002.

[47] P. Malberti and M. Ciappa, "A Power-Cycling-Induced Failure Mechanism of IGBT Multichip Modules," *International Symposium on Testing and Failure Analysis*, 1995.

[48] V. Mehrotra, J. He, M. S. Dadkhah, K. Rugg, and M. C. Shaw, "Wirebond Reliability in IGBT-Power Modules: Application of High Resolution Strain and Temperature Mapping," *International Society of Power Semiconductor Devices and ICs (ISPSD)*, 1999.

[49] V. A. Sankaran, C. Chen, C. S. Avant, and X. Xu, "Power Cycling Reliability of IGBT Power Modules," *IEEE Industry Applications Society Annual Meeting (IAS)*, 1997.

[50] B. J. Baliga, *Power Semiconductor Devices*. Boston, MA: PWS Publishing Company, 1996.

[51] D. W. Brown, M. Abbas, A. Ginart, I. N. Ali, P. W. Kalgren, and G. J. Vachtsevanos, "Turn-Off Time as an Early Indicator of Insulated Gate Bipolar Transistor Latch-up," *IEEE Transactions on Power Electronics*, vol. 27 (2), pp. 479–489, 2012.

[52] H. Delambilly, and H. Keser, "Failure Analysis of Power Modules – A Look at the Packaging and Reliability of Large IGBTs," *IEEE Transactions on Components Hybrids and Manufacturing Technology*, 1993.

[53] W. Wu, M. Held, P. Jacob, P. Scacco, and A. Birolini, "Investigation on the Long Term Reliability of Power IGBT Modules," *Proceedings of International Symposium on Power Semiconductor Devices and IC's (ISPSD)*, pp. 443–448, 1995.

[54] J. Harris and M. Matthews. "Selecting Die Attach Technology for High-Power Applications," *Power Electronics Technology*, vol. 35 (11), pp. 18–23, 2009.

[55] T. Aoki, "A Discussion on the Temperature-Dependence of Latch-Up Trigger Current in Cmos/Bicmos Structures," *IEEE Transactions on Electron Devices*, vol. 40 (11), pp. 2023–2028, 1993.

[56] W. Wu, M. Held, P. Jacob, P. Scacco, and A. Birolini, "Thermal Stress Related Packaging Failure in Power IGBT Modules," *Proceedings of the 7th International Symposium on Power Semiconductor Devices and ICs*, 1995.

[57] W. Wu, M. Held, P. Jacob, P. Scacco, and A. Birolini, "Investigation of the Long Term Reliability of Power IGBT Modules," *International Symposium on Power Semiconductor Devices and IC's*, 1995.

[58] R. Degraeve, G. Groeseneken, R. Bellens, M. Depas, and H. E. Maes, "A consistent model for the thickness dependence of intrinsic breakdown in ultra-thin oxides," *International Electron Devices Meeting*, 1995.

[59] H. Lin, D. Lee, T. Wang, and 6, "New Insights into Breakdown Modes and their Evolution in ultra-thin Gate Oxide," *International Symposium on VLSI Technology, Systems, and Applications*, Hsinchu, Taiwan, 2001.

[60] R. Rodriguez, J. H. Stathis, and B. P. Linder, "Modeling and Experimental Verification of the Effect of Gate Oxide Breakdown on CMOS Inverters," *IEEE International Reliability Physics Symposium*, 2003.

[61] C. Hu and Q. Lu, "A Unified Gate Oxide Reliability Model," *IEEE International Reliability Physics Symposium*, San Diego, CA, 1999.

[62] T. Tomita, H. Utsunomiya, Y. Kamakura, and K. Taniguchi, "Hot Hole Induced Breakdown of Thin Silicon Dioxide Films," *Applied Physics Letters*, vol. 71 (25), pp. 3664–3666, 1997.

[63] D. DIMARIA and J. STASIAK, "Trap Creation in Silicon Dioxide Produced by Hot-Electrons," *Journal of Applied Physics*, vol. 65 (6), pp. 2342–2356, 1989.

[64] J. Sune and E. Wu, "A New Quantitative Hydrogen-based Model for Ultra-thin Oxide Breakdown," *Transactions of the IRE Professional Group on Audio*, pp. 97–98, 2001.

[65] K. Um, "IGBT Basic II," Fairchild Semiconductor, Apr 2002. Available: http://ieeexplore.ieee.org/xpls/Abs_all.jsp?arnumber=1042815.

[66] J. Suehle and P. Chaparala, "Low Electric Field Breakdown of Thin SiO_2 Films under Static and Dynamic Stress," *IEEE Transactions on Electron Devices*, vol. 44 (5), pp. 801–808, 1997.

[67] A. Berman, "Time-Zero Dielectric Reliability Test by a Ramp Method," *Reliability Physics Symposium*, pp. 204–209, 1981.

[68] D. L. Crook, "Method of Determining Reliability Screens for Time Dependent Dielectric Breakdown," *Reliability Physics Symposium*, pp. 1–7, 1979.

[69] J. W. McPherson and D. A. Baglee, "Acceleration Factors for Thin Gate Oxide Stressing," *Reliability Physics Symposium*, pp. 1–5, 1985.

[70] J. McPherson and H. Mogul, "Underlying Physics of the Thermochemical E Model in Describing Low-Field Time-Dependent Dielectric Breakdown in SiO_2 Thin Films," *Journal of Applied Physics*, vol. 84 (3), pp. 1513–1523, 1998.

[71] H. Liu, Y. Hao, and J. Zhu, "A Thorough Investigation of Hot-Carrier-Induced Gate Oxide Breakdown in Partially Depleted N- and P-Channel SIMOX MOSFETs," *Microelectronics Reliability*, vol. 42 (7), pp. 1037–1044, 2002.

[72] C. T. Wang, *Hot Carrier Design Considerations for MOS Devices and Circuits*. London: Tavistock Publications, 1992.

[73] F. Crupi, B. Kaczer, G. Groeseneken, and A. De Keersgieter, "New Insights into the Relation between Channel Hot Carrier Degradation and Oxide Breakdown Short Channel nMOSFETs," *IEEE Electron Device Letters*, vol. 24 (4), pp. 278–280, 2003.

[74] W. K. Henson, N. Yang, and J. J. Wortman, "Observation of Oxide Breakdown and its Effects on the Characteristics of Ultra-Thin-Oxide nMOSFETs," *IEEE Electron Device Letters*, vol. 20 (12), pp. 605–607, 1999.

[75] T. Pompl, H. Wurzer, M. Kerber, R. C. W. Wilkins, and I. Eisele, "Influence of Soft Breakdown on NMOSFET Device Characteristics," *IEEE International Reliability Physics Symposium* 1999.

[76] N. Patil, D. Das, K. Goebel, and M. Pecht, "Identification of Failure Precursor Parameters for Insulated Gate Bipolar Transistors (IGBTs)," *International Conference on Prognostics and Health Management (PHM)*, 2008.

[77] N. Patil, J. Celaya, D. Das, K. Goebel, and M. Pecht, "Precursor Parameter Identification for Insulated Gate Bipolar Transistor (IGBT) Prognostics," *IEEE Transactions on Reliability*, vol. 58 (2), pp. 271–276, 2009.

[78] J.-H. Fabian, S. Hartmann, and A. Hamidi, "Analysis of Insulation Failure Modes in High Power IGBT Modules," *IEEE Industry Applications Society Meeting*, vol. 2 pp. 799–792, 2005.

[79] N. Malik and A. Al-Arainy, *Electrical Insulation In Power Systems*. Boca Raton, FL: CRC Press, 1998.

[80] S. L. Nunes and M. T. Shaw, "Water Treeing in Polyethylene – A Review of Mechanisms," *IEEE Transactions on Electrical Insulation*, (6), pp. 437–450, 1980.

[81] H. Matsuba and E. Kawai, "Water Tree Mechanism in Electrical Insulation," *IEEE Transactions on Power Apparatus and Systems*, vol. 95 (2), pp. 660–670, 1976.

[82] A. Ashcraft, "Water Treeing in Polymeric Dielectrics," *World Electrical Congress*, Moscow, 1977.

[83] T. Yoshimitsu and T. Nakakita, "New Findings on Water Tree in High Polymer Insulating Materials," *1978 IEEE International Conference on Electrical Insulation*, Philadelphia, PA, 1978.

[84] R. Patsch and J. Jung, "Water Trees in Cables: Generation and Detection," *IEEE Proceedings-Science Measurement and Technology*, vol. 146 (5), pp. 253–259, 1999.

[85] R. Ohl, "Light-sensitive electric device including silicon," U.S. Patent Patent 2,402,662, 1948.

[86] "Solar Progress Report: Advancing Toward a Clean Energy Future," United States Department of Energy, May 2014. Available: http://www.whitehouse.gov/sites/default/files/docs/progress_report–advancing_toward_clean_energy_future.pdf.

[87] E. Ela, V. Diakov, E. Ibanez, and M. Heaney, "Impacts of variability and uncertainty in solar photovoltaic generation at multiple timescales," National Renewable Energy Laboratory. NREL/TP-5500-58274, 2013. Available: http://www.nrel.gov/docs/fy13osti/58274.pdf.

[88] NERC Resources Subcommittee Technical Report, *Balancing and Frequency Control*. Atlanta, GA: North American Electric Reliability Corporation, Jan 26, 2011.

[89] S. M. Chalmers, M. M. Hitt, J. T. Underhill, P. M. Anderson, P. L. Vogt, and R. Ingersoll, "The Effect of Photovoltaic Power Generation on Utility Operation," *IEEE Transactions on Power Apparatus and Systems*, (3), pp. 524–530, 1985.

[90] M. Thomson and D. G. Infield, "Impact of Widespread Photovoltaics Generation on Distribution Systems," *IET Renewable Power Generation*, vol. 1 (1), pp. 33–40, 2007.

[91] M. A. Eltawil and Z. Zhao, "Grid-Connected Photovoltaic Power Systems: Technical and Potential Problems—A review," *Renewable and Sustainable Energy Reviews*, vol. 14 (1), pp. 112–129, 2010.

[92] K. Knorr, B. Zimmermann, D. Kirchner, *et al.*, "Kombikraftwerk 2 Final Report," 2014. Available: http://www.kombikraftwerk.de/mediathek/abschlussbericht.html (in German).

[93] J. D. Glover, M. Sarma, and T. Overbye, *Power Systems Analysis and Design*, 4th edition. Thomson Learning, 2008.

[94] International Electrotechnical Commission (IEC) Technical Report 61850-90-7, "Communication networks and systems for power utility automation – Part 90-7: Object models for power converters in distributed energy resources (DER) systems." Geneva, Switzerland, 2013.

[95] N. G. Hingorani and L. Gyugyi, *Understanding FACTS*. New York: Wiley-IEEE Press, 2000.

[96] Y. G. Rebours, D. S. Kirschen, M. Trotignon, and S. Rossignol, "A Survey of Frequency and Voltage Control Ancillary Services—Part II: Economic Features," *IEEE Transactions on Power Systems*, vol. 22 (1), pp. 358–366.

[97] Y. G. Rebours, D. S. Kirschen, M. Trotignon, and S. Rossignol, "A Survey of Frequency and Voltage Control Ancillary Services—Part I: Technical Features," *IEEE Transactions on Power Systems*, vol. 22 (1), pp. 350–357.

[98] "Q at Night: Reactive power outside of feed-in operation," Dec 22 2014. Available: http://www.sma.de/fileadmin/Partner/SMA_Connect/WP_QATNIGHT.AEN132110W.pdf.

[99] "DC Loading of Distributed PV Systems Utilizing AE 3TL Inverters," Advanced Energy Application Note ENG-AE3TL-DCLoading-260-02, 2014. Available: solarenergy.advanced-energy.com/upload/File/Application Notes/ENG-AE3TL-DCLoading-260-02.pdf.

[100] B. Vermeersch, G. De Mey, M. Wójcik, *et al.*, "Chimney Effect on Natural Convection Cooling of Electronic Components," *IMAPS Advanced Technology Workshop on Thermal Management*, Palo Alto, CA, pp. 1–6, 2008.

[101] B. Gebhart, *Buoyancy-Induced Flows and Transport*. Milton Park: Taylor & Francis, 1988.

[102] W. M. Rohsenow, J. P. Hartnett, and E. N. Ganić, *Handbook of Heat Transfer Fundamentals*. New York: McGraw-Hill, 1985.

[103] *The Energy Performance of In-roof PV*. Cambridge: Viridian Solar, 2014.

Chapter 15
Reliability of power supplies for computers

Yaow-ming Chen[1]

Many factors can affect the reliability of a computer power supply. It is necessary to establish a consistent methodology to verify the performance of a power supply under different situations during the design stage and manufacturing process to ensure the product's reliability. This chapter will describe the commonly used methods for assuring the reliability of power supplies for computers.

15.1 Purpose and requirements

The power supplies mentioned in this chapter are either fixed or mobile. They include, but are not limited to internal or external power supplies for desktop computers, DC-DC converters, voltage regulation modules (VRMs), and external AC adapters for notebooks.

Before delivering products to a client, a power supply manufacturer needs to run different reliability tests. Usually, the power supply manufacturer needs to propose a reliability-oriented project that needs to be proved from the client side to ensure the quality and reliability of their product. This is usually referred to as a power system reliability qualification plan with the aim of reducing the failure risk in a controlled manner under analysis and test supervision.

Usually, the power reliability qualification plan includes reliability engineering activity, report deliverables, due date requirements, and so on. An example of a power reliability qualification plan is shown in Table 15.1. More items for reliability engineering activity, such as capacitor life analysis or fan tests, may be added to the plan upon the agreement between the client and the power supply manufacturer. Also, a detailed reliability qualification schedule should be submitted to and approved by the client at the beginning of each qualification project. It should be mentioned that testing alone may not reveal design margins or deficiencies. Therefore, the reliability qualification plan needs a balanced combination of analysis and testing, including electrical stress and environmental factors.

[1]Department of Electrical Engineering, National Taiwan University, Taiwan

Table 15.1 An example of a power reliability qualification plan

Reliability engineering activity	Report deliverables	Due date requirement	Remark
1. Reliability qualification schedule	Detailed schedule showing all reliability qualification activities, including when, where, and by whom activities were conducted	45 days prior to EVT	
2. Thermal profile analysis	Report that summarizes component thermal data based on worst-case maximum operating ambient temperature on multiple sources of components	10 days after thermal profile activity	EVT and DVT
3. De-rating analysis	Report of the electrical component stress and durability levels based on worst-case maximum operating ambient temperature and applicable to multiple sources of components	15 days after de-rating analysis activity	EVT and DVT
4. Highly accelerated life test	Report that details operational and destruct temperature and vibration limits of power systems.	15 days after highly accelerated life test activity	EVT and DVT
5. Shock and vibration test	Report on the power system's mechanical and electrical design issues caused by shock and vibration.	15 days after highly accelerated life test activity	DVT

Note: EVT – engineering verification test; DVT – design verification test.

In each reliability test report, the following items should be included:

- description of tests, procedures, and calibrated equipment
- photos with legible labels on the test samples, both external and with the cover removed
- detailed measured and recorded data with executive summary, observations, and recommendations
- any pass or fail assessment (including any recommendations for alternative design)
- adequate references for any non-standard processes used
- all referenced data files, specifications, and documents to enable complete review.

15.1.1 Design failure modes and effects analysis

The design failure modes and effects analysis (DFMEA) is a technique to find the weaknesses in a design before the design is realized [1,2]. It is a preventive form of problem-solving and has been applied to different engineering disciplines, including computer power supply manufacturing. The DFMEA is a bottom-up approach that examines the effects of single-point failures on overall system performance.

Three major tasks of DFMEA are

1. Recognize and evaluate the potential failure modes of each component in a product and its effects on the product.

2. Identify actions that could eliminate or reduce the chance of a potential failure occurring.
3. Document the process for improvement of future designs.

In a DFMEA process, a ranking chart with severity, occurrence, and detection should be established first. A commonly used DFMEA ranking chart is shown in Table 15.2.

A risk priority number (RPN), which is the product of severity, occurrence, and detection rank numbers, provides a relative measure of the importance of each effect [2–5]

$$\text{RPN} = \text{Severity} \times \text{Occurrence} \times \text{Detection} \tag{15.1}$$

For example, a failure mode can be found to have low severity, moderate occurrence, and absolute uncertain detection. According to Table 15.2, its RPN = $2 \times 3 \times 5 = 30$. The higher the RPN, the higher the priority of its associated failure mode that needs to be resolved. It should be noted that the failure mode with a safety hazard (severity ranking of 5) must be eliminated even though its RPN is low.

Since DFMEA is a weakness-oriented team-based systematic approach, a worksheet for knowledge sharing and action discussion is required. An example of the DFMEA worksheet used by a power supply company is shown in Table 15.3.

Table 15.2 A commonly used DFMEA ranking chart

		Severity
Ranking	**Description**	**Interpretation**
5	Hazardous	Safety-related catastrophic failure
4	High	Product is totally inoperable
3	Moderate	Product is operable, but at a reduced level of performance
2	Low	Product is operable, but comfort or convenience item(s) are inoperable or at a reduced level of performance
1	None	No effect
		Occurrence
Ranking	**Description**	**Interpretation**
5	Very High	Failure is almost inevitable, <1 in 3
4	Frequent	Repeated failure, <1 in 8
3	Moderate	Occasional failure, <1 in 80
2	Infrequent	Relatively few failures, <1 in 150,000
1	Rare	Failure is unlikely, <1 in 1,500,000
		Detection
Ranking	**Description**	**Interpretation**
5	Absolute uncertainty	Undetectable until catastrophic failure occurs or there is no design control
4	Remote	Remote chance that the design control will detect a potential cause of failure or subsequent failure mode
3	Low	Low chance that the design control will detect a potential cause of failure or subsequent failure mode
2	Moderate	Moderate chance that the design control will detect a potential cause of failure or subsequent failure mode
1	High	High chance that the design control will detect a potential cause of failure or subsequent failure mode

Table 15.3 Sample DFMEA worksheet

Item	Function (PSU function description)	Potential failure mode	Potential effect(s) of failure	Severity (S)	Potential cause(s)/ mechanism of failure	Occurrence (O)	Current design controls prevention	Current design controls detection	Detection (D)	RPN
Fuse1	Protect component for AC input	OPEN	No output	4	Inrush current, over-current, lower AC voltage, or heat	1	Total power protection, AC sense	Turn on test	2	**8**
		SHORT	No effect	1	Solder bridge	1	N/A	Visual inspection/ turn on test	2	**2**
Relay	The switch for the standby circuit	Pin1 to Pin2, Pin3 to Pin4 OPEN	No output	4	Over-stress of component, heat, and component break	1	Check component stress and power rating	Turn on test	2	**8**
		Pin1 to Pin4, Pin2 to Pin3 SHORT	Due to input short, huge current will go through input to damage capacitors and other components	4	Solder bridge	1	N/A	Visual inspection	3	**12**
Q52	Main switching MOSFETs of PFC	OPEN	The PFC circuit is disabled	4	Poor solder or missing component	1	N/A	Turn on test	2	**8**
		SHORT	**No output**	**4**	Over-stress of component, heat, and component break	**1**	**Check component stress and power rating**	**Turn on test**	**2**	**8**

(Continues)

M1	The PFC circuit control IC	Pin3, Pin4, Pin5, Pin6, Pin7, Pin8, Pin9, Pin11, Pin12, Pin13, Pin18 OPEN	The PFC control IC cannot work normally	4	Poor solder or missing component	1	N/A	Turn on test	2	**8**
		Pin15 short to Pin16	The PFC control IC cannot work normally	4	Over-stress of component, heat, and component break	1	Check component stress and power rating	Turn on test	2	**8**
T5	Main transformer of DC to DC	Pin1,2,A,B OPEN	No output	4	Over-stress of component, heat, and component break	1	Check winding stress and power rating	Turn on test	2	**8**
		Pin 1 to 2, Pin A to B SHORT	Cause Due to input short, huge current will go through input to damage capacitors and other components	4	Solder bridge	1	N/A	Turn on test	2	**8**
D704	Snubber circuit of SR circuit	OPEN	V_{ds} spike over-rating	4	Poor solder or missing component	1	N/A	De-rating test	4	**16**
		SHORT	No output	4	Over-stress of component, heat, and component break	1	Check component stress and power rating	Turn on test	2	**8**

Usually, the power converter maker needs to provide a summarized report to its client before the design is completed. It should be mentioned that if the effect of failure is not acceptable to the client, the design must be modified to eliminate the failure mode even though the RPN value is not high.

15.2 Thermal profile analysis

The objective of thermal profile analysis is to measure and record case temperatures of critical electronic components under worst-case conditions for all internal or external power systems being qualified. Measurements of alternative component types from all potential suppliers should be included. The design engineer must use sound engineering judgment to ensure that all operational configurations are accounted for, including standby mode and minimum airflow condition, to capture the worst-case conditions. The thermal test data should be used in other qualification procedures, such as de-rating analysis, capacitor life calculation, and reliability prediction.

To run the thermal profile analysis, different numbers of samples for both EVT and DVT should be determined. All planned component suppliers for each reference designator on the bill of materials (BOM) should be included for the thermal profile analysis.

The temperature is measured by using thermocouples. The measuring thermocouple junction must be placed where the temperature is to be measured. If the thermocouple is connected to parts of different polarity, it may be appropriate to place additional electrical insulating sleeving over the conductors (not the thermocouple junction). Thermocouples usually are secured by tying, cementing, or soldering without affecting the thermal contact with the part being measured.

Tying with thread is used primarily for round items such as wire insulation. Cementing provides a better thermal bond and reduces the surface area exposed to air at a temperature different from that being measured. Soldering can provide better thermal conductivity and greater mechanical security for metal surfaces to which solder will adhere. Currently, cementing the thermocouple onto the component using high thermal conductivity glue is the most popular way to measure the device's temperature. In addition, the attachment point of the thermocouple should be able to reflect the true temperature of the component, and some components may need more than two measurement points. Figure 15.1 shows the recommended thermocouple attachment, within the marked circle, for key components such as a capacitor, MOSFET, and inductor.

The thermal profile analysis should include the methodology and results of temperature measurements. The pictures of the device under test and its corresponding thermal images should be placed side-by-side in the thermal profile for easy comparison. An example of side-by-side pictures of a test sample and its thermal image is shown in Figure 15.2. It should be mentioned that the thermal image only provides limited information about the thermal test since, for many components, it is not possible to measure their temperature directly. Therefore,

(a) (b) (c)

Figure 15.1 Thermocouple attachments for different components: (a) capacitor, (b) MOSFET, and (c) inductor

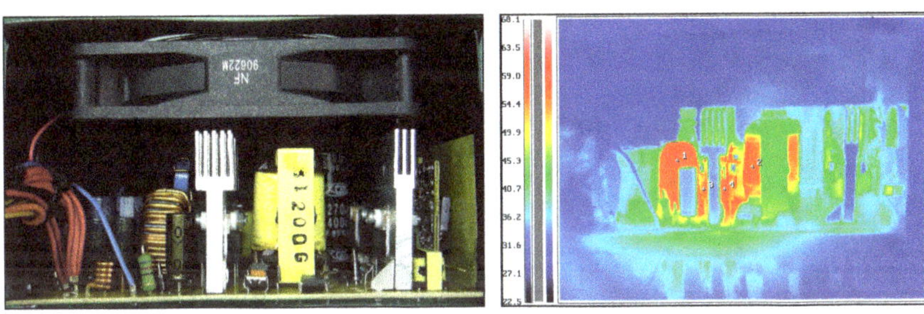

Figure 15.2 Example of side-by-side pictures showing a test sample and its thermal image

using thermocouples to measure the required temperature readings for different components is necessary.

Thermal measurement must be recorded by data loggers with a sampling rate agreed to by the client's reliability engineering personnel. Thermocouple-type information should be approved by the client. Information about the mounting locations, selected components for measurement, and airflow diagrams must be provided before the measurement procedure begins. Temperatures for components that are not directly measured should be estimated with a method agreed upon by the client's reliability engineering personnel. Thermal measurements need to be taken at specified operating temperatures and standby condition temperatures. The temperature must stabilize at each condition for a minimum of 30 min with ±1 °C change maximum. Figure 15.3 shows the picture of a power supply with many thermocouples attached to different components. The thermocouples and their wires should be well arranged so that they will not affect the normal air flow of the power supply. Usually, temperature readings of different components will be continuously recorded during the test. Figure 15.4 shows the temperature curves of

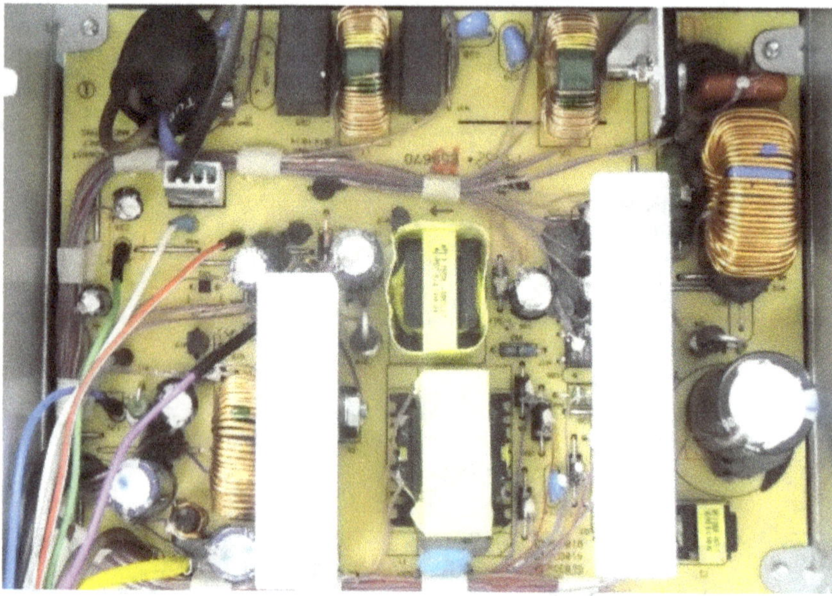

Figure 15.3 A power supply with many thermocouples attached to different components

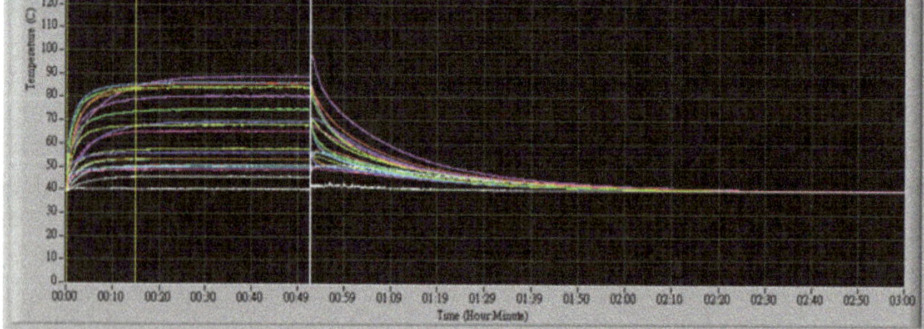

Figure 15.4 Temperature curves of different components through the entire test

different components on the same monitor throughout the entire test. A commonly seen thermal measurement summary, including information on the part number, component description, temperature limits, and judgment, is shown in Table 15.4.

Table 15.4 A sample thermal measurement summary

NO	Location	Vendor	Model	Rated temp.	De-rating limit	Judgment	Remark
2	Q1	FAIRCHILD	FQP13N50C	150	135	Pass	
28	D71	ON	MBR2045CTG	175	157.5	Pass	
32	C35	LTEC	TKD1HM4R7 D11A-5*11-P5	105	95	Pass	
43	C73	TEAPO	SE025M0047A5F-P0511-P5	105	95	Pass	
49	RT2	THINKING	TTC05103KSYA02	125	125	Pass	
60	T1:Wire	ACBEL	API P/N R25F06-0001I	130	110	Pass	
62	T2	FE-TRONIC	API P/N R25C65-0001I	130	110	Pass	
69	L8:Croe	ACBEL	API P/N R21D59-0001I)	130	90	Pass	

15.3 De-rating analysis

The objective of de-rating analysis is to verify that component ratings are not exceeded under either steady-state operation or transient conditions [6]. Proper component de-rating will lower failure rates through reduced stress. It can also reduce the impact of material, manufacturing, and operational variability, as well as enable continued operation with part parameter shifts. In order to ensure a reliable power conversion product, the power supply manufacturer needs to provide details on the de-rating methods, conditions, and results to the client.

De-rating analysis is comprehensive and is a part of the design process. De-rating considerations shall be considered before the components are selected. Every critical parameter, including voltage, current, temperature, and other stress factors, shall be considered. Sometimes, mechanical considerations should be included. For instance, mechanical bending can cause cracks in multi-layer ceramic capacitors (MLCCs) and lead to electrical failure.

The de-rating level, or stress factor, for different types of components, plays a key role in the de-rating analysis. The stress factor is defined as the applied level divided by rating

$$\text{Stress Factor (SF)} = \frac{\text{Applied Level}}{\text{Rating}} \times 100\% \tag{15.2}$$

For example, a 250 V-rated part used at 200 V has an 80% stress factor. The stress factor for some parameters (such as temperature) is expressed as a quantity (such as 25 °C), not a percentage number. A sample de-rating chart is shown in Table 15.5.

The parameters listed in Table 15.5 reflect the main causes of a component's failure. It also provides the design guidelines for circuit designers. For example,

Table 15.5 An example of de-rating chart

Device type	Parameter	10-Year stress factor	5-Year stress factor
Resistors			
Carbon composition	Power dissipation	≤ 60%	≤ 70%
	Maxi working voltage	≤ 60%	N/A
	Below max temp. Limit	≥ 25%	≥ 25%
Zero ohm	Current	≤ 85%	≤ 85%
Fixed film	Power dissipation	≤ 60%	≤ 70%
	Maxi working voltage	≤ 70%	≤ 70%
	Below max temp. limit	≥ 25%	≥ 25%
	SMD max body temp.	100 °C	100 °C
Capacitors			
Fixed	DC voltage	≤ 80%	≤ 80%
ceramic-MLCC	Temp. below max limit	≥ 10%	≥ 10%
Fixed aluminum	DC voltage	≤ 80%	≤ 85%
electrolytic	Ripple current	≤ 70%	N/A
	Temp. below max limit	≥ 10%	≥ 10%
	Life/endurance (using life cap. equations)	≥ 10 years at 25 °C, full load	≥ 5 years at 25 °C, full load

electrolytic capacitor failure is due to excess voltage, reverse voltage, ripple current, operating temperature, and electrolyte evaporation. However, power dissipation and ambient temperature are the main concerns for resistor failure. It should be mentioned that components with certified safety agent approvals may be used up to 100% of their approved ratings. 100% is the maximum voltage that the component would see based on rigorous worst-case analysis. Some examples to illustrate the stress ratio calculation are presented below:

Example 1: A 12 $V_{dc}\underline{+}3\%$ output terminal needs a capacitor with less than an 80% stress factor. The selected capacitor should satisfy: $(12\times1.03)/V_{rated} \times 100\% < 80\%$. Therefore, the rated capacitor voltage should be higher than 15.45 V.

Example 2: A capacitor's rated temperature is 105 °C. If it is used for a de-rating guideline with "Temp. Below Max. Limit >25 °C," the maximum operating temperature that meets the de-rating guideline is: 105–25 °C = 80 °C.

Example 3: A diode's maximum junction temperature resistor is specified as 130 °C, and its thermal resistance from junction to case is 3 °C/W. If the power dissipation is 2.5 W and the case temperature is 80 °C, the junction temperature is calculated as follows: 80 °C + (3 °C/W × 2.5 W) = 87.5 °C. Therefore, the stress factor for the diode under this specified operation condition is: (87.5 °C/ 130 °C) × 100% = 67.3%.

The stress factor is an operation-condition-oriented value and highly depends on the client's demands. As the stress factor decreases, the rating of the used component should increase. The power supply manufacturer needs to follow the test procedure and criteria provided by the client. Usually, primary and alternate component sources need to be included, and measurements at different loading levels, such as 100% and 80%, are required. It is usually a requirement that the supplier provide the worst-case component stress in the de-rating analysis report using a worst-case application. The worst case for voltage parameters may occur with the maximum input voltage, but the one for current parameters occurs when the minimum input voltage is applied. For components where electrical measurements are not feasible, justification for using calculated results under application conditions is allowed.

15.4 Capacitor life analysis

It is well known that electrolytic capacitors have the shortest lifetime among all components in a power converter [7]. Also, the failure of electrolytic capacitors is caused by many factors such as voltage, temperature, or ripple current. The purpose of capacitor life analysis is to ensure that all capacitors, especially electrolytic ones, have adequate predicted life for the usage environment of the power converter. Data required for this analysis is gathered from the preceding thermal profile analysis and de-rating analysis. The life analysis should include all capacitors from multiple (primary and alternate) suppliers for each reference designator. Usually, the client will provide a preferred supplier list for the power supply manufacturer to ensure a more reliable component supply. The capacitor lifetime is referred to as the worst-case operating condition, which is the maximum system operation temperature with 24 hours a day, 7 days a week operation.

The determination of the capacitor lifetime is affected by many factors, including materials, temperature, voltage, and the de-rating factor. The most comprehensive failure rate of electrolytic capacitors can be found in the military handbook MIL-HDBK-217:

$$\lambda_c = \lambda_b \cdot \pi_T \cdot \pi_C \cdot \pi_V \cdot \pi_{SR} \cdot \pi_Q \cdot \pi_E \text{ failure}/10^6 \text{ h} \tag{15.3}$$

where λ_b is the basic failure rate depending on the capacitor structure and materials, π_T is the ambient temperature factor, π_C is the capacitance factor, π_V is the voltage stress factor, π_{SR} is the series resistance factor, π_Q is the quality factor, and π_E is the environmental factor. Although (15.3) shows that the capacitor's failure rate depends on many different factors: two of them play the most important roles—operation temperature and voltage stress.

Different computer power supply companies have their own lifetime equations for different types of capacitors. In addition, capacitor suppliers may have their own ways of estimating their products' lifetimes. Therefore, the capacitor lifetime equations should be discussed between the power supply manufacturer and its client. In general, the client will provide the equations for capacitor lifetime analysis. Commonly used life formulas for two types of capacitors are given as examples below.

15.4.1 Aluminum electrolytic capacitors

A commonly used lifetime equation for the aluminum electrolytic capacitor's expected lifetime, λ_c, is [7–10]

$$\lambda_c = \lambda_b \cdot \pi_{T,ext} \cdot \pi_{T,\,\text{int}} \cdot \pi_V \cdot \pi_Q \tag{15.4}$$

where λ_b is the base lifetime provided by the capacitor supplier, $\pi_{T,ext}$ is the function related to external temperature, $\pi_{T,int}$ is the function related to internal temperature, π_v is the voltage stress-related function, and π_q is the function relative to quality.

Functions included in (15.4) could be varied under different applications and requirements. A commonly used simplified $\pi_{T,ext}$ is

$$\pi_{T,ext} = 2^{\left(\frac{T_{\max} - T_a}{10}\right)} \tag{15.4a}$$

where $T_{\max}$ is the capacitor's rated temperature, and T_a is the case temperature in the application. A commonly used simplified $\pi_{T,int}$ is

$$\pi_{T,\text{int}} = 2^{\left(\frac{\Delta T_O - \Delta T_i}{\Delta T_O}\right)} \tag{15.4b}$$

where ΔT_o is the core temperature rise caused by the internal heating due to the rated ripple current, ΔT_i is the core temperature rise caused by the ratio of ripple current in the application, i_{rapp}, and the rated ripple current, i_{ro}. Usually, most capacitor manufacturers use a fixed ΔT_o to determine its rated ripple current by holding the capacitor at its maximum temperature and voltage and then increasing the ripple current. The expression for ΔT_i is

$$\Delta T_i = \Delta T_o \left(\frac{i_{rapp}}{i_{ro}}\right)^2 \tag{15.4c}$$

Commonly adopted functions for π_V and π_Q are

$$\pi_V = 1.05 - \frac{\left(\frac{V_a}{V_r} - 0.6\right)^2}{2} \tag{15.4d}$$

$$\pi_Q = \frac{\ln(\lambda_b \times T_{\max})}{13.1} \tag{15.4e}$$

where V_r is the rated voltage and V_a is the applied voltage.

15.4.2 Os-con type capacitors

The os-con type capacitor is an aluminum solid capacitor with highly conductive organic semiconductor electrolyte material. It has the advantages of low ESR and long lifespan and has been adopted for computer powers to reduce the size and ripple voltage as well as to increase the reliability of the power supply system. Because of the material characteristics, the lifetime equation for an os-con capacitor is different from the one for an electrolytic capacitor. A commonly adopted

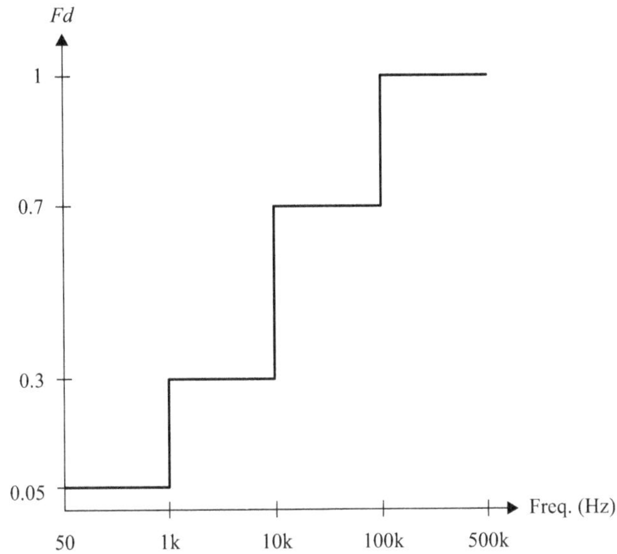

Figure 15.5 The typical frequency de-rating factor curve for os-con capacitors

lifetime equation for an os-con capacitor, $\lambda_{\text{ocos-con}}$, is

$$\lambda_{\text{os-con}} = \lambda_b \times 10^{\left(\frac{T_{\max} - T_a}{20}\right)} \times F_d \tag{15.5}$$

where F_d is the frequency de-rating factor with a typical curve given in Figure 15.5. Equation (15.5) reveals that the ocos-con capacitor is not suitable for line frequency regulation, but it is good for high-frequency switching applications.

15.5 Fan life

Fans, which are sometimes called blowers, are important components for power converters to provide heat dissipation through airflow. It is a relatively cheap component inside the power converter, but its failure may lead to malfunction or even severe damage to the system. The fan's reliability is commonly overlooked. Usually, reliability deliverables include fan specifications, lead-free reliability qualification, de-rating, design verification tests, and third-party reliability tests [11–13].

Almost every fan used for computer power is driven by a DC motor. Therefore, fans can be categorized by their control type: PWM-controlled, voltage-controlled, or thermistor controlled. The fan specification must specify its dimension, voltage, power, control type, RPM, airflow, noise level, expected life, and bearing/ lubricant types.

It should be noted that the fan is the only component with rotating parts inside a computer power system, so its mechanical-based lifetime may affect the system's reliability. One important factor in a fan lifetime is the bearings. Bearing life is usually called the L10 (or B10) life. It is defined as the life at which 10% of the bearings are expected to fail under classical fatigue failure. Or, it is the life at which 90% of the bearings are still operating. The L10 life is a theoretical life and may not represent the service life of the bearing. Other factors, such as load, temperature, materials defects, or contamination, can have a significant effect on bearing life. Other than bearing life, impeller unbalance and lubricant evaporation must be used for failure mode determination.

Unlike other electronic components, the non-functional (storage) test is included in the DVT to ensure that the fan can meet the specified storage conditions. Normally, the temperature/humidity ranges for a storage test are wider than for an operation test. The commonly adopted operation test conditions are 0–70 °C and 10%RH–90%RH, while the ones for the storage test are −40–70 °C and 5% RH–95%RH. The so-called four-corner cycling test process for both storage and operation tests is required for fan analysis. During these testing processes, numbers of randomly selected production-level fans will be soaked inside an environmental chamber with controlled temperature and humidity. The client will provide the duration for each step and the temperature/humidity ramp for testing.

A typical temperature/humidity storage test flow is shown in Figure 15.6(a), while the one for an operation test is shown in Figure 15.6(b). The testing starts at point O with room conditions (25 °C, 40%RH). Then, the test condition goes to points A, B, C, and D and returns to O eventually. The numbers inside the parentheses at a specific point or between two points indicate the testing step and the duration hours. For example, step 1 goes from point O to point A in 4 hours and step 2 is to stay at point A for 24 hours. It should be mentioned that the soaking points and temperature/humidity ramps may vary, and they are usually defined by the client.

Thermal shock testing is also required for fans. Tens of thermal shock cycles with temperature between two extremes (–40 and 70 °C) are commonly adopted. The minimum dwell times, such as 30 min, should be specified for the thermal shock test.

The fan life analysis includes a sequence of performance measurements and different tests. The performance measurements are speed (RPM), current, and noise (dBA). The test samples are measured first before any test. Then, the four-corner storage test, the four-corner operation test, and the thermal shock test will be performed in order. After each test, the measurements for all test samples need to be conducted. All four measurements in different stages should be included in the analysis report.

In addition to the design verification tests mentioned above, the fan supplier must conduct a reliability demonstration test for a minimum of 6 weeks using the client-defined fan accelerated life test, which includes power cycling and parametric testing. Since different types of fans have different specifications and control parameters, their parametric testing is different. For example, RPM at different duty cycles (such as 0%, 10%, 50%, and 100%) needs to be measured for PWM

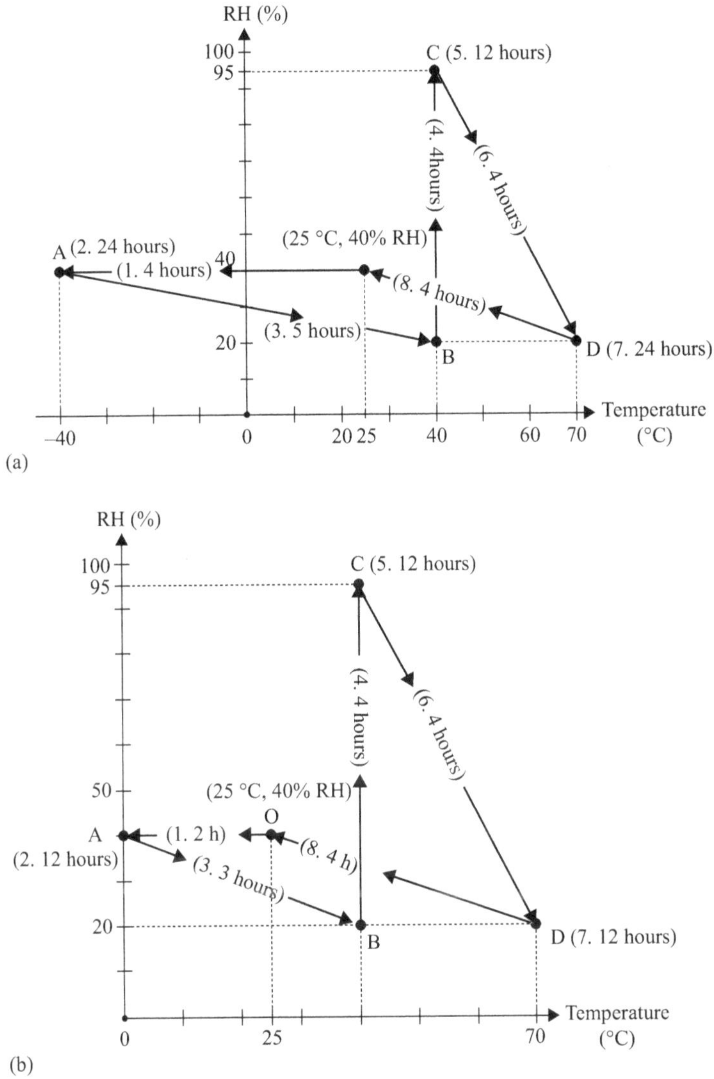

Figure 15.6 Four corners cycling test process for fan analysis: (a) the storage test flow and (b) the operation test flow

fans, while the voltage-controlled fans need to measure RPM at different applied voltage levels (such as maximum, minimum, and midpoint).

The fan fails when it does not perform one or more of its required functions within the limits mentioned in its specifications. Common failure criteria are

- Fail to spin
- Fail to rotate with the RPM tolerances

- Draw more current than a specified value
- Abnormal noise
- Mechanical cracking
- Grease leakage
- Housing warpage
- Impeller touching the housing
- Smoke from the fan

15.6 High accelerated life test

To ensure the robustness of power converters in today's competitive marketplace, the design phase becomes a perfect opportunity to find a product's maximum limitation. Traditional verification testing approaches (such as burn-in) require many cycles (days or weeks) when a new product is operated under normal operation conditions. Accelerated stress testing (AST) provides a series of operational overstresses to effectively uncover the product's weak point by accelerating fatigue. Different aspects of AST are developed and applied during different stages of computer power manufacturing.

The highly accelerated life test (HALT), highly accelerated stress screening (HASS), and the highly accelerated stress audit (HASA) are collectively referred to as AST [14–18]. The HALT is a design test used to improve the robustness and reliability of a product through a test-fail-fix process where the applied stresses are beyond the specified operating limits. The HASS is also a test-fail-fix process for all manufactured units, where the applied stresses are beyond the specified operating limits. It provides for the early discovery of manufacturing process issues. The HASA has the same testing process as HASS but is only performed on selected samples. Since the HALT is performed during the design stage and the HASS/HASA is for the manufacturing process, this section will focus on the explanation of HALT.

The testing of HALT is performed in a so-called HALT environmental chamber, which can simultaneously provide controlled temperature and vibration to the device under test. Pictures of the HALT chamber with a device under test are shown in Figure 15.7. The appearance of the HALT chamber is shown in Figure 15.7(a), while Figure 15.7(b) shows a device under test that has been placed inside the chamber. A picture of the automatic test equipment (ATE), which provides the required power source and various load conditions for the tested device, is shown in Figure 15.7(c).

The HALT chamber must be calibrated periodically. It should be able to produce the desired rapid thermal change rate (such as 50 °C/min) and maintain thermal stability. Also, it should be capable of generating rigorous random shock vibration with six degrees of freedom (X, Y, and Z simultaneous with pitch, roll, and yaw) and with a broadband vibration spectrum (such as 10 Hz–4 kHz). In addition, the calibrated thermocouples and accelerometers with associated instruments, such as meters and the spectrum analyzer, are required to monitor the test

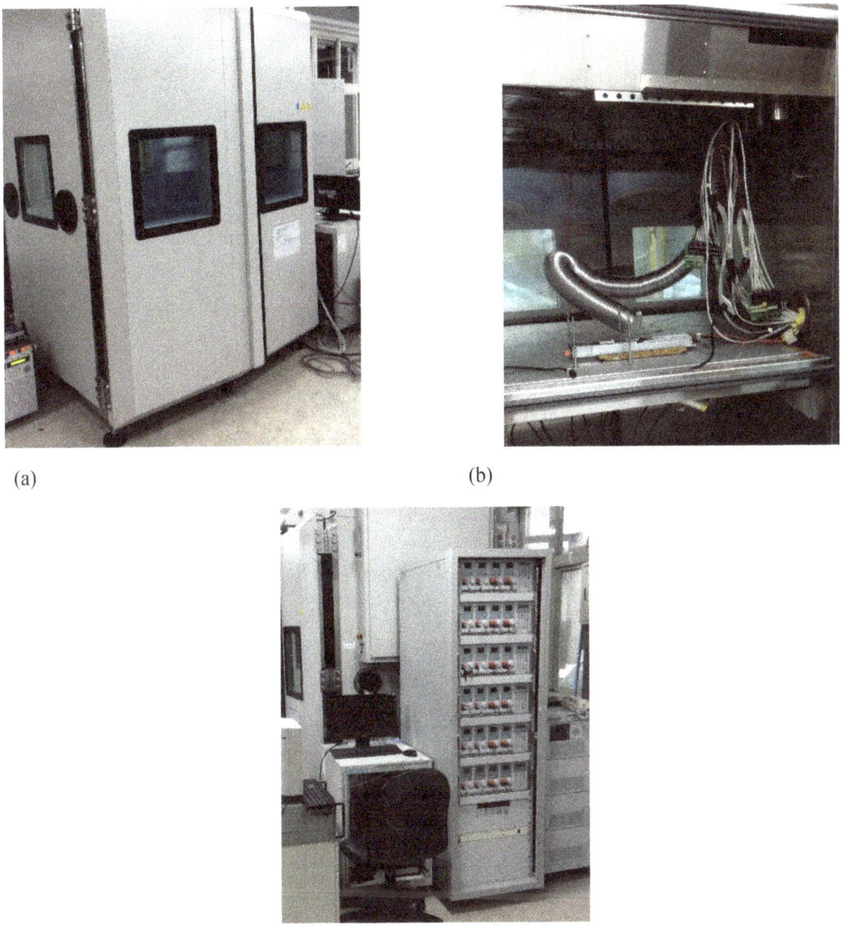

(a) (b) (c)

Figure 15.7 *Pictures of HALT chamber with the device under test: (a) the appearance of the HALT chamber, (b) a device under test inside the HALT chamber, and (c) the automatic test equipment for HALT chamber*

and to prepare the testing report. Since the device is operating during the HALT, programmable voltage sources, electronics loads, and appropriate cables and connectors are necessary.

Before the HALT, all samples under test should be able to work functionally within the specifications. The thermocouples should be attached to thermally neutral points away from high thermal dissipating components or air flow to ensure that the whole sample reaches the thermal set point. The accelerometers should be small enough to be mounted in a central location on the sample, and light enough that their mass does not significantly impact the vibration dynamic characteristics

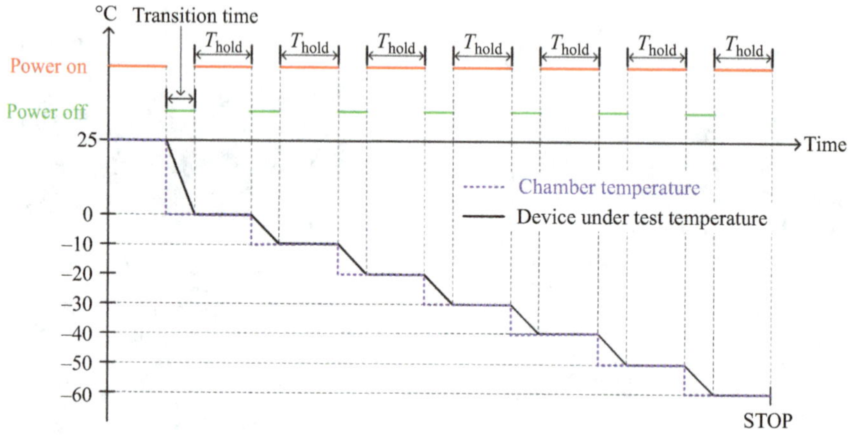

Figure 15.8 Example of a low-temperature stress test graph

of the sample. The recommended order of applying thermal and vibration stresses during HALT is as follows:

1. Low-temperature stress
2. High-temperature stress
3. Vibration stress
4. Combined temperature vibration stress

15.6.1 Low-temperature stress

An example of a low-temperature stress test graph is shown in Figure 15.8. The low-temperature test starts at 0 °C, and the lowest testing temperature is usually set to –60 °C, although the sample may fail earlier at a higher temperature. At the beginning, all samples should operate correctly according to all of their specifications at room temperature. Usually, a diagnostic testing program is developed to perform a complete function test for all specifications.

Before each temperature stress step test, all samples are powered off. Then, the chamber temperature decreases to 0 °C. Once a stable temperature is reached, all samples are powered on. All sample functions should be verified by running the diagnostic test program. The chamber temperature needs to be held at 0 °C for at least T_{hold} minutes. After the diagnostic test program is over, all samples are powered off, and the temperature stress step test is complete.

Then, the chamber temperature is decreased by 10 °C to a new step temperature. Again, once the stable temperature is achieved, all samples are powered on to run the diagnostic test program within T_{hold} minutes. Once all samples pass the diagnostic test program, they are powered off, and the temperature stress step is completed. The steps continue until the minimum temperature (–60 °C) is reached or any sample fails the testing function.

If the sample fails at a specific step temperature (–X °C), then the temperature is increased by 5 °C, and the test program is run again. If all samples pass the test program, then the low operating limit temperature becomes –(X–5) °C. Otherwise, the low operating limit temperature is –(X–10) °C.

15.6.2 High-temperature stress

The testing process for high-temperature stress is similar to that for the low-temperature stress test, except that the starting temperature is 60 °C, and the maximum temperature is usually 120 °C. An example of high-temperature step test graph is shown in Figure 15.9. The high-temperature stress is also used to determine the high operating limit temperature. If the sample fails at a specific step temperature (Y °C), then decrease the temperature by 5 °C and run the test program again. If all samples pass the test program, then the high operating limit temperature becomes (Y–5) °C. Otherwise, the high operating limit temperature is (Y–10) °C.

15.6.3 Vibration stress

The chamber temperature is kept at 25 °C, and all samples are powered on with a diagnostic test program running during the vibration stress test. Usually, the starting point of the vibration stress test is 10 G_{rms}, and the highest vibration level is 50 G_{rms}. It should be mentioned that the maximum vibration level depends on the need of the application from the client side. During each vibration stress level, the testing time must last for T_{hold} minutes. A typical step chart of the vibration stress test is shown in Figure 15.10.

If a failure of the diagnostic test program occurs, the vibration test should be stopped so that the problem can be fixed. If the failure cannot be fixed without

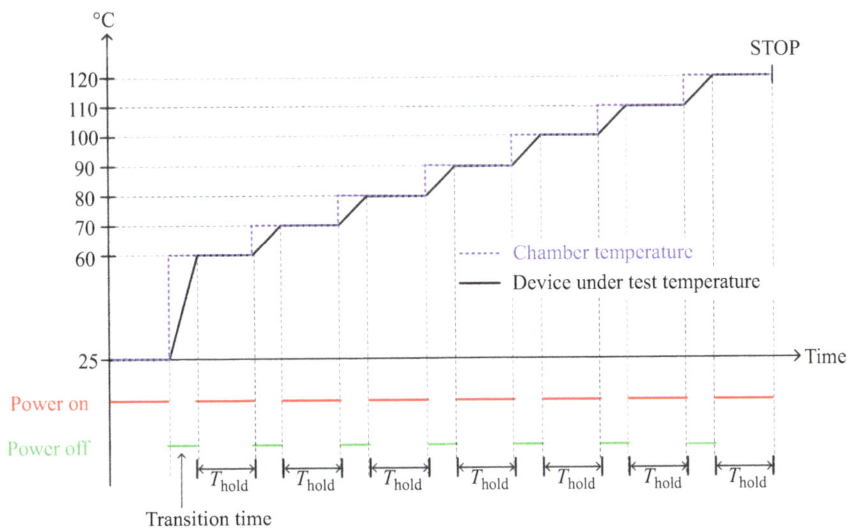

Figure 15.9 Example of a high-temperature step test graph

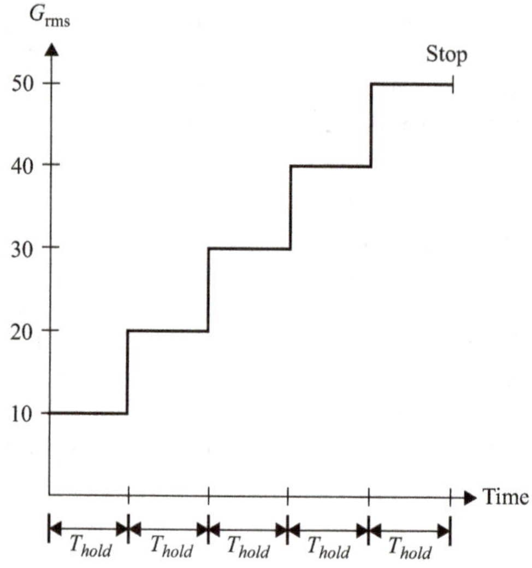

Figure 15.10 A typical step chart of the vibration stress test

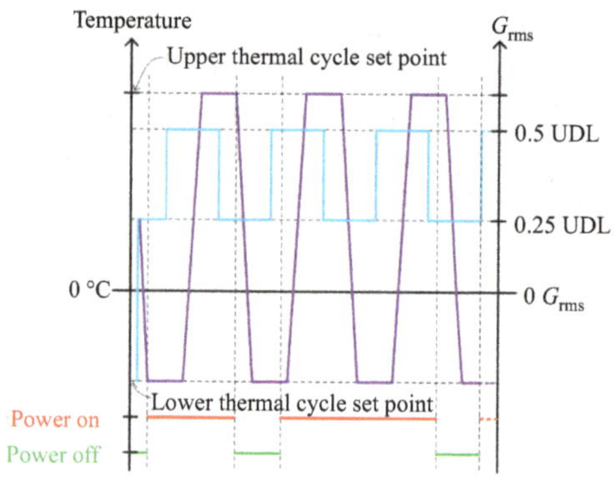

Figure 15.11 A typical combined temperature–vibration stress test

vibration, then the last vibration level will be documented as the upper destruct limit. If failure is not present, the last vibration level achieved will be documented as the upper operating limit.

15.6.4 Combined temperature vibration stress

The environmental chamber should be able to provide a combined thermal and vibration stress test. Since this is a combined stress test, the temperature and vibration levels are slightly less than the operation limits obtained by the temperature or vibration stress tests. A typical combined temperature vibration stress test profile is shown in Figure 15.11.

Thermal cycling should have a very steep temperature ramp (such as 60 °C/min) and run between two set points. Usually, the upper thermal cycle set point is 10 °C below the high operating temperature limit, and the lower one is 10 °C above the low operating temperature limit. The dwell time of each temperature set point should be long enough to complete the diagnostic program as well as under different levels of vibration stress.

The vibration test has two levels. Usually, vibration levels of 25% and 50% of the upper destruct limit (UDL) obtained from the vibration stress test are adopted. The dwell time of each vibration level is dependent on the dwell time, T_{hold}, of the thermal cycling. Each vibration level, either 25% or 50%, should be performed during half of T_{hold}.

A specified number of power supply samples are randomly selected for the combined stress test. They are powered on and off during the combined test, and a diagnostic test program should be performed to verify the function of the power supply. At least a specified number of thermal cycles should be run to complete the combined temperature and vibration stress test. If any failure occurs, the testing process should stop, and the failed sample should be removed from the chamber to perform the failure analysis to determine the cause. Commonly seen mechanical failures of the HALT procedure are loose screws or rivets, cracks in plastics, misalignment, or permanent deformation of components. Unusual noise or any failure issued by the diagnostic test program are called functional failures.

15.7 Vibration, shock, and drop test

The purpose of the vibration and shock test is to ensure that the module, components, and solder joints are sufficiently rugged. It can also merit further power converter design qualification. Two random vibration tests, operating and survival, and one sinusoidal-wave vibration test need to be performed [19].

15.7.1 Vibration test

The random vibration is in direct contrast to the definition of periodic (e.g., sinusoidal) vibration. The random vibration has all frequencies, within given limits, occurring at the same time. The amplitude of any discrete frequency is a function of a Gaussian curve within the capabilities of the test equipment availability.

The usual way to describe a random motion is in terms of its power spectral density. The unit of power spectral density is g^2/Hz, where g is the gravitational acceleration and can be represented by metric units, m/s^2. The energy of the random vibration is specified by the integration of the power spectral density over the frequency range and is represented as G_{rms}. G_{rms} is the root mean square (rms) value of the power spectral density. For a pure sinusoidal function, the peak value is its rms value times the square root of 2: $X_{peak} = (\sqrt{2})\, X_{rms}$. However, the rms value of a signal is equal to the standard deviation, assuming a zero mean. Therefore, random vibration is very complicated and has no simple relationship between its

Table 15.6 Vibration spectral density for operating vibration test

Frequency (Hz)	Power spectral density (g^2/Hz)	Slope (dB/octave)
5–350	0.015	0.0
350–500	N/A	–6.0
500	0.0074	0.0

Table 15.7 Vibration spectral density for survival test

Frequency (Hz)	Power spectral density (g^2/Hz)	Slope (dB/octave)
20	0.01	0.0
20–80	N/A	3.0
80–350	0.04	0.0
350–2000	N/A	–3.0
2000	0.007	0.0

peak and rms values. The peak value of a stationary random time history is typically three or four times the rms value.

For the operating vibration test, a specified number of samples (usually 3) shall be tested, and each one is operating at full load conditions throughout the entire test. The commonly used vibration level is 2.6 G_{rms}, with the spectral densities shown in Table 15.6. The tested sample should be able to withstand the random vibration for at least 10 min per axis on all three axes. After the operating mode vibration test, all samples should be able to meet all required specifications.

Conversely, the survival vibration test has a higher vibration level, usually up to 6.06 G_{rms}, with the spectral densities shown in Table 15.7. Again, all tested samples should be able to withstand the random vibration for at least 10 min per axis on all three axes. All samples should be able to meet all required specifications after the random vibration test.

The sinusoidal-wave vibration has a vibration level of 0.75 G_{rms} for frequency ranging from 5 to 450 Hz at the rate of one octave per minute. After the sinusoidal-wave sweep, the four largest amplitude resonating frequencies should be noted. Then, the tested samples need to undergo a vibration test for at least 10 min at each of those frequencies on all three axes. This is a non-operating test, and all test samples should survive with permanent damage.

15.7.2 Shock and drop test

A complete shock test includes four different types: operational, non-operational, trapezoidal, and drop. Usually, at least three samples are required for each test.

For the operational shock test, the test samples shall operate at full load and perform within specifications throughout the test. Those samples are subjected to a one-half sinusoidal-wave pulse of 3.0 ms duration with an effective acceleration of 20g.

For the non-operational shock test, the effective acceleration is increased to 80*g* with the same pulse and duration. Furthermore, this test should be performed on each of the sample's six faces. Also, these tested samples should perform within specification after the non-operational shock test.

The trapezoidal shock test is also a non-operational test. The trapezoidal shock has an effective acceleration of 50*g* with a velocity change of 0.746 m/s. The trapezoidal shock shall be applied to each of the sample's six faces at least three times. As in previous tests, these tested samples shall perform within specification after the shock.

For the drop test, the sample free-falls from a height of 1 m and strikes a concrete surface. Since each sample has 6 faces and 8 corners, the drop test should be performed a total of 28 times on the same sample. That is 2 drops, operational and non-operational, for each face or corner. The sample should be able to perform within specifications throughout the drop test.

15.8 Manufacturing conformance testing

The power supply manufacturer needs to test its products continuously during the manufacturing stage. The objective of this testing, called manufacturing conformance testing, is to identify any possible causes of failures, including workmanship, during different production processes.

In-circuit tests and automatic optical or high-resolution X-ray inspection are applied to all products before final assembly and functional testing. This stage of testing is performed on all fabricated printed circuit boards to assure workmanship quality and detect soldering or component misplacement faults. After this early stage test, the assembly of a complete power converter device will be performed. Many types of tests, including the functional test, stress test, safety test, quality assurance, and reliability testing, will be applied to the product during or after the manufacturing process.

The ATE is required to perform functional tests of every manufactured unit. Each power converter should pass every parameter specified in the functional specification. The power supply manufacturer should have an agreement with the client on the functional testing method.

Stress tests are performed after the manufactured power converters pass the functional test. Based on the client's request, the burn-in test may be required on 100% or a selected sample of the products. HASS on 100% of the products or highly accelerated stress audit HASA on a low percentage of samples are performed in this manufacturing stage.

All power conversion devices shall pass the safety tests as defined by functional specifications and by applicable regulatory documents such as EMC requirements or IEC standards. Quality assurance (QA) should be performed to assure conformance to functional, mechanical dimensions, and cosmetic requirements.

15.8.1 The ongoing reliability testing

Besides tests conducted during different manufacturing stages, ongoing reliability testing (ORT) is performed after the first shipment of the product [20]. ORT is performed on a sample basis to assure that no process or component degradation has affected the long-term reliability of the product. ORT is used to identify any variability in the production process before any customer experiences abnormal production performance. Normally, ORT may be further divided into two parts: long-term testing and short-term testing.

For long-term testing, ORT will continue to run until the end of the product's lifetime. Devices that complete a long-term ORT are not salable and must be properly disposed of. The required number of samples is usually determined from the client's side. Not every product requires long-term ORT. Short-term ORT should run long enough to detect any changes in early life failures. For most power supply manufacturers, the so-called ORT is referred to as the short-term ORT. Devices that complete a short-term ORT may be cycled back through a re-testing process and sold as new if the combined stress level and time effects do not degrade the long-term reliability. However, most power supply manufacturers simply discard their ORT samples.

The ORT should be performed on products from each production site on a calendar basis during the life of the product. It should be mentioned that ORT is not a way to verify product design or to catch trivial issues. The sample sizes for ORT are determined by the following factors: first customer shipment date, risk factor, power supply complexity, ship volumes, and acceptable quality level sampling plan.

The risk factor, high or low, plays an important role in the ORT. For the first 6 months after the first customer shipment date, all power supplies are considered "high risk." However, if any ORT failure appears in the previous 20 weeks, or if manufacturing yields do not meet the set targets, the products remain at "high risk" after 6 months. Conversely, all power supplies are considered "low risk" if there were no ORT failures in the previous 20 weeks and if the manufacturing yields met the set target after six months. The risk factor will affect the required sample sizes of the products. Table 15.8 shows the recommended weekly sample size for ORT under different risk factors.

Samples for ORT are taken from the production line, and they must be able to uniformly represent one week's products. By doing so, the results from the ORT samples are associated with all power supplies manufactured in the corresponding week.

2-week ORT, samples are pulled daily until the sample size is complete in 1 week. Samples are tested and pulled out after the 2-week test period. Therefore, after the first week, there are always two sets of ORT samples under testing.

Table 15.8 Recommended weekly sample size for ORT

Lot size	High risk	Low risk
<25	4	2
26–100	6	4
101–250	10	6
251–500	14	12
501–1200	20	16
1201–3200	24	18
>3200	26	20

Table 15.9 Burn-in with the recommended specifications for ORT

Line voltage	Split population, half at 100–120 V_{ac} and half at 220–240 V_{ac}
Burn-in temperature	Chamber temperature 50–55 °C
Burn-in hours	14 days (or to be specified)
Power cycling (T_{on} − T_{off})	45 min on, 5 min off, 10 min quick power cycle
Load	95% ± 5% of rated power
Quick power cycle	30 s on, 30 s off for 10 cycles (10 min)

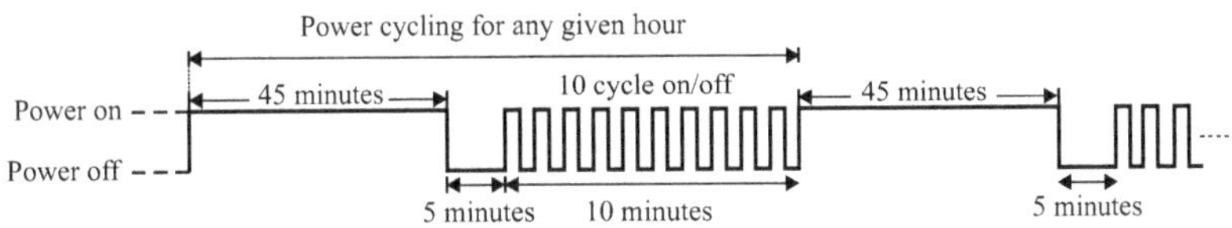

Figure 15.12 A typical input power cycling chart

It should be mentioned that the test period may be varied depending on the actual requirements of the client. For example, if a particular model of power supply has been regularly and stably manufactured for a period of time, then the test period may be reduced to one week. Conversely, for a newly developed product, the test period may be increased to 3 weeks or more upon the client's request.

The test methodology of the ORT is a burn-in with the recommended specifications shown in Table 15.9. Before and after the ORT, all samples must pass all specifications at full load. During the ORT, samples are split into two groups, with half for 100–120 V_{ac} line voltage and the other half for 220–240 V_{ac}. The testing chamber temperature should be maintained between 50 and 55 °C. Forced air circulation is allowed but should not directly blow on the samples. A typical input power cycling chart for any given hour of burn-in is shown in Figure 15.12. The input power is on for 45 min and turned off for 5 min. Then, the procedure enters the quick power cycles for 10 min. During a quick power cycle, the input power is turned on and off for 30 s, respectively. In total, 10 cycles will performed in 10 min. Sometimes, the client will request different power cycles based on different application demands.

15.9 Conclusions

This chapter presents the basic concept of the reliability of power supplies for computers. In order to ensure the reliability of power supplies for computers, many approaches, including work planes, parameter/component analysis, and variety tests, need to be performed by the power supply manufacturer. Many tasks need to be done during the design and manufacture stages. The performance of the power supply at all specifications needs to be verified under different operation situations.

To assure and improve the power supply's reliability, the testing should continue even after the products have been shipped out.

Acknowledgments

The author would like to express his sincere appreciation to AcBel Polytech Inc. for providing photographs and helpful information for the preparation of this chapter.

References

[1] Z. Zhong, C. Zhou, and P. Cao, "AMT failed mode analysis by integration of hardware-in-loop and DFMEA," *CECNet* 2012, pp. 134–137.

[2] D. Ling, H.-Z. Huang, W. Song, Y. Liu, and M.J. Zuo, "Design FMEA for a diesel engine using two risk priority numbers," *RAMS* 2012, pp. 1–5.

[3] J.-M. Cai; X.-M. Li, and G.-H Yang, "The risk priority number methodology for distribution priority of emergency logistics after earthquake disasters," *MSIE* 2011, pp. 560–562.

[4] Y. Zhao, G. Fu, B. Wan, and C. Pei, "An improved cost-based method of risk priority number," *IEEE PHM* 2012, pp. 1–4.

[5] Y. Zhao, G. Fu, and B. Wan, "An improved risk priority number method based on AHP," *RAMS* 2013, pp. 1–7.

[6] Z. Huang, "De-rating design and analysis for safety and reliability," *RAMS* 2014, pp. 1–7.

[7] B. Abdi, R. Ghasemi, and S. M. M. Mirtalaei, "The effect of electrolytic capacitors on SMPS's failure rate," *International Journal of Machine Learning and Computing*, Vol. 3, No. 3, pp. 300–304, 2013.

[8] M. L. Gasperi, "Life prediction model for aluminum electrolytic capacitors," *IEEE Industry Applications Conference* 1996, pp. 1347–1351.

[9] A. Dehbi, W. Wondrak, Y. Ousten, and Y. Danto, "High temperature reliability testing of aluminum and tantalum electrolytic capacitors," *Microelectronics Reliability*, Vol. 42, No. 6, pp. 835–840, 2002.

[10] R. Jano, and D. Pitica, "Accelerated ageing tests for predicting capacitor lifetimes," *IEEE SIITME* 2011, pp. 63–68.

[11] X. Tian, "Cooling fan reliability: failure criteria, accelerated life testing, modeling and qualification," *RAMS* 2006, pp. 380–384.

[12] S. Narasimhan, G. Shankaran, and S. Basak, "Modeling of fan failures in networking enclosures," *IEEE SEMI-THERM* 2012, pp. 249–254.

[13] O. Hyunseok, M. H. Azarian, M. Pecht, C. H. White, R. C. Sohaney, and E. Rhem, "Physics-of-failure approach for fan PHM in electronics applications," *PHM* 2010, pp. 1–6.

[14] A. Barnard, "Ten things you should know about HALT and HASS," *RAMS*, 2012, pp. 1–6.

[15] M. Silverman, "Summary of HALT and HASS results at an accelerated reliability test center," *RAMS* 1998, pp. 30–36.

[16] M. Silverman, and J. Hofmeister, "The useful synergies between prognostics and HALT and HASS," *RAMS* 2012, pp. 1–5.
[17] R. Schmidt, and C. Spindler, "Failure assessment and HALT test of electrical converters," *RAMS* 2012, pp. 1–6.
[18] H. McLean, "From HALT results to an accurate field MTBF estimate," *RAMS* 2010, pp. 1–5.
[19] Y. Liu, F. Sun; H. Zhang, Z. Zhou, and Y. Qin, "A comparison of two board level mechanical tests-drop impact and vibration shock," *APM* 2011, pp. 199–203.
[20] F. Schenkelberg, "Establishing effective ORT requirements," *RAMS* 2012, pp. 1–6.

Chapter 16
High-power converters

Rik W. De Doncker[1], Nils Soltau[2], Hanno Stagge[1] and Marco Stieneker[1]

16.1 High-power applications

This chapter presents an introduction to the main applications and some general guidelines for high-power converters. Afterwards, new thyristor-based devices are introduced that may be used in future medium-voltage power converters. Following that, a short introduction to high-power inverters and dc–dc converters is given.

16.1.1 General overview

High-power, medium-voltage converter systems usually operate at low switching frequencies in the range of 300–700 Hz. Today, multilevel converters have become more prominent in applications as they offer a lower total harmonic distortion using a higher number of switching states. With these converters, the size of passive filter components can be reduced, and the disadvantage of the low switching frequency of individual medium-voltage silicon-based power electronic switches can be overcome. With soft-switching dc–dc converters, the operating frequency may be increased into the kilohertz range, thereby reducing the size and weight of the passive components.

With the smaller size of the passive components, the size and weight of the converter system can be reduced. Moreover, the costs of the filter components are reduced. In applications such as industrial drives, traction drives, or wind turbines, these factors play a decisive role. Also, since power electronics has become cheaper with further development and increased use, it is economically desirable to use less copper and steel for passive components and increase the share of semiconductors in converter systems.

High-power converters are used in a number of different fields of electrical engineering. Below, an overview of applications is given, separated into drive and grid applications. Independent of the application, the main requirements of high-power converter systems are robustness, reliability, efficiency, and life-cycle cost.

[1]E.ON Energy Research Center, Institute for Power Generation and Storage Systems, RWTH Aachen University, Germany
[2]Mitsubishi Electric Europe B.V, Ratingen, North Rhine-Westphalia Germany

16.1.1.1 Drive applications

The main drive applications of high-power converters are in various areas of industry, renewable power sources (wind), and traction drive systems. Megawatt industrial drive systems control the speed of compressors, pumps, and mills in the power, oil and gas, metal forging, chemical, paper, and mining industries. Trains, locomotives, heavy-duty off-highway vehicles, and marine propulsion systems are the main applications in the traction sector. Since industrial drives are usually stationary, small size and weight are of secondary importance.

Power converters for drive systems in general consist of two stages: the machine-side converter and the grid-side converter. If only uni-directional power flow is required, e.g., for motor applications, the grid-side stage may utilize a diode rectifier. In some applications, thyristor rectifiers are used. Especially for the wind power industry, insulated-gate bipolar transistor (IGBT)-based grid-side converters are more prominent, which can also participate in grid services.

Wind turbine systems have power ratings in the low megawatt range. With increasing power, medium-voltage drives have become attractive for reducing the conduction losses in the cables connecting the converter system in the nacelle of the wind tower to the transformer that is usually placed on the ground level. The collector field of a wind farm in the future may be designed using dc with a converter station for connection to a high-voltage dc (HVDC) line or grid. These stations then convert the power of a complete wind farm at ratings of a few hundred megawatts.

16.1.1.2 Grid applications

Increasingly, power electronic converters are being used in electric grids to assist stable operation and control power flow. Flexible AC transmission systems are used to provide reactive power and enhance grid stability during faults. As grid codes have become more demanding, even more power electronic systems will be required in the future. Active filtering and HVDC transmission are other applications that require power electronics. As dc in general has a number of advantages over ac, in the future, entire grid sections also in distribution grids may be designed using dc. As the passive transformer cannot be used to convert dc, active power stations with power electronics are required for these systems.

16.2 Thyristor-based high-power devices

After its invention in the late 1950s, the thyristor became the dominant power-electronic device for high-power applications for many years. High voltage and current ratings and low forward on-state voltage have enabled, for the first time, efficient high-power electrical energy conversion.

Also today, thyristor-based power-electronics remain important. Due to their low conduction losses, fast switching transients, high surge-current ratings, and their capability to be light triggered, they offer decisive advantages over other power-electronic switches. Moreover, they are still the devices with the highest voltage and current ratings.

From the original thyristor many enhanced devices have been developed. Four of these devices are introduced in the following.

16.2.1 Integrated gate-commutated thyristor (IGCT)

In general, a thyristor turns off when the current reverses and falls below the holding current. Only special thyristors are capable of being turned off by force. These thyristors are the gate turn-off thyristor (GTO) and the gate-commutated thyristor (GCT).

They are turned off by a negative gate current, $-I_g$. The required current is determined by the turn-off gain

$$\beta = \frac{I_a}{I_g} \tag{16.1}$$

where I_a is the anode current to be turned off [12,23,27].

While for GTOs the turn-off gain is in the range of 3–5 [24], GCTs are operated with unity turn-off gain. Consequently, the negative gate pulse equals the anode current at turn-off. By this, the GCT has improved switching performance, and it does not necessarily require snubber circuits like the GTO. Today, the GCT has almost completely replaced the GTO in new converter designs.

The unity-gain turn-off requires a large negative gate pulse. This results in very high demands on the gate drive unit (GDU) of the GCT. Furthermore, the connection between GDU and GCT has to ensure a very small parasitic resistance, R_s and inductance, L_s in order to achieve a steep current pulse. Therefore, GDU and GCT are usually integrated into one unit, called an IGCT.

As depicted in Figure 16.1, the GDU is directly connected to the GCT. Moreover, a large area of the IGCT is used by electrolytic capacitors and MOSFETs.

The main purposes of the GDU (turning on and off the GCT) and their implementation are explained briefly. The turn-on unit is depicted in Figure 16.2 (a). Using the switches S_1, S_2, and S_3, a current is built up in L_1 and L_2, respectively. Opening S_2 or S_3 the current is driven into the gate, firing the GCT.

Figure 16.1. IGCT consists of GCT in press-pack housing (left) and GDU (right)

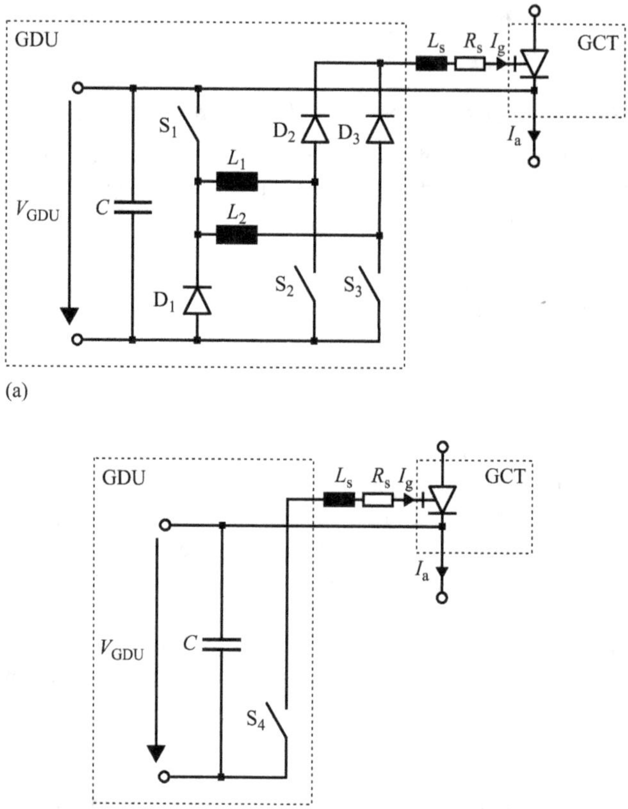

Figure 16.2. Schematic of a GCT GDU: (a) turn-on circuit and (b) turn-off circuit

The elementary schematic of the turn-off unit is depicted in Figure 16.2(b). Through the switch S_4, a capacitor charged with the supply voltage of the GDU (in the range of 20) is reversely connected to the gate. The negative gate voltage results in a large current pulse to turn off the GCT.

In the following, the reliability and the failure modes of thyristor-based devices are discussed. One of the failure modes of GTOs is filament formation at high turn-off gains [2,34]. IGCTs are more reliable due to their lower turn-off gain and a lower amount of passive components. Compared to the IGCT, GTO and IGBT failure rates are 1.6 or 2.3 times higher, respectively [33]. Studies have shown that a 3MVA three-phase IGCT inverter has a mean time between failure higher than 45 years. The failure distribution is [34]

- GCT: 13
- Clamp circuit*: 8

*Additional passive snubber components [34].

- GDU: 53
- GDU supply: 26.

It is evident that, compared to the GCT and the auxiliaries, the GDU fails more often. According to Steimer *et al.* [34], the main contributions to failure come from the fiber optics and the logic circuits.

16.2.2 Internally commutated thyristor (ICT)

The large number of capacitors and MOSFETs shown in Figure 16.1 belongs to the turn-off unit of the IGCT. They make the IGCT large and bulky. The concept of the ICT minimizes the size of the turn-off unit by integration into the press-pack housing of the GCT. The GCT wafer itself remains unchanged [15,16].

As depicted in Figure 16.3(b), the capacitors and MOSFETs (represented by single devices in the schematic) are integrated into the press-pack housing. Therefore, the commutation inductance is reduced significantly, and, hence, the required capacitance is reduced. Consequently, ceramic capacitors can be used, which additionally have a lower equivalent series inductance. In total, the amount of needed capacitance is only about 10 compared to the external electrolytic capacitors of an IGCT [15].

The major advantages of the ICT compared to the IGCT are increased safe-operating area and reliability. It has been demonstrated that the maximum turn-off current is 35 higher when the GCT wafer is operating as an ICT. Moreover, the part count for the GDU is more than halved, and its power consumption is reduced [16].

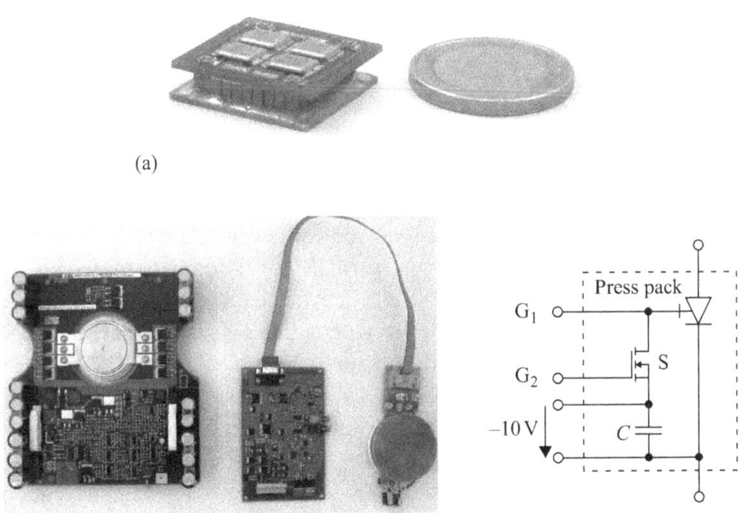

Figure 16.3. Internally commutated thyristor: (a) turn-off unit, (b) IGCT (left) versus ICT (right), and (c) equivalent circuit

16.2.3 Dual-ICT

Already in the 1990s, experts proposed to connect a fast switching device as a MOSFET in parallel to a device with low conduction losses, like an IGBT or IGCT [14]. This utilizes both devices' advantages, namely low switching losses, and low conduction losses. Similarly, two different GCTs can be used in parallel, where one GCT is designed for good conduction and the other one for fast switching performance. This can be realized by adjusting the charge carrier lifetime properly [17].

Using discrete devices, however, the parasitic inductance and resistance between the devices impede an optimal function. In the Dual-ICT, two GCTs are coaxially diffused onto one wafer (cf. Figure 16.4(a)) [17]. The inner GCT is optimized for low on-state voltage, which is referred to as "conduction GCT." When the Dual-ICT conducts, the current mainly flows through this GCT, leading to low conduction losses. The GCT located in the outer ring of the wafer is optimized for low switching losses, and, hence, is referred to as "switching GCT." Its purpose is to provide optimal performance when the Dual-ICT turns off. The integration on one wafer minimizes the parasitic inductance between the GCT structures, which optimizes the dynamic current sharing.

Figure 16.4(c) shows the turn-off process of the Dual-ICT. Initially, both GCTs are in the on-state. However, the conduction GCT carries a high share of the total current. The sequence is initiated by turning off the conduction GCT first. This GCT turns off against the on-state voltage of the switching GCT. Since the on-state voltage is comparably low, the losses of this turn-off sequence are negligible. Subsequently, the switching GCT is turned off. Since it is specially designed for this purpose, the turn-off losses are reduced compared to a conventional IGCT with equivalent on-state losses.

Compared to a single GCT with the same silicon area, the Dual-ICT achieves improved efficiency, especially in hard-switched applications. Alternatively, considering the same loss density, the current density can be increased by 56 or the switching frequency by a factor of 4.8 [38].

16.2.4 ETO/IETO

The emitter turn-off thyristor (ETO) is a GCT, which is controlled through two MOSFETs [21,22]. One MOSFET is located in series with the cathode of the GCT, i.e., cascode arrangement, while a shunt MOSFET is connected to the gate, as shown in Figure 16.5(b). Both MOSFETs are not exposed to high voltages. The maximum voltage is determined by the gate-cathode forward voltage of the PN junction plus the voltage drop of the cathode switch, S_2. For a 4000 ETO, for example, the transient voltage of the cathode switch stays below 60 V [22].

Compared to an IGCT, the ETO features a smaller GDU with less power consumption and MOS-controlled switching (similar to an IGBT).

In order to turn the device off, switch S_1 is closed first. When subsequently S_2 is opened, the anode current is forced to commutate to the gate, leading to the unit-gain turn-off of the GCT [4]. However, in order to achieve unity turn-off gain, especially during over-currents, the stray inductance in the cathode-gate commutation loop is crucial [22].

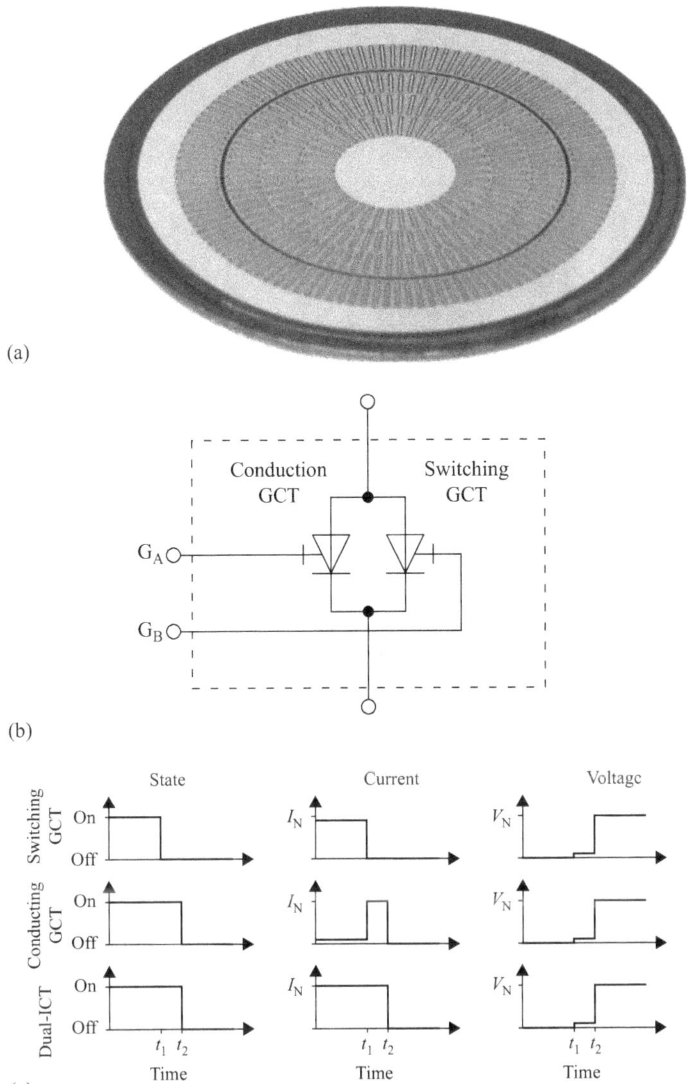

Figure 16.4. Dual-ICT: (a) two GCTs diffused on one wafer, (b) equivalent circuit, and (c) turn-off sequence

Stray-inductance minimization, which maximizes the turn-off capability, motivates the integration of the entire MOS structure into the press-pack housing, as shown in Figure 16.5(a) [3–5]. The equivalent circuit of this integrated emitter turn-off thyristor (IETO) is depicted in Figure 16.5(b).

As for the ETO, the IETO features a small GDU. In fact, the GDU of an IETO is only the size of a credit card (cf. Figure 16.5(b)), and its power consumption is

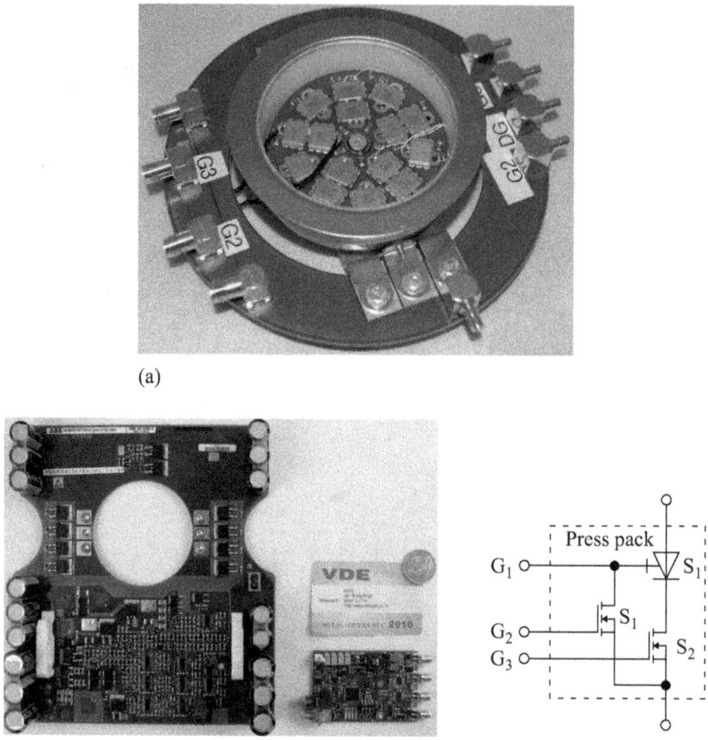

(a)

(b) (c)

Figure 16.5. Integrated emitter turn-off thyristor

less than one-sixth of an IGCT GDU. Moreover, the turn-off capability is increased through a low-inductive commutation path. With a commercial 520 GCT wafer, under hard-switching conditions, the IETO enables one to increase the maximal turn-off current to 1600 [5].

16.2.5 Reliability of thyristor-based devices

As described in chapter 16.2.1, the IGCT itself is a very reliable device compared to an IGBT, for example. If an IGCT fails, the plasma is contained inside the hermetically sealed press-pack housing. Nevertheless, the bulky GDU with its electrolytic capacitors and numerous components fails most often.

The ICT, dual-ICT, and IETO integrate the turn-off unit into the press-pack housing, resulting in a reduction of capacitance and components. Therefore, instead of electrolytic capacitors, ceramic capacitors can be used to provide a lower series resistance, improved stability for low temperatures, and longer lifetime.

The reduction of capacitance also reduces the total amount of components for the GDUs. The GDU of a conventional IGCT rated for 5.5 and 520 requires 300 parts. In contrast, the GDU of the ICT, dual-ICT, and IETO require 131, 259, and 124 components, respectively.

Finally, the integration of the turn-off unit enables a cable connection between the GCT and GDU. This allows for a larger distance between the GDU and the device, leading to a lower temperature of the GDU. Moreover, the load current stays inside the (hermetically sealed) press-pack housing, increasing the safety of the device and reducing the thermal stress of the GDU.

On the one hand, thyristor-based power-electronic devices are very efficient and reliable. One the other hand, it is worthwhile to further develop IGCTs. In any case, thyristor-based devices are still an enabler when it comes to high-power conversion.

16.3 High-power inverter topologies

16.3.1 Two-level converters

In the medium to high power range, the most commonly used circuit for power conversion from ac to dc or vice versa is the three-phase full-bridge topology or two-level voltage source converter, which is depicted in Figure 16.6. Note that only a single phase leg is shown in Figure 16.6. However, in principle, an arbitrary number of phases can be used in multi-phase inverters. Grid-connected converters, as well as drive converters, typically have three phases.

The output voltage, u_{AN}, of this circuit can only take the values $+U_{dc}/2$ or $-U_{dc}/2$. Due to this low number of voltage levels, i.e., switching states, the harmonic content of the output signal is high, which makes filtering expensive. This represents one of the main disadvantages of the two-level converter topology in medium-voltage grid-connected applications. Due to the maximum current and voltage of modern power electronic devices, a single two-level converter is limited to power ratings in the low megawatt range.

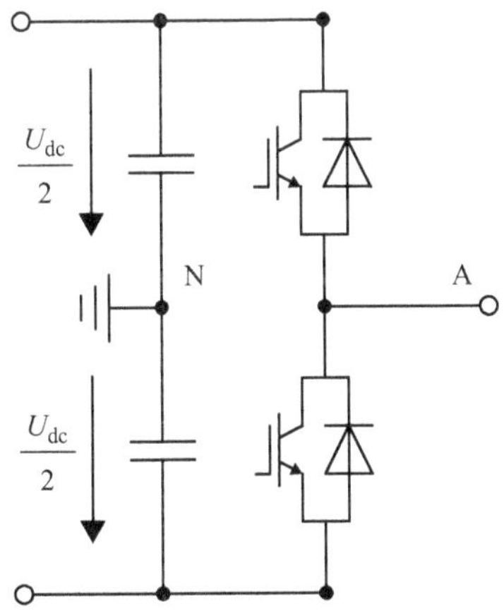

Figure 16.6. Phase leg of a two-level voltage source converter

A parallel connection of devices yields higher output power; however, it requires additional measures to ensure proper current sharing between semiconductor switches. Moreover, scaling output power by increasing the current rating causes higher losses in cables, passive components, and connectors. A more frequently used approach is to increase system voltage by connecting devices in series. An example of a two-level converter with a series connection of two IGBTs per switch position is shown in Figure 16.7. To ensure homogeneous voltage sharing across all series-connected devices (particularly under dynamic conditions), snubber circuits or active voltage balancing control (in the case of transistors) have to be employed. These lead to additional losses and costs.

16.3.2 Multi-level converters

The aforementioned drawbacks of the two-level circuit can be overcome with multi-level converter topologies. As the name indicates, the output of a multi-level converter can take more than two discrete values. In general, an n-level converter creates a phase-to-neutral output voltage waveform with n different values. Consequently, the line-to-line voltage in a three-phase system has $2n + 1$ levels. Among the main advantages of multi-level topologies, compared to their two-level counterpart, are significantly lower harmonic distortion of the output voltage (at a similar switching frequency). With increasing levels, the achievable voltage (and thus the maximum obtainable apparent power) increases while the harmonic distortion at the output terminals decreases. For an n-level inverter, the effective output switching frequency is $n - 1$ times higher than the switching frequency of the single semiconductor devices. At the same time, the maximal voltage steps are limited, and with it, the stress on the insulation of inductive loads.

A generalized structure of a multi-level converter that was introduced by Peng [29] is illustrated in Figure 16.8. Again, only a single phase-leg is shown. For a

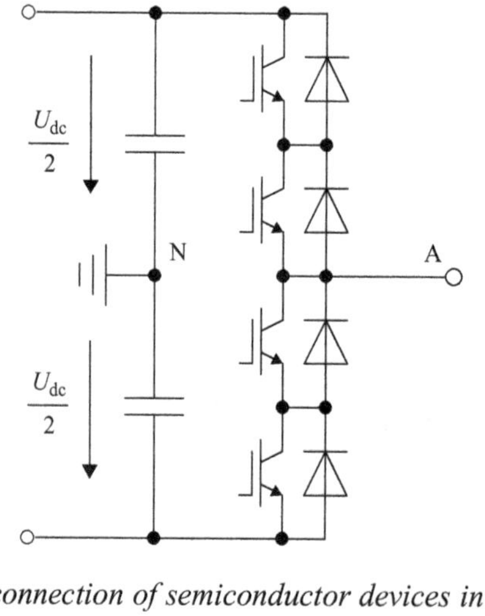

Figure 16.7. Series connection of semiconductor devices in a two-level converter

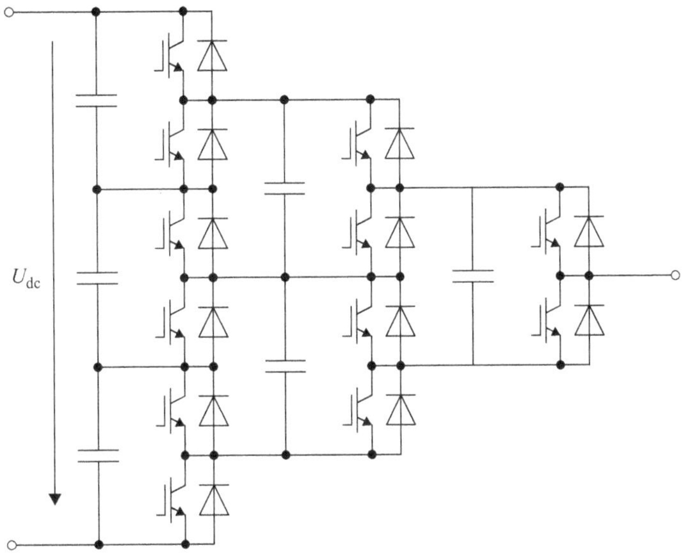

Figure 16.8. Generalized four-level converter

three-phase configuration, three structures of Figure 16.8 are connected to the dc link. The topology shown in Figure 16.8 has four output levels. Compared to other circuits, the component count of this basic topology is rather high. For n levels, it consists of $\sum_{i=1}^{n-1} i$ basic cells, each comprising two switches, two diodes in anti-parallel, and a capacitor. All capacitors are charged to a voltage $U_{dc}/(n-1)$, with U_{dc} being the full dc-link voltage across the input terminals of the converter. All semiconductor devices have to be rated for the same voltage.

This converter features many degrees of freedom to balance semiconductor losses and capacitor voltages. However, due to the high number of components, it is more suitable for the general analysis of multi-level converter topologies than for a real product. Hence, the generalized multi-level converter topology can be used as a starting point for the analysis of converter topologies that are geared to real-world application needs. By removing individual switches, diodes and capacitors from the circuit, other multi-level topologies, e.g., the neutral-point-clamped converter (NPC), the active-NPC (ANPC), or the flying capacitor inverter (FLC), can be deduced from this converter.

For more information on multi-level converters, the reader is referred to [1] or [18], which give overviews and more details on different topologies, their advantages, and applications.

16.3.2.1 NPC converters

Starting from the generalized multi-level converter in Figure 16.8, removing all switches except the outer ones in each vertical line, removing all capacitors except those that are connected to the dc link, and cross-connecting the diodes yields the four-level NPC converter, also called the diode-clamped converter

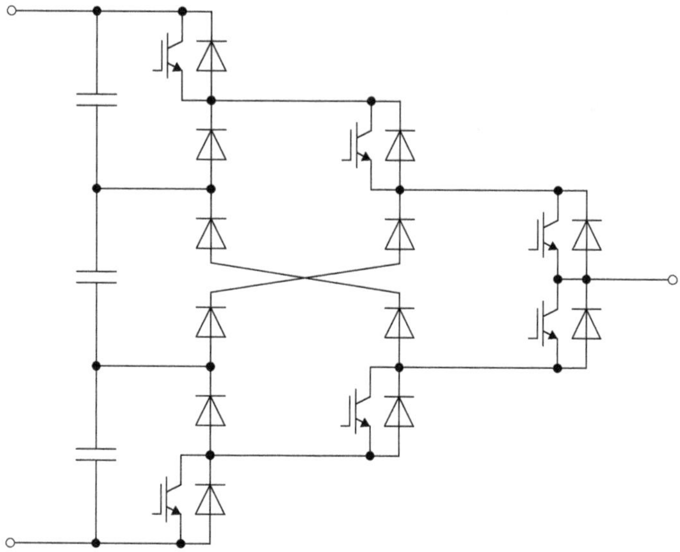

Figure 16.9. Four-level NPC converter derived from the generalized multi-level circuit

(Figure 16.9). The principle of operation of this topology will be explained in the following paragraphs using the example of a three-level NPC inverter introduced by [28].

The three-level version of the NPC converter (3L-NPC) is one of the most commonly used multi-level topologies for industrial applications. The equivalent circuit diagram is given in Figure 16.10. As can be seen, the converter structure is very similar to the conventional two-level converter with a series connection of two devices per switch position (cf. Figure 16.7). However, the 3L-NPC features two additional diodes that clamp the midpoints of the upper and the lower switch position to the midpoint of the dc link.

The NPC consists of two commutation cells (S_1–S_3 and S_2–S_4). Thus, to control the 3L-NPC inverter, two switches are always in the on-state, while the other switches are in the blocking state. Three different states are used (Figure 16.11) to create three different values of the output voltage, u_{AN}. These states are given in Table 16.1, where "1" and "0" indicate whether the corresponding switch is turned on and off, respectively.

A disadvantage of the three-level NPC is that the device stress is not evenly shared between the inner (S_2, S_3) and the outer (S_1, S_4) switches. Moreover, capacitor voltage balancing becomes an issue for high-level counts. Therefore, commercial NPC converters typically have three levels. A better loss distribution between the inner and the outer switches can be achieved by adding additional active turn-off devices (e.g., IGBTs) in parallel with the clamping diodes. The resulting topology is called an ANPC [6].

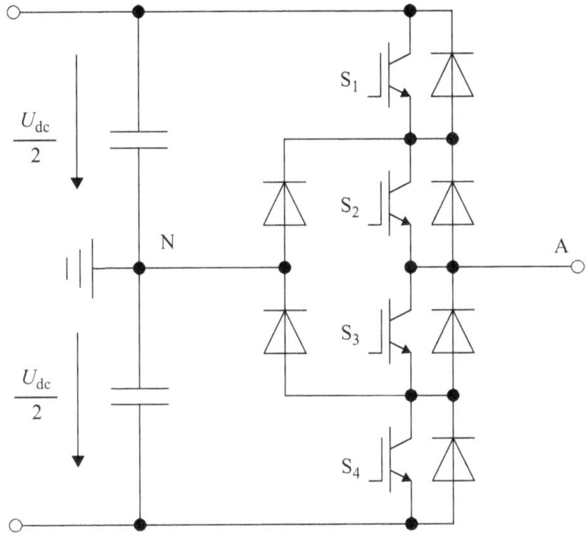

Figure 16.10. Three-level NPC inverter

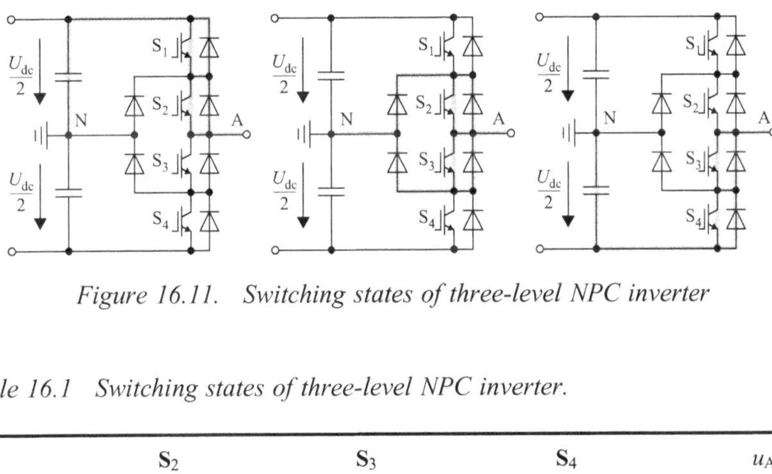

Figure 16.11. Switching states of three-level NPC inverter

Table 16.1 Switching states of three-level NPC inverter.

S_1	S_2	S_3	S_4	u_{AN}
1	1	0	0	$+U_{dc}/2$
0	0	1	1	$-U_{dc}/2$
0	1	1	0	0

16.3.2.2 Flying-capacitor converter

Removing all inner switches and diodes from the generalized multi-level converter resulted in the FLC, first introduced by Meynard and Foch [26]. The FLC does not suffer from capacitor voltage balancing problems like the NPC, as redundant switching states allow controlled charging and discharging of the capacitors. Thus, building converters that have more than three levels is possible without difficulty. Multi-megawatt four-level FLCs are also commercially available. Operation with auxiliary resonant-commutated poles (ARCPs) is possible to decrease switching losses [37].

16.3.2.3 Series-connected H-bridge converter

The series-connected H-bridge converter introduced by Hammond [13] is also known as the cascaded H-bridge or the cascaded cell converter. It consists of n series-connected power cells per phase. The three-phase legs are star-connected with a floating neutral. All cells are isolated from each other and comprise a six-pulse diode rectifier as the input stage, a capacitive dc link, and an H-bridge as the output stage. The cells are powered by a transformer with an appropriate number of isolated secondary windings. Redundancy is easily implemented by providing redundant power cells and a switch that bypasses a defective power cell.

16.3.2.4 Modular multi-level converter

The modular multi-level converter (MMC or M2C) proposed by Lesnicar and Marquardt [20] consists of an upper and a lower arm, each having a certain number of identical power modules. A power module comprises IGBTs, diodes, and a capacitor in a half-bridge or H-bridge configuration. Combining the switching states of all modules in a phase leg in a suitable way produces a multi-level waveform at the output. To ensure proper balancing of capacitor voltages within one arm, the voltages across the capacitors in each module are continuously measured and compared to each other. This allows for an appropriate selection of cells to be turned on or off.

The main advantages of the MMC are modularity and scalability together with greatly improved ac-side power quality [18].

16.4 High-power dc–dc converter topologies

The three-phase dual-active bridge (DAB) dc–dc converter [7] is a well-suited topology for high-power medium-voltage applications. The galvanic isolation provided by the transformer offers advantages for applicable semiconductor devices and operation security. Improved control strategies [9,10] lead to highly dynamic and precise operation. Bidirectional power flow capability allows utility applications. Also, commercially available power electronic building blocks can be used to construct the dc–dc converter.

Within this chapter, the analysis and characteristics of a DAB are presented. Afterwards, the design of a DAB-based modular dc–dc converter system is discussed.

16.4.1 DAB converter

As shown in Figure 16.12, the three-phase DAB dc–dc converter [7] consists of two full-bridge converters and a medium or high-frequency transformer. The primary-side full-bridge converts the dc voltage, V_p into ac. The ac voltage is transformed

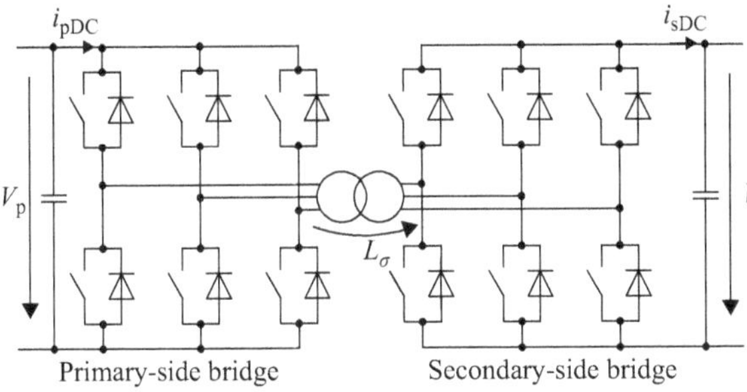

Figure 16.12 Schematic of a three-phase DAB dc–dc converter

according to the winding ration, n, of the transformer. The secondary-side full-bridge actively rectifies the ac-link voltage into the dc voltage, V_s.

Since the transformer does not require a purely sinusoidal voltage, the full-bridge converter can be operated with fundamental-switching frequency instead of pulse-width modulation (PWM). Hence, the frequency of the ac link is equal to the switching frequency, f_{sw}, of the power electronics converter. In comparison to PWM converters, the switching losses are reduced significantly.

f_{sw} has a significant influence on the characteristic of the transformer. The higher the frequency, the smaller the size and weight and the lower the core losses of the transformer. However, the semiconductors of the full-bridge converter limit the maximum switching frequency for a certain power rating.

As shown in Soltau *et al.* [32], a five medium-voltage dc–dc converter operating with $f_{sw} = 1$ and $V_p = V_s = 5$ has been developed.

16.4.1.1 Power transfer

As shown in Figure 16.13, a fundamental-frequency switching pattern results in the six-step waveform voltages applied to the transformer terminals. v_{1a}, v_{1b}, and v_{1c} are the voltages on the primary side, and v_{2a}, v_{2b}, and v_{2c} are the values on the secondary side. Introducing a phase shift or load angle, φ, between the voltage modulated on the primary and secondary terminals of the transformer results in a voltage across its stray reactance, X_σ. These voltages lead to phase currents i_{1p}, i_{2p}, and $i3p$ on the primary side that are equal to the currents on the secondary side i_{1s}, i_{2s}, and i_{3s} considering a unity winding ratio of the transformer ($n = 1$). Figure 16.13 also shows the plot of the dc input current, i_{pDC}, and of the dc output current, i_{sDC}. The power transfer of the DAB P_{DAB} can be controlled by varying the load angle, φ. Neglecting the losses of the power electronic converters and of the transformer, P_{DAB} can be calculated with (16.2)

$$P_{\text{DAB}} = \frac{V_{\text{p}}{}^2}{\omega L_{\text{s}}} \cdot d \cdot \begin{cases} \dfrac{2\varphi}{3} - \dfrac{\varphi^2}{2\pi} & \text{for } 0 \le \varphi \le \pi/3 \\ \varphi - \dfrac{\varphi^2}{\pi} - \dfrac{\pi}{18} & \text{for } \pi/3 \le \varphi \le 2\pi/3 \end{cases} . \tag{16.2}$$

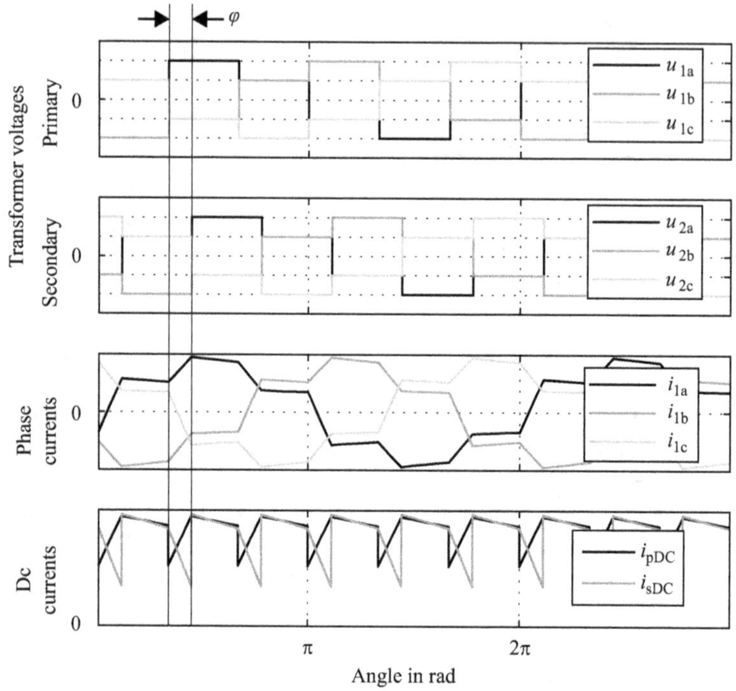

Figure 16.13 Characteristic waveforms of a three-phase DAB

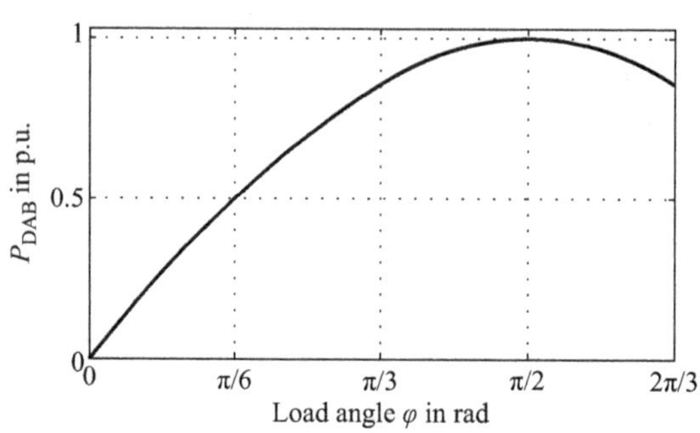

Figure 16.14 DAB power transfer over load angle φ

where L_s is the stray inductance of the transformer, and $\omega = 2 \cdot \pi \cdot f_{sw}$ is the angular frequency of the ac-link voltage. $d = V_p V_s{}'$ gives the dynamic voltage ratio of the primary-side and secondary-side dc voltage. Therefore, $V_s{}' = V_{s/n}$ is the secondary-side dc voltage V_s related to the primary side considering the turn ratio of the transformer n. As can be seen in Figure 16.14, maximum power transfer can be achieved with φ equal to $\pi/2$.

16.4.1.2 Stray inductance design

According to (16.2), the stray inductance of the transformer, L_σ, influences the power transfer of the DAB, P_{DAB}. Increasing L_σ leads to a lower rise of current and also to a lower maximum P_{DAB}. Hence, L_σ has to be designed related to the intended power rating of the dc–dc converter

$$L_{s,\max} = \frac{V_p{}^2}{\omega P_{nom}} \cdot d \cdot \begin{cases} \dfrac{2\varphi}{3} - \dfrac{\varphi^2}{2\pi} & \text{for } 0 \leq \varphi_{nom} \leq \pi/3 \\ \varphi - \dfrac{\varphi^2}{\pi} - \dfrac{\pi}{18} & \text{for } \pi/3 \leq \varphi_{nom} \leq 2\pi/3 \end{cases}. \tag{16.3}$$

At the nominal load angle, φ_{nom}, rated power P_{nom} is transferred. Setting φ_{nom} to $\pi/3$ ensures a safety margin since the maximum power transfer is reached at $\varphi = \pi/2$. However, L_σ can also be set to a smaller value than $L_{\sigma,max}$.

Allowing variations of V_p and V_s, an optimum value for L_σ can be found such that the maximum ac-link current is minimized [30] in all operation points the DAB is designed for. Hence, the current stress on the semiconductor devices and transformer is reduced, and the efficiency of the system is improved.

16.4.1.3 Power devices

For high-power applications, IGBTs and IGCTs with suitable voltage and current ratings are available on the market. The advantage of IGBTs is the relatively low switching loss, whereas the conduction losses dominate applications with DAB. The conduction performance of IGCTs is superior to IGBTs, but the switching losses are higher if no further measures are taken. In total, it was found that the use of IGCTs leads to a higher efficiency of high-power DABs [32] since soft-switching operation and lossless snubbers reduce the switching loss significantly.

16.4.1.4 Soft-switching operation

Soft-switching operation reduces the switching losses of the power semiconductors considerably. In the case of the DAB converter, the power devices are turned on while the anti-parallel diode carries the current. Equation (16.4) gives the conditions for soft switching of the primary-side bridge, whereas soft-switching operation is ensured if (16.5) is fulfilled. i_{pk} andi_{sk} are the currents in the kth phase of the primary-side bridge and secondary-side bridge, respectively.

$$i_{pk}((k-1) \cdot 2 \cdot \pi/3) < 0 \;\Leftrightarrow\; i_s k(0) > 0 \tag{16.4}$$

$$i_{pk}((k-1) \cdot 2 \cdot \pi/3 + \varphi) > 0 \;\Leftrightarrow\; i_s k((k-1) \cdot 2 \cdot \pi/3 + \varphi) < 0 \tag{16.5}$$

In other words, soft switching is guaranteed when the phase current crosses zero between the switching instances of the primary and secondary switches of that phase.

In soft-switching operation, the voltage across the active switch is nearly zero during turn-on (zero-voltage switching, ZVS), and the turn-on losses are virtually eliminated. If ZVS is ensured for the first phase of a full-bridge converter due to symmetry, this is also the case for the second and third converter phases.

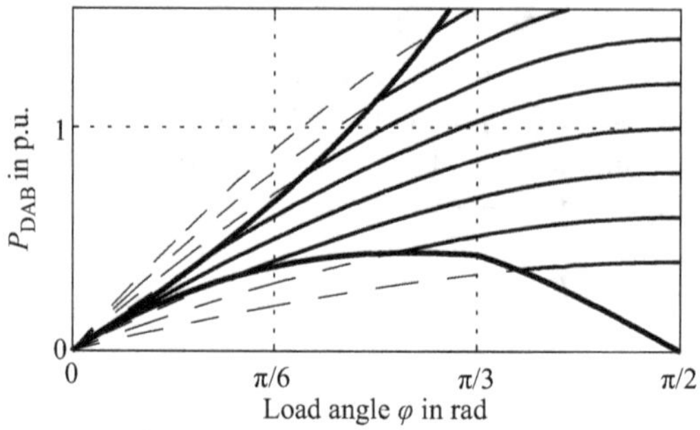

Figure 16.15. Soft-switching operation area in dependency of φ, P_{DAB}, and d

However, the achievement of ZVS mainly depends on the dynamic voltage ratio, d, and the transferred power. The conditions for soft switching of the primary-side bridge are

$$d < \begin{cases} \dfrac{1}{1-\dfrac{3\varphi}{2\pi}} & \text{for } 0 \leq \varphi \leq \pi/3 \\ \dfrac{1}{\dfrac{3}{2}-\dfrac{3\varphi}{2\pi}} & \text{for } \pi/3 \leq \varphi \leq 2\pi/3 \end{cases}. \tag{16.6}$$

and for the secondary-side bridge

$$d < \begin{cases} 1-\dfrac{3\varphi}{2\pi} & \text{for } 0 \leq \varphi \leq \pi/3 \\ \dfrac{3}{2}-\dfrac{3\varphi}{2\pi} & \text{for } \pi/3 \leq \varphi \leq 2\pi/3 \end{cases}. \tag{16.7}$$

The solid lines in Figure 16.15 indicate the soft-switching operation points in dependency of P_{DAB}, φ, and d. Operating the DAB with $d = 1$ allows soft-switching over the entire range of φ and hence P_{DAB}. However, if d varies from unity, either the primary-side or the secondary-side bridge operates hard-switched at smaller load angles. For increasing d, the primary-side bridge is hard-switched, while for a d below 1, the secondary-side bridge leaves the soft-switching operation range. It can be seen that with the decrease of φ, the borders to hard-switched operation become closer. Hence, deviating $d = 1$, where soft switching is ensured, becomes smaller.

16.4.1.5 Lossless snubber

Lossless snubbers can be applied to increase the efficiency of the DAB dc–dc converter. Capacitors connected in parallel to the power switches (Figure 16.12) reduce the voltage rise during turn-off. This results in a significant reduction of the switching loss. The application of a snubber with a capacitance of 1 μF lowers the switching losses by up to 70% [19].

However, it has to be ensured that the snubber capacitors are discharged completely before the parallel-connected semiconductor is supposed to be turned on. Otherwise, the capacitor is short-circuited by the power switch, and the high short-circuit current may lead to the destruction of the semiconductor due to the high thermal stress.

This can be avoided by operating the DAB within the soft-switching operation range. The ac load current discharges the snubber capacitor before it commutates to the anti-parallel diode. Meanwhile, the snubber capacitor related to the opposite switch within the converter leg is charged [30]. This process has to be finished within the switching dead time between the upper and lower semiconductor that avoids short-circuiting of the dc link.

In soft switching, the direction of the current ensures recharging of the snubber capacitors. Additionally, the amplitude of the ac-link current has to be sufficiently high during the dead time. Hence, soft-switching operation is also restricted to a minimum P_{DAB}.

Alternatively, an ARCP [25] can be applied to allow operation beyond the soft-switching area [8]. Also, the ac load current can be boosted with an ARCP to permit lower power transfer [31].

16.4.2 Modular dc–dc converter system

The interconnection of dc transmission and distribution grids as well as the grid integration of, e.g., wind parks and PV fields with dc collector grids, requires utility-scale dc–dc converters. These converters enable the adaption of different voltage levels, control the power flow, and provide galvanic isolation.

However, the dc–dc converter has to be designed according to the different required specifications. Instead of designing a dc–dc converter for every individual application scenario, a modular approach can be used to reduce engineering and production costs.

A DAB, as the smallest unit or building block in a modular dc–dc converter system, can be developed once. According to the use, the voltage rating is adapted by series-connection, and the power rating by parallel-connection of these modules [11,35]. Figure 16.16 shows a schematic of an example.

This chapter considers only a DAB with one full-bridge converter on the primary and secondary sides of the transformer.

Hence, the total number of DABs within the converter system, N_{tot}, is equal to the number of installed primary-side bridges, N_p, or secondary-side bridges, N_s ($N_p = N_s$).

16.4.2.1 System design

As shown in Figure 16.16, the center of the dc-link capacitors can be grounded. This bipolar approach halves the voltage against the ground, resulting in less insulation effort.

The primary-side voltage, $V_{p,sys}$, and secondary-side voltage, $V_{s,sys}$, of the dc–dc converter system depend on the number of series-connected full-bridge converters:

$$V_{p,sys} = N_{p,series} \ V_p \tag{16.8}$$

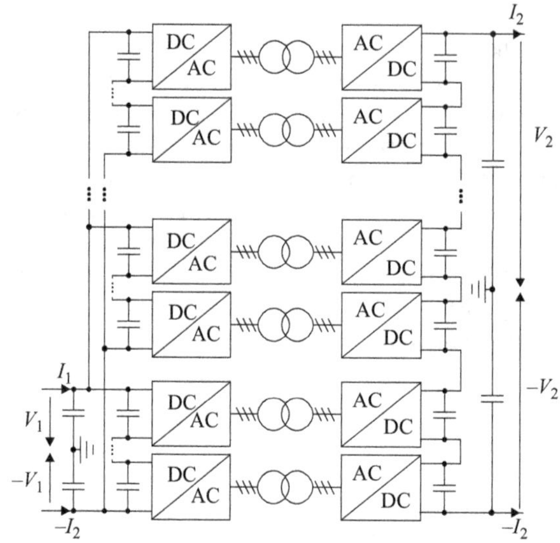

Figure 16.16. Schematic of a modular dc–dc converter system for connecting medium-voltage DC (MVDC) (left) with HVDC (right)

$$V_{s,sys} = N_{s,series}\ V_s \tag{16.9}$$

$N_{p,series}$ and $N_{s,series}$ are the number of series-connected full-bridge converters on the primary and secondary sides, respectively, and can differ from each other. However, it is mandatory that N_p be equal to N_s

$$N_p = N_s \tag{16.10}$$

Equation (16.10) can be fulfilled if several series-connections of primary-side and secondary-side converters are connected in parallel. In this case, N_p and N_s can be calculated with (16.11) and (16.12), respectively.

$$N_p = N_{p,series} \cdot N_{p,par} \tag{16.11}$$

$$N_s = N_{s,series} \cdot N_{s,par} \tag{16.12}$$

Inserting (16.11) and (16.12) into (16.10) leads to (16.13), which can be used to estimate the required amount of parallel-connected converter strings:

$$N_{p,series} \cdot N_{p,par} = N_{s,series} \cdot N_{s,par} \tag{16.13}$$

The design of a modular dc–dc converter system according to the required voltage levels leads to a nominal power of the system, P_{sys}, that is determined by the total amount of installed DAB

$$P_{sys} = N_{tot} \cdot P_{DAB} \tag{16.14}$$

However, the power rating can be increased by additional DABs connected in parallel. Hence, the modular dc–dc converter system can be adapted easily to the specific application.

16.4.2.2 Redundancy

Depending on the application scenario, high system reliability can be a very important criterion. This can also be taken into account within the design process of the dc–dc converter system.

When a single DAB fails, the voltage contribution of this converter is lost. If the system is supposed to continue operation, the remaining DABs have to compensate for the voltage loss. Hence, the DAB applied in the converter system has to be designed with a certain safety margin related to the voltage rating. During a fault, the operation voltage of each DAB is increased equally until the original value at the system terminals is reached again. The faulty DAB is bypassed in that case [36].

However, the system and each single DAB can be designed to allow the bypassing of the faulty dc–dc converters. Besides additional investment costs for spare DABs, the efficiency of a redundant system is decreased during normal operation. This results from the voltage-blocking capability that is reserved for compensating faults. The semiconductor devices are not optimally utilized.

Furthermore, redundancy can be implemented by additional parallel-connected DABs. This allows the loss of whole series connections of DABs while the rated power transfer is ensured. In contrast to redundancy with additional series-connected DAB, this approach does not lead to reduced efficiency but to considerably higher investment costs.

The modularity of utility-scaled dc–dc converter systems leads to cost reduction in design, production, maintenance, and repair. The converter system can be adapted to every application scenario easily since the DAB, being the smallest unit of the system, has to be developed once. It is also beneficial if the system production can be based on a single converter design. Also, the maintenance strategies and repair processes are equal for all converter systems in operation. Technicians do not have to be trained for every individual converter system. Logistics and stock strategies of spare parts can be improved.

Figures 16.6, 16.7, 16.8, 16.9, 16.10, and 16.11 are taken from the book "Lecture Notes Power Electronics Control, Synthesis, Application", ISBN: 978-3-943496-01-7, © ISEA, RWTH Aachen University.

References

[1] H. Abu-Rub, J. Holtz, J. Rodriguez, and G. Baoming. "Medium-voltage multilevel converters – State of the art, challenges, and requirements in industrial applications". In: *IEEE Transactions on Industrial Electronics* 57. 8 (2010), pp. 2581–2596.

[2] H. Bleichner, K. Nordgren, M. Rosling, M. Bakowski, and E. Nordlander. "The effect of emitter shortings on turn-off limitations and device failure in

GTO thyristors under snubberless operation". In: *IEEE Transactions on Electron Devices* 42.1 (1995), pp. 178–187.

[3] M. Bragard. "The integrated emitter turn-off thyristor. An innovative MOS-gated high-power device". *PhD thesis*. Institute for Power Generation and Storage Systems, RWTH Aachen University, 2012.

[4] M. Bragard, M. Conrad, H. van Hoek, and R.W. De Doncker. "The integrated emitter turn-off thyristor (IETO) – An innovative thyristor-based high power semiconductor device using MOS assisted turn-off". In: *IEEE Transactions on Industry Applications* 47.5 (2011), pp. 2175–2182.

[5] M. Bragard, H. van Hoek, and R.W. De Doncker. "A major design step in IETO concept realization that allows overcurrent protection and pushes limits of switching performance". In: *IEEE Transactions on Power Electronics* 27.9 (2012), pp. 4163–4171.

[6] T. Bruckner and S. Bemet. "Loss balancing in three-level voltage source inverters applying active NPC switches". In: *IEEE 32nd Annual Power Electronics Specialists Conference, 2001 (PESC 2001)*, vol. 2, 2001, pp. 1135–1140.

[7] Doncker, R. W. De, D. M. Divan, and M. H. Kheraluwala. "A three-phase soft-switched high-power-density DC/DC converter for high-power applications". In: *IEEE Transactions on Industry Applications*, vol. 27, 1991, pp. 63–73.

[8] Doncker, R. W. De, and J. P. Lyons. "The auxiliary resonant commutated pole converter". In: *Industry Applications Society Annual Meeting, 1990, Conference Record of the 1990 IEEE*. 1990, pp. 1228–1235.

[9] S. P. Engel, N. Soltau, H. Stagge, and De Doncker, R. W. "Dynamic and balanced control of three-phase high-power dual-active bridge DC–DC converters in DC-grid applications". In: *IEEE Transactions on Power Electronics*,28 (2013), pp. 1880–1889.

[10] S. P. Engel, N. Soltau, H. Stagge, and Doncker, R. W. De. "Improved instantaneous current control for high-power three-phase dual-active bridge DC–DC converters". In: *IEEE Transactions on Power Electronics* 29 (2014), pp. 4067–4077.

[11] S. P. Engel, M. Stieneker, N. Soltau, S. Rabiee, H. Stagge, and Doncker, R. W. De. "Comparison of the modular multilevel DC converter and the dual-active bridge converter for power conversion in HVDC and MVDC grids". In: *IEEE Transactions on Power Electronics* 30 (2015), pp. 124–137.

[12] J. M. Goldey, I. M. Mackintosh, and I. M. Ross. "Turn-off gain in p–n–p–n triodes". In: *Solid-State Electronics* 3.2 (1961), pp. 119–122.

[13] P.W. Hammond. "A new approach to enhance power quality for medium voltage AC drives". In: *IEEE Transactions on Industry Applications* 33.1 (1997), pp. 202–208.

[14] Y. M. Jiang, G. C. Hua, E. X. Yang, and F. C. Lee. "Soft-switching of IGBTs with the help of MOSFETs". In: VPEC Seminar. Ed. by Virginia Power Electronics Center. 1992.

[15] P. Koellensperger. "The Internally Commutated Thyristor. Concept, Design and Application". *PhD thesis*. Institute for Power Electronics and Electrical Drives, RWTH Aachen University, 2011.

[16] P. Koellensperger and R. W. De Doncker. "The internally commutated thyristor – A new GCT with integrated turn-off unit". 2006 4th International Conference *on* June 2006, pp. 1–6.

[17] P. Kollensperger, M. Bragard, T. Plum, and R.W. De Doncker. "The dual GCT – A new high-power device using optimized GCT technology". In: *IEEE Transactions on Industry Applications* 45.5 (Sept. 2009), pp. 1754–1762.

[18] S. Kouro, M. Malinowski, K. Gopakumar, *et al.* "Recent advances and industrial applications of multilevel converters". In: *IEEE Transactions on Industrial Electronics* 57.8 (2010), pp. 2553–2580.

[19] R. Lenke, H. van Hoek, S. Taraborrelli, R. W. De Doncker, J. San Sebastian, and I. Etxeberria Otadui. "Turn-off behavior of 4.5 kV asymmetric IGCTs under zero voltage switching conditions". In: *Proceedings of the 2011 – 14th European Conference on Power Electronics and Applications (EPE 2011)*. 2011, pp. 1–10.

[20] A. Lesnicar and R. Marquardt. "A new modular voltage source inverter topology". In: *European Conference on Power Electronics and Applications (EPE)*. 2003.

[21] Yuxin Li, A.Q. Huang, and F.C. Lee. "Introducing the emitter turn-off thyristor (ETO)". In: *Thirty-Third IAS Annual Meeting Industry Applications Conference, 1998. The 1998 IEEE*. vol. 2. 1998, pp. 860–864.

[22] Yuxin Li, A.Q. Huang, and K. Motto. "Experimental and numerical study of the emitter turn-off thyristor (ETO)". In: *IEEE Transactions on Power Electronics* 15.3 (2000), pp. 561–574.

[23] R. H. van Ligten and D. Navon. "Base turn-off of p–n–p–n switches". In: *Records of the Western Electronic Show and Convention (IRE WESCON)*. Ed. by Institute of Radio Engineers. 1960, pp. 49–52.

[24] K. Lilja and H. Gruning. "Onset of current filamentation in GTO devices". In: 21st Annual IEEE Power Electronics Specialists Conference, 1990. PESC '90 Record. 1990, pp. 398–406.

[25] W. McMurray. "Resonant snubbers with auxiliary switches". In: *IEEE Transactions on Industry Applications*,29 (1993), pp. 355–362.

[26] T.A. Meynard and H. Foch. "Multi-level conversion: High voltage choppers and voltage-source inverters". In: *23rd Annual IEEE Power Electronics Specialists Conference, 1992. PESC'92 Record*, vol. 1, 1992, 397–403.

[27] D.R. Muss and C. Goldberg. "Switching mechanism in the n–p–n–p silicon controlled rectifier". In: *IEEE Transactions on Electron Devices* 10.3 (1963), pp. 113–120.

[28] A. Nabae, I. Takahashi, and H. Akagi. "A new neutral-point-clamped PWM inverter". In: *IEEE Transactions on Industry Applications* IA-17.5 (1981), pp. 518–523.

[29] Fang Zheng Peng. "A generalized multilevel inverter topology with self voltage balancing". In: *IEEE Transactions on Industry Applications* 37.2 (2001), pp. 611–618.

[30] Robert Ulrich Lenke. "A contribution to the design of isolated DC–DC converters for utility applications". *PhD thesis*. Aachen: RWTH Aachen University, 2012.

[31] N. Soltau, J. Lange, M. Stieneker, H. Stagge, and De Doncker, R. W. "Ensuring soft-switching operation of a three-phase dual-active bridge DC–DC converter applying an auxiliary resonant-commutated pole". In: *2014 16th European Conference on Power Electronics and Applications (EPE'14-ECCE Europe)*. 2014, pp. 1–10.

[32] N. Soltau, H. Stagge, De Doncker, R. W., and O. Apeldoorn. "Development and demonstration of a medium-voltage high-power DC–DC converter for DC distribution systems". In: *2014 IEEE 5th International Symposium on Power Electronics for Distributed Generation Systems (PEDG)*. 2014, pp. 1–8.

[33] P. Steimer, O. Apeldoorn, E. Carroll, and A. Nagel. "IGCT technology baseline and future opportunities". In: *2001 IEEE/PES Transmission and Distribution Conference and Exposition*, vol. 2, 2001, pp. 1182–1187.

[34] P.K. Steimer, H.E. Gruning, J. Werninger, E. Carroll, S. Klaka, and S. Linder. "IGCT – A new emerging technology for high power, low cost inverters". In: *IEEE Industry Applications Magazine*, 5.4 (1999), pp. 12–18.

[35] M. Stieneker, N. R. Averous, N. Soltau, H. Stagge, and R. W. De Doncker. "Analysis of wind turbines connected to medium-voltage DC grids". In: *2014 16th European Conference on Power Electronics and Applications (EPE'14-ECCE Europe)*. 2014, pp. 1–10.

[36] M. Stieneker and R. W. De Doncker. "System efficiency estimation of redundant cascaded-cell converters in applications with high-power battery energy storage systems". In: *2012 International Conference on Renewable Energy Research and Applications (ICR-ERA)*, 2012, pp. 1–6.

[37] C. Turpin, L. Deprez, F. Forest, F. Richardeau, and T.A. Meynard. "A ZVS imbricated cell multilevel inverter with auxiliary resonant commutated poles". In: *IEEE Transactions on Power Electronics* 17.6 (2002), pp. 874–882.

[38] E. Van Brunt, A.Q. Huang, T. Butschen, and R.W. De Doncker. "Dual-GCT design criteria and voltage scaling". In: *2012 IEEE Energy Conversion Congress and Exposition (ECCE)* 2012, pp. 2596–2603.

Index